Shelf Life and Food Safety

Shelf Life and Food Safety

Edited by
Basharat Nabi Dar
Manzoor Ahmad Shah
Shabir Ahmad Mir

CRC Press
Taylor & Francis Group
Boca Raton London New York

CRC Press is an imprint of the
Taylor & Francis Group, an **informa** business

First edition published 2022
by CRC Press
6000 Broken Sound Parkway NW, Suite 300, Boca Raton, FL 33487-2742

and by CRC Press
2 Park Square, Milton Park, Abingdon, Oxon, OX14 4RN

CRC Press is an imprint of Taylor & Francis Group, LLC

Library of Congress Cataloging-in-Publication Data
Names: Dar, Basharat Nabi, editor. I Shah, Manzoor Ahmad, editor. I Shah, Shabir Ahmad, editor.
Title: Shelf life and food safety/edited by Basharat Nabi Dar, Manzoor Ahmad Shah, Shabir Ahmad Shah.
Description: First edition. I Boca Raton: Taylor and Francis, 2022. I Includes bibliographical references and index.
Identifiers: LCCN 2021052582 (print) I LCCN 2021052583 (ebook) I ISBN 9780367540173 (hardback) I ISBN 9780367550370 (paperback) I ISBN 9781003091677 (ebook)
Subjects: LCSH: Food—Storage. I Food—Shelf-life dating. I Food—Safety measures. I Food—Quality control.
Classification: LCC TP373.3 .S54 2022 (print) I LCC TP373.3 (ebook) I DDC 664—dc23/eng/20211108
LC record available at https://lccn.loc.gov/2021052582
LC ebook record available at https://lccn.loc.gov/2021052583

ISBN: 978-0-367-54017-3 (hbk)
ISBN: 978-0-367-55037-0 (pbk)
ISBN: 978-1-003-09167-7 (ebk)

DOI: 10.1201/9781003091677

Typeset in Times
by KnowledgeWorks Global Ltd.

Contents

**Sourav Garg, Syed Mansha Rafiq, Balaji G., Sristi Mundhada,
Syed Insha Rafiq, and Nikunj Sharma**

**Insha Zahoor, Nadira Anjum, Meenakshi Trilokia, Arshied Manzoor,
Haroon Rashid Naik, Anjum Ayoub, and Farhana Mehraj Allai**

Chetan Sharma, Damir D. Torrico, and Sandip Singh

Arjuara Begum, Shagufta Rizwana, and H.A. Makroo

**Syed Insha Rafiq, Sourav Garg, Balaji G., Irshaan Syed,
Hikhmathunnisa Sultan Noordeen, and Syed Mansha Rafiq**

Michael V. Joseph and Chetan Sharma

Thoithoi Tongbram, Jinku Bora, and H.A. Makroo

**Darakshan Majid, Sajad Ahmad Sofi, Aabida Jabeen, Farhana Mehraj Allai,
H.A. Makroo, and Shahnaz Parveen Wani**

Preface

The quality and safety of the food we eat elicits a great deal of attention and is a priority for consumers. Shelf life of food products is an important concept and is of interest to everyone in the supply chain from producer to consumer. Various factors including microbiological changes, chemical changes, and packaging interactions influence the shelf life and stability of food. Shelf life studies can provide important information to manufacturers and consumers to ensure a high-quality food product. Various evaluation methods are used for shelf life determination and are usually performed at the manufacturer level. Moreover, various techniques are used throughout the food chain to enhance the shelf life of food products. There is a lack of scientific information on the shelf life and safety under one umbrella. This sensitive issue is reviewed in this well-timed book, which brings together a group of subject experts to represent up-to-date and objective outlines on a broad range of subject matter including food spoilage and safety, preservation, packaging, and sensory aspects. The use of traditional and innovative technologies for enhancing food safety and/or increasing shelf life, along with the assessment and prediction of food safety and shelf life, are addressed in this book. *Shelf Life and Food Safety* is the first of its kind to focus on issues related to evaluation techniques for shelf life, determinants, and techniques for shelf life enhancement. This book is appropriate for students, researchers, scientists, and professionals in food science and technology. It is also a handy source of information for people involved in the food industry, processing sector, product development, marketing, and other associated fields.

Basharat Nabi Dar

Editors

Basharat Nabi Dar, PhD, is a graduate in agricultural sciences and received his MS and PhD in food technology. Dr. Dar started his professional career in 2012 as Assistant Professor of Food Technology at IUST, Awantipora, India. He was a Visiting Scientist at the Institute of Food Science, Cornell University, in the United States under the CV Raman Fellowship of the University Grants Commission, India. He is the recipient of the UGC Research Award 2014–2016 in the field of agricultural sciences. Dr. Dar is on the expert panel of the Joint FAO/WHO Expert Meeting on Microbiological Risk Assessment (JEMRA) for the prevention and control of microbiological hazards in fresh fruits and vegetables. He is the coordinator of the Food Testing Laboratory sponsored by the Ministry of Food Processing and Industries, India. He is also associated with several research projects as a PI/Co-PI member. Recognition of his contributions has provided ample opportunity to collaborate, present talks, and interact with professionals, and Dr. Dar has an active presence in public policy discourse through his position as a technical expert to WHO/FAO and the GOI. He has edited 5 books and has more than 90 publications in his field.

Manzoor Ahmad Shah, PhD, works as an Assistant Professor at the Government Degree College for Women, Anantnag, Jammu and Kashmir, India. He received an MSc degree from the Islamic University of Science & Technology, Awantipora, India, and a PhD (in food technology) from Pondicherry University, Puducherry, India. The author was awarded with the gold medal for outstanding performance during his MSc degree program. Dr. Shah has more than 40 publications in internationally reputed journals and books. He has edited four books related to food science and technology. Dr. Shah is an active reviewer for several international scientific journals of repute. He has attended several national and international conferences, workshops, and seminars.

Shabir Ahmad Mir, PhD, obtained his PhD in food technology from Pondicherry University, Puducherry, India. Dr. Mir is presently an Assistant Professor at the Government College for Women, Srinagar, India. He received the best PhD Thesis Award for outstanding research in the field of food technology. He has organized several conferences and workshops in food science and technology. Dr. Mir has published numerous international papers and book chapters and edited 10 books.

Contributors

Syed Afiya
Division of Food Science and
  Technology
Sher-e-Kashmir University of
  Agricultural Sciences and Technology
  of Kashmir
Srinagar, India

Saghir Ahmad
Department of Post-Harvest Engineering &
  Technology
Faculty of Agricultural Services
Aligarh Muslim University
Aligarh, India

Sajid Ali
Department of Horticulture
Faculty of Agricultural Sciences and
  Technology
Bahauddin Zakariya University
Multan, Pakistan

Farhana Mehraj Allai
Department of Post-Harvest Engineering &
  Technology
Faculty of Agricultural Sciences
Aligarh Muslim University
Aligarh, India
and
Department of Food Technology
Islamic University of Science and
  Technology
Awantipora, India

Tawheed Amin
Division of Food Science and
  Technology
Sher-e-Kashmir University of
  Agricultural Sciences and Technology
  of Kashmir
Srinagar, India

Muhammad Akbar Anjum
Department of Horticulture
Faculty of Agricultural Sciences and
  Technology
Bahauddin Zakariya University
Multan, Pakistan

Nadira Anjum
Division of Food Science and Technology
Sher-e-Kashmir University of Agricultural
  Sciences and Technology (SKUAST) Jammu
Chatha, India

Anjum Ayoub
Division of Food Science and Technology
Sher-e-Kashmir University of Agricultural
  Sciences and Technology (SKUAST) Jammu
Chatha, India

Omar Bashir
Division of Food Science and Technology
Sher-e-Kashmir University of Agricultural
  Sciences and Technology of Kashmir
Srinagar, India

Arjuara Begum
Department of Food Engineering and Technology
Tezpur University
Tezpur, India

Anju Bhat
Division of Food Science and Technology
Sher-e-Kashmir University of Agricultural
  Sciences and Technology (SKUAST) Jamma
Chatha, India

Jinku Bora
Department of Food Technology
Jamia Hamdard
New Delhi, India

Nitamani Choudhury
Food Processing Technology
Udalguri Polytechnic
Bhergaon, India

Basharat Nabi Dar
Department of Food Technology
Islamic University of Science and Technology
Awantipora India

Pranjal Pratim Das
Department of Chemical Engineering
IIT Guwahati
Guwahati, India

Shaghef Ejaz
Department of Horticulture
Faculty of Agricultural Sciences and
    Technology
Bahauddin Zakariya University
Multan, Pakistan

Shemilah Fayaz
Department of Food Technology
Islamic University of Science and
    Technology
Awantipora, India

Balaji G.
Anna University
Chennai, India
and
Indian Institute of Technology Guwahati
Assam, India

Neeraj Gandhi
Department of Food Science and Technology
Punjab Agriculture University
Ludhiana, India

Sourav Garg
National Institute of Technology Rourkela
Rourkela, India

Mahmood Ul Hasan
Institute of Horticultural Science
University of Agriculture
Faisalabad, Pakistan

Hikhmathunnisa Sultan Noordeen
B.S. Abdur Rahman Crescent Institute of
    Science and Technology
Chennai, India

Baby Z. Hmar
Department of Food Technology
Mizoram University
Aizawl, India

Sajjad Hussain
Department of Horticulture
Faculty of Agricultural Sciences and
    Technology
Bahauddin Zakariya University
Multan, Pakistan

Farheena Iftikhar
Department of Food Science and Technology
University of Kashmir
Srinagar, India

Abida Jabeen
Department of Food Science and Technology
SKUAST-K
Srinagar, India

Bisma Jan
Department of Food Technology
School of Interdisciplinary Science and
    Technology
Jamia Hamdard University
New Delhi, India

Nusrat Jan
Division of Food Science and Technology
Sher-e-Kashmir University of Agricultural
    Sciences and Technology of Kashmir
Srinagar, India

Smruthi Jayarajan
School of Agriculture
Lovely Professional University
Jalandhar, India

Michael V. Joseph
Prestage Department of Poultry Science
North Carolina State University
Raleigh, North Carolina

Swati Kapoor
Punjab Agriculture University
Ludhiana, India

Gurkirat Kaur
Department of Food Science and Technology
Punjab Agriculture University
Ludhiana, India

K. Rama Krishna
Department of Horticulture
School of Life Sciences
Central University of Tamil Nadu
Thiruvarur, India

Darakshan Majid
Department of Food Technology
Islamic University of Science and Technology
Awantipora, India

Ishrat Majid
Department of Food Technology
Islamic University of Science and Technology
Awantipora, India

H.A. Makroo
Department of Food Technology
Islamic University of Science and Technology
    Awantipora
Awantipora India

Arshied Manzoor
Department of Post-Harvest Engineering &
    Technology
Faculty of Agricultural Sciences
Aligarh Muslim University
Aligarh, India

Spurti Morab
Department of Dairy Microbiology
ICR-NDRI
Karnal, India

Sristi Mundhada
Anna University
Chennai, India

Aarifa Nabi
Plant Physiology Section
Department of Botany
Aligarh Muslim University
Aligarh, India

Haroon Rashid Naik
Department of Food Technology
Islamic University of Science and Technology
Awantipora, India

Vikas Nanda
Department of Food Engineering and Technology
Sant Longowal Institute of Engineering and
    Technology
Longowal, India

Farah Naqash
Department of Food Technology
Islamic University of Science and Technology
Awantipora, India

Aamir Nawaz
Department of Horticulture
Faculty of Agricultural Sciences and
    Technology
Bahauddin Zakariya University
Multan, Pakistan

Jyoti Nishad
Department of Food Technology
SRCASW
University of Delhi
New Delhi, India

Shivani Pathania
Food Industry Development Department
TEAGASC
Dublin, Ireland

Syed Insha Rafiq
ICAR–National Dairy Research Institute
    Karnal
Karnal, India

Syed Mansha Rafiq
Department of Food Technology
National Institute of Food Technology
    Entrepreneurship and Management
Sonepat, India

Mahmoud Said Rashed
Food Industry Development Department
TEAGASC
and
School of Food Science and Environmental
    Health
Technological University Dublin
Dublin, Ireland
and
Food Science and Technology Department
Faculty of Agriculture
Alexandria University
Alexandria, Egypt

Sumira Rashid
Amity Institute of Food Technology
Amity University Uttar Pradesh
Noida, India

Qurat Ul Eain Hyder Rizvi
Department of Food Technology
Eternal University
Sirmour, India

Shagufta Rizwana
Department of Food Technology
Islamic University of Science and Technology
   Awantipora
Awantipora, India

Muhammad Shafique
University of Agriculture
Faisalabad, Pakistan

Chetan Sharma
Department of Wine, Food, and Molecular
   Biosciences
Lincoln University
Christchurch, New Zealand

Nikunj Sharma
National Institute of Food Technology
   Entrepreneurship and Management
Sonepat, India

Mohd Aaqib Sheikh
Department of Food Engineering and
   Technology
Sant Longowal Institute of Engineering and
   Technology
Longowal, India

Rayees Ahmad Shiekh
Department of Chemistry
Government Degree College Pulwama
Pulwama, India

Haamiyah Sidiq
Division of Food Science and Technology
Sher-e-Kashmir University of Agricultural
   Sciences and Technology of Kashmir
Shalimar, India

Sandip Singh
Wave Beverages Pvt. Ltd.
Punjab, India

Mabrouk Sobhy
School of Food and Biological Engineering
Jiangsu University
Zhenjiang, China
and
Food Science and Technology Department
Faculty of Agriculture
Alexander University
Alexandria, Egypt

Sajad Ahmad Sofi
Department of Food Technology
Islamic University of Science and Technology
Awantipora, India

Preetisagar Talukdar
Department of Chemical Engineering
IIT Guwahati
Guwahati, India

Mamta Thakur
Department of Food Technology School of
   Science
ITM University
Gwalior, India

Thoithoi Tongbram
Department of Food Engineering and Technology
Tezpur University
Tezpur, India

Damir D. Torrico
Department of Wine, Food and Molecular
   Biosciences
Lincoln University
Christchurch, New Zealand

Meenakshi Trilokia
Division of Food Science and Technology
Sher-e-Kashmir University of Agricultural
   Sciences and Technology (SKUAST) Jammu
Chatha, India

Abira Umam
School of Vocational Studies and Applied
   Sciences
Gautam Budha University
Greater Noida, India

Sajad Mohd Wani
Division of Food Science and Technology
Sher-e-Kashmir University of Agricultural
Sciences and Technology of Kashmir
Shalimar, India

Shahnaz Parveen Wani
Department of Food Science and Technology
SKUAST-K
Srinagar, India

Sabreena Yousuf
Division of Food Science and Technology
Sher-e-Kashmir University of Agricultural
Sciences and Technology of Kashmir
Shalimar, India

Insha Zahoor
Department of Post-Harvest Engineering &
Technology
Faculty of Agricultural Sciences
Aligarh Muslim University
Aligarh, India

Aiman Zehra
Division of Food Science and Technology
Sher-e-Kashmir University of Agricultural
Sciences and Technology of Kashmir
Shalimar, India

CHAPTER 1

Food Spoilage and Safety

**Sourav Garg, Syed Mansha Rafiq, Balaji G., Sristi Mundhada,
Syed Insha Rafiq, and Nikunj Sharma**

CONTENTS

Email: Syed Mansha Rafiq, mansharafiq@gmail.com

DOI: 10.1201/9781003091677-1   1

1.1 INTRODUCTION

Food provides nutrition, thus it is susceptible to many biochemical reactions and is highly prone to various classes of microorganisms, which are inherently prone to deterioration and spoilage with time (Pitt and Hocking, 2009). This is the very basic definition:

> Food spoilage can be considered as any change which renders a product unacceptable for human consumption; it may arise from insect damage, physical damage (bruising, freezing, drying, etc.), indigenous enzyme activity in animal or plant tissues, and chemical changes (usually involving oxygen) or microbial contamination. Therefore, spoilage is a complex phenomenon, involving physical, chemical, microbiological, and biochemical changes.
>
> (Bevilacqua et al., 2016)

Physical spoilage of food is generally a result of a significant gain of moisture by excessively dried foods or by excessive dehydration of moist foods, which leads to their texture disruption. Chemical spoilage of food is a result of various chemical reactions occurring between different food components (or by reaction of the individual food components with the ambient environmental conditions), which alter its sensory attributes. Microbial spoilage of food is caused by the growth of undesirable microorganisms in the food system, which produces unwanted enzymes and by-products (Benner, 2014).

The problem of food spoilage worsens when it leads to food safety hazards, making the food unfit for consumers, and can lead to severe illness. Other cases of food spoilage can degrade the color, flavor, aroma, or texture of the product. In a few cases, it can lead to alteration of the nutrients present in the product so that it can no longer meet the declared nutritional values. The time required for a food to reach one or more of these spoilage conditions is commonly called its shelf life, which helps the consumer decide how long a particular product can be stored before usage. A food manufacturer needs to know what kind of spoilage can occur in a particular food, how to detect it, and how to effectively reduce its occurrence. Although spoilage of food or its quality deterioration cannot be stopped, reducing the rates of these processes by proper handling, various processing methods, modified formulations, appropriate packaging, and storage is always desirable (Singh and Anderson, 2004). To discuss this problem in a detailed manner, it is necessary to understand various types of food spoilage, their causes, and the factors that can lead to major food spoilage.

1.2 TYPES OF FOOD SPOILAGE

Food spoilage can broadly be categorized as physical, chemical, and microbial spoilage.

1.2.1 Physical Spoilage

Physical spoilage includes physical damage, moisture loss/gain, texture disruption, and other physical changes in the visual attributes of the food as a result of mishandling foods during storage, transportation, and distribution. In addition, fresh fruits and vegetables are also susceptible to physical damage due to poor respiration and chilling injury during storage. Migration of moisture from or into food systems can lead to spoilage of many foods. For instance, bakery products (cakes and breads) are prone to moisture loss upon storage in a dry environment, which changes their texture from moist and chewy to hard and brittle and can lead to staling (Roudaut et al., 2002). Due to the low moisture content in powdered foods, they are highly susceptible to gain moisture from their storage environment, which makes their structure amorphous and makes the particles stick together and forms lumps (Ozmen and Langrish, 2002).

Another physical spoilage that can readily occur in food systems is crystal growth. For instance, the formation of ice crystals due to slow freezing (or multiple freeze-thaw cycles) of ice cream resulting in a grainy texture of the final product (Ozmen and Langrish, 2002). Emulsion breakdown is another major category of physical food spoilage that commonly occurs in salad dressings, mayonnaise, and similar products (Depree & Savage, 2001). In case of mayonnaise, this occurs due to the auto-oxidation of fats caused by the presence of oxygen (Parreño and Carandang, 2021).

1.2.2 Chemical Spoilage

Spoilage of food products can occur as a result of the reaction between its chemical components or due to their breakdown by several factors such as temperature, water activity, pH, and storage environment. An important class of chemical spoilage in foods is the spoilage of carbohydrates, e.g., gelatinization of starch granules, which easily swell up by absorbing water and lose their crystalline structure. Starches are also susceptible to the retrogradation process that results from migration of moisture or due to high exposure of starch granules to multiple freeze-thaw cycles (Manzocco et al., 2002). The spoilage caused by the interaction of sugars and proteins can be classified as enzymatic and non-enzymatic reactions. Enzymatic reactions in fresh fruits and vegetables can result in the softening of tissues and the formation of brown pigments, as shown in Figure 1.1. The spoilage of lipids in foods may also occur due to oxidation, enzymatic degradation, or other hydrolytic reactions. In hydrolytic rancidity, the fatty acids separate from triglycerides due to the reaction of lipolytic enzymes in the presence of water. The resulting smaller fatty acids have lesser flavor and can sometimes acquire a rancid odor. However, these reactions mostly occur at elevated temperatures

Figure 1.1 Chemical spoilage of some food products.

above 60°C and can be inhibited by minimizing the interaction of lipid molecules with moisture (Kilcast and Subramaniam, 2000).

1.2.3 Microbial Spoilage

Microbial spoilage is one of the most common and significant categories of spoilage in all types of foods. It accounts for nearly 25% of the total food loss worldwide and various foodborne diseases (Bondi et al., 2014). Numerous bacteria are known to be responsible for causing food spoilage and major foodborne illness. Common examples include *Escherichia coli* O157:H7, *Listeria monocytogenes*, *Salmonella* spp., *Bacillus cereus*, *Campylobacter jejuni*, and *Vibrio* spp. The rate of food spoilage caused by bacteria is quite high and is most predominant in high-protein foods like fish, dairy, and meat products. Although yeasts are an important class of microorganisms for the production of fermented foods, several of them like *Candida* spp., *Saccharomyces* spp., *Dekkera* spp., and *Zygosaccharomyces* spp. are responsible for major food spoilage (Petruzzi et al., 2017).

Among the molds causing food spoilage, *Rhizopus* spp., *Aspergillus* spp., *Penicillium* spp., and *Fusarium* spp. are the predominant ones with some species of *Aspergillus* producing aflatoxins that are responsible for foodborne illness (Ellis et al., 1991). Although viruses cannot grow in the food, they can easily be sustained once they get inside the system. Such viruses can enter the food product either through a previously infected plant or animal, by cross-contamination during handling and processing, or due to infestation by insects and rodents. These viruses evidently cause foodborne illnesses such as hepatitis A, rotavirus, mad cow disease, and Norwalk virus (Singh and Anderson, 2004).

1.3 FACTORS AFFECTING FOOD SPOILAGE

Food spoilage generally occurs by several intrinsic characteristics of the food as well as by various environmental factors. The factors affecting physical and chemical spoilage of foods has briefly been discussed in the previous section. In this section, the focus will be on the major factors that affect microbial spoilage in food systems.

1.3.1 Structure and Composition of Food

Raw foods are naturally resistant to microbial infestation due to the rigid structure of their coverings (fruit and vegetable peels, shells, bran, etc.). However, these products become damaged physically during various processing operations and can easily get infested by microbes and become vulnerable to spoilage due to enzymes (Petruzzi et al., 2017). In general, solid foods have a tendency to degrade first from outer surfaces followed by interior spoilage. Liquid and semi-liquid foods are more prone to spoilage due to the higher mobility of microbes within them (Modi, 2009).

The nutritional composition of the food plays a vital role in affecting the type and rate of microbial spoilage. Sugar-rich foods are generally vulnerable to fermentative microorganisms, whereas protein-rich and high-fat foods are more prone to proteolytic and lipolytic microbes, respectively (Modi, 2009). Many of the spoilage reactions are also catalyzed by the presence of various metal ions and other trace elements. On the other hand, the presence of several antimicrobial compounds (that are found naturally in foods) can inhibit or reduce the rate of such spoilage and can extend the shelf life of the product (Davidson and Critzer, 2012). For instance, lysozyme present in egg albumin ruptures the cell walls of Gram-positive bacterial species and prevents degradation of proteins (Davidson and Critzer, 2012). Other antimicrobial food components that hinder the growth of spoilage microbes are nisin, sorbates, citrate, butylated hydroxyanisole, benzoates, sulfur dioxide, etc. (Jeantet et al., 2016). The presence of artificial sweeteners like aspartame, acids, and other hydrocolloids can also inhibit spoilage caused by the growth of *Zygosaccharomyces bailii* (Campos et al., 2014).

1.3.2 Water Activity

Water activity (a_w) is the ratio of the vapor pressure of food to that of pure water under similar conditions. The most desirable range for growth of most of the microorganisms is between 0.995 and 0.98. A slight reduction in water activity of the food system (by a reduction in moisture content or by addition of salt, sugar, etc.) can greatly decrease the growth rate of the microbes present in it and can enhance its shelf life to a large extent. In general, bacteria require higher water activity for their growth and cannot be sustained easily in dry food products (Modi, 2009). It is worthy to note that several bacterial species like *Brevibacterium linens* produce osmoprotectants like ectoine, which helps them to store water inside their cytoplasm and can be used for sustenance in dry foods (Jeantet et al., 2016).

1.3.3 pH

pH is a measure of the hydrogen ion concentration of the product, which is related to its acidity level. Every microorganism has an optimum range of pH for its growth and metabolic activities. Although most of the bacteria grow rapidly in the neutral pH range (which is similar to the pH of their cytoplasm, and favorable for enzymatic activities), there are a few exceptions that can withstand extreme ranges (Jeantet et al., 2016). On the other hand, yeasts can easily grow in the pH range of 4.5–7.0, and some can even grow in the 2.0–2.5 pH range (Modi, 2009). Higher growth rates of yeasts at lower pH levels help them compete with and dominate other classes of microbes, including bacteria (Howell, 2016). A few proteolytic bacterial species can easily grow in a higher pH range by producing amines and buffering the higher pH conditions (Davidson and Critzer, 2012).

1.3.4 Temperature

Every microorganism grows within a particular range of temperature with the maximum growth rate at the optimum temperature. Therefore, the microorganisms can be categorized as psychrophiles, mesophiles, and thermophiles with their optimum growth range of 12–15°C, 30–45°C, and 52–75°C, respectively (Modi, 2009). Generally, the lag phase for the growth of microorganisms tends to decrease with an increase in temperature, which aids in better growth during a subsequent period of time. At the same time, a higher temperature also affects other enzymatic activities and synthesis of proteins and, therefore, shelf life of the product. Yeasts and molds, on the other hand, can grow easily at or below room temperature and can be accountable more often for spoilage of products stored in cold and chilled conditions (Modi, 2009). Bacterial and fungal spores can be considered to be the most heat-tolerant organisms, with a few of them surviving very high temperatures above 100°C (Jeantet et al., 2016).

1.3.5 Gas Tension

The oxidative tension and redox potential of food highly influence the type of microbes that can grow within the system. For instance, the growth of aerobic microbes and spoilage caused by them is most dominant at the outer surface of foods, whereas facultative microbes tend to grow both on the surface level and inside the food, and anaerobic microbes tend mostly to grow only inside the system (Modi, 2009).

1.4 CONTROL OF FOOD SPOILAGE

The spoilage of food indicates the degradation or deterioration of nutritional and sensory qualities along with the safety of the food for human consumption. The first step to control spoilage is to implement knowledge about the spoilage microorganisms and use of various systems, practices, and processes. At the industry level, implementation of the Hazard Analysis Critical Control Point (HACCP) plan along with a good manufacturing process and, finally, risk assessment is done to control or detect spoilage (Panisello and Quantick, 2001). Food preservation by heat, packaging, processing, and storage reduces the probability of food spoilage factors in its environment. The key preservative strategies are discussed below.

1.4.1 Preservation by Application of Heat

Enzymes and microorganisms are destroyed at high temperature. The extent of damage to spoilage microbes and enzymes depends on the temperature, exposure time, and of course the heat resistance of the given medium. Exposure to high temperatures is not just destructive to microorganisms and enzymes, it also speeds up a multitude of chemical reactions that lead to changes in the texture, taste, appearance, color, digestibility, and nutritional value of the product (Tewari and Juneja, 2008).

1.4.2 Preservation by Removal of Heat

At low temperatures, the activity of microorganisms and enzymes and the rate of chemical reactions are suppressed. Low temperature, unlike heat, does not dramatically kill enzymes and microorganisms, it merely depresses their activity. Cold storage requires two distinct technical processes: cooling (keeping the food well above the freezing point of the liquid) and freezing (cooling it below the freezing point). A significant part of food preservation by freezing effect is due to the phase change, from liquid to solid, resulting in a decrease in molecular mobility. The transition from liquid to the ice of part of the water results in a significant decrease of the water activity, which in turn reduces the microbial and enzyme activities (Wu et al., 2014).

1.4.3 Preservation by Reduction of Water Activity

Microorganisms are susceptible to low water activity as their metabolic movement is hindered by restricted water activity. Enzyme activity is also dependent upon water activity. Dehydration and adding solutes (sugar, salt) are techniques of preservation focused on the reduction of water movement. The glass transition phenomenon and the subsequent decline of molecular mobility are also a consequence of the combined impact of low water activity and low temperature, thereby preventing food spoilage (Amit et al., 2017).

1.4.4 Preservation by Radiation

Ionizing radiation has the potential to kill microorganisms and inactivate enzymes. The process of ionizing radiation disrupts the cellular homeostasis, thereby generating radiolytic products that generate irreversible changes in microorganisms and enzymes. It also controls the spoilage of grains by pest infestation (Karel and Lund, 2003).

1.4.5 Preservation by Chemical Methods

The effect of salt and smoke chemicals on microorganisms is centered on two of the oldest food preservation techniques, namely salting and smoking. Many pathogens do not grow at low pH;

hence, acids can act as preservatives for food. The pH of foods can be reduced either by adding acids (acetic, citric, lactic, etc.) or by producing acids in situ (mainly lactic) by fermentation. Although today it would be absurd to define wine as "preserved grape juice," it may be assumed that alcoholic fermentation was a type of chemical preservation at its origin. Many spices and herbs are used to avoid spoilage of food or to delay it. Many non-natural chemical compounds (e.g., sulfur dioxide, benzoic acid, and sorbic acid) are effective in preventing or retarding other forms of food spoilage (preservation against microbial spoilage, and oxidative degradation, or stabilization against unwanted changes in texture and structure, etc.) (Lück, 1985).

1.4.6 Novel Preservation Techniques

The growing consumer demand for nutritious, fresh-like foods has initiated significant research and development activities towards creating alternative preservation processes for the production of these foods. These processes can be classified into thermal and non-thermal technologies. The non-thermal technologies include high hydrostatic pressure, pulsed electric fields, irradiation, ultrasound, cold plasma, and ozone supercritical water, whereas thermal technologies include microwave, radiofrequency, and ohmic and inductive heating (Hameed et al., 2018).

1.4.7 Hurdle Technology

Most of the shelf-stable foods available in the market have undergone a combination of different processing, preservation, and packaging strategies, which is known as hurdle technology. Employment of different hurdles that targets water activity, redox potential, relative humidity, etc., will generate huge stress on the microbes, thereby rendering them metabolically inefficient to cope, leading to cell death and preventing food spoilage. As an example, one can assume that dry sausage's shelf life and safety is provided by the combined effect of low water activity, comparatively low pH, smoking, salt, cold temperatures, and perhaps spices, but each of these "hurdles" are not large enough to deliver the intended preservation when acting alone (Leistner, 2000).

1.4.8 Packaging

With an increased demand for an extended shelf life of food without the addition of preservatives, packaging is the most successful alternative to control the spoilage of food. Packaging has the main purpose of establishing a protective barrier between food and the environment. Food packaging preservation functions include preventing microbial contamination, regulating the exchange of materials (water vapor, oxygen, odorous substances) between the food and its atmosphere, light safety, etc. Recently, packaging materials have been produced that contain preservatives such as antioxidants and antimicrobial compounds (Mathlouthi, 2013). New techniques like active packaging have beneficial interactions with the internal packaging environment and food to improve shelf life while preserving the nutritional quality, inhibiting infestations by spoilage and pathogenic microorganisms, and preventing contaminant migration. Antimicrobial packaging has earned tremendous praise because it prevents the spoilage of organisms and prolongs the food product's shelf life. Food products such as fish, poultry, meat, cheese, fruit, and vegetables are packaged extensively in antimicrobial containers as they are prone to rapid microbial infestation due to their high water activity (Pradhan et al., 2015).

1.5 DETECTION AND IDENTIFICATION METHODS OF SPOILAGE

Detection and identification of spoilage is one of the important parameters in modern food industries. The health risks associated with consuming spoiled or contaminated foods are microbially debilitating and potentially lethal. Food products such as soda, salad dressing, etc., are often contaminated by *Zygosaccharomyces* and *Torulaspora* species, whereas beer and wine are vulnerable to *Brettanomyces*, *Saccharomyces*, and a variety of other genera. Spoiled cheese and meats often have high counts of *Rhodotorula*, *Yarrowia*, and *Debaryomyces*. Various methods used for the detection of spoilage microorganisms in different food products are discussed in the following sections.

1.5.1 Biosensors

A biosensor is an analytical tool that combines a biological component with a physicochemical detector for the identification of a chemical substance. Recent techniques used enzymatic reactor systems with amperometric electrodes to sense levels of diamine and thus assess the consistency of beef, fish, and poultry. Postmortem enzymatic activity also spoils muscle food. Biosensors dependent on DNA or RNA, mass spectroscopy, bioluminescence, and similar systems are being developed to detect bacteria in foods (Freitas et al., 2011); however, in the field of mold detection little has been achieved.

1.5.2 Microbial Metabolites

Microbial metabolites are potential food spoilage detection methods in meat-related products. Once bacteria have used glucose at surface levels, they will then metabolize other substrates consecutively, such as free amino acids. To achieve this, many bacteria secrete a variety of proteolytic enzymes, with Gram-negative bacteria predominantly secreting amino peptidases at refrigeration temperatures on the meat (Davies et al., 1998). Several researchers have proposed that this involvement of enzymes could be used rapidly with the use of enzyme assays to measure meat stability in terms of bacterial numbers. The production of amines, ammonia, indoles, and other pH-reducing compounds also increases the spoilage indicated as a result of amino acid utilization.

1.5.3 ELISA

Currently, many newer techniques have been developed such as the use of the enzyme-linked immunosorbent assay (ELISA), which is an immunological method for the detection of food spoilage. This method uses enzyme labeling to target the identification of by-products of spoilage organisms (such as *E. coli* O157: H7, toxins developed by *Staphylococcus aureus* and proteases from the *Pseudomonas* genus of food spoilage) related to meat, fish, and poultry spoilage (Jabbar and Joishy, 1999). Ergosterol is the main sterol present in most fungi (molds and yeasts) but not in plants and it absorbs at ultraviolet (UV) ranges of 240–300 nm; thus, it has been used to estimate total fungal biomass in soil, plants, and cereals. The application of the use of ergosterol and ELISA has been performed in many research projects. Nonspecific ELISA further detected 6 types of molds in the mixture (*Aspergillus*, *Cladosporium*, *Fusarium*, *Geotrichum*, *Mucor*, and *Penicillium*) as well as 9 other genera (*Byssochlamys*, *Eurotium*, *Leptosphaerulina*, *Monascus*, *Neosartorya*, *Talaromyces*, *Trichoderma*, *Trichothecium*, and *Verticillium*).

1.5.4 Polymerase Chain Reaction

Another tool developed is the use of nucleic acid-based detection of spoilage microbes known as polymerase chain reaction (PCR). This tool detects different gene sequences that specifically help identify the spoilage microorganism. Because of the use of PCR-based typing techniques such as fingerprinting by microsatellite PCR and random amplified polymorphic DNA RAPD analysis (Williams, 1989), there is now an opportunity to differentiate individual spoilage strains, whether of the same or different species, and to identify their sources of origin (Couto et al., 1996). A listing for this kit for the identification of fungi in foods by DNA fingerprinting is available in Fung (2002);

however, no current activity is reported on its recognition. With further work, a DNA-based method for the detection of molds in foods could become commercially available in future.

1.5.5 Quantitative Analysis

The current quantitative methods available to enumerate microorganisms in foods are based on adenosine triphosphate (ATP) bioluminescence, electrical phenomena, or microscopic methods of measurements. These methodologies are based on quantifying the amount of ATP in the cell culture of the bacteria, which in turn can supply the data for the number of cells. Also, these approaches are efficient in quantifying pollution on industrial machines. Microscopic approaches include the application of direct epifluorescent filter techniques (Eed et al., 2016).

1.5.6 Organoleptic Detection

Spoilage can be detected by changes in organoleptic properties such as odor change, visible micro-flora colonies, meat surface stickiness, and flavor variation. The changes in the organoleptic odor are now computed by electronic noses and have been very successful in meat and fish testing (Gardner and Bartlett, 2000). The detection of volatiles as a measure of the growth of molds was used in the production of electronic noses. Söderström et al. (2003) experimented with measuring mold growth in malt extract solution using an electronic tongue similar to an electronic nose in that it contains multiple metal sensors that sense nonvolatile compounds in liquid.

1.5.7 Fourier Transform Infrared Spectroscopy

Fourier transform infrared (FT-IR) spectroscopy involves observing molecules that are excited by an infrared beam, resulting in an infrared absorbance spectrum that is a "fingerprint" feature of any chemical or biochemical substance. In the case of meat products, the infrared study of muscle foods focused mainly on authenticity and adulteration studies, such as the distinction among beef, ox liver, and pork; identification of offal beef adulteration (Al-Jowder et al., 2002); and attempts to classify species such as pork, chicken, and turkey. Near-infrared spectroscopy (NIRS) and linear discriminate analytics software can sort the wheat kernels into scab-damaged, mold-damaged, and sound kernels groups with 95% accuracy. Wang et al. (2004) reported similar accuracy percentages by NIRS for sorting mold-damaged soybeans combined with the analysis of partial least squares (PLS) and neural network data. FT-IR photo acoustic spectroscopy, which samples the surface of solid particles, has been used to detect *Aspergillus flavus* in maize, and an artificial neural network was developed to differentiate between contaminated and non-contaminated maize. Reflectance spectroscopy as figured in NIR was combined with a hyperspectral imaging device to establish a better imaging technique for detecting defects on the surface of an apple including mold rots.

1.5.8 Denaturing Gradient Gel Electrophoresis

Denaturing gradient gel electrophoresis (DGGE) is a promising technique used in foods and beverages to classify organisms and quantify yeast populations. This technique relies on separating DNA fragments of different nucleotide sequences (e.g., species specific) in a polyacrylamide gel containing a linear gradient of DNA denaturants (a combination of urea and formamide) by decreasing electrophoretic mobility of liquified double-stranded DNA amplicons. A related technology is temperature gradient gel electrophoresis (TGGE), in which a temperature gradient replaces the gel gradient of DGGE (Muyzer and Smalla, 1998). Recent DGGE applications include yeast recognition and population dynamics in sourdough bread (Meroth et al., 2003), in coffee fermentation (Masoud et al., 2004), and in wine grapes (Prakitchaiwattana et al., 2004).

1.6 FOOD SAFETY

The term "food safety" represents different perspectives to different audiences. It is generally referred to as preparation, handling, and storage with intent to minimize the risk of an individual from foodborne illness (Hentges et al., 2005). Food safety is a global concern as it covers most of the essential areas of everyday life. Food encompasses many hazards that result in illness among humans (Hussain, 2016). The major contributors for foodborne illness start from the farm, which includes excessive use of agrochemicals, using contaminated water for field irrigation, chemical migration from industries and their surrounding sites, etc. There are various other contributors that pose food safety hazards including microbiological contamination, food adulteration, inappropriate storage, etc. The 21st century witnessed an increased awareness among consumers regarding safe and wholesome food (Kilcast and Subramaniam, 2011). This has impacted world trade and demand for a safe food supply. Thus, analysis of risks associated with food has become more essential. Different identification techniques to detect different hazards are given in Table 1.1.

1.6.1 Hazards in Food

The Codex Alimentarius Commission defines hazard "as a biological, chemical or physical agent in, or condition of, food with the potential to cause an adverse health effect." Therefore, food hazards can be categorized into the following categories: physical, chemical, biological, and allergens.

1.6.1.1 Physical Hazards

Any extraneous inanimate substance found in food that poses a hazard to humans is known as a physical hazard or contaminant. The most common physical hazards include glass pieces, wooden pieces, hair, iron nails, stone, metal insulation plastics, etc. These hazards might lead to internal bleeding, choking, cuts, broken teeth, etc. In some cases, the removal of these materials from the body requires surgery (Wallace et al., 2018).

1.6.1.2 Chemical Hazards

Naturally occurring substances or process-induced chemicals that cause foodborne illness are known as a chemical contaminants or chemical hazards. They originate from various sources such as soil, environment, disinfectant, packaging materials, personal care products, air, water, etc. (Tang et al., 2009). Naturally occurring toxins include biological toxins and mycotoxins such as aflatoxin, ochratoxin, etc. There has been emerging documentation of the accumulation of numerous persistent environmental contaminants in animals, poultry, fish, and dairy products. Pesticide residues such as organophosphates and organochlorine compounds are always found in agricultural commodities such as fruits, vegetables, etc. Compounds such as mercury and perchlorate sometimes occur naturally and enter the food chain posing a threat. The risk associated with these chemical contaminations is dose dependent, which depends on the level of the chemical in the food and the amount of contaminated food consumed.

1.6.1.3 Biological Hazards

Any living organism or compound produced by an organism that causes a threat and health ailments to humans are termed biological hazards or biological contaminants. These contaminants include substances generated by humans, rodents, bugs, and microbes (Chatterjee and Abraham, 2018). Biological contamination is the main source of foodborne diseases and food intoxication (Schirone

Table 1.1 Identification Techniques to Detect Different Hazards

Hazard	Detection Method	Principle	Target	Reference
Physical	Metal detectors	Electromagnetic (EM) detection creates an electromagnetic field through which a food product passes. The output signal is recorded by the signal receiver that records any disturbance in the EM field, which occurs when there is the presence of metal.	This equipment is suitable for foods that conduct electricity, i.e., frozen foods and packaged foods. It can detect the ferrous material contaminants	Batt (2016)
	Optical sorting	The product is made to pass through a broad spectrum of lasers under the scan zone. The signals reflected while passing the scan zone are evaluated against the standard input. If found defective or foreign, the object is removed by a jet at high speed	Useful for items such as fruits, vegetables, nuts, seeds, etc.	Lan et al. (2017)
	X-ray detectors	X-ray detectors work on the principle of adsorption. When X-rays pass through food, a part gets absorbed and reflected. The foreign objects are detected by the image processing software	This technique can also be employed for image analysis, which increases its versatility in contaminant detection, for example, filling level control, control of missing items or product, fat analysis, broken parts control, or mass evaluation	Demaurex and Sallé (2014)
	Thermal imaging	Uses thermal scanning of heated food to detect the foreign object based on the thermographic image generated by the product	Applied to products that require heating and cooling such as roasting of nuts, seeds, etc.	Vadivambal and Jayas (2011)
Chemical	Chromatographic techniques	The major chromatographic techniques used for the identification of contaminants include gas chromatography and liquid chromatography	Detection of most of the chemical components like pesticide residues, biological toxins, perfluorochemicals, bisphenols, polychlorinated biphenyls (PCBs), organophosphate pesticides, etc.	Medana (2020)
	Immunoassay	The most common method for the immune assay is ELISA. Studies show that high-affinity immune detection methods allow one to find the aflatoxins, pesticide residues, and even veterinary drug residues if probed against the particular antibody	The detection of naturally occurring toxins and drug residues	Raeisossadati et al. (2016)
	Biosensor	The principle of lateral flow technology is based on the response of liquid movement through an adsorption membrane. This movement triggers the detachment of target moieties in the test strip	This type of assay could be employed for most types of chemical contaminants like mycotoxins, fish toxins, agrochemicals, veterinary drugs, unconventional chemical hazards like melamine, environmental contaminants like PCB, dioxins, etc.	Zhang et al. (2015)
	Aptasensing technology	Application of nano-materials like carbon nanotubes, magnetic nanoparticles, graphene, quantum dots, and nano-rods along with fluorescent, colorimetric, electro-chemiluminescent, chemiluminescent, for detection of food contaminants	Many reagents label-free one-step analytical methods have been developed with nanomaterial-based optical apta sensor technology for a wide range of contaminants	Lan et al. (2017)

(Continued)

Table 1.1 Identification Techniques to Detect Different Hazards (*Continued*)

Hazard	Detection Method	Principle	Target	Reference
Biological	Electronic nose (eNose)	Enose is mainly intended to identify the odor and smell of simple to complex flavors	For detection of contaminants in dairy products. Studies show that this device successfully detected the presence of *Escherichia coli*, *Enterobacter aerogenes*, and *Pseudomonas aeruginosa*, which are potent spoilage microbes responsible for the spoilage of milk	Ghasemi-Varnamkhasti et al. (2018)
	Polymerase chain reaction (PCR)	PCR is capable of amplifying a specific fragment of DNA. It has been used in pathogen diagnostics. With the increasing amount of sequencing data available, it is possible to design quantitative PCR (qPCR) assays for every microorganism (groups and subgroups of microorganisms, etc.) of interest	A wide range of pathogenic microbes, along with their behavior could be studied by including its viable factors such as its environment, food components, temperature, pH, aerobic and anaerobic conditions, etc.	Chatterjee and Abraham (2018)
	Lateral flow immunoassay	The principle of lateral flow technology is based on the response of liquid movement through an adsorption membrane	Lateral flow immunoassay has been successfully employed in the identification of pathogenic bacterial strains such as *Listeria* sp., *E. coli*, *Bacillus anthracis, Staphylococcus aureus, Streptococcus* sp., *and Yersinia pestis*, along with different strains of *Salmonella and Campylobacter*	Raeisossadati et al. (2016)

et al., 2017). Symptom onset of foodborne diseases does not start immediately after consumption. Food intoxication happens when an individual expends food intoxicated with a toxin. The manifestations show up quickly once the food is devoured. There are six sorts of microorganisms that can cause foodborne ailments: bacteria, viruses, parasites, protozoa, fungi, and prions.

Common bacteria causing foodborne illness include *Salmonella* sp., enterohaemorrhagic *E. coli*, *L. monocytogenes*, *S. aureus*, *Clostridium botulinum*, *Campylobacter* sp., and *Pseudomonas* sp. (Rawat, 2015). Common yeast and mold causing foodborne infections include *Candida* sp., *Aspergillum* sp., and *Helicosporium*. Common viruses causing foodborne infections include norovirus, rotavirus, adenovirus, and western Nile virus (Pinu, 2016).

1.6.1.4 Allergens

An allergen is any substance that is harmless but can trigger an immune response that leads to an allergic reaction in a vulnerable person. These allergens are mostly proteins, which are certain additives added in food that can trigger an allergic response in an individual. Common allergens present in food include gluten, lactose, peanut, egg and egg products, fish and fish products, crustaceans, etc. Handling of allergens lies in the appropriate labeling of the product. While manufacturing products in the same line (i.e., normal and special products such as allergen-free products), it must be strictly noted that beginning from the raw materials to product storage everything must be placed separately to prevent cross-contamination. One cannot prevent allergens, but the consumer can be informed by proper information on the packaging indicating the presence of an allergen in legible terms (Metcalfe et al., 2011).

1.7 REGULATIONS TO CONTROL FOOD SPOILAGE

Food safety implies the assurance that the consumed food will not cause any illness to an individual. Hence, food industries around the world adhere to various regulations that inscribe quality programs based on assessment and testing of food materials for potential hazards. The legitimate prerequisites for food safety and food quality have been set up by numerous national governments to protect consumers and guarantee that food is fit for consumption. These necessities are contained in food laws and regulations, the extent of which fluctuates starting with one nation then onto the next. Some of the world food standards are created by

- Codex Alimentarius Commission
- United States Food and Drug Administration (USFDA)
- Food safety and Standards Authority of India
- European Food Safety Agency

Although various standard agencies prevail across the globe, they are all bound by uniform standards and minimum requirements as stated by the Codex Alimentarius Commission, which is a separate statutory body created by the Food and Agricultural Organization and World Health Organization. The following regulations propose basic hygiene practices and certain quality requirements:

- Good Manufacturing Practices (GMP)
- Hazard Analysis and Critical Control Point (HACCP)
- Food Safety Management System (FSMS)

1.7.1 Good Manufacturing Practices

The minimum mandatory prerequisites for sanitary and phytosanitary conditions, as directed by the regulatory authority for food processing conditions, are collectively known as Good Manufacturing

Practices (GMP). GMP form the basis for the implementation of the food safety program within the food processing industry. The principle objective of this program is to meet the legal regulations and standards of the government to guarantee the production of safe food (Jarvis, 2014). Various companies use GMP standards as a significant tool for the consideration of their procurement of raw material, packaging materials, etc. There are many standards available for GMP across different regulations across the globe:

- Codex General Principles of Food Hygiene
- USFDA: current Good Manufacturing Practices (cGMP)
- British Retail Consortium (BRC)
- Schedule 4 of Indian Food Safety and Standards Regulation 2011

GMP encompass all mandatory guidelines as discussed in the Codex Code of Hygiene Practices.

1.7.2 Hazard Analysis and Critical Control Point

The HACCP system is a framework of science-based methodology that identifies specific threats and measures to control it while guaranteeing the safety of food (Wallace et al., 2018). It is a preventive way to examine and address hazards at each progression to produce a safe and quality product employed throughout the stages of the food supply chain. The effective implementation of HACCP in a company requires concrete support and involvement from top management and the workforce.

HACCP framework significantly relies upon the basic foundations of prerequisite programs (PRP) such as GMP and Good Hygiene Practices (GHP). These PRPs provide fundamental ecological and working conditions that are vital to the production of safe and wholesome food. The Codex Alimentarius General Principles of Food Hygiene portray the essential conditions and practices required for foods intended for worldwide trade. The application of HACCP is compatible with the implementation of quality management systems, such as the ISO 9000 series, ISO 22000, and FSSC 22000 and is the system of choice in the management of food safety within such systems.

HACCP rests on the following seven principles (Corlett, 1998),which have been universally acknowledged by government bodies, trade associations, and the food industry around the globe (Al-Zuhairi, 2016):

1. Conduct hazard analysis
2. Determine critical control points
3. Establish critical limits
4. Establish monitoring process
5. Establish corrective actions
6. Establish verification process
7. Establish documentation and record-keeping

The entire HACCP system largely relies on the type of product and process that a food industry manufactures. It varies for different types of products that have different processes (Jan et al., 2016). The Codex Alimentarius Commission formulated a model comprising 12 sequential steps for the development and implementation of HACCP in an industry. This is collectively known as the logical sequence for the application of HACCP. The first and foremost duty of a food establishment is to form HACCP team which comprises different sections of the industry such as production, stores, quality control, research and development, supply chain, etc., such that the hazard at each step is identified and monitored (Mortimore, 2000). These people should be well aware of the product and process for which the HACCP plan has to be implemented. The success of HACCP greatly depends on the commitment provided by top management (Wallace et al., 2018). The HACCP system involves continuous monitoring, verification, record keeping, and updating as required.

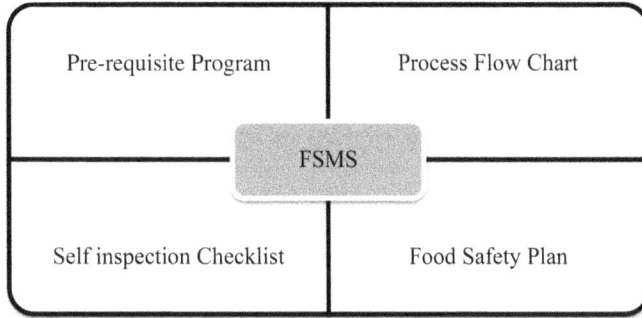

Figure 1.2 Structure of FSMS implementation.

1.7.3 Food Safety Management System

An FSMS is a system of interrelated components that adjoin to ensure food safety and prevent adverse human health effects caused by food spoilage (Figure 1.2). These components are included in all quality programs, project plans, government and regulatory policies, trade policies, process control, employee roles and responsibilities, documents, records, etc. (Mensah and Julien, 2011). FSMS underlines the incorporation of elements of GMP and HACCP (Untermann, 1999). In 2005, the International Organization for Standardization published the ISO 22000:2005 standard, which are FSMS requirements, widely pertinent to all establishments that operate in the food chain (Mensah and Julien, 2011).

The FSMS has five basic key elements:

1. PRPs such as GMP and GHP
2. HACCP
3. Management
4. Statutory and regulatory requirements
5. Communication

1.8 RELATIONSHIP BETWEEN FOOD SAFETY AND FOOD SPOILAGE

Despite interventions from governments, regulatory bodies, and various food safety programs, foodborne outbreaks are on the rise worldwide. There has been increased speculation of new threats because of alterations in the characteristics of microorganisms, their evolution, production strategies, environmental influence, etc. These risks arise as we foster new technologies in food processing such as genetically modified organisms (GMOs), proteomics, and the intervention of nanotechnology. As we discover new advancements in the field, analyzing the pace for new risks associated with it is clearly very important. Thus, the new challenge for food safety points out microbiological food safety. The efficient way to keep track of this threat is to develop a systematic risk assessment program, which includes demands for the safety of food, an evidence-based scientific approach, and well-designed risk assessment and risk management protocols. The most challenging risk among microbiological food safety is the possible significance of spoilers created by opportunistic pathogens and the evolution of antibiotic resistance within the system of microbes.

The food industry currently focuses on food safety by solely following conventional methods; instead, it should focus on conventional reactive methods to achieve complete safety by upgrading its assessment tools and methods. Microbiological safety could be mitigated by regular assessment of target organisms for their mutations, antibiotic resistance, internal and extraneous genetic modifications, etc., thereby alternative safety procedures could be developed and employed.

1.9 CONCLUSION

Foodborne diseases pose a significant threat to the community, individual, and food industries. There are numerous factors, such as lack of quality of raw materials, poor personal hygiene, and improper handling of materials that are responsible for food spoilage. To ensure the safety and authenticity of food delivered to the consumer, stringent enforcement of standards and codes of practices such as GMP. GHP, HACCP, and FSMS were established and followed. Current concerns about the various hazards to food in all stages of the supply chain could be prevented by the enforcement of these standards. This chapter gave a brief outlook about the various types of spoilage factors, detection and prevention of hazards, and regulations that are being widely discussed. A range of advanced tools and continuing research efforts should continue to be applied to address all food quality issues and to ensure brand and consumer protection.

REFERENCES

Al-Jowder O., Kemsley E. & Wilson R. H. (2002). Detection of adulteration in cooked meat products by mid-infrared spectroscopy. *Journal of Agricultural and Food Chemistry.* 50:1325–1329.

Al-Zuhairi A. J. (2016). Hazard analysis and critical control point HACCP system. *Iraqi National Journal of Chemistry.* 16:172–185.

Amit S. K., Uddin M., Rahman R., Islam S. M. & Khan M. S. (2017). A review on mechanisms and commercial aspects of food preservation and processing. *Agriculture & Food Security,* 6(1), 1–22.

Batt C. A. (2016). Chemical and physical hazards in food. *Reference Module in Food Science, 1* (2). DOI:10.1016.B978-0-08-100596-5.03437-5.

Benner Jr R. (2014). Organisms of concern but not foodborne or confirmed foodborne: Spoilage microorganisms. *Encyclopedia of Food Safety.* Elsevier; Vol. 2, 245–250.

Bevilacqua A., Corbo M. R. & Sinigaglia M. (2016). *The Microbiological Quality of Food: Foodborne Spoilers.* Woodhead Publishing.

Bondi M., Messi P., Halami P.M., Papadopoulou C. & de Niederhausern S. (2014). Emerging microbial concerns in food safety and new control measures. *Biomed International.* 2014:251512.

Campos C.A., Gliemmo M.F. & Castro M.P. (2014). Strategies for controlling the growth of spoilage yeasts in foods. *Microbial Food Safety and Preservation Techniques.* CRC Press; 516–531.

Chatterjee A. & Abraham J. (2018). Microbial contamination, prevention, and early detection in food industry. *Microbial Contamination and Food Degradation.* Elsevier; 21–47.

Corlett D. A. (1998). *HACCP User's Manual.* Springer Science & Business Media.

Couto M.B., Hartog B., In't Veld J.H., Hofstra H. & Van der Vossen J. (1996). Identification of spoilage yeasts in a food-production chain by microsatellite polymerase chain reaction fingerprinting. *Food Microbiology.* 13:59–67.

Davidson P. M. & Critzer F. M. (2012). Interventions to inhibit or inactivate bacterial pathogens in foods. *Microbial Food Safety.* Springer; 189–202.

Davies A.R., Board R. & Board R. (1998). *Microbiology of Meat and Poultry.* Springer Science & Business Media.

Demaurex G. & Sallé L. (2014). Detection of physical hazards. *Food Safety Management.* Academic Press; 511–533.

Depree J. & Savage G. (2001). Physical and flavour stability of mayonnaise. *Trends in Food Science & Technology.* 12:157–163.

Eed H.R., Abdel-Kader N.S., El Tahan M. H., Dai T. & Amin R. (2016). Bioluminescence-sensing assay for microbial growth recognition. *Journal of Sensors.* 2016:1492467.

Ellis W., Smith J., Simpson B., Oldham J. & Scott P.M. (1991). Aflatoxins in food: occurrence, biosynthesis, effects on organisms, detection, and methods of control. *Critical Reviews in Food Science & Nutrition.* 30:403–439.

Freitas F., Alves V.D. & Reis M.A. (2011). Advances in bacterial exopolysaccharides: from production to biotechnological applications. *Trends in Biotechnology.* 29:388–398.

Fung D.Y. (2002). Predictions for rapid methods and automation in food microbiology. *Journal of AOAC International.* 85:1000–1002.

Gardner J.W. & Bartlett P.N. (2000). Electronic noses. *Measurement Science and Technology.* 11:1087.

Ghasemi-Varnamkhasti M., Apetrei C., Lozano J. & Anyogu A. (2018). Potential use of electronic noses, electronic tongues and biosensors as multisensor systems for spoilage examination in foods. *Trends in Food Science & Technology.* 80:71–92.

Hameed F., Ayoub A. & Gupta N. (2018). Novel food processing technologies: an overview. *International Journal of Chemical Studies.* 6(6):770–776.

Hentges D. L., Schmidt R. H. & Rodrick G. E. (2005). Safe handling of fresh-cut produce and salads. *Food safety handbook,* 425–442.

Howell K. (2016). Spoilage: Yeast spoilage of food and beverages. *Encyclopedia of Food and Health.* Elsevier; 113–117.

Hussain M. A. (2016). Food contamination: major challenges of the future. *Foods.* 5(2):21.

Jabbar H. & Joishy K. (1999). Rapid detection of *Pseudomonas* in sea foods using protease indicator. *Journal of Food Science.* 64:547–549.

Jan T., Yadav K. & Borude S. (2016). Study of HACCP Implementation in Milk Processing Plant at Khyber Agro Pvt. Ltd in Jammu & Kashmir. *Journal of Food Processing & Technology.* 7(2):610–615.

Jarvis B. (2014). Good manufacturing practice. *Encyclopedia of Food Microbiology* (2nd edn.). Elsevier; 106–115.

Jeantet R., Croguennec T., Schuck P. & Brulé G. (2016). *Handbook of Food Science and Technology 1: Food Alteration and Food Quality.* John Wiley & Sons.

Karel M. & Lund D. B. (2003). *Physical Principles of Food Preservation: Revised and Expanded.* CRC Press.

Kilcast D. & Subramaniam P. (2000). *The Stability and Shelf-Life of Food.* Woodhead Publishing.

Kilcast D. & Subramaniam P. (2011). *Food and Beverage Stability and Shelf Life.* Elsevier.

Lan L., Yao Y., Ping J. & Ying Y. (2017). Recent progress in nanomaterial-based optical aptamer assay for the detection of food chemical contaminants. *ACS applied materials &Interfaces.* 9:23287–23301.

Leistner L. (2000). Basic aspects of food preservation by hurdle technology. *International Journal of Food Microbiology.* 55:181–186.

Lück E. (1985). Chemical preservation of food. *Zentralblatt fur Bakteriologie, Mikrobiologie und Hygiene. 1. Abt. Originale B, Hygiene.* 180:311–318.

Manzocco L., Nicoli M. C. & Labuza T. P. (2002). Study of bread staling by x-ray diffraction analysis. *Italian Journal of Food Science.* 14:235–245.

Masoud W., Bjørg C. L., Jespersen L. & Jakobsen M. (2004). Yeast involved in fermentation of *Coffea arabica* in East Africa determined by genotyping and by direct denaturating gradient gel electrophoresis. *Yeast.* 21:549–556.

Mathlouthi M. (2013). *Food Packaging and Preservation.* Springer Science & Business Media.

Medana C. (2020). *Analysis of Chemical Contaminants in Food.* Multidisciplinary Digital Publishing Institute.

Mensah L. D. & Julien D. (2011). Implementation of food safety management systems in the UK. *Food Control.* 22:1216–1225.

Meroth C. B., Hammes W. P. & Hertel C. (2003). Identification and population dynamics of yeasts in sourdough fermentation processes by PCR-denaturing gradient gel electrophoresis. *Applied and Environmental Microbiology.* 69:7453–7461.

Metcalfe, D. D., Sampson, H. A. & Simon, R. A. (Eds.). (2011). *Food Allergy: Adverse Reactions to Foods and Food Additives.* John Wiley & Sons.

Modi H. (2009). *Microbial Spoilage of Foods.* Aavishkar Publishers.

Mortimore S. (2000). An example of some procedures used to assess HACCP systems within the food manufacturing industry. *Food Control.* 11:403–413.

Ozmen L. & Langrish T. (2002). Comparison of glass transition temperature and sticky point temperature for skim milk powder. *Drying Technology.* 20:1177–1192.

Panisello P. J. & Quantick P. C. (2001). Technical barriers to hazard analysis critical control point (HACCP). *Food Control.* 12:165–173.

Parreño Jr R. P. & Carandang M. B. (2021). Effect of modified cassava starch in reduced-fat mayonnaise by correlating emulsion stability with anti-oxidation reaction using gas chromatography–mass spectrometry (GC-MS). *Philippine Journal of Science.* 150(4):753–763.

Petruzzi L., Corbo M. R., Sinigaglia M. & Bevilacqua A. (2017). Microbial spoilage of foods: Fundamentals. *The Microbiological Quality of Food.* Elsevier; 1–21.

Pinu F. R. (2016). Early detection of food pathogens and food spoilage microorganisms: application of metabolomics. *Trends in Food Science & Technology.* 54:213–215.

Pitt J. I. & Hocking A. D. (2009). The Ecology of Fungal Food Spoilage. *Fungi and Food Spoilage.* Springer; 3–9.

Pradhan N., Singh S., Ojha N., Shrivastava A., Barla A., Rai V. & Bose S. (2015). Facets of nanotechnology as seen in food processing, packaging, and preservation industry. *BioMed Research International.* 1–17.

Prakitchaiwattana C. J., Fleet G. H. & Heard G. M. (2004). Application and evaluation of denaturing gradient gel electrophoresis to analyse the yeast ecology of wine grapes. *FEMS Yeast Research.* 4:865–877.

Raeisossadati M. J., Danesh N. M., Borna F., Gholamzad M., Ramezani M., Abnous K. & Taghdisi S. M. (2016). Lateral flow based immunobiosensors for detection of food contaminants. *Biosensors and Bioelectronics.* 86:235–246.

Rawat S. (2015). Food spoilage: microorganisms and their prevention. *Asian Journal of Plant Science and Research.* 5:47–56.

Roudaut G., Dacremont C., Pàmies B. V., Colas B. & Le Meste M. (2002). Crispness: a critical review on sensory and material science approaches. *Trends in Food Science &Technology.* 13:217–227.

Schirone M., Visciano P., Tofalo R. & Suzzi G. (2017). Biological hazards in food. *Frontiers in Microbiology.* 7:2154.

Singh R. & Anderson B. (2004). The major types of food spoilage: An overview. *Understanding and Measuring the Shelf-life of Food.* Woodhead Publishing; 3–23.

Söderström C., Borén H., Winquist F. & Krantz-Rülcker C. (2003). Use of an electronic tongue to analyze mold growth in liquid media. *International Journal of Food Microbiology.* 83:253–261.

Tang Y., Lu L., Zhao W. & Wang J. (2009). Rapid detection techniques for biological and chemical contamination in food: a review. *International Journal of Food Engineering.* 5: Art. 2.

Tewari G. & Juneja V. (2008). *Advances in Thermal and Non-Thermal Food Preservation.* John Wiley & Sons.

Untermann F. (1999). Food safety management and misinterpretation of HACCP. *Food Control.* 10:161–167.

Vadivambal R. & Jayas D. S. (2011). Applications of thermal imaging in agriculture and food industry—a review. *Food and Bioprocess Technology.* 4:186–199.

Wallace C. A., Sperber W. H. & Mortimore S. E. (2018). *Food Safety for the 21st Century: Managing HACCP and Food Safety Throughout the Global Supply Chain.* John Wiley & Sons.

Wang D., Dowell F., Ram M. & Schapaugh W. (2004). Classification of fungal-damaged soybean seeds using near-infrared spectroscopy. *International Journal of Food Properties.* 7:75–82.

Williams A. (1989). Methodological developments in food mycology. *Journal of Applied Bacteriology.* 67:61s–67s.

Wu C.-H., Yuan C.-H., Ye X.-Q., Hu Y.-Q., Chen S.-G. &Liu D. (2014). A critical review on superchilling preservation technology in aquatic product. *Journal of Integrative Agriculture.* 13(12):2788–2806.

Zhang Y., Zuo P. & Ye B.-C. (2015). A low-cost and simple paper-based microfluidic device for simultaneous multiplex determination of different types of chemical contaminants in food. *Biosensors and Bioelectronics.* 68:14–19.

CHAPTER **2**

Shelf Life and Food Safety
Analysis, Methods, and Techniques

**Insha Zahoor, Nadira Anjum, Meenakshi Trilokia, Arshied Manzoor,
Haroon Rashid Naik, Anjum Ayoub, and Farhana Mehraj Allai**

CONTENTS

2.1 INTRODUCTION

Consumers are increasingly demanding high-quality food and have expectations that such quality will be maintained during the period between purchase and consumption. Depending on the physical and chemical properties of food, there will come a point when its quality will be unacceptable to consumers and, at this point, it has reached the end of its shelf life. Shelf life is concerned with all food commodities and is considered an important common attribute in raw materials (supplied),

Email: Insha Zahoor, insha.zahor@gmail.com

DOI: 10.1201/9781003091677-2

19

ingredients, and the final manufactured products. The time duration during which a food product remains safe is when it maintains its original and desired physical, chemical, microbial, and sensory characteristics that bestow consumer acceptance on to the product (Earle and Earle, 2007). It is also the time duration ensuring the acceptability of the food products by consumers due to their safe consumption nature and quality. The time to determine the shelf life starts once the product is manufactured, and the period that is highlighted on the product by a label is a function of various factors such as type of raw ingredients, the process used for manufacturing, nature of packaging, and storage conditions. The label provides the consumer with an idea of the time after which the product starts deteriorating. However, studies have shown ways to extend the shelf life of food products through various procedures including high-temperature and low-temperature incorporation, which provide the growth retardation or even the elimination of many spoilage and pathogenic microorganisms (bacteria, yeasts, and molds) and any toxin formation. Some bacteria resistant to cold treatment may grow and increase in number during storage, making the food unsafe and affecting the product's shelf life and safety. The guidelines set by enforcement authorities or agencies, current industry best practice, relevant food legislation, procedures that are framed by professional bodies (independent; e.g., Institute of Food Technologists [IFT]), assessing end point, and the market conditions are the main factors deciding a product's shelf life.

The spoilage and deterioration of a variety of foods are explained by the following mechanisms:

- Gain or loss due to moisture or water vapor transfer
- Changes occurring in the foods due to exposure to sunlight or artificial light
- Changes in foods due to chemical or biochemical reactions
- Changes occurring due to microbial interventions
- Changes occurring due to the transfer (physical in nature) of substances including flavors and odors

In other words, shelf life is regarded as multidimensional characteristics that the food manufacturers and processors, including consumers, will accept. It is considered that food quality along with food safety are two important aspects of an acceptable shelf life that are intimately linked together. Therefore, food safety and food deterioration cannot be viewed side by side, i.e., a deteriorated food cannot be safe for consumption. The quality of a food product depends on the nature of the product and the manufacturing process. However, similar or even competing products might have different characteristics and shelf life. Hence, the prime concern of the shelf life is to monitor and understand how these characteristics change during storage and eventually lead to consumer rejection. We can see the manufacturer's label on a food product with a shelf life shorter than it actually has and this can be revealed through the commercial and marketing evaluation (Steele, 2004).

Evaluation of the shelf life of foods and food products is recommended only after the correct processing, packaging, and storage have been done and requires the basic knowledge of the science and technology of foods, processing of foods, analysis and packaging of foods, and statistical techniques used during the process (Steele, 2004). High-quality maintenance prevents unwanted changes in sensory quality and the safety of foods is necessary due to the increasing demand of consumers for the same. Labeling on food packages by food manufacturers reflects the quality and life of that food product during which it is safe to consume. Two factors, namely type of food and total shelf life of the food packed, determine the date coding on the label. For example, in the United Kingdom, the "use by date" is put on the package label of foods prone to high microbial perishability. Similarly, "best before" or "best before end" labels are put on the packages of foods with a shelf life of 1.5 years or more. A "use by" date is necessary for safety reasons because food could become unsafe during its shelf life. Food products having passed the use by date should not be consumed and sold because of safety concerns. In the case of best before, one can sell the food products that have reached their best before date, as it is not illegal. The best before date is not used for foods with high perishability, and specific instructions for storage must be included on the label, which should be followed by the seller to ensure the safety of the food product during the labeled shelf life to consumers. Likewise, the "expiry date" is essential to mention on the label of food products with longer shelf life and is critical to nutrition. The expiry date is a date usually listed by a food producer that conveys to customers the end day a product will be safe to consume. Moreover, the circumstances necessary for the food storage must be easily met during distribution, time duration in the retail systems, and storage in the home. The growth of pathogenic microorganisms and their production of toxic substances may not provide visible signs of the unsafe nature of food. These kinds of food safety concerns can be eliminated through the addition of substances such as antioxidants helping to prevent color change, prevention of light exposure to avoid changes in food products, and by following the prescribed storage instructions. However, the potential for chemical toxicity developing in food has been mitigated to a large extent through modern technological improvements in food processing, ingredients, packaging, and storage. Generally, for foods with a short shelf life, microbiological changes are of prime concern, whereas for medium- to long-life products, chemical and sensory changes determine the food quality.

In the case of foods that are frozen, enzymes are the big concern, resulting in quality changes and flavor changes in meat and poultry and non-blanched fruits and vegetables due to various enzymes such as lipoxygenases found in the fruits and vegetables, because microbes generally cannot grow in a freezing environment. The presence of nitrogenous substances (non-protein in high levels), lipid fractions (unsaturated), high water content, and microbial metabolic activities coupled with the oxidative reactions occurring during storage makes the aquatic species highly prone to perishability. Hence, shelf life and quality determination remain an important aspect of food products. Selecting the quality characteristics (deteriorating rapidly) and mathematical modeling are the two most important steps involved in the process of shelf life testing. Furthermore, testing needs professional judgment and depends on the type of food product.

2.2 EXPIRY DATE AND BEST BY DATE

The expiry date is a previously determined date of consumable products like food, medicine, or supplements after which they should not be consumed. Going past the expiry date may cause certain changes (spoilage, damage, product is not-effective, etc.) in the consumable product. In addition, the expiry date is a date that is usually listed by the producer of the product that conveys to customers the end day a product will be safe to consume. Later, the consumable product will not have the same safety and quality attributes expected by customers (Alongi et al., 2018).

It is essential to mention the expiry date on the label of the food product with a longer shelf life and it is critical to nutrition, for example, in meal replacements, nutritional supplements, infant formulas, formulated liquid diets, etc. These should not be used after the expiry date has passed.

The expiry date is important because it

- Is focused on the normal, expected handling of the consumable product and its exposure related to temperature
- Tells how much longer (end date) the consumable product retains its nutritional, microbial, and physical stability
- Indicates the freshness of the consumable product

2.2.1 Best by Date

Best by or best before refers to a date that signifies the end-day usage of the consumable product at specified storage conditions (Food Safety and Standards [Packaging and Labelling] Regulations, 2011). During that period, the consumable product retains its quality. These dates are only advisory and listed on the label of the packaged food product, for example, frozen foods, dried fruits and bread, etc. The provision of listing best by date or best before date on the label is a legal requirement.

Claims have been made that the food product is not necessarily unsafe beyond the best by date. However, the food product may perhaps lose its quality in terms of optimum flavor, texture, and freshness.

The importance of best by date is to

- Keep the consumer safe before buying and consuming and
- To show the specified period a food product will be of greatest flavor and best quality.

2.3 FACTORS AFFECTING FOOD SHELF LIFE AND SAFETY

There are various factors that influence the shelf life and safety of food. These are categorized into intrinsic and extrinsic factors.

1. *Intrinsic factors:* They influence the food shelf life internally and are generally related to the composition and properties of the food itself. These factors cannot be controlled and changed. They include pH, moisture content or water activity (a_w), nutritional composition, availability of oxygen, the addition of preservatives (such as antioxidants, spices, salt, sugar, etc.), oxidation/reduction potential and product formulation, etc.
   - *pH:* This is defined as a measure of acidity or alkalinity. It varies according to the composition and formulation of a food product and might differ during its shelf life. pH needs to be controlled as it will greatly influence the safety and shelf life of the food product (Man, 2002).
   - *Water activity (a_w):* This is defined as the vapor pressure of water in a substance compared with the vapor pressure of pure water at the same temperature. Water activity values have been widely used to indicate food stability and are helpful in predicting microbial survival and chemical and biochemical changes (Labuza and Hyman, 1998) that can reduce the shelf life of foods (Robertson, 2010).
   - *Nutritional composition:* This is an important factor that influences the shelf life and safety of food, for example, margarine with a particular fat content (minimum 80%) limits microbial growth (pathogens and spoilage organisms) (Delamarre and Batt, 1999).
   - *Availability of oxygen:* The amount of oxygen available in the environment has a great deal of influence on the type of microorganisms that can grow in certain foods. It also causes browning, rancidity, and flavor changes resulting in food deterioration.
   - *Addition of preservatives:* Preservatives (such as antioxidants, spices, salt, sugar, etc.) influence the food shelf life and safety by inhibiting microbial growth.
   - *Oxidation/reduction potential:* The oxidation and reduction potential of a food product will influence the kind of organism and chemical changes produced in it, for example, aerobes (*Bacillus* spp.) will grow in high oxidation potential and anaerobes (*Clostridium perfringens*) cannot survive in the presence of oxygen (Man, 2002).
   - *Product formulation:* This includes the use of ingredients, type and amount of preservatives, and type and amount of enzymes present. All things that are used during the formation of a food product have a strong influence on food safety and on shelf life, because any changes in the amount of ingredients (substituting or removing ingredients) may allow some microbial growth that decreases the shelf life. Moreover, in case of multicomponent and composite products, when contact between different components occurs it leads to moisture, color, and flavor transference; for example, moisture migration occurs in cereal flakes mixed with dried fruits (Howarth, 2000).
2. *Extrinsic factors:* These are factors that influence the food shelf life externally during transportation, handling, and storage. They can be controlled and changed and include temperature, gases present, light exposure, relative humidity, mechanical handling during distribution and storage, processing, type of packaging material and system, etc.
   - *Temperature:* This is a key factor responsible for the rates of deteriorative reactions. Also, in some cases it plays an important role as it can slow down the growth of microorganisms, which is important to food safety and quality.
   - *Gases present:* Gases like O_2 and CO_2 are used to control the growth of microorganisms in foods. For example, Controlled Atmospheric Packaging (CAP), Modified Atmospheric Storage

(MAP), and vacuum packaging have a beneficial effect that retards microbial growth, extending the shelf life of foods.

- *Light exposure:* The length and the intensity of light exposure to food influences food safety and sometimes leads to the production of flavor defects in some packaged foods.
- *Relative humidity:* This is the ratio of the partial pressure of water vapor to the equilibrium vapor pressure of water at a given temperature. It is the second most important environmental factor that directly affects the moisture and water activity content of food. It also affects the rate of food deterioration reactions (Labuza et al., 1969).
- *Mechanical handling during distribution and storage:* Mishandling of foods during distribution and storage will reduce shelf life.
- *Processing:* Processing includes several operations, such as mixing, salting, fermenting, heating, freezing, pasteurization and sterilization, etc., to which a food may be exposed. All of these have a considerable effect on microflora and physical, chemical, nutritional, and sensory properties of the food product and, hence, food shelf life (Man, 2002). Similarly, the selection of process can alter the shelf life of the final product; for example, commercially sterile canned foods have a shelf life of 2 years or more (Shakuntalamanay and Shadaksharaswamy, 2008).
- *Type of packaging material and system:* Packaging is one of the important factors that influence the shelf life and safety of food. The type of packaging material and system selected for a food product can act as a barrier against light, gaseous exchange, and moisture vapor transmission. Moreover, it protects against many deteriorative changes that limit the shelf life of a food, for example, metal cans used for canned foods (Man, 2002).

Both intrinsic and extrinsic factors play a vital role in influencing food shelf life and safety interactively or independently. These factors are beneficial in creating new strategies for extending the shelf life of food products.

2.4 FACTORS AFFECTING SHELF LIFE TESTS

The shelf life test of a food product usually describes exactly how much longer food will maintain its quality during storage. The following factors need to be considered while determining shelf life.

2.4.1 Understand the Nature of the Product

The nature of the food product plays a vital role in the shelf life determination. Before setting up a shelf life test, it is essential for the evaluator to thoroughly know the product (Man, 2002). Food can be categorized into three categories depending on the nature of the changes that mostly occurs during the storage period: (1) perishable foods (very short life products), (2) semi-perishable foods (short to medium shelf life products), and (3) non-perishable foods (medium to long shelf life products) (Robertson, 2006).

- *Perishable foods (very short life products):* These are foods that deteriorate rapidly and can be held at room temperature for a limited amount of hours or 1 or 2 days before deterioration. Perishable foods require refrigeration and need to be stored at chill or freezer temperatures, i.e., 0°C to 7°C or −12°C to −18°C, respectively, for example, milk, fresh flesh foods, minimally processed foods, and many fresh fruits and vegetables. (Robertson, 2010). These foods are mostly subjected to microbiological and/or enzymatic deterioration. Therefore, measurement needs to be taken every day for shelf life determination (Phimolsiripol and Suppakul, 2016).
- *Semi-perishable foods (short to medium shelf life products):* These foods do not need immediate refrigeration, although they have a restricted shelf life. Semi-perishable foods may contain natural inhibitors such as root vegetables, eggs, and some cheeses (Robertson, 2010) and received some mild preservation treatment, for example, pasteurized milk, cheese, smoked meat, and some bakery products. For these products, measurements are made every week for shelf life determination (Phimolsiripol and Suppakul, 2016).

- *Non-perishable foods (medium to long shelf life products):* These foods are kept for months or years without spoiling unless handled and stored improperly. Generally, this type of food remains unaffected by microbes because they have less moisture content (cereal grains, nuts, and some confectionery products). This also includes food that have experienced a thermal process or were sustained in specific conditions, for example, canned foods, soft drinks, and cake mixes. For these products, measurement is observed weekly or monthly for shelf life determination (Phimolsiripol and Suppakul, 2016).

2.4.2 Understand the Factors Affecting Food Product Quality

Each product has its own aspects that will possibly limit its shelf life or prolong the shelf life (NZFSA, 2005). Accordingly, before setting up for a shelf life test, the evaluator must understand all of the factors that affect food product quality. These factors can be categorized into compositional and environmental factors.

- *Compositional factors*: These are factors that define the composition and characteristics of the final product. The factors include water activity; pH value; total acidity; type of acid; redox potential; availability of oxygen, nutrients, natural microflora, and survival microbial counts; natural biochemistry of the product formulation (enzymes, chemical reactants); use of preservatives in product formulation (e.g., salt) (Kilcast and Subramaniam, 2000); and concentration of reactant, inhibitor, and catalyst.
- *Environmental factors*: These include certain factors the final product faces as it moves across the food chain. These factors include time–temperature profile, the pressure required in the headspace, temperature control, and relative humidity; light exposure (ultraviolet and infrared), environmental microbial counts while processing, and storage and distribution; atmospheric composition inside packaging; subsequent heat treatment (e.g., reheating or cooking before consumption); and distributor, retailer, and consumer handling (Phimolsiripol and Suppakul, 2016).

2.4.3 Identify the Critical Quality Based on Several Guidelines

The evaluator must follow these guidelines:

- a. *Government laws:* As the critical quality is primarily for microbial safety.
- b. *Customer standards:* Customer standards may be higher than the government standards.
- c. *Competitors:* The critical quality of the product can be correlated to a competitor's product.
- d. *Consumers:* Assigned the best judgment of the critical quality for each food product.

2.4.4 Understand the Concerted Series

It comprises biochemical or physicochemical reactions that need to be sharply interpreted along with finding several mechanisms liable for spoilage or ruin the desirable characteristics of the food product. Various deteriorative changes arise from reactions such as rancidity development, enzymatic activity (Kilcast and Subramaniam, 2000), microbiological spoilage, and moisture and/or other vapor migration (Phimolsiripol and Suppakul, 2016).

2.5 METHODS FOR DETERMINATION OF SHELF LIFE AND SAFETY

2.5.1 Direct Method

This approach includes the storage of the product in circumstances that are quite similar to the actual atmosphere and what it actually has to face during storage. The direct method consists of real-time studies to monitor the evolution of a product's storage in regular intervals. This method has an advantage over other methods in that it helps to accurately estimate the time taken by the food product to deteriorate. It does not take into consideration the factor instability of the product during storage conditions, which results in a long time duration in evaluation studies.

1. *Challenge test:* The prime concern of the challenge test is to make sure that the food product is exposed to the conditions that it will face in real life during storage. For this, the pathogens or microorganisms are introduced experimentally in the food during the production process. The main disadvantages of this method involve the difficulty in its implantation, the complexity of the method, and not analyzing all the factors that will influence the product at the same time during storage. This method of testing the growth potential of *Listeria monocytogenes* in ready-to-eat foods is the most commonly used in laboratories. The advantage of the method is that it is very easy to implement compared with other methods. The main disadvantage is that the results obtained during this test are exclusively applicable to the foods stored in the challenge test conditions. Hence, each time the test is done we have to change the formulation or the process (Beaufort, 2011).

2. *Predictive microbiology:* This is a rapid method used to estimate quantitative microbial growth in food by using mathematical models. Using this method, food products are exposed to different environmental conditions and the microbial responses or behavior to these conditions are predicted through mathematical and statistical models. The advantage of this method is that it considers the conditions of a product that will change during the storage period, which guarantees its potential application in developing new products. However, it is more complex for the manufacturer with inaccurate results, i.e., these results should not be relied on completely but considered as tools to support decisions. This method helps in the improvement of experimental design that leads to the interpretation of experimental results. In addition, it helps to prevent wasting time on useless things and helps to decide which tests to perform and which ones to omit. Predictive microbiology results in the quantitative prediction of future developments during the storage life of the product through a series of events. Predictive microbiology is used to evaluate the results of estimating the occurrence of *Bacillus cereus* in milk.

3. *Accelerated shelf life tests:* The procedure involving sell by or use by labeling, introduced by many countries for retail outlets with least or no delay for new products, is possible only for short shelf life products. However, due to possible delays, this method would not work for products to be used for a long time in the future, which require complete information including the storage conditions. This drawback has paved the way for a new method called accelerated shelf life procedures, which can be used when the storage life is long and the relationship between the storage characteristics and the quality characteristics under an ambient storage condition is known. Accelerated shelf life testing refers to the evaluation of the stability of the product that the expert obtains in a shorter time duration (shelf life) than the product possesses (Steele, 2004). The prime concern of this method includes the estimation of a product's shelf life in a shorter time duration and includes the following concepts:
   - The higher temperature in which the food is stored can affect the food adversely during storage, leading finally to a change in shelf life in a short storage time.
   - The curve obtained based on an extrapolation by using data (obtained from accelerated determination) helps in the shelf life determination in normal storage conditions.

The basic principle that an accelerated test consists of physical and chemical processes causing deterioration is increased by changing the conditions of food storage. Due to this, the relationship of ambient storage conditions with the shelf life is determined if the processes causing the deterioration in the two conditions are considered almost the same. The Arrhenius model can also be used to determine the accelerated deterioration through an increase in storage temperature and is applicable only for simple chemical systems. Several accelerated shelf life tests have been introduced to date. The rate of reaction is singled out to increase the rate of oxidation because the increase in absolute temperature speeds up the rate of reaction exponentially. A typical accelerated shelf life test has the following steps:

1. The first step is the selection of a method suitable for testing the food product of which shelf life is to be determined.
2. After this step, the induction period (the time taken to approach the end point) of a sample put under the same conditions of the test is measured.

3. The final step is a quite difficult one that includes the translation of the value of a period of induction obtained in the real shelf life (months) of the product during storage. An arbitrary factor is used for the translation process that is based on previous experience. These methods are mostly used for the evaluation of the effectiveness of antioxidants in highly oxidizable foods with added antioxidants.

Numerous shelf life testing approaches related to the accelerated method with procedures to obtain the data of deterioration within a shorter period have been introduced. These also help to determine which model and how a model is to be used in predicting the product's shelf life (actual). Among all, the two approaches commonly used in accelerated shelf life testing are the initial rate approach and kinetic model approach.

a. *Initial rate approach*

This is considered the simplest approach in accelerating the process of shelf life testing applicable where it is easy to monitor the deterioration process by using a method that is highly sensitive and accurate. In this method, the knowledge of the procedure for the evaluation of shelf life and its dependence on time (given by the order of a chemical reaction) are prime factors required for the actual shelf life prediction of the food product.

b. *Kinetic model approach*

The steps involved in the most commonly used method (kinetic model approach) of accelerated shelf life testing include the following:

1. The first step involves the selection of desired factors that are active kinetically and cause acceleration in the deterioration process.
2. Select levels of accelerating factors that favor fast deterioration, and carrying out kinetic studies at these levels.
3. The data of normal storage are extrapolated, thereby the kinetic model parameters are evaluated.
4. Finally, the product's actual storage shelf life is predicted by using the kinetic model extrapolated data.

2.6 SURVIVAL METHOD

The survival method is different from other methods and is based on the consumer's opinion regarding the product's characteristics (physical). This method is based on how people (consumers) react to the same product with different manufacturing dates. Evaluation is based on whether the consumer accepts the products for consumption or not and results in the establishment of a relationship between the perception of consumers and the product's shelf life. Hence, it requires the determination of this relationship in deciding the best product for consumption.

2.7 TECHNIQUES FOR SHELF LIFE PREDICTION

Currently, people are progressively demanding high-quality food, and they are expecting that the quality of their food product will remain at the elevated point from the period of the buying until utilization. Time span or shelf life of usability is a significant property of all foods, including raw materials, manufactured products, and ingredients. It is the time during which the food item will stay safe and hold various desired physical, chemical, microbiological, and sensory attributes; consent or comply with any label presentation or nutritional information; and remain adequate to the buyer (Earle and Earle, 2007). To calculate the shelf life of food products, the analyst is required to comprehend the fundamental innovation technology of food science, including food analysis, packaging of foods, processing of foods, and various statistical techniques of statistics. Shelf life analysis should be done just when the processing of foods is effectively done, including packaging of foods, storing of foods, and readying for purchase and utilization (Steele, 2004). The determination of the end of the time span or shelf life can be from (1) applicable legislation of foods, (2) rules provided by agencies or implementation specialists, (3) special guides given by autonomous bodies (for example, IFT) that are expert in the concerned field, (4) best trending modern practices, and (5) information or knowledge about the market.

2.7.1 Kinetic Reactions

The kinetic reaction approach is the easiest and simplest method for testing shelf life. To forecast the time span, there must be an understanding of how the degradation process acts as a function of time, which can be accessed from the kinetic data. This provides a quantification of the quality or characteristic of products or food items dependent on the response alteration in the reaction (Corradini and Peleg, 2007). The kinetic equation can be shown as:

$$rA = d[a]/dt \qquad (2.1)$$

The change or transition in mass A of an ingredient of interest is observed. The characteristic factors of quality $[A]$ are generally quantifiable or perceptible by sensory, microbiological, chemical, or physical parameters, like degradation or nutrient loss or development of bad odor or off-flavor. The time to reach the quality index value at a condition that is specified (t_s, i.e., time span or the shelf life) is proportional but inversely to the rate constant at this condition.

$$t_s = fq(Ats) / k \qquad (2.2)$$

The quality function types of food for a possible zero-, first-, second-, or nth-order reaction are shown in Figure 2.1 and Table 2.1, which depict the various patterns of orders of the reactions. The rate of the reaction does not depend on the concentration or mass of the reactant as far as zero-order reactions are concerned. In the case of the first-order reaction, it depends on only one reactant, so one is its exponent value. In the case of the second-order reaction, the reaction rate could be proportional to the product of the two concentrations or squared single concentration (Labuza and Riboh, 1982). The reactions that cause the loss of the shelf life based on physicochemical, microbial index, or chemical characteristics (e.g., Millard browning, overall quality of frozen food) and first-order reactions (e.g., microbial growth, oxidative colorless, vitamin loss) are depicted in Figure 2.1 and Table 2.1. To explain the temperature dependence, a great deal of published research has been applied to the kinetic model. Chemical properties like peroxide value in extra virgin olive oil

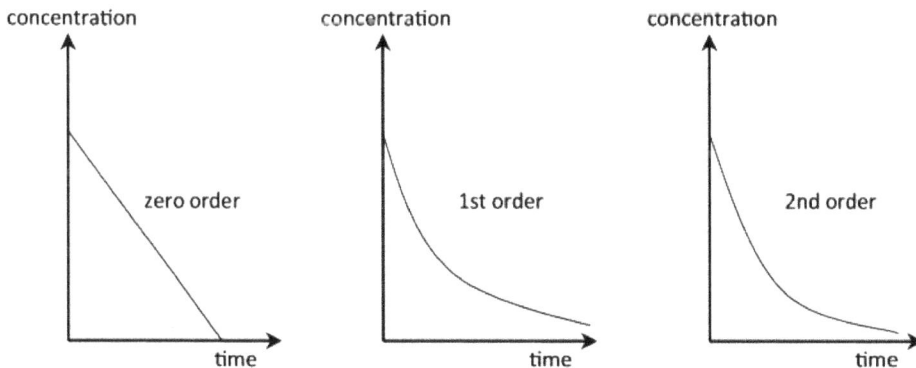

Figure 2.1 Process of changes in concentration in different orders of reaction.

Table 2.1 Quality Function of the Order of the Reaction

Apparent Reaction Order	Quality Function
0	$A_t - A_0$ or A_t/A_0
1	$\ln(A_t - A_0)$ or $\ln(A_t/A_0)$
2	$1/A_t - 1/A_0$
n $(n \neq 1)$	$(1/n - 1) \times (A_t^{1-n} - A_0^{1-n})$

(Calligaris et al., 2006), degradation of vitamins in citrus juice concentrate like vitamin C (Burdurlu et al., 2006), and hydrogen ion in coffee liquids (Manzocco and Nicoli, 2007) can be parameters of quality, as well as various physical properties like loss of color in fresh-cut asparagus (Sothornvit and Kiatchanapaibul, 2009), weight loss in frozen bread dough (Phimolsiripol et al., 2011), and sensory attributes like sensory properties in shrimp that is frozen (Tsironi et al., 2009).

Generally, the most straightforward and often utilized technique to calculate the order of the reaction is the integration process (van Boekel, 2008). This procedure begins by first assuming the order of the reaction followed by integration and then linearize by linear regression. This is followed by plotting the experimental data or information in a setup that is linearized. Finally, if the information or data fit a straight line, it shows that our assumption is right; if it shows deviation then we do the procedure once again.

2.7.2 Accelerated Shelf Life Simulation

A very short time is required by the food industries to decide the time span or shelf life of food items. As far as practical reasons are concerned, when the time of storage is long, the food industries generally use quick or rapid test techniques that reduce the processing time for obtaining the essential experimental data. Rapid or accelerated shelf life techniques are defined as any procedure that is capable of calculating the product stability that is based on the data, which are obtained in a smaller time span than the exact shelf span of the food item (Steele, 2004). To reduce the time and, when needed, to calculate the shelf lifetime, the procedure of the accelerated or rapid shelf life simulation or technique includes the following:

1. The prediction is that when food is stored at temperatures that are higher than any adverse or deteriorating effect on its behavior of storage, the final appearance of the time span or shelf life could be done in a shorter time.
2. Extrapolation can be used to estimate the time span or shelf life in normal storage conditions with the help of data that are obtained from rapid determination.

2.7.3 The Arrhenius Model

The rate of a chemical reaction can be related to the changes in the temperature with the help of the Arrhenius model. It is one of the classical models that are usually and widely used in various storage tests and processing tests usually affected by temperature (Corradini and Peleg, 2007; Phimolsiripol et al., 2011). The model is shown as:

$$K = k_0 \exp(-Ea)/RT \qquad (2.3)$$

where T shows absolute temperature (Kelvin, K), R shows gas constant (8.3144 J mol^{-1} K^{-1}, 1.9872 cal mol^{-1} K^{-1} or 8.3144 J mol^{-1} K^{-1}), Ea is the energy of activation, and k_0 is the rate constant. By using the natural logarithm, Equation (2.3) can be put in the form of the standard slope intercept.

$$\ln k = \ln A - Ea = RT \ (\text{or})$$

$$\ln k = \ln A - (Ea = R) \times 1/T$$

$$\Updownarrow\Updownarrow\Updownarrow$$

$$y = b + ax$$

(2.4)

Various steps of shelf life determination of the Arrhenius model are as follows:

- Follow the kinetic reaction concept as stated in the Kinetic Reactions section to find the order of the reaction.
- Secondly, follow the Arrhenius relationship.
- Then plot the Arrhenius relationship.
- With the help of the linear regression, fit the curve.
- Then Ea/R is the slope of the plot between $\ln k$ versus $1/T$.

2.7.4 The Bracket Method

Accelerated stability data are used to estimate the activation energy, but when the energy of activation is already known, the degradation rate at the temperature of storage may be assumed or predicted from data that are collected at only a single elevated temperature. This method is most favored and usually preferred in industries because it helps to reduce the time and size of accelerated stability tests. Experience demonstrates that few pharmaceutical analytes have activation energy within the range of 10–20 kcal/mol, but it is not certain that you will have exact data or you will be able to make assumptions about the energy of activation of a particular product (Anderson and Scott 1991). This method is clear utilization of the Arrhenius equation that is used when the value of the activation energy is known (Anderson and Scott 1991). There should not be any confusion by bracketing in the bracket method; it is an experimental design that permits us to test the least number of samples. It also helps to test at the extremes of certain factors, like container fill, container size, and strength. The assumption of bracketing is that the intermediate level stability is shown by the extreme stability and the testing or analysis at those extremes is conducted at all points of time (Tsong, 2003).

2.7.5 Q-Rule

The Q-rule states that when the temperature is lowered by a certain amount of degrees the degradation rate decreases by a constant factor. Typically the value of the Q is set at 2, 3, or 4 and this value or the factor is proportional to the change in temperature as Q^n in which n is equal to the alteration or change in the temperature in °C divided by 10°C. So 10°C is the baseline temperature, and sometimes this Q-rule is called Q10. To explain the various applications of the Q-rule, let us say the product stability at 50°C is 32 days. The temperature recommended for the storage is 25°C and $n = (50 - 25)/10 = 2.5$. Now let us enter a value that is intermediate, $Q = 3$. In this manner, $Q^n = (3)^{2.5} = 15.6$. So the anticipated lifetime is 32 days × 15.6 = 500 days. When lower values of Q are used this methodology is more conservative (Anderson and Scott, 1991).These rules for the bracket method and the Q-rule are the rough approximation of the stabilities. They could be effectively and successfully used to plan the duration of testing as well as the elevated temperature levels in the accelerated stability testing protocol.

2.8 CONCLUSION

Shelf life is considered an important characteristic of food products and the materials used for their preparation. Shelf life actually is a period up to which a particular food remains safe for consumption and retains the physical, chemical, sensory, and microbiological attributes to the desired level. The shelf life of food is affected by several intrinsic as well as extrinsic factors. When a food product meant for human consumption is processed and stored, its shelf life must be evaluated using a proper procedure or technique. Sensory evaluation and physicochemical and other dispersive techniques are useful in identifying the detrimental changes in the food. Apart from these techniques, some kinetic models are also used to understand the proper mechanism of the deteriorative reactions in food products. A shelf life evaluation technique can be considered ideal when it is nondestructive and warns about the detrimental changes in the food as early as possible.

REFERENCES

Alongi, M., Silani, S., Lagaizo, C., & Manzocco, L. (2018). Effect of expiry date communication on acceptability and waste of fresh-cut lettuce during storage at different temperatures. *Food Research International*, *116*, 1121–1125.

Anderson, G., & Scott, M. (1991). Determination of product shelf life and activation energy for five drugs of abuse. *Clinical Chemistry*, *37*(3), 398–402.

Beaufort, A. (2011). The determination of ready-to-eat foods into *Listeria monocytogenes* growth and no growth categories by challenge tests. *Food Control*, *22*(9), 1498–1502.

Burdurlu, H. S., Koca, N., & Karadeniz, F. (2006). Degradation of vitamin C in citrus juice concentrates during storage. *Journal of Food Engineering*, *74*(2), 211–216.

Calligaris, S., Sovrano, S., Manzocco, L., & Nicoli, M. C. (2006). Influence of crystallization on the oxidative stability of extra virgin olive oil. *Journal of Agricultural and Food Chemistry*, *54*(2), 529–535.

Corradini, M. G., & Peleg, M. (2007). Shelf-life estimation from accelerated storage data. *Trends in Food Science & Technology*, *18*(1), 37–47.

Delamarre, S., & Batt, C. A. (1999). The microbiology and historical safety of margarine. *Food Microbiology*, *16*, 327–333.

Earle, M., & Earle, R. (Eds.). (2007). *Case studies in food product development*. Elsevier.

Food Safety and Standards (Packaging and Labelling) Regulations. (2011). 1.1.2: Compendium packaging labelling regulations, Food Safety and Standards Authority of India.

Howarth, J. A. K. (2000). Ready to eat breakfast cereals. In: *Shelf-life evaluation of foods*, 2nd ed. Aspen Publishers, 182–196.

Kilcast, D., & Subramaniam, P. (2000). *The stability and shelf-life of food*. Woodhead Publishing, 1–344.

Labuza, T. P., & Hyman, C. R. (1998). Moisture migration and control in multi domain foods. *Trends in Food Science and Technology*, *9*, 47–55.

Labuza, T. P., & Riboh, D. (1982). Theory and application of Arrhenius kinetics to the predication of nutrient losses in foods. *Food Technology*, *36*(10), 66–74.

Labuza, T. P., Tannenbaum, S. R., & Karel, M. (1969). Water content and stability of low moisture and intermediate moisture foods. *Food Technology*, *24*, 543–550.

Man, D. (2002). *Food industry briefing series: shelf life*. Blackwell Science, 1–129.

Manzocco, L., & Nicoli, M. C. (2007). Modeling the effect of water activity and storage temperature on chemical stability of coffee brews. *Journal of Agricultural and Food Chemistry*, *55*(16), 6521–6526.

NZFSA. (2005). A Guide to calculating the shelf life of foods, information booklet for the food industry. *New Zealand Food Safety Authority*, 1–32.

Phimolsiripol, Y., Siripatrawan, U., & Cleland, D. J. (2011). Weight loss of frozen bread dough under isothermal and fluctuating temperature storage conditions. *Journal of Food Engineering*, *106*(2), 134–143.

Phimolsiripol, Y., & Suppakul, P. (2016). Techniques in shelf life evaluation of food products. *Reference Module in Food Science*, *1*, 1–8.

Robertson, G. L. (2006). *Food packaging principles and practice*, 2nd ed. CRC Press, 291.

Robertson, G L. (2010). *Food packaging and shelf life: a practical guide.* CRC Press, 1–408.

Shakuntalamanay, N., & Shadaksharaswamy, M. (2008). *Food facts and principles*, 3rd ed. New Age International, 1–490.

Sothornvit, R., & Kiatchanapaibul, P. (2009). Quality and shelf-life of washed fresh-cut asparagus in modified atmosphere packaging. *LWT-Food Science and Technology*, *42*(9), 1484–1490.

Steele, R. (Ed.). (2004). *Understanding and measuring the shelf-life of food.* Woodhead Publishing.

Tsironi, T., Dermesonlouoglou, E., Giannakourou, M., & Taoukis, P. (2009). Shelf life modelling of frozen shrimp at variable temperature conditions. *LWT-Food Science and Technology*, *42*(2), 664–671.

Tsong, Y. (2003). Recent issues in stability study: introduction. *Journal of Biopharmaceutical Statistics*, *13*(3), vii–ix.

Van Boekel, M. A. (2008). Kinetic modeling of food quality: a critical review. *Comprehensive Reviews in Food Science and Food Safety*, *7*(1), 144–158.

CHAPTER 3

Sensory Methods for Shelf Life Assessment of Foods

Chetan Sharma, Damir D. Torrico, and Sandip Singh

CONTENTS

3.1 INTRODUCTION

Each product has a specific time frame under which it is acceptable for consumers, and this time frame may range from minutes, such as brewed tea (in a loose sense), to months or years, such as potatoes, onions, carbonated beverages, milk powder, wine, or honey, respectively. Generally, this time frame is comprehended under the term *shelf life*, and a variety of definitions of the same can be mined from past literature (Manzocco, Calligaris, & Nicoli, 2010; Young & O'Sullivan, 2011; Tanner, 2016; ASTM E2454-20, 2020). An assortment of changes, including physical (Sharma, 2012; Joseph, 2016), chemical (Sharma, 2012; Singh, Sharma, & Sharma, 2017), microbiological, or sensorial (Chanadang & Chambers, 2019; Tano-Debrah, Saalia, Ghosh, & Hara, 2019), appear in food throughout and after this acceptable time frame, which may collectively result in a deterioration of the product, irrespective of nature. Although all these physical, chemical, or microbiological changes directly affect the sensory properties, once the hurdles, irrespective of physicochemical or microbiological, have been ascertained, the remaining barrier to ensure the final quality depends very much

Email: Chetan Sharma, chetan.sharma@lincoln.ac.nz

DOI: 10.1201/9781003091677-3

on the sensory properties of the product. Among these hurdles, some have a more predominant role than others in specific product categories, such as the microbiological integrity in dairy and other perishables. In contrast, sensory aspects will dominate in food categories that do not tend to suffer from microbiological changes, such as biscuits, oil, roasted peanuts, honey, ghee, and coffee (Hough, Langohr, Gómez, & Curia, 2003; Lawless & Heymann, 2010; Tano-Debrah et al., 2019). However, there is no consensus on this, and many new protocols based on sensory performance indicators appear even in highly perishable categories, such as fish (Freitas, Vaz-Pires, & Câmara, 2019; Joshy et al., 2020). Hough and Garitta (2012) claimed that the number of products that establish their shelf life on sensory properties is far greater than those whose shelf life depends on microbiological or nutritional properties. Furthermore, some product properties can be just too complicated to measure objectively, such as perceived freshness in salads, and physicochemical measurements alone will not be adequate to indicate the perception-based subtle changes in the product and its final acceptability.

The Ministry for Primary Industries (MPI), Aotearoa, New Zealand, defined shelf life as "the period established under intended conditions of distribution, storage, retail, and use, that the food would remain safe and suitable". During this period, the product should provide consumers with its intended sensory experience, performance, and benefits (ASTM E2454-20, 2020). Food companies are required by Aotearoa New Zealand law to attribute shelf life to their products under defined storage conditions. Food sold in Aotearoa New Zealand follows two typologies, namely, *use by* and *best before*. According to the MPI, the use by date is required for health and safety reasons, just like the European Union (EU), whereas best before is more related to food quality. Often food quality is beheld into objective indices related to microbiological, nutritional, or physicochemical characteristics (Cardello, 1995), but this is not what best before is implying. MPI (Ministry for Primary Industries, New Zealand, 2016) defined *quality attributes* as color, taste, texture, and flavor, as well as any specific qualities for which one makes express or implied claims, such as the freshness of the food. Motivated by these quality attributes, the author(s) would like to invoke the definition of sensory science here as a bridge to fill the gap between these quality attributes and best before date: "a scientific method used to evoke, measure, analyze, and interpret those responses to products as perceived through the senses of sight, smell, touch, taste, and hearing" (Lawless & Heymann, 2010). By connecting the quality attributes with sensory science, sensory shelf life methods can be used to provide the best eating experience and to provide economic importance to consumer desires.

One can view the sensory shelf life testing as no particular category of sensory testing but a program of repeated testing using accepted sensory methods (Lawless & Heymann, 2010). Sensory shelf life testing may employ any of the three major kinds of sensory tests, descriptive, discrimination, or affective, depending on the objectives of the study (Lawless & Heymann, 2010). Proceeding with the foundations of these tests, shelf life can be broadly viewed as an *expert opinion*, *conformance*, or *consumer appeal* (Lawless, 1995), which will be eventually covered in this chapter in the form of analytic and affective tests for the sensory shelf life determination. Along with the methods, the importance of sample storage conditions and designs has been discussed in detail for their relevance and better logistics arrangement. The sample storage designs were pictorially presented to engage readers, help them comprehend, and give them a step-by-step procedure to follow. Accelerated storage is discussed for its relevance with shelf life and its possible advantages and disadvantages concerning actual shelf life testing. Finally, recommendations and future opportunities are summarized to conclude the course.

3.2 DESIGN OF SAMPLE STORAGE

Test samples for sensory shelf life testing can be stored by two different approaches, *basic* and *reversed/flipped* (Hough, 2010). The basic approach is a *typical* sample storage design, which involves the storage of all test samples at prescribed conditions and periodically removes them for sensory analysis. Periodic removal of samples for sensory shelf life testing demands the periodic gathering of trained assessors or consumers (Lopane, 2018; Chanadang & Chambers, 2019), which can be a costly enterprise. Hence, this approach to storage has been criticized for having low efficiency concerning time, panel handling, and resources (Lawless & Heymann, 2010). Additionally, it may pose a risk of drift in the evaluation criteria if assessors or consumers somehow realize that they are participating in a sensory shelf life study. An effective identity control and the addition of fresh samples at each time point may help in minimizing these biases (Giménez, Ares, & Ares, 2012).

Another approach, termed reversed design (Figure 3.1), allows all samples of different storage times to be evaluated in one single instance. For example, if one has to study lettuce shelf life at 4°C, the first batch of lettuce stored at 4°C would correspond to the longest storage time, whereas the second batch harvested 2 weeks later would correspond to the second longest storage time, and this would continue until all storage times have been included (Figure 3.1) (Araneda, Hough, & De Penna, 2008). This staggering production approach is suitable for products where batch-to-batch variability remains small, such as carbonated beverages, biscuits, and others. This approach found its application in apple baby food shelf life testing (Gámbaro, Ares, & Gimenez, 2006). Simultaneously, confusion with storage times and batches may occur in some cases.

A variation on this is to store products of a single batch under desired storage conditions, and once the elapsed time for sampling is reached, take them out and keep them under conditions that essentially stop all aging processes, such as via freezing (Figure 3.2a) (Lawless & Heymann, 2010). For example, Samotyja (2015) used this design to have all samples of different storage times available on the same day. She stored commercial refined oil samples for other storage points and later froze them at −18°C until evaluation. The crux of the on/off aging approach is that not all products can qualify for on/off aging, such as lettuce, and those qualified indeed should not introduce sensory changes during the thawing process. High hydrostatic pressure-processed avocado pulp was tested for shelf life by this design approach (aging was stopped by storing at −80°C) (Jacobo-Velázquez & Hernández-Brenes, 2011). Likewise, fish burgers were refrigerated for 30 days and were subsequently frozen immediately on storage time points until the last storage point was reached (Marques, Lise, de Lima, & Daltoé, 2020).

Another variation of this procedure is to store the product under conditions that essentially stop all aging processes and remove them at each selected sampling time to keep them under normal

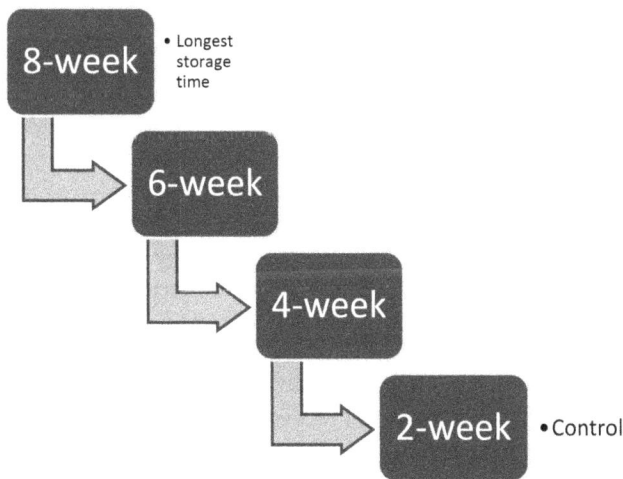

Figure 3.1 Reversed storage design.

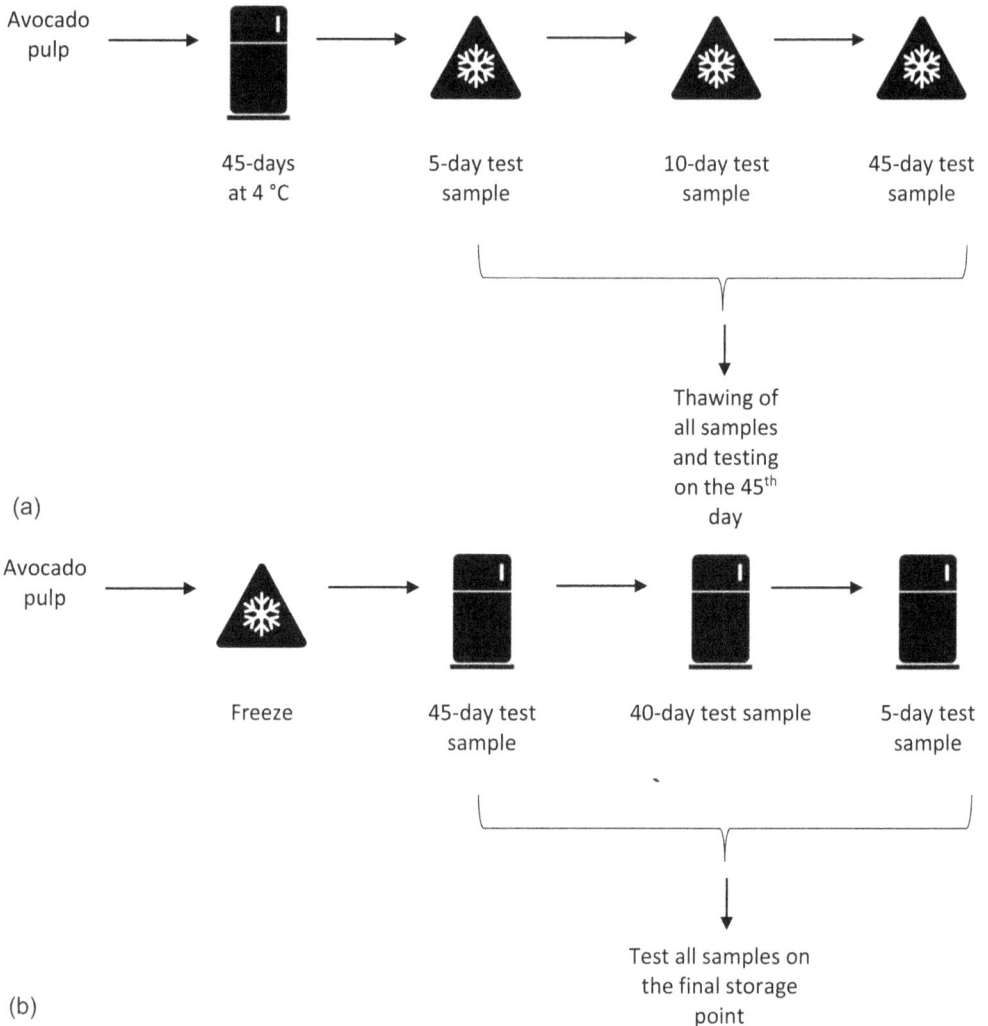

Figure 3.2 (a) Additional variation of reversed storage design with an avocado pulp test sample; (b) additional variation of reversed storage design with an avocado pulp test sample.

conditions until their evaluation (Figure 3.2b). To avoid the difficulty of repeatedly assembling a large group of consumers throughout shelf life studies, Hough et al. (2003) and others (Hough, Garitta, & Gómez, 2006) presented samples in one session using reversed design shelf life storage. Overall, the flipped design approach overcomes the main drawbacks of basic storage design, but selecting storage conditions that halt all deterioration might be a challenging task due to associated changes in the test sample.

The frequency of testing during storage has significant implications in the accurate prediction of shelf life. Generally, equally spaced time intervals are used to test the shelf life, but Hough (2010) gave the example of ultra-pasteurized milk where equally spaced time points captured minimal changes, especially in early storage days. For example, Hough (2010) found no significant differences in ultra-pasteurized milk up to the first 9 days of storage. For the same reasons it could be better, in some cases, to have more storage points toward the end of the shelf life period.

3.3 METHODS OF SENSORY SHELF LIFE ESTIMATION

Various methods exist for sensory shelf life testing, but they can be broadly categorized into two main categories, analytical and affective. Analytical testing involves trained to semi-trained assessors, whereas affective testing involves untrained, naïve participants. In general, descriptive analysis, difference-from-control tests, and quality rating methods are considered analytical, whereas consumer tests or acceptability limit methodology and survival analysis are considered affective or hedonic tests. Considering the objective nature of analytical tests, shelf life limits established based on trained assessors may lead to undesired financial losses due to the overrepresentation of the more sensitive portion of the population. The low *recognition threshold* of trained assessors does not represent the general public, and specifications based on them may be very conservative and contribute to more food losses.

In contrast, the higher risk of false positives in consumer-based sensory shelf life testing demands a higher number of participants to accommodate the likelihood of large variability in each consumer's sensitivity (Stocking, Suffet, McGuire, & Kavanaugh, 2001). In all instances, sensory methods of shelf life testing are destructive; hence, sufficient samples must be stored and available throughout the study. Irrespective of the method type, all techniques require some sort of sensory end point or a failure criterion to test against, beyond which the sensory quality is no longer acceptable. The selection of the test could be based on a previously determined sensory end point criterion or product type. Each method type has some implications, such as economic, logistical, or environmental, and the successful implications need careful method consideration.

3.3.1 Descriptive Analysis

Descriptive analysis is regarded as an analytical technique of high sensitivity, precision, and accuracy. For the same reasons, it is used extensively to discern the composition and detect small deviations relevant to the stakeholder, such as incipient rancidity in oil or sourness in milk. In the bearing of sensory shelf life estimation, it can be used to obtain a detailed description of the nature and range of the sensory changes a product undergoes during storage. The descriptive analysis relies on *trained assessors* to receive such precise and accurate information. Trained assessors have a long history since the inception of sensory science as a field. ASTM described trained assessors as having an established degree of sensory acuity, experience with the test procedure, and an established ability to make consistent and repeatable sensory assessments (ASTM E253-20, 2020). This method's power is its *panel* (assessor), and the training of the panel can be adjusted to meet the specific goals. Several training methods are available (Lawless & Heymann, 2010) depending on the intended purpose. Still, one thing is the same in all those methods that the panelists are trained to operate in unison, as an instrument, and each panelist serves as a function analogous to an individual sensor on an instrument (Drake, 2007). A few variations do exist in selecting assessors for sensory shelf life estimation; for example, Lopane (2018) used *experts*, pretrained coffee professionals from coffee houses, for the determination of the cold-brew coffee's shelf life. On the other side, Chanadang, Koppel, & Aldrich (2016) and Jacobo-Velázquez & Hernández-Brenes (2011) used *trained assessors* to evaluate pet food and avocado pulp shelf life, respectively. Trained assessors either individually (ballot method) or collectively (consensus method) select the appropriate attributes based on fresh control or stored samples and subsequently use some of those agreed-on attributes for description via discussion with the panel leader. Ballot and consensus are two variants of descriptive analysis and more information on them can be obtained from other sources (Chambers, 2018). Attribute identification based on stored samples is termed *critical* attributes, i.e., a sensory defect appearing from prolonged storage, and is responsible for consumers' rejection of the sample (Giménez et al., 2012). Once critical attributes are identified, follow-up training on those selected attributes should be set out. Previously developed lexicons can also be used for initial attribute selection (Ari Akin, Miller, Jaffe, Koppel, & Ehmke, 2019; Kumar & Chambers, 2019; Sharma et al., 2020). For instance, Chanadang and Chambers (2019) used the dry dog food lexicon for attribute selection while Lopane (2018) used the coffee lexicon (Chambers et al., 2016).

Sometimes, the selection of critical attributes for shelf life estimation is *experiment based* and at other times it is *experience based*. For example, Garitta, Hough, & Sánchez (2004) identified *critical* attributes from a list of attributes through accelerated storage sample testing, while Chanadang and Chambers (2019) considered *rancid* and *painty* attributes as key surrogates of consumer disliking and used them for shelf life cutoff points. The former, experiment-based strategy is more appropriate and could be of enormous significance in known and unknown product category testing. For instance, the selection of critical attributes was performed via open discussion with the panel leader following the assessment of four lettuce samples stored at different storage times of 0, 7, 10, and 17 days (Lareo et al., 2009). In contrast to the discussion-based selection, a ballot variant was used to identify the critical attributes that changed most frequently over accelerated storage time (Garitta et al., 2004). Similarly, Ares et al. (2009) used stored samples corresponding to five storage times for the critical attribute selection, and Jacobo-Velázquez and Hernández-Brenes (2011) used stored samples of up to 35 days for the same task. Gauchez et al. (2020) selected attributes for sourdough bread based on a preliminary storage study. Hence, critical attributes should always be set through preliminary storage studies, which can be accelerated or real-time studies. It is suggested that readers refer to the accelerated storage section of this chapter before selecting the accelerated storage study.

Following attribute selection, a pretrained panel is calibrated on those selected critical attributes with reference standards and dilutions, respectively. For instance, Jacobo-Velázquez and Hernández-Brenes (2011) calibrated seven pretrained assessors with citric acid for sourness and hexanal for rancidity by using 0, 6, 20, and 100% w/w concentration of each critical attribute. Likewise, different concentrations of critical attributes were used to calibrate the pretrained assessors on different intensities (Garitta et al., 2004). Unstructured or structured line scales are generally used to rate the intensity, and for the same reason, sometimes this method is also known as *intensity* measurement. For instance, Center for Sensory Analysis and Consumer Behavior (CSACB) at Kansas State University typically usually uses a 15-point scale with 0.5 increments (Chanadang et al., 2016; Chanadang & Chambers, 2019). Assessors are trained on the intensity scale usage with the help of standards of different concentrations (Jacobo-Velázquez & Hernández-Brenes, 2011). It is advised to use a minimum of three reference standards for near-zero, medium, and extreme scale points (Sharma, 2019). Hough et al. (2002) calibrated the assessors by asking them to correctly match unknown stimuli (labeled as K) with the corresponding attribute and score the different concentrations of an attribute, labeled as A, B, and C, to achieve a consensus score. Following calibration, intensities are assigned to different concentrations after discussing with the panel leader (Garitta et al., 2004; Jacobo-Velázquez & Hernández-Brenes, 2011). After calibration, the performance of the panel can be validated by some kind of tests, such as by previously developed consensus scores for each reference (Hough et al., 2002) or by correctly matching the reference standard with a blind-coded intensity (Garitta et al., 2004), or by other tests (Tomic, Forde, Delahunty, & Næs, 2013). Once the panel performance is considered reliable by the panel leader, it is time to start the actual sample evaluation. Hough et al. (2002) removed the stored samples of milk powder every 3 weeks of the 3-month storage, reconstituted, and served to the trained assessors paired with a fresh control sample. Assessors recorded the appearance and flavor attributes that differentiated stored milk samples from the fresh control. Many other products have been tested for sensory shelf life by the intensity measurement property of this method, such as whole milk powder (Hough et al., 2002), ostrich steaks (Fernández-López et al., 2008), pork (Blixt & Borch, 2002), and others. The average intensity of duplicate or triplicate evaluations, calculated over each storage time point, is regressed over time to create a scatter plot of attribute versus time as shown in Figure 3.3.

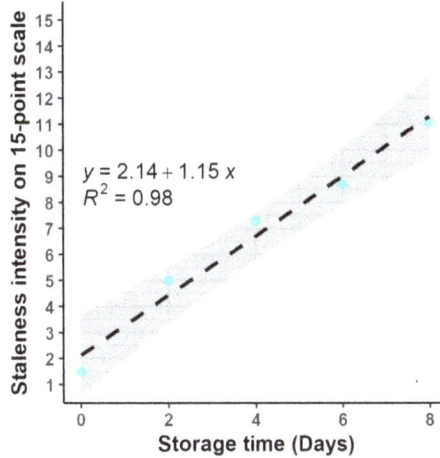

Figure 3.3 Typical scatterplot example of critical attribute intensity rating by trained assessors at different storage times with a fitted regression line and 95% confidence intervals.

Sensory shelf life can be determined as the time point when the intensity of a sensory attribute reaches the predetermined criterion, known by failure criterion or cut off point. It is advised to avoid selecting an arbitrary failure criterion for attribute intensity and refer to the later section of this chapter for criteria selection. Descriptive analysis is one of the most popular methods used for sensory shelf life determination and has been used extensively (Garitta et al., 2004; Hough, Subramaniam, Narain, & Beeren, 2013; Chanadang et al., 2016; Lopane, 2018; Chanadang & Chambers, 2019). A simplified flow diagram of the process is presented in Figure 3.4.

In this approach, the intensity of *selected* sensory attributes determines the shelf life. Hough (2010) echoed the use of a limited number of critical sensory attributes for shelf life studies, generally ranging from three to six. A relatively small number of attributes helps to reduce the time and resources needed to maintain a well-trained and calibrated panel. For instance, Garitta (2004) used four critical attributes for shelf life testing. Concurrently, the number of assessors and replications are as crucial as the number of critical attributes. For example, Chanadang and Chambers (2019) evaluated samples in triplicate. This method's primary disadvantage is its higher maintenance cost concerning time, price, and human resources. An extensive training requires a battery of relevant and up-to-date reference standards, which in turn demands several helping hands behind the curtain to plan, prepare, and serve those standards during sensory sessions. Also, the availability of a facility in which the evaluations can be performed reliably is also an added challenge to the experiments' logistics. Practitioners are advised to follow recommended guidelines of descriptive analysis methods as it is not uncommon to find published research where best practices were not followed and which may compromise the accuracy and reliability of sensory shelf life estimations. It is also essential to understand here that the analytical approach of trained assessors may have a very low *recognition threshold* than ordinary untrained assessors or consumers, which may not be an ideal situation for establishing a cutoff point.

Assuming an uncomplicated simplest design type, such as completely randomized design (CRD) can be used to assign the treatment factor (storage) levels (range of storage days, such as 5, 10, and 15 days) to the experimental units (test sample), one can start with preliminary exploratory analysis or graphical analysis followed by one-way analysis of variance (ANOVA) to establish that differences or a relationship among factor levels are of sufficient magnitude to qualify for *statistical significance*. For CRD, the levels of the primary factor are randomly assigned to the experimental units. For instance, select sample sizes $n_1, n_2, \ldots, n_g$ with $n_1 + n_2 + \cdots + n_g = N$ (total units) and then choose n_1 units at

Test sample

Basic Design		Reversed Design

Basic Design

↓

Assemble previously trained
assessors (8 to 15) at each
storage time

↓

Storage time 0 − Identification
of critical attributes (5 to 6) ◄── through preliminary shelf-life ──► of critical attributes (5 to 6)

Critical attributes can be found

testing of samples with

different storage times or from

published lexicons and product

category experience

Reversed Design

↓

Assemble previously trained
assessors (8 to 15) at once

↓

Storage time 0 − Identification
of critical attributes (5 to 6)

↓

Train assessors on selected
critical attributes with the help
of references and definitions

↓

Train assessors on scale usage

↓

Check panel performance

↓

Test samples for selected
critical attributes and record

↓

Storage time 1 − Repeat testing

Storage time 2 − Repeat testing

Storage time n − Repeat testing

↓

Train assessors on selected
critical attributes with the help
of references and definitions

↓

Train assessors on scale usage

↓

Check panel performance

↓

Test samples for selected
critical attributes and record

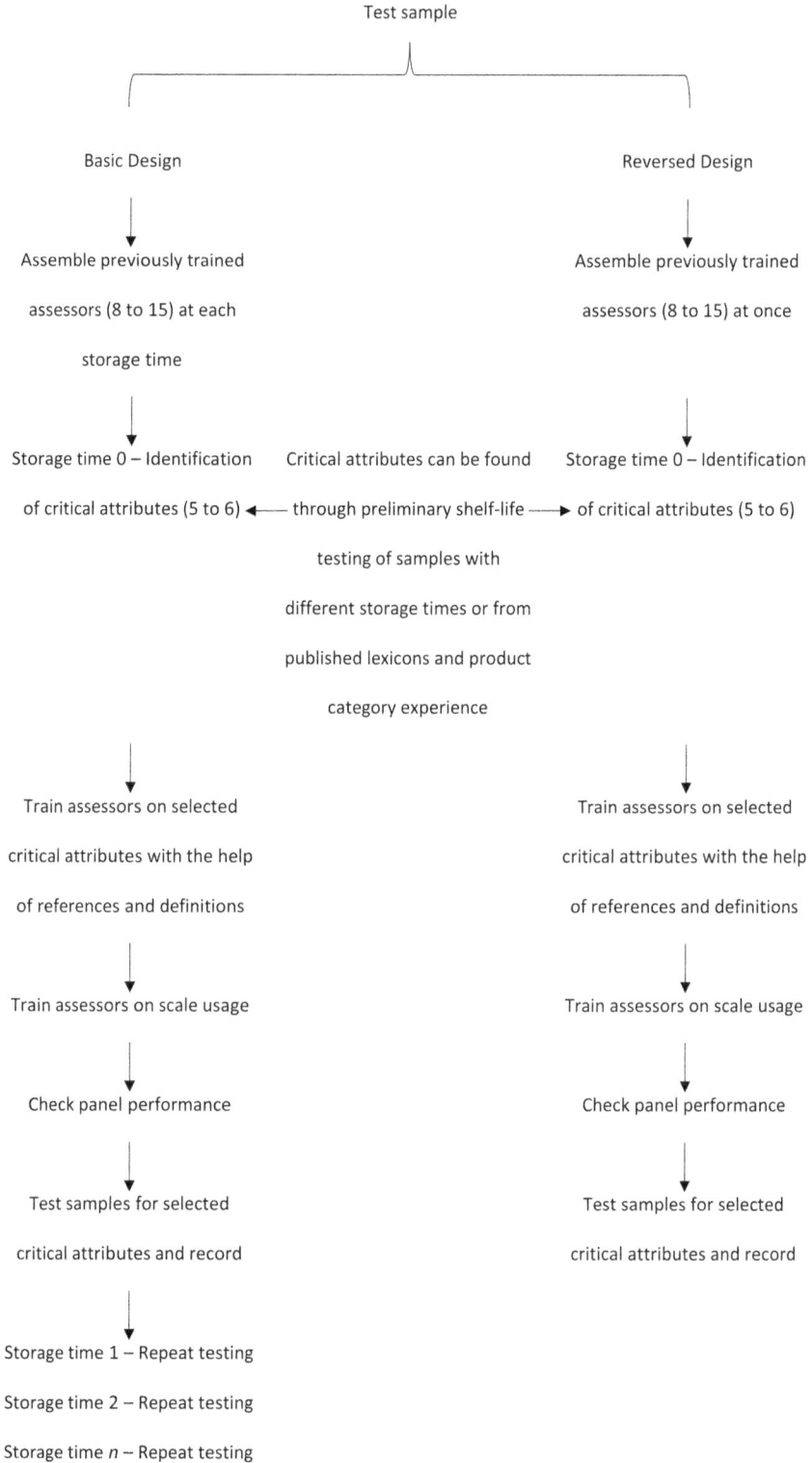

Figure 3.4   Flow diagram of the descriptive analysis process concerning basic and reversed designs.

random to receive treatment 1. Likewise, choose n_2 units at random to receive treatment 2 and so on, or store the samples in required conditions and randomly take out samples at the required storage point, say after 5 days, and so on. It is important with this design that there should be similar or comparable variability within the units. Readers are advised to check assumptions ahead of running an ANOVA or any other tests, such as pairwise comparisons and focused procedures based on contrasts. Even though an ANOVA can give us an indication that not all treatment groups have the same mean response, it does not, by itself, tell us which treatments are different or in what ways they differ. To do this, we need to look at the treatment means or the treatment effects. One method to examine treatment means is called *multiple comparisons*. A huge variety of multiple comparison tests exist, and based on the *error rate* one wants to control, an appropriate test type can be selected (Oehlert, 2010).

3.3.2 Difference-from-Control Test

One of the most popular approaches of sensory shelf life estimation is the *difference-from-control* test, which measures the *overall difference* and *magnitude* of the difference between stored samples and control via trained or semi-trained assessors using discriminative tests or intensity scales (Lawless & Heymann, 2010; Leong, Kasamatsu, Ong, Hoi, and Loong, 2016; Tano-Debrah et al., 2019). This said, the implementation of this method varies widely from only rating the *overall difference* to rating the *directional* differences on *specific attributes* to acquire the diagnostic information on the *nature* of the difference (Meilgaard, Carr, & Civille, 2015). The magnitude of difference between stored samples and the control is modeled as a function of storage conditions (of time, temperature, relative humidity, and so on), and the shelf life is estimated as the time at which the product reaches a *predetermined difference* from the control product. A key challenge in this test type is finding the appropriate storage condition for the *control* sample to stop or delay changes and keep it as fresh as possible. Water, if not the most important ingredient, is an important ingredient of food stability and shelf life. In the same vein, it is critical to understand the science of *freezing* and the role of water in the context of perishable produce. For instance, rapid freezing favors smaller ice-crystal formation by stopping water migration to seed crystals (Kwak et al., 2013), whereas slow freezing favors larger ice-crystal formation. Jacobo-Velázquez and Hernández-Brenes (2011) used −80°C to quickly freeze the pulp sample and prevent the large ice-crystal formation in the shelf life study. Thereby this process minimizes the texture deterioration and has been used previously to test cooked rice texture (Kwak et al., 2013). Likewise, the optimum temperature for the frozen storage of meat has been reported to be −40°C, as only a tiny percentage of water is unfrozen at this point (Leygonie, Britz, & Hoffman, 2012). Simultaneously, time is also as crucial as the freezing method or condition. For example, the ricotta cheese control sample stored at $2 \pm 1°C$ was replaced with a new sample (to minimize the difference between control samples) every 7 days (Hough, Puglieso, Sanchez, & da Silva, 1999). In such cases where it is not possible to keep the same control sample throughout the whole period, the replacement of the control sample should be validated using discriminative tests. Hough et al. (1999) used 15 trained assessors to ensure that the fresh batch was not statistically different from the previous batch via a triangle test.

Another critical challenge in this test type is the selection and training of participants. The participants' training objectives are to familiarize themselves with the developed terminology and differences between basic or complex stimuli at varying levels and execute those differences by scales. However, in general, training of the panel to use this method contains *less involvement* than the training for the traditional descriptive analysis (Meilgaard et al., 2015). Also, simplicity, high flexibility, and ease of panel training have been outspoken benefits of this method. In a hybrid form of difference-from-control test, Leong et al. (2016) screened naïve participants for taste acuity, and they were trained later in six training sessions of 1.5 hours on a developed terminology to acclimatize for *directional* differences rather than *global* or *overall*. By directional author means to specify the sensory attribute, for instance sweetness, in which the samples differ. Acknowledging the later

Table 3.1   A Typical Example of Discrimination Test Findings Reporting

Methods	Correct/Total Response	Statistical Analysis	z-Value	Number for Significance at 0.05	D/n
Triangle	6/17	n.s.	−0.09	10	0.03
Dual standard	11/17	n.s.	0.97	13	0.29
3-AFC	12/17	s.	3.00	13	0.56
2-AFC	13/17	s.	1.94	13	0.53

Abbreviations: n.s., not significant; s., significant.

(overall differences) property of this test, Donovan et al. (2016) echoed this test's use for complex test stimuli, such as salads. Simultaneously, Hough et al. (1999) used a trained panel for testing the *global* differences in the ricotta cheese samples.

For each sampling time, assessors receive the control sample labeled as K, stored samples labeled with three-digit code numbers, and a blind control labeled with a three-digit code number. From here onward, two different approaches can be used to answer two different questions, such as "whether the stored sample and a fresh control sample are perceptibly different" or "to what degree they are different." In the first approach, discrimination tests such as paired comparison, triangle, tetrad, or duo-trio tests can be used, whereas in the second approach, degree-of-difference scale types, in the context of a difference-from-control trial, can be used. Just to vigilant the practitioners here, forced-choice discrimination tests are more discriminating than scale type tests to measure subtle differences (Hough, 2010). However, if preliminary testing indicates that samples have very large and obvious differences, then it may be more appropriate to use scaling techniques to indicate the exact magnitude of the difference between the samples. Discrimination tests are commonly used to determine whether small changes in ingredients, processing, packaging, or products by themselves have any effect on the sensory properties of a particular product. This approach has been used for estimating the sensory shelf life of *Deglet Nour dates* by performing a triangle test with 24 semi-trained panelists (Nabily, Nabili, Namsi, Majdoub, & Azzouna, 2020). In general, practitioners should consider using a large sample size when power needs to be high, for instance, 50 or greater (Lawless & Heymann, 2010). Also, readers are advised that not all discrimination test methods perform equally well in detecting the small differences between products. For example, the paired comparison test tends to be more sensitive to differences than the duo-trio test (Table 3.1). Therefore, the methods should be selected wisely (Lawless, 2012). This methodology, either in original or modified form, is frequently used in industrial environments rather than the degree-of-difference scale type under the difference-from-control context.

From the data analysis point, the percentage of correct responses for each test method is determined as shown in Table 3.1. Use the following *binomial approximation* of normal distribution, with a critical z-value of 1.645 for significance:

$$z = \frac{\left(P_{obs} - P_{chance}\right) - \left(\frac{1}{2n}\right)}{\sqrt{\frac{pq}{n}}} \tag{3.1}$$

where P_{obs} is the correct proportion identified by the subject in the test, P_{chance} is the chance probability for each test, $q = 1 - p$, and n is the number of subjects. Check the results against the tables of "minimum number of corrected judgments" found in Lawless and Heymann (2010) to make these determinants. The *estimated proportion of discriminators* (*D/n*) for each method is calculated by the following formula:

$$C = D + p(n - D)$$

where C = number of correct responses, D = number of discriminators, n = number of respondents, and p = chance performance level of the test.

There are situations, especially those encountered during shelf life, when it is more important to know the size of differences than the mere existence of a difference; it is where the degree-of-difference scale type tests found their worth. A variety of scales have found their use in this approach, such as the 5-point scale (Tano-Debrah et al., 2019), 7-point scale (Hough et al., 1999), 7-point bipolar scale (Gauchez et al., 2020), 9-point scale (Dong, Wrolstad, & Sugar, 2000), 11-point bipolar scale (Leong et al., 2016), and others. Some of the types are shown in Figure 3.5. Hough et al. (1999) used a 7-point structured unidirectional scale, where 0 = no difference, 1 = very slight, 2 = slight, 3 = moderate, 4 = moderately big, 5 = big, and 6 = very big, whereas Tano-Debrah et al. (2019) used a 5-point scale, where 5 meant that the product was as good as the control, 3 referred to the product acceptability limit, and scores ranging from 0 to 2.9 referred to products that were not acceptable. The scale also can be designed to provide directional information related to the attribute being analyzed, for example, 4 = "Extremely Weaker," 0 = "Not at All Different," to 4 = "Extremely Stronger" (Meilgaard et al., 2015).

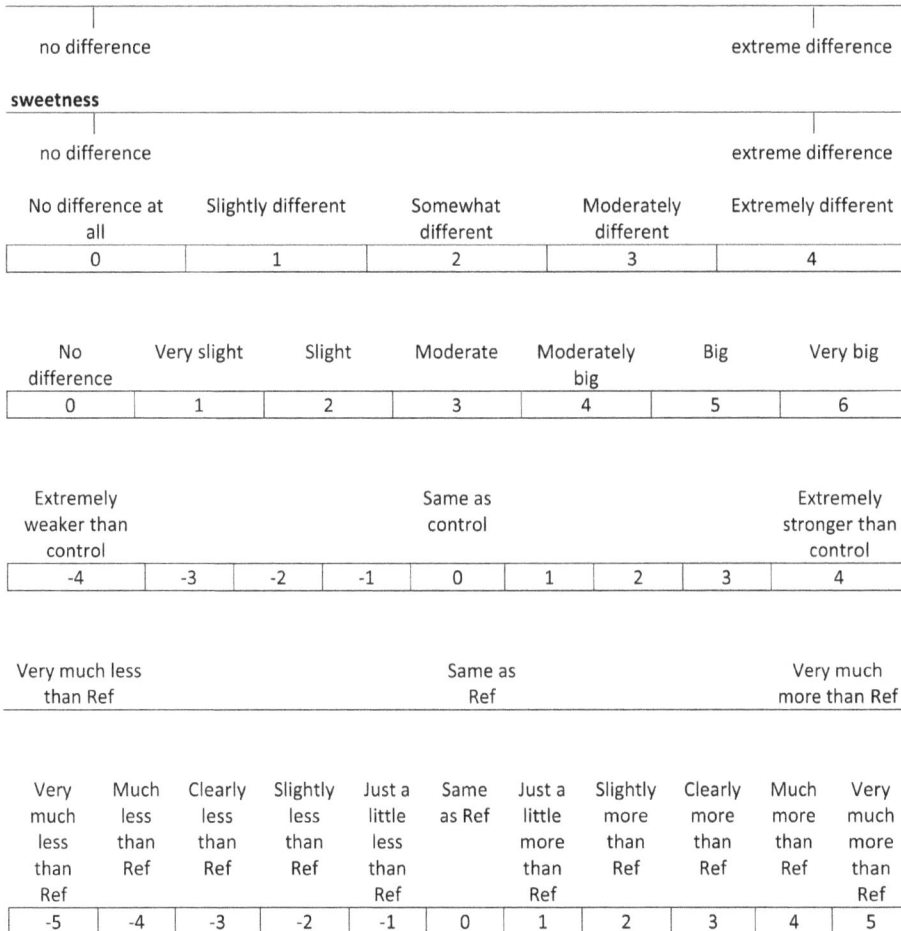

Figure 3.5 Examples of unidirectional and bidirectional degree-of-difference scale types under difference-from-control test type.

Table 3.2   A Typical Example of Data Set Obtained from Assessors (Excerpt)

	Storage Time							
	t_1		t_2		t_3		t_n	
Assessors	Sample	Control	Sample	Control	Sample	Control	Sample	Control
1	1.2	0.5	1.4	0.6	1.4	0.5	1.8	0.4
2	1.1	0.4	1.5	0.7	1.3	0.5	1.7	0.5
3	0.8	0.8	1.4	0.7	1.4	0.4	1.7	0.4
–	1.1	0.8	1.6	0.6	1.4	0.4	1.7	0.4
N	1.1	0.7	1.3	0.5	1.3	0.5	1.7	0.4
Average	1.06	0.64	1.44	0.62	1.36	0.46	1.72	0.42
Sample score	1.06 – 0.64 = 0.42		1.44 – 0.62 = 0.82		1.36 – 0.46 = 0.9		1.72 – 0.42 = 1.3	

A sensory specification or cutoff score between stored and fresh control samples should be set in advance to establish the end of the product's shelf life. According to the product or situation, several criteria have been used previously, such as Tano-Debrah et al. (2019) who used 3 on a 5-point scale as an acceptability limit, whereas Hough et al. (1999) used ≥1.5 on a 7-point scale as the cutoff point to define failure. The sample score was obtained by subtracting the mean of the blind control from the mean of the stored sample (Table 3.2), and this score was further regressed over the storage time to estimate the shelf life of the product (Figure 3.6). Still, practitioners are advised to avoid arbitrary selection of the failure criterion; they should choose consumer perception responses for the cutoff point identification.

The main disadvantage of this method is that the overall difference rating does not provide specific guidance on the *nature* of the difference used to identify the cause. Even the inclusion of

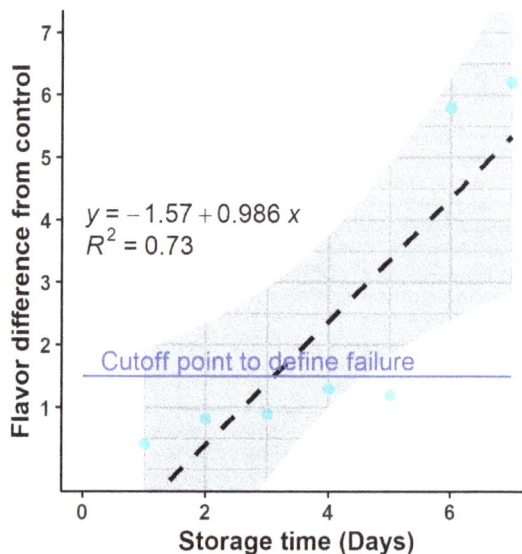

Figure 3.6   Typical scatterplot representation of mean sample score data regressed over storage time to estimate the shelf life with 95% confidence intervals. *Note:* The cutoff point is arbitrarily selected, which is not a recommended practice.

several attribute scales does not guarantee that all sources of difference will be identified (Meilgaard et al., 2015). Another disadvantage is that the control sample must always be available, which means that it should be stored for extended periods with minimal accompanying sensory changes or easy to produce.

From a statistical analysis point of view, data generated from the difference-from-control test can be analyzed using an ANOVA test, followed by multiple comparison tests, such as the Tukey test (Dong et al., 2000) or Dunnett test (Donovan et al., 2016; Leong et al., 2016). It is important to recognize here that in some situations, we do not want to make all pairwise comparisons; instead, we want to make comparisons between a control and the other treatments. In those situations, a Dunnett test is recommendable. Yet, for most of the sensory shelf life studies, there is the assumption that the stored sample would be directionally different from control, either on the decreasing freshness side or on increasing off-flavor; hence, one-sided tests should be selected.

3.3.3 Quality Rating Methods

Quality rating methods involve a direct, *overall sensory quality* measurement throughout storage. Hitherto *quality* is a multifaceted and complex term to define, and Cardello (1995) indeed equated quality with the "consumer-based perceptual construct," but for the sake of this method, we would align ourselves with the objective and expert-based definition of Molnár (1995), whose theoretical model for overall quality was unitary and dependent on direct expert measurements. By direct measurement, we mean integrating different characteristics of the test sample into a single score or grade, or index (Molnár, 1995). To provide valid and reliable information, researchers should define quality beforehand, its specifications, and its tolerance limits.

The quality index method (QIM), a univariate theoretical model, is widely used in the fisheries (Freitas et al., 2019; Joshy et al., 2020) and is an excellent example of the application of quality rating methods. Established by the Tasmanian Food Research Unit, QIM is a fast, simple, nondestructive, and descriptive *grading* system of *freshness* evaluation in seafood. This method integrates the differences between fish species through an objective assessment of fish attributes, such as body surface or gill odor. This method measures the degree of change in critical sensory attributes using a scoring system of 0 (fresh) to 3 (spoiled) demerit points; the longer the time, the higher is the punctuation. However, it is uncommon to discover that all attributes scored the maximum demerit points because the changes during the storage period may not be significant to achieve such scores. Demerit scores of critical sensory attributes, provided by trained assessors, are summed up to a total to give an overall evaluation index, the so-called Quality Index (QI) for shelf life prediction. The lowest value of the cumulative total is 0, representing the best quality, whereas any values above 0 indicated the increasing deterioration of quality (Joshy et al., 2020). This prediction is possible if a linear correlation between QI and storage time exists. Extending the idea of Molnár (1995), Joshy et al. (2020) developed a demerit score-based Fish Quality Index (FQI), which gives different weight to important attributes via the Delphi method. Readers are encouraged to consult this work for more information on this method as we are focusing only on QIM in this chapter.

QIM demands a precise description of quality parameters or attributes describing the product under a panel leader's guidance. In the first place, descriptive analysis methods, such as Quantitative Descriptive Analysis (QDA™) or others, can be used to define the attributes in identifying the maximum storage time of the test sample. QDA was used for the previously mentioned purposes (Sveinsdottir, Hyldig, Martinsdottir, Jørgensen, & Kristbergsson, 2003). In the course of determining maximum shelf life, the panel detects *characteristic* and *spoilage* attributes, which may together develop into a profile and can be used as a reference profile for future studies. Following the selection of the critical attributes, assessors are trained in the QIM evaluation of these attributes. To develop QIM, Sveinsdottir et al. (2003) used three previously experienced QIM assessors, whereas Bonilla,

Sveinsdottir, and Martinsdottir (2007) used only two, possibly untrained assessors, to observe and register the sensory changes in the form of the so-called *preliminary* scheme, from day 0 until spoiled. Although Bonilla et al. (2007) rated the preliminary scheme's attribute on a 3-point scale, it was not clear from their research if they already had an existing attribute or if the new attribute was developed on the spot. Although no such rating was reported by Sveinsdottir et al. (2003) in the preliminary scheme, they did run six 1-hour sessions to establish the QIM. Such variations are common and may be confusing sometimes. Following the preliminary scheme development, 9–12 previously trained assessors (Sveinsdottir et al., 2003; Bonilla et al., 2007; Ari Akin et al., 2019), could become QIM assessors of the different product categories or in general sensory assessors for the product's evaluation. The panel leader explained how to use the previously developed preliminary scheme and evaluate each quality parameter. All suggestions of improvements are made in the scheme during the training sessions, and the QIM scheme is finalized before the last training session.

Among other quality rating methods, the in/out or pass/fail method is quite a popular choice at the business level and has found applications in fish inspection and carbonated beverage product categories, to name a few. This method involves semi-trained assessors' evaluations and classification of samples as either "in" or "out" of the acceptable range of variation related to the standard product (Civille & Carr, 2016). In quality control settings, this method is mainly used to identify products that show apparent deviations, such as off-notes or other defects. This method found its applications in the products that exhibit variation in only a small number of sensory attributes. It requires the least amount of panel training, and simplicity is its primary advantage. Finally, the panel of this method requires careful monitoring and ongoing maintenance because the panelists' ratings are directly related to the disposition of the product. Constant care must be exercised to ensure that the panelists are objectively assessing the products and are not succumbing to pressure to meet production quotas or other non-quality related criteria. A snippet of the evaluation sheet is presented in Figure 3.7 for assessing defects during the storage of carbonated beverages. Interested readers are encouraged to stretch and try reading the book, *Sensory Evaluation in Quality Control*, for detailed information on this method.

Quality rating methods have also found their applications in other product categories, such as dairy and meat products. Among other advantages, they offer a reliable and standardized

	Test samples					
	Ts1	Ts2	Ts3	Ts4	Ts5	Ts6
IN	☐	☐	☐	☐	☐	☐
OUT						
Fruity	☐	☐	☐	☐	☐	☐
Burnt caramel	☐	☐	☐	☐	☐	☐
Sulfurous	☐	☐	☐	☐	☐	☐
Medicinal	☐	☐	☐	☐	☐	☐
Earthy	☐	☐	☐	☐	☐	☐
Metallic	☐	☐	☐	☐	☐	☐

Figure 3.7   In/Out sheet.

methodology that involves instructions and easily understandable illustrational material. It is well suited to teach inexperienced people and train or monitor panelists' performances. Despite the wide usage for sensory shelf life estimation, many published studies do not comply with recommendations for best practices in sensory evaluation. Many unpublished sources also confirm this. The biggest problem observed is the *inadequate* number of trained assessors for the rating tasks. For instance, Lopez-Galvez et al. (2013) used only three trained assessors. It is widely recommended to use 8–20 trained assessors to get reliable results, except in a few conditions where the laboratory has a long history of boarding trained assessors and are specialized sensory labs. Another common problem observed in dealing with quality rating systems is the lack of appropriate panel training. Although this is the most critical step in determining the results' validity, many studies do not specify the procedures used, whereas others used some arbitrary training methods (Sharma, 2012; Lopez-Galvez et al., 2013). Also, inadequate training sessions, such as only one training session by Siripatrawan and Noipha (2012), were used to merely meet the symbol, i.e., *trained*, rather than the concept to follow. Likewise, in some cases, hedonic information was gathered from trained assessors as an index of product quality (Siripatrawan & Noipha, 2012; Lopez-Galvez et al., 2013). It is highly recommended that trained assessors should not measure liking since their perception is not representative of the perception of a naïve consumer. It is essential to advance acceptable practices to increase the validity of the sensory shelf life estimations.

3.3.4 Consumer Test

The use of *consumers* for the sensory shelf life estimation has a long history, de facto our descendants have been exercising this method for centuries, obviously not by scale usage, but by being actual consumers. This same age-old practice of estimating shelf life has been advanced to include a scale for comparison and objective purposes. It was thought to be the most appropriate tool to determine when a food product reaches the end of its shelf life (Hough et al., 2002). However, simultaneously the repeated assembly of consumers for the multiple measurements needed during shelf life stages would be impractical and expensive. A typical consumer method, the *acceptability limit methodology*, is used to estimate the shelf life by the consumer liking data using a 9-point hedonic scale. This scale is the most widely used in sensory science, and was developed at the United States Army Food and Container Institute. This method's primary prerequisite is to recruit the target consumer population who intends to be the *actual consumer* of the test sample. There is no such thing as universal or all-purpose consumers. The selection should be based on the "target population," which means that consumers who would actually purchase and use the product. Hough et al. (2002) screened consumers for dairy product consumption and selected consumers with a frequency of at least once a week. Measuring sensory acceptability with a reduced number of subjects is not recommended (Araneda et al., 2008). A minimum of 100 consumers has been recommended for obtaining reliable data (Hough et al., 2006). In this test type, participants are presented with a set of samples with different storage times and are asked to score their overall liking. The product's shelf life is defined as the storage time at which its overall acceptability falls below a previously set value. Figure 3.6 shows a typical linear decrease of the overall acceptability score with storage time.

Many criteria have been used previously to set the acceptability limit, such as Sharma (2012) used the score of 7 on a 9-point scale as a cutoff point (Figure. 3.8), whereas others proposed a score of 6 (Giménez et al., 2012). However, the use of one universal acceptability limit as a cutoff point across product categories has certain disadvantages. For example, considering the lower overall acceptability score of fresh produce, such as of 5 to 6 on a 9-point hedonic scale for potatoes (Sharma, 2019), selecting a universal acceptability limit as a cutoff point is not the right approach. A closeness of overall acceptability score, such as enumerated above, to the cutoff point may lead to too conservative shelf life estimations that limit their commercial life. Although both Sharma (2012) and Giménez et al. (2012) used 7 and 6 (on a 9-point scale) as a cutoff criterion, neither

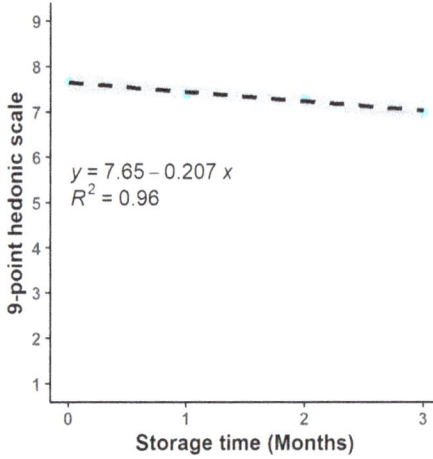

$$y = 7.65 - 0.207\,x$$
$$R^2 = 0.96$$

Figure 3.8 Redeveloped plot of rice-mung bean extrudates. (From Sharma 2012.)

of them was able to tell what percentage of consumers would stop eating by this point. Recently, Gauchez et al. (2020) found that the middle of the scale corresponds to 50% consumer rejection for sourdough bread. This may be a good approximation to assume, but this finding was only tested with one test sample and needs further research with other product categories.

The *first significant difference* in overall acceptability via statistical testing, from that of the fresh product, has also been used for the cutoff point; Gauchez et al. (2020) identified the first significant difference in the overall liking of sourdough bread on a 84- to 96-hour for sample B via Dunnett test. Gauchez et al. (2020) simply plotted overall liking as a function of storage time and estimated the shelf life when overall liking became significantly (statistically) different from that of the control time slot. In a sense, this seems quite a straightforward test, but practitioners should keep one thing in mind: not all consumers would stop consuming the product at one specific point and statistical significance is indeed not a real consumer opinion of whether they will stop consumption, but more of a change in an attribute. For the same reason, practitioners often consider other approaches. One such approach is rejection rate, such as a 25 or 50% rejection rate, which suggests the *percentage* of consumers who would stop eating the test sample in a real-life situation. By considering the economic implications of percent rejection, such as those related to the withdrawal of samples from the marketplace when they are still fresh for some segment of consumers, this approach holds a special position in business strategy. Generally, a question stated as "would you continue to consume this sample" or "would you normally consume it" is used to follow percent consumers rejecting the stored sample along the way and later use the obtained data either with survival analysis (Hough, 2010) or with their own (Gauchez et al., 2020). Gauchez et al. (2020) found a 50% rejection rate @ 64 hours with linear regression for sample B compared with 84–96 hours by the first significant difference method.

In line with cutoff points, Ramírez et al. (2007) used a criterion where the *minimum size of a significant difference* in consumer liking score, responsible for statistical significance among storage times, obtained via Fisher's least significant difference (LSD) or obtained through any other *post hoc* tests, such as the Student-Neuman-Keuls (SNK) test (Fritsch, Hofland, & Vickers, 1997) or Dunnett test (Makhoul, Ghaddar, & Toufeili, 2006), was used to establish a cutoff point. By doing so, researchers were providing a conservative estimate of the shelf life. For example, the cutoff point for the shelf life of sunflower kernels was determined by subtracting the *minimum value* (assuming 0.7 here; Fritsch et al. 1997 found 0.5), obtained via SNK test, from the average acceptability score (assuming 6.5; Fritsch et al. 1997 found 5 on a 9-point hedonic scale) of the fresh control sample, as:

Minimum acceptable liking score = Average acceptability/liking score of control sample − LSD

$$= 6.5 - 0.7 = 5.8$$

The previously mentioned criterion can also be rewritten as shown in the following equation:

$$\text{If } LSD = Z\alpha\sqrt{\frac{2 \times MSE}{N}}$$

$$\text{or } S = F - Z\alpha\sqrt{\frac{2 \times MSE}{N}} \tag{3.2}$$

where S is the minimum tolerable acceptability of stored sample; F is average acceptability of fresh or control sample; Z_α is a one-tailed coordinate of the standard curve for an α significance level, such as 1.645 for 5%; MSE is mean square of the error derived from ANOVA of the consumer data; and N is the number of consumers. *Note:* LSD can be calculated by either z or t **distributions**, so readers are advised to consult a statistician for the appropriate choice to make. In general, the t distribution is preferred over z because it is a more conservative approach and the practitioner would be less likely to commit a type I "false-positive" error. In addition, if the sample size is large enough, i.e., more than 30, the t-quantile will be close to the z-quantile.

The obtained value, i.e., 5.8, can be used as a cutoff point to determine the *sensory failure point* on the trained assessor's data via regression through the origin (Fritsch et al., 1997) or simple regression, as shown in Figure 3.9. Confidence intervals can also be obtained concerning shelf life, as shown in Figure 3.9. Hough and Garitta (2012) echoed the use of confidence intervals in sensory shelf estimations.

This *sensory failure point* (3.98 or 4, in our case) can be further used to determine the shelf life by regressing the trained assessor's data on storage time. This method combines the analytical descriptive analysis with consumer responses and can be regarded as a combination of *intensity*

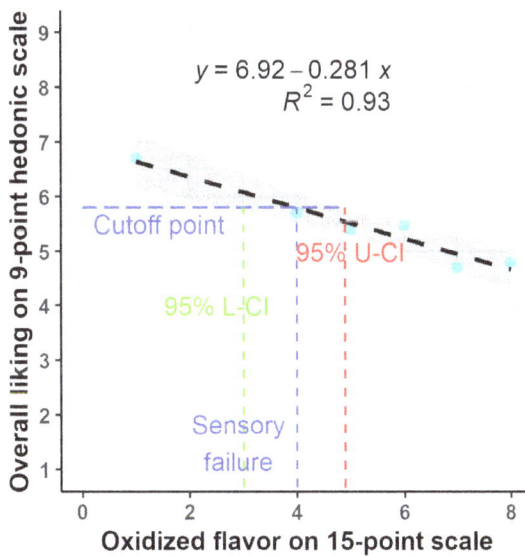

Figure 3.9 A typical scatter plot obtained with the consumer's sensory failure equation with 95% confidence intervals, both upper and lower (U-CI and L-CI, respectively).

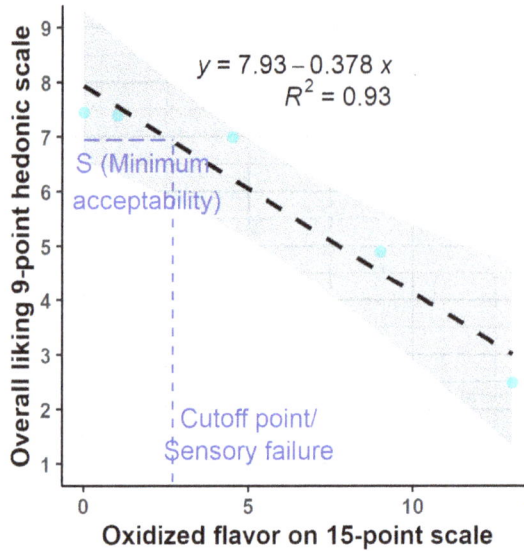

$$y = 7.93 - 0.378\,x$$
$$R^2 = 0.93$$

Figure 3.10 Typical scatter plot obtained from the first step of the cutoff point (cutoff point on the graph) estimation. Regression of mean overall liking score as a function of oxidized flavor intensity from trained assessors to locate the cutoff point is seen.

measurement and *acceptability limit methodology*. An integral approach to combine both perspectives avoids the *bias* caused by *arbitrary* choices based only on the analytic point of view. Ultimately, it is the consumer who decides what the cutoff point (Hough & Garitta, 2012).

Demonstrating how it works, Marques et al. (2020) reported the minimum tolerable acceptability (*S*) of stored samples = 6.95 on a 9-point scale and used it subsequently to locate a cutoff point on intensity scale data, as shown in Figure 3.10.

Using a simple linear regression, the cutoff point is determined as the intensity of the defect that corresponds to a predetermined overall liking score. Linear regressions have been reported previously (Garitta et al., 2004; Ramírez et al., 2007). Minimum acceptability of 6.95 was found corresponding to 2.7 on off-flavor intensity (Figure 3.10). Following cutoff point identification, this cutoff point was used to estimate the shelf life for oxidized flavor intensity, as shown in Figure 3.11.

A major limitation of this approach is the limited number of critical attributes used by practitioners to estimate shelf life. Hence, Giménez et al. (2012) advocated using more critical descriptors to establish a sensory cutoff limit. Following up on this, Jacobo-Velázquez and Hernández-Brenes (2011) used two critical attributes as a failure criterion, whereas Marques et al. (2020) used only one. This methodology has been used to estimate the sensory shelf life of powdered milk (Hough et al., 2002), apple baby food (Gámbaro et al., 2006), dulce de leche (Garitta et al., 2004), avocado paste (Jacobo-Velázquez & Hernández-Brenes, 2011), fish burgers (Marques et al., 2020), etc.

An essential step in this methodology is identifying critical attributes with increasing levels of sensory defects with time. Critical attributes can be already known from category experience or can be identified by evaluating samples with different storage times. Gámbaro et al. (2006) ran a preliminary study to identify critical attributes of apple baby food storage. Trained assessors evaluate selected critical attributes of test samples for intensity while simultaneously samples are evaluated by consumers for acceptability ratings using the 9-point hedonic scale. A wide range of *consumers* were used, such as 40 consumers (Gámbaro et al., 2006), 50–52 consumers (Araneda

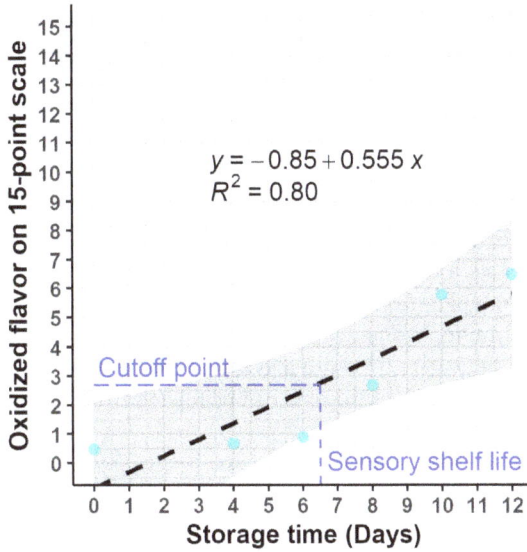

Figure 3.11 Typical scatter plot of sensory shelf life estimation based on the cutoff point obtained from the first step.

et al., 2008), and 80 consumers (Marques et al., 2020). This is quite apparent in the literature, but it is advised to use the recommended number of consumers, such as a minimum 100–120, for acceptability testing.

Giménez et al. (2012) criticized this criterion for being overly conservative. A similar finding was reported previously by Marques et al. (2020). This criterion mirrors when a consumer perceives a difference in the stored sample, which may fall short of the actual real-life shelf life settings by considering the wider consumption thresholds. Previously, it was found that consumers have higher thresholds for rejection to *consumption* than *purchase* (Ares, Giménez, & Gámbaro, 2008; Giménez et al., 2007). A slight difference does not necessarily relate to rejection, and Giménez et al. (2007) concluded that a relatively low percentage of consumers who actually rejected the bread samples reached the acceptability limit through Equation (3.2). Marques et al. (2020) found a shelf life of about 17 days with Equation (3.2), whereas 21 days (50% rejection rate) was found with survival analysis.

ANOVA can be used to establish the statistical significance, followed by multiple comparison tests, such as the Tukey honest significant difference (HSD) (Ares et al., 2009). Strictly speaking, the HSD is only applicable in an equal sample size situation. For the unequal sample size case, the approximate HSD, often called the Tukey-Kramer, is advised. Gauchez et al. (2020) used the Dunnett test to compare overall liking for storage time slots, whereas Ramírez et al. (2007) used Fisher's LSD to separate the means at a 5% significance level. A two-way ANOVA was used to accommodate consumers and samples as variation factors (Ramírez et al., 2007).

3.3.5 Survival Analysis

The survival analysis method of shelf life estimation is a statistical method widely used in clinical, epidemiologic, biologic, sociologic, and reliability studies (Hough et al., 2003). Hough et al. (2003) were the first to apply this method with some modifications to the sensory shelf life estimation. Originally, methods of survival analysis were developed to evaluate *times* until an event of interest occurs, often called *survival times* (Hough et al., 2003). To estimate the sensory shelf life by the survival analysis method, consumers are asked to try a set of samples with different storage times and answer *yes*, or *no* to the question "Would you normally consume this product?." Likewise, Olivares-Tenorio et al. (2017) asked each consumer, "are you willing to consume this fruit?" and recorded their response as merely *yes* or *no*, according to their willingness to consume, based only on the general appearance. In general, consumers are instructed to answer the question considering that they have already bought the product or the product is being served at home. It is essential to consider that each consumer tries all samples with different storage times. Each consumer answers the question by *yes* (accepts the sample) or *no* (rejects the sample) for each storage time. The exact moment the consumer rejects a sample (T) will never be observed due to the discrete nature of storage time points. For example, if one denies a sample on storage time 3 or t_3 (day 10), the time to rejection would be $<t_3$. Limited resources and other considerations are the causes for preventing the subject from testing in between t_2 (5 days) and t_3 (10 days) time slots with samples stored at 6, 7, 8, and 9 days. This incompleteness of the data can be related to the usual censoring definitions in survival analysis (Hough et al., 2003).

Left censoring: A compassionate consumer would likely reject the sample much before other consumers. For example, a sample was denied on first storage in Table 3.2 by *N* participant. Hough et al. (2003) questioned the credibility of the subject who rejected fresh samples and echoed not to use the data from those subjects. Likewise, Marques et al. (2020) excluded those rejecting fresh samples and were inconsistent at 23 of 80.

Interval censoring: If, for example, a consumer accepts samples stored at times 1 and 2 months but rejects the sample stored at 3 months, the time to rejection is $2 < T \leq 3$ months. Thus, what we know is that the consumer would start rejecting sometime between 2 and 3 months (Table 3.2). Inconsistent rejections may also occur, such as rejecting a storage time, say t_3, accepting t_4, and rejecting t_5. In those situations, censoring can be interpreted differently, such as Hough et al. 2003 interpreted a similar type of data as interval data with t_2-t_5, but the same data can be treated as t_2-t_3 by ignoring the subject's answers after the first rejection.

Right censoring: If an event of interest, such as off-odor here, does not occur during the storage period, this observation is right censored. For example, if a consumer accepts all samples, we will say that time to rejection is >4 months (Table 3.3).

Based on reversed sample storage design, either all samples can be evaluated in a single session, or specific samples of the basic storage design can be assessed in multiple sessions. Marques et al. (2020) used reverse design to estimate the shelf life of fish burgers. A variant of survival analysis, the so-called current-status survival analysis, is used in situations where consumers evaluate a single sample at each storage time. This topic is discussed in the forthcoming section. The survival analysis method recommends a higher number of consumers, such as close to 120, in compliance with other consumer-oriented methods (Hough, Luz Calle, Serrat, & Curia, 2007).

Table 3.3 Typical Data Matrix of Survival Analysis Method for One Replication

Consumer	Storage Time (Months)				Censoring
	1	2	3	4	
1	Yes	Yes	Yes	Yes	Right: >4
2	Yes	Yes	No	Yes	Interval: 2–4
–	Yes	No	No	No	Interval: 1–2
N	No	No	No	No	Left: <1

Abbreviations: No, rejection; Yes, acceptance.

In survival analysis, the survival function $S(t)$ is defined as the probability of a consumer accepting a product beyond time t, i.e., $S(t) = P\,(T > t)$ (Hough et al., 2003). The likelihood function, a nonparametric estimation, is used to estimate the survival function (Klein & Moeschberger, 2006):

$$L = \prod_{i \in R} S(r_i) \prod_{i \in L} (1 - S(l_i)) \prod_{i \in I} (S(l_i) - S(r_i)) \tag{3.3}$$

where r is the set of right-censored data, l is the set of left-censored data, and i is the set of interval-censored data. However, parametric models can furnish more precise estimates of the survival function than nonparametric estimates by assuming either the log-linear distribution or Weibull distribution for the survival times. Usually, survival times are not normal; instead, their distribution is often right-skewed. Readers are encouraged to look for Hough et al.'s (2003) publication for more information on this function. If the log-linear distribution is chosen for T, the survival function is given by:

$$S(t) = 1 - \phi\left(\frac{\ln(t) - \mu}{\sigma}\right) \tag{3.4}$$

where $\phi\,(\cdot)$ is the standard normal cumulative distribution function, and μ and σ are model's parameters.

If the Weibull distribution is chosen, the survival function is given by:

$$S(t) = S_{sev}\left(\frac{\ln(t) - \mu}{\sigma}\right) \tag{3.5}$$

where $S_{sev}\,(\cdot)$ is the survival function of the smallest extreme value distribution: $S_{sev}\,(w) = \exp\,(-e^{w})$, and μ and σ are the model's parameters.

The parameters of the log-linear model are obtained by maximizing the likelihood function via substituting $S(t)$ in Equation (3.2). Once the likelihood function is formed for a given model, specialized software can estimate the parameters (μ and σ) for the experimental data. Hough et al. (2003) used S-PLUS software to calculate the survival function nonparametrically. Codes for RStudio have been provided by Hough (2010). Later the previously mentioned parameters are used to plot the proportion of consumers rejecting the product as a function of storage time (Figure 3.12). To estimate the shelf life, a probability level of consumers' rejections must be chosen. A shelf life value considering 50% rejection has been recommended previously (Hough et al., 2003). Gauchez et al. (2020) used 50% of rejection as the failure criterion to estimate sourdough bread's shelf life. Others have used 25% rejection, which is more conservative but more practical (Jacobo-Velázquez & Hernández-Brenes, 2011). This methodology has been used to estimate the shelf life of yogurt (Hough et al., 2003), apple baby food (Gámbaro et al., 2006), fish burgers (Marques et al., 2020), sourdough bread (Gauchez et al., 2020), and other products.

3.3.5.1 Current-Status Survival Analysis

Current-status survival analysis is a variation of the survival analysis method, beneficial in situations where each consumer evaluates only one sample conforming to one storage time rather than testing all samples of different storage times (Araneda et al., 2008). In contrast to the original, only left and right censoring exists in this method since subjects evaluate only one sample stored for the time t, and one needs to test at least two samples to determine the interval. If a consumer rejects a sample on day 11 (t_2), then this consumer would be regarded as left-censored, corresponding to

(B)

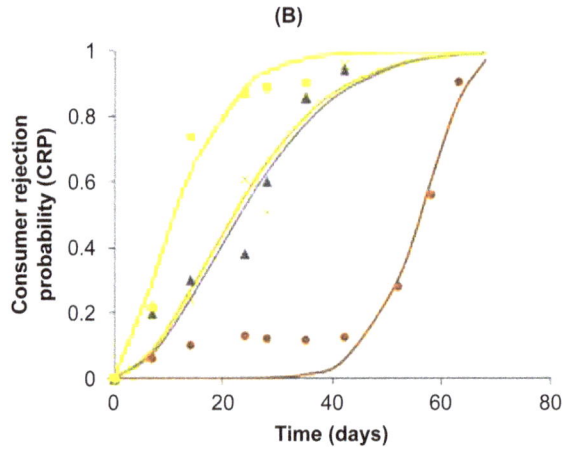

Figure 3.12 Plot of consumer rejection proportion as a function of time. Lines represent Weibull model and markers represent experimental data. (Reproduced from *LWT*, 80, Mary-Luz Olivares-Tenorio, Matthijs Dekker, Martinus A.J.S. van Boekel, Ruud Verkerk, Evaluating the effect of storage conditions on the shelf-life of cape gooseberry (*Physalis peruviana* L.), 523–530, (2017) with permission from Elsevier.)

exact rejection time (T) is shorter than $t_3 = 16$ days (T $< t_3$), whereas other consumers who accept the same sample on day 11, would be regarded as right-censored corresponding to storage time (T $> t_2$). Araneda et al. (2008) described this type of data as current-status or quantal-response data and models used to analyze this data as are similar to survival analysis method with the exception that likelihood function is replaced by the following expression:

$$L = \prod_{ifR}(1 - Fr_i)\prod_{ifL}F(l_i)$$ (3.6)

where r is the set of right-censored observations and l is the set of left-censored observations.

3.4 ACCELERATED STORAGE

Accelerated storage is used to fasten the aging and shorten the storage time. The idea behind this storage is that chemical reactions will proceed in a predictable manner at higher temperatures, according to simple kinetic models; thus, the shelf life at long time intervals can be simulated by shorter intervals at higher temperatures (Lawless & Heymann, 2010). However, MPI, New Zealand, sees accelerated storage and, hence, accelerated shelf life tests as an "indirect" method of shelf life testing, which is also less accurate (Ministry for Primary Industries, New Zealand, 2016). Lawless and Heymann (2010) stated that problems occur when changes in the product are not as similar as in real time at average temperatures. Chanadang and Chambers (2019) reported a few discrepancies in the shelf life estimation of fortified extruded blends with the accelerated storage compared with real-time storage. Others also echoed the concerns related to accelerated storage and advised to use real-time testing in parallel with accelerated storage, especially for new products (Chanadang & Chambers, 2019; Lawless & Heymann, 2010). Accelerated storage could be of worth in critical attribute identification (Garitta et al., 2004), provided the changes are similar to products with extended shelf life, such as extruded snacks, breakfast cereals, and others.

3.5 AN AFTERTHOUGHT AND RECOMMENDATIONS

Sensory science can contribute significantly to provide the highest eating qualities, but the success of the business in the current context requires its alignment with other initiatives as well, such as zero-waste, and this requires a complete trace down of eating qualities to pinpoint exact shelf life. Sensory shelf life methods prove their worth in ensuring that consumers do not find an unacceptable eating quality within its intended shelf life. The inclusion of *after pack opening* shelf life in intended shelf life is gaining importance in business. Sensory science could also play a more significant role in this upcoming *secondary shelf life* issues, which are more concerned with quality and consumer perception after pack opening (Nicoli & Calligaris, 2018). A wide variety of dried and semi-dried foods with a water activity (a_w) below 0.6 undergo modifications that are not of much safety concern but are a quality concern. Most of those concerns are related to packing atmosphere condition and usually lead to structural changes, oxidation, rancidity, and flavor profile depletion once opened. Together, they can cause the rejection of the product *during usage* and may eventually contribute to food wastage. Nicoli and Calligaris (2018) argued that the likelihood of repeated purchase could be reduced by addressing the sensory defects after pack opening. Indeed, they are much less researched topics, which may be due to a lack of regulations. However, in the view of decreasing food wastage, secondary shelf life issues may gain attention in the near future.

Food products do not have sensory shelf lives of their own, rather they depend on the interaction of the food with the consumer (Araneda et al., 2008), and for this reason, consumers should be regarded as the most appropriate tool for determining the failure criterion and sensory shelf life determination. By considering the shelf life as a food-consumer interaction (Samotyja, 2015), practitioners should focus on the consumer attitude toward quality changes and percent rejection. Earlier, sensory shelf life was focused on ensuring minimum differences in the sensory characteristics, but this may lead to unnecessarily high amounts of food loss because consumers keep consuming a stored product until their mental cutoff point is reached. In the same vein, intended sensory characteristics may vary from person to person and minimal changes occurring during storage should not be taken as the only standard for shelf life estimations. Also, Ares et al. (2008) showed that consumers have a higher threshold for rejection to *consume* than a rejection to *purchase*, implying that *before-purchase* counter/display-cabin shelf life should be estimated differently than *after-purchase* shelf life. Many different failure criteria were discussed in the chapter, but more practical information is required to refine these methodologies. By considering the growing importance of best before dates and secondary shelf life issues, we see a bright future in sensory science in shelf life studies. Also, sensory characteristics can have a significant role in cross-cultural differences where food represents a group of people, their ingredients, their place, and their taste.

3.6 CONCLUSION

In this chapter, consideration was given to enumerating the major sensory science-based methods used in shelf life estimation. Occupied with this, we discussed descriptive analysis, difference-from-control tests, quality rating, acceptability limit, and survival method in great depth. Good practices concerning design type, number of participants, number of critical attributes, selection of critical attributes, procedures to follow, and statistical analyses were emphasized throughout all these methods. Later in the chapter, accelerated storage was discussed from the perspective of its applications and limitations in shelf life studies. For the readers' perspective, figures, flow diagrams, examples, and notes were provided for clear and easy comprehension of the concepts.

REFERENCES

Araneda, M., Hough, G., & De Penna, E. W. (2008). Current-status survival analysis methodology applied to estimating sensory shelf life of ready-to-eat lettuce (*Lactuca sativa*). *Journal of Sensory Studies, 23*(2), 162–170. https://doi.org/10.1111/j.1745-459X.2007.00140.x

Ares, G., Barrios, S., Lareo, C., & Lema, P. (2009). Development of a sensory quality index for strawberries based on correlation between sensory data and consumer perception. *Postharvest Biology and Technology, 52*(1), 97–102. https://doi.org/10.1016/j.postharvbio.2008.11.001

Ares, G., Giménez, A., & Gámbaro, A. (2008). Sensory shelf life estimation of minimally processed lettuce considering two stages of consumers' decision-making process. *Appetite, 50*(2), 529–535. https://doi.org/10.1016/j.appet.2007.11.002

Ari Akin, P., Miller, R., Jaffe, T., Koppel, K., & Ehmke, L. (2019). Sensory profile and quality of chemically leavened gluten-free sorghum bread containing different starches and hydrocolloids. *Journal of the Science of Food and Agriculture, 99*(9), 4391–4396. https://doi.org/10.1002/jsfa.9673

ASTM E2454-20 (2020). *Standard guide for sensory evaluation methods to determine sensory shelf life of consumer products*. ASTM International. https://doi.org/10.1520/E2454-20

ASTM E253-20 (2020). *Standard terminology relating to sensory evaluation of materials and products*. ASTM International. https://doi.org/10.1520/E0253-20

Blixt, Y., & Borch, E. (2002). Comparison of shelf life of vacuum-packed pork and beef. *Meat Science, 60*(4), 371–378. https://doi.org/10.1016/S0309-1740(01)00145-0

Bonilla, A. C., Sveinsdottir, K., & Martinsdottir, E. (2007). Development of quality index method (QIM) scheme for fresh cod (*Gadus morhua*) fillets and application in shelf life study. *Food Control, 18*(4), 352–358. https://doi.org/10.1016/j.foodcont.2005.10.019

Cardello, A. V. (1995). Food quality: relativity, context and consumer expectations. *Food Quality and Preference, 6*(3), 163–170. https://doi.org/10.1016/0950-3293(94)00039-X

Chambers, E. IV. (2018). Consensus methods for descriptive analysis. In S. E. Kemp, J. Hort, & T. Hollowood (Eds.), *Descriptive analysis in sensory evaluation* (pp. 211–236), Wiley Blackwell. https://doi.org/10.1002/9781118991657.ch6

Chambers, E. IV., Sanchez, K., Phan, U. X., Miller, R., Civille, G. V., & Di Donfrancesco, B. (2016). Development of a "living" lexicon for descriptive sensory analysis of brewed coffee. *Journal of Sensory Studies, 31*(6), 465–480. https://doi.org/10.1111/joss.12237

Chanadang, S., & Chambers, E. IV. (2019). Sensory shelf life estimation of novel fortified blended foods under accelerated and real-time storage conditions. *Journal of Food Science, 84*(9), 2638–2645. https://doi.org/10.1111/1750-3841.14758

Chanadang, S., Koppel, K., & Aldrich, G. (2016). The impact of rendered protein meal oxidation level on shelf-life, sensory characteristics, and acceptability in extruded pet food. *Animals, 6*(8), 44. https://doi.org/10.3390/ani6080044

Civille, G. V., & Carr, B. T. (2016). Sensory evaluation in quality control (QC/sensory). In G. V. Civille, & B. T. Carr (Eds.), *Sensory evaluation techniques* (pp. 493–512), CRC Press.

Dong, X., Wrolstad, R. E., & Sugar, D. (2000). Extending shelf life of fresh-cut pears. *Journal of Food Science, 65*(1), 181–186. https://doi.org/10.1111/j.1365-2621.2000.tb15976.x

Donovan, J. D., Keller, K. L., & Tepper, B. J. (2016). A brief task to assess individual differences in fat discrimination. *Journal of Sensory Studies, 31*(4), 296–305. https://doi.org/10.1111/joss.12212

Drake, M. (2007). Invited review: sensory analysis of dairy foods. *Journal of Dairy Science, 90*(11), 4925–4937. https://doi.org/10.3168/jds.2007-0332

Fernández-López, J., Sayas-Barberá, E., Munoz, T., Sendra, E., Navarro, C., & Pérez-Alvarez, J. (2008). Effect of packaging conditions on shelf-life of ostrich steaks. *Meat Science, 78*(1–2), 143–152. https://doi.org/10.1016/j.meatsci.2007.09.003

Freitas, J., Vaz-Pires, P., & Câmara, J. S. (2019). Freshness assessment and shelf-life prediction for *Seriola dumerili* from aquaculture based on the quality index method. *Molecules, 24*(19), 3530. https://doi.org/10.3390/molecules24193530

Fritsch, C., Hofland, C., & Vickers, Z. M. (1997). Shelf life of sunflower kernels. *Journal of Food Science, 62*(2), 425–428. https://doi.org/10.1111/j.1365-2621.1997.tb04018.x

Gámbaro, A., Ares, G., & Gimenez, A. (2006). Shelf-life estimation of apple-baby food. *Journal of Sensory Studies, 21*(1), 101–111. https://doi.org/10.1111/j.1745-459X.2006.00053.x

Garitta, L., Hough, G., & Sánchez, R. (2004). Sensory shelf life of dulce de leche. *Journal of Dairy Science*, *87*(6), 1601–1607. https://doi.org/10.3168/jds.S0022-0302(04)73314-7

Gauchez, H., Loiseau, A., Schlich, P., & Martin, C. (2020). Impact of aging on the overall liking and sensory characteristics of sourdough breads and comparison of two methods to determine their sensory shelf life. *Journal of Food Science*, *85*(10), 3517–3526. https://doi.org/10.1111/1750-3841.15410

Giménez, A., Ares, F., & Ares, G. (2012). Sensory shelf-life estimation: A review of current methodological approaches. *Food Research International*, *49*(1), 311–325. https://doi.org/10.1016/j.foodres.2012.07.008

Giménez, A., Varela, P., Salvador, A., Ares, G., Fiszman, S., & Garitta, L. (2007). Shelf life estimation of brown pan bread: a consumer approach. *Food Quality and Preference*, *18*(2), 196–204. https://doi.org/10.1016/j.foodqual.2005.09.017

Hough, G. (2010). *Sensory shelf life estimation of food products* (1st ed.), CRC Press.

Hough, G., & Garitta, L. (2012). Methodology for sensory shelf-life estimation: A review. *Journal of Sensory Studies*, *27*(3), 137–147. https://doi.org/10.1111/j.1745-459X.2012.00383.x

Hough, G., Garitta, L., & Gómez, G. (2006). Sensory shelf-life predictions by survival analysis accelerated storage models. *Food Quality and Preference*, *17*(6), 468–473. https://doi.org/10.1016/j.foodqual.2005.05.009

Hough, G., Langohr, K., Gómez, G., & Curia, A. (2003). Survival analysis applied to sensory shelf life of foods. *Journal of Food Science*, *68*(1), 359–362. https://doi.org/10.1111/j.1365-2621.2003.tb14165.x

Hough, G., Luz Calle, M., Serrat, C., & Curia, A. (2007). Number of consumers necessary for shelf life estimations based on survival analysis statistics. *Food Quality and Preference*, *18*(5), 771–775. https://doi.org/10.1016/j.foodqual.2007.01.003

Hough, G., Puglieso, M. L., Sanchez, R., & da Silva, O. M. (1999). Sensory and microbiological shelf-life of a commercial ricotta cheese. *Journal of Dairy Science*, *82*(3), 454–459. https://doi.org/10.3168/jds.S0022-0302(99)75253-7

Hough, G., Sánchez, R. H., Garbarini de Pablo, G., Sánchez, R. G., Calderón Villaplana, S., Giménez, A. M., & Gámbaro, A. (2002). Consumer acceptability versus trained sensory panel scores of powdered milk shelf-life defects. *Journal of Dairy Science*, *85*(9), 2075–2080. https://doi.org/10.3168/jds.S0022-0302(02)74285-9

Hough, G., Subramaniam, P., Narain, C., & Beeren, C. (2013). Collecting samples from the shelf: Does this contribute to shelf-life knowledge? *Journal of Sensory Studies*, *28*(1), 47–56. https://doi.org/10.1111/joss.12022

Hough, G., Wakeling, I., Mucci, A., Chambers, E. IV., Gallardo, I. M., & Alves, L. R. (2006). Number of consumers necessary for sensory acceptability tests. *Food Quality and Preference*, *17*(6), 522–526. https://doi.org/10.1016/j.foodqual.2005.07.002

Jacobo-Velázquez, D., & Hernández-Brenes, C. (2011). Sensory shelf-life limiting factor of high hydrostatic pressure processed avocado paste. *Journal of Food Science*, *76*(6), S388–S395. https://doi.org/10.1111/j.1750-3841.2011.02259.x

Joseph, M. V. (2016). Extrusion, physico-chemical characterization and nutritional evaluation of sorghum-based high protein, micronutrient fortified blended foods. Doctoral dissertation, Kansas State University, Manhattan, Kansas. http://hdl.handle.net/2097/32907

Joshy, C., Ninan, G., Panda, S., Zynudheen, A., Kumar, K. A., & Ravishankar, C. (2020). Development of demerit score-based fish quality index (FQI) for fresh fish and shelf life prediction using statistical models. *Journal of Aquatic Food Product Technology*, *29*(1), 55–64. https://doi.org/10.1080/10498850.2019.1693463

Klein, J. P., & Moeschberger, M. L. (2006). *Survival analysis: techniques for censored and truncated data*, Springer Science & Business Media. https://doi.org/10.1007/b97377

Kumar, R., & Chambers, E. IV. (2019). Lexicon for multiparameter texture assessment of snack and snack-like foods in English, Spanish, Chinese, and Hindi. *Journal of Sensory Studies*, *34*(4), e12500. https://doi.org/10.1111/joss.12500

Kwak, H. S., Kim, H. G., Kim, H. S., Ahn, Y. S., Jung, K., Jeong, H. Y., & Kim, T. H. (2013). Sensory characteristics and consumer acceptance of frozen cooked rice by a rapid freezing process compared to homemade and aseptic packaged cooked rice. *Preventive Nutrition and Food Science*, *18*(1), 67–75. https://doi.org/10.3746/pnf.2013.18.1.067

Lareo, C., Ares, G., Ferrando, L., Lema, P., Gambaro, A., & Soubes, M. (2009). Influence of temperature on shelf life of butterhead lettuce leaves under passive modified atmosphere packaging. *Journal of Food Quality*, *32*(2), 240–261. https://doi.org/10.1111/j.1745-4557.2009.00248.x

Lawless, H. (1995). Dimensions of sensory quality: a critique. *Food Quality and Preference*, *6*(3), 191–199. https://doi.org/10.1016/0950-3293(94)00023-O

Lawless, H. T. (2012). Comparison of discrimination test methods. In H. T. Lawless (Ed.), *Laboratory exercises for sensory evaluation* (pp. 27–31), Springer Science & Business Media. https://doi.org/10.1007/978-1-4614-5713-8

Lawless, H. T., & Heymann, H. (2010). *Sensory evaluation of food: principles and practices* (2nd ed.), Springer Science & Business Media. https://doi.org/10.1007/978-1-4419-6488-5

Leong, J., Kasamatsu, C., Ong, E., Hoi, J. T., & Loong, M. N. (2016). A study on sensory properties of sodium reduction and replacement in Asian food using difference-from-control test. *Food Science & Nutrition*, 4(3), 469–478. https://doi.org/10.1002/fsn3.308

Leygonie, C., Britz, T. J., & Hoffman, L. C. (2012). Impact of freezing and thawing on the quality of meat: Review. *Meat Science*, 91(2), 93–98. https://doi.org/10.1016/j.meatsci.2012.01.013

Lopane, S. N. (2018). An investigation of the shelf life of cold brew coffee and the influence of extraction temperature using chemical microbial and sensory analysis. Master's thesis, Clemson, University, Clemson, South Carolina. https://tigerprints.clemson.edu/all_theses/2899

Lopez-Galvez, F., Ragaert, P., Palermo, L. A., Eriksson, M., & Devlieghere, F. (2013). Effect of new sanitizing formulations on quality of fresh-cut iceberg lettuce. *Postharvest Biology and Technology*, 85, 102–108. https://doi.org/10.1016/j.postharvbio.2013.05.005

Makhoul, H., Ghaddar, T., & Toufeili, I. (2006). Identification of some rancidity measures at the end of the shelf life of sunflower oil. *European Journal of Lipid Science and Technology*, 108(2), 143–148. https://doi.org/10.1002/ejlt.200500262

Manzocco, L., Calligaris, S., & Nicoli, M. (2010). Methods for food shelf life determination and prediction. In M. Hu, & C. Jacobsen (Eds.), *Oxidation in foods and beverages and antioxidant applications* (pp. 196–222), Elsevier. https://doi.org/10.1533/9780857090447.1.196

Marques, C., Lise, C. C., de Lima, V. A., & Daltoé, M. L. M. (2020). Survival analysis and cut-off point to estimate the shelf life of refrigerated fish burgers. *Food Science and Technology*, 40(1), 171–177. https://doi.org/10.1590/fst.36918

Meilgaard, M. C., Carr, B. T., & Civille, G. V. (2015). *Sensory evaluation techniques* (5th ed.), CRC Press.

Ministry for Primary Industries, New Zealand. (2016). How to determine the shelf life of food. (No. 2), Wellington, New Zealand.

Molnár, P. J. (1995). A model for overall description of food quality. *Food Quality and Preference*, 6(3), 185–190. https://doi.org/10.1016/0950-3293(94)00037-V

Nabily, M., Nabili, A., Namsi, A., Majdoub, H., & Azzouna, A. (2020). Shelf life prediction and storage stability of Deglet Nour dates (*Phoenix dactylifera* L.): microbiological and organoleptic properties. *Chemistry Africa*, 3(1), 189–197. https://doi.org/10.1007/s42250-019-00108-4

Nicoli, M. C., & Calligaris, S. (2018). Secondary shelf life: An underestimated issue. *Food Engineering Reviews*, 10(2), 57–65. https://doi.org/10.1007/s12393-018-9173-2

Oehlert, G. W. (2010). *A first course in design and analysis of experiments* (1st ed.), W.H. Freeman and Company.

Olivares-Tenorio, M. L., Dekker, M., van Boekel, M. A., & Verkerk, R. (2017). Evaluating the effect of storage conditions on the shelf life of cape gooseberry (Physalis peruviana L.). *LWT*, 80, 523–530. https://doi.org/10.1016/j.lwt.2017.03.027

Ramírez, G., Hough, G., & Contarini, A. (2007). Influence of temperature and light exposure on sensory shelf-life of a commercial sunflower oil. *Journal of Food Quality*, 24(3), 195–204. https://doi.org/10.1111/j.1745-4557.2001.tb00602.x

Samotyja, U. (2015). Consumer acceptability studies in shelf-life estimation of rapeseed and sunflower oils. *Acta Alimentaria*, 44(1), 60–67. https://doi.org/10.1556/aalim.44.2015.1.5

Sharma, C. (2012). Development of extruded snacks utilizing broken rice and mung bean. Master's thesis, Punjab Agricultural University, Ludhiana, Punjab, India. http://krishikosh.egranth.ac.in/handle/1/5810025059

Sharma, C. (2019). Sensory and consumer profiling of potatoes grown in the USA. Doctoral dissertation, Kansas State University, Manhattan, Kansas. http://hdl.handle.net/2097/40044

Sharma, C., Chambers, E. IV., Jayanty, S. S., Sathuvalli Rajakalyan, V., Holm, D. G., & Talavera, M. (2020). Development of a lexicon to describe the sensory characteristics of a wide variety of potato cultivars. *Journal of Sensory Studies*, 35(4), e12577. https://doi.org/10.1111/joss.12577

Singh, B., Sharma, C., & Sharma, S. (2017). Fundamentals of extrusion processing. In V. Nanda, & S. Sharma (Eds.), *Novel food processing technologies* (pp. 1–46), New India Publishing Agency.

Siripatrawan, U., & Noipha, S. (2012). Active film from chitosan incorporating green tea extract for shelf life extension of pork sausages. *Food Hydrocolloids*, *27*(1), 102–108. https://doi.org/10.1016/j.foodhyd.2011.08.011

Stocking, A. J., Suffet, I. H., McGuire, M. J., & Kavanaugh, M. C. (2001). Implications of an MTBE odor study for setting drinking water standards. *Journal AWWA*, *93*(3), 95–105. https://doi.org/10.1002/j.1551-8833.2001.tb09156.x

Sveinsdottir, K., Hyldig, G., Martinsdottir, E., Jørgensen, B., & Kristbergsson, K. (2003). Quality index method (QIM) scheme developed for farmed Atlantic salmon (*Salmo salar*). *Food Quality and Preference*, *14*(3), 237–245. https://doi.org/10.1016/S0950-3293(02)00081-2

Tanner, D. (2016). Impacts of storage on food quality. *Reference module in food science*, Elsevier. https://doi.org/10.1016/B978-0-08-100596-5.03479-X

Tano-Debrah, K., Saalia, F. K., Ghosh, S., & Hara, M. (2019). Development and sensory shelf-life testing of KOKO plus: A food supplement for improving the nutritional profiles of traditional complementary foods. *Food and Nutrition Bulletin*, *40*(3), 340–356. https://doi.org/10.1177/0379572119848290

Tomic, O., Forde, C., Delahunty, C., & Næs, T. (2013). Performance indices in descriptive sensory analysis – A complimentary screening tool for assessor and panel performance. *Food Quality and Preference*, *28*(1), 122–133. https://doi.org/10.1016/j.foodqual.2012.06.012

Young, N., & O'Sullivan, G. (2011). The influence of ingredients on product stability and shelf life. In D. Kilcast, & P. Subramaniam (Eds.), *Food and beverage stability and shelf life* (pp. 132–183), Elsevier. https://doi.org/10.1533/9780857092540.1.132

CHAPTER 4

Modeling of the Shelf Life of Foods

Arjuara Begum, Shagufta Rizwana, and H.A. Makroo

CONTENTS

4.1 INTRODUCTION

The time for any food to be edible without deterioration is called shelf life. It acts as an indicator of expiry for the purchaser. There is no clear or general definition for the term shelf life because it varies from product to product and there are different procedures to calculate. These procedures are based on food regulatory bodies for shelf life assessment, extension, and dating. Thus, shelf life is defined as a definite time length after production or packaging until which a food product retains the desired qualities at well-defined storage conditions (Nicoli, 2012).

There are enormous numbers of food products with different physical and chemical properties. Based on their properties, they require different processing, packaging, and storage conditions. So, while figuring out the shelf life of food, we have to take quite a number of factors into consideration

Email: H.A. Makroo, hilalmakroo@gmail.com

DOI: 10.1201/9781003091677-4

like water activity, the permeability of the packaging material, storage temperature and humidity, and so on. The factors affecting the shelf life of food are as follows (Galić et al., 2009):

- *Processing conditions:* Temperature, time of treatment
- *Product properties:* Moisture content, physiochemical properties
- *Packaging materials:* Barrier (water vapor, O_2, CO_2, aroma), mechanical characteristics
- *Storage conditions:* Temperature, humidity, light

Therefore, mathematical models are required to estimate the shelf life considering all the factors by taking their mathematical terms as numeric values.

Modeling is important to understand the fundamental mechanism of a process as well as to optimize or control the operating conditions. A number of models for describing the kinetics of food materials have been developed; they are generally classified into empirical and mechanistic models. The shelf life rate equations are mostly semi-empirical models. The data need to be generated experimentally.

For example, for moisture-sensitive foods, shelf life calculations are based on sorption isotherm models in which water activity (a_w) is expressed as an explicit function of moisture content. The rate of change of moisture content with respect to time is expressed as a function of change in relative humidity and water activity keeping other terms like surface area and permeability of packaging material constant. The time required to reach critical moisture content is considered as the shelf life of the product. In such case, shelf life calculation involves fitting data into moisture sorption models, solving the ordinary differential rate equation, and interpolating the solved values to obtain shelf life (Das, 2005).

4.2 EXPERIMENTAL CONDITIONS

The experimental design of a shelf life experiment is carried on by

- Considering the factors affecting shelf life,
- Quality testing at different time intervals, and
- Observing the rate of change quality with regard to the factors.

So, before designing an experiment the guidelines of storage are to be known. According to food safety laws, the storage conditions for better shelf life are shown in Table 4.1.

4.3 THEORETICAL MODELS

The advancement of technology and the development of microelectronics have made computation quite available and easy. The theoretical models that require mathematical skills are easy to solve with the help of computation. Theoretical models are those which are based on fundamental

Table 4.1   General Storage Conditions Based on Food Types

Food Items	Storage Temperature (°C)	Storage Conditions
Fruits and vegetables	10 to 15	Dry and cool
Dairy products	2 to 4	Refrigeration
Meat, poultry, and seafood	−18 to 4	Freezing

principles. The most predictive model for shelf life estimation is the kinetic model. Because our food is a biochemical system, the principles of chemical kinetics apply to it. Saguy and Karel (1980) studied the deterioration of food quality during processing and storage based on the composition factors C_i and environmental factors E_j. The rate of quality (dQ/dt) can be expressed as a first-order equation:

$$dQ/dt = F(C_i, E_j) \qquad (4.1)$$

The composition factors are the concentration of reactive compounds, inorganic catalysts, enzymes, reaction inhibitors, pH, water activity, and microbial populations. Temperature, relative humidity, total and partial pressure of different gases, and light and mechanical stresses are the most considered environmental factors. For simplification, the environmental factors are considered to be constant and the composition factors affecting the food are expressed as a function of time. Therefore Equation (4.1) can be expressed as

$$dQ/dt = kC_i^n \qquad (4.2)$$

where n is the order of the reaction, $n = 0, 1, 2, \ldots.$, and k is the rate constant.

Because in most practical cases environmental factors are not stable, the inclusions of such factors are necessary to predict an accurate model. The reliable and easy approach is to model the factor E_j into an apparent reaction rate constant as E_j: $k = k(E_j)$. In the case of food storage, temperature (T) is considered to be the strongest factor influencing other factors (Taoukis et al., 1997). The widely used equation is the Arrhenius equation.

$$\frac{\delta lnk}{\delta\left(\frac{1}{T}\right)} = -\frac{\Delta E^0}{R} \qquad (4.3)$$

where E^0 is the activation energy and R is the universal gas constant.

For a stationary food system, considering the Arrhenius model to a great extent gives a very satisfying prediction. In the case of a dynamic system, the integral factor of the Arrhenius equation is used.

Dynamic shelf life dating is quite helpful in critical analysis of the dependent on factors. The experimental data after quality analysis are subjected to the computational method for modeling. A type of shelf life testing method known as accelerated life testing (ALT) introduces a stress integral factor into kinetic models. ALT usually brings down the mean time to failure time in long storage experiments. Expressing the quality change under dynamic temperature is expressed in Equation (4.4) (Martins et al., 2008).

$$C^n = C_0^n - (1-n)k_0 \int_{t_0}^{t} e^{\left[\frac{E_a}{R}\left(\frac{1}{T}-\frac{1}{T}\right)\right]^0} dt - \int_{t_0}^{t} w(t)dt \qquad (4.4)$$

where $w(t)$ is the noise function. As $t \rightarrow \infty$, $\int w(t)dt$ tends to 0.

Microbial growth spoilage can be explained using mechanistic models like Huang's model and the model of Baranyi and Roberts (Lee, 2014). Huang's model can be expressed as the change in cell numbers (N) with respect to time (t) as shown in Equation (4.5). The latter is written in the form of two differential equations (Equations 4.6 and 4.7).

$$\frac{dN}{dt} = \frac{\mu_{max}\left(1 - N/N_{max}\right)}{1 + \exp[-25(t - \lambda)]} \qquad (4.5)$$

where μ_{max}, λ and N_{max} are maximum specific growth rate, lag time, and maximum cell density, respectively.

$$\frac{dq}{dt} = \mu_{max} q \tag{4.6}$$

$$\frac{dN}{dt} = \mu_{max} \left(\frac{q}{1+q} \right) \left(1 - \frac{N}{N_{max}} \right) N \tag{4.7}$$

where q is a hypothetical physiological state of the cell population representing a normalized concentration of unknown substance critically needed for cell growth.

There are a few more mechanistic models based on the type of food for predictability, but they are mostly avoided because they are very complex to solve.

4.4 EMPIRICAL MODELS

Chemical degradation and microbial growth can be explained using ad hoc empirical models based on the factor's dependency on temperature (Corradini and Peleg, 2007).

The time at which a consumer rejects a product can be expressed in terms of illumination factor (in the case of light-sensitive foods) and temperature. Light is taken as a binomial response and temperature as an Arrhenius term (Garitta et al., 2018). A logarithmic equation taking light and temperature as accelerating factors is expressed in Equation (4.8):

$$Y = \ln(Time_R) = \beta_0 + \frac{E_a}{R} \times \frac{1}{Temp} + \beta_1 \cdot I + \beta_2 \cdot I \cdot \frac{1}{Temp} + \sigma W \tag{4.8}$$

where $Time_R$ = storage time at which a consumer rejects a sample; β_0, β_1, and β_2 are regression coefficients; E_a = activation energy (cal/mol); R = gas-law-constant = 1.98 cal/(mol·K); $Temp$ = storage temperature, K; I = illumination condition, 0 for no-illumination, and 1 for with-illumination; σ is the scale parameter; and W is the error distribution.

For products with less oxidative stability, predictive models for both high and low temperature using Rancimat-based and chemical composition-based models works very well with an accuracy of more than 90%. Farhoosh and Hoseini-Yazdi (2013) formulated the chemical compound change over a temperature range as a pseudo zero-order rate equation. For example, peroxide value (PV) change is expressed in (Equation 4.9):

$$PV = PV_0 + k \times IP \tag{4.9}$$

where PV_0 is the initial PV, k is the rate constant, and IP is the induction period.

Highly perishable food is very prone to microbial degradation. Microbial spoilage can be determined by empirical models based on reciprocal shelf life (RS) (Dabadé et al., 2015). These models include Equations (4.10) (exponential model), (4.11) (Ratkowsky model), and (4.12) (Arrhenius model). Accordingly, the RS is expressed as a factor of storage temperature.

$$RS = b_1 \times e^{a \times T} \tag{4.10}$$

$$RS = b_2^2 \times (T - T_{min})^2 \tag{4.11}$$

$$RS = b_3 \times e^{\left(\frac{-E_a}{R \times K} \right)} \tag{4.12}$$

where T is the storage temperature; R is the gas constant 8.314 (J/mol K); K is storage in Kelvin; and b_i, a, T_{min}, and E_a are coefficients to be estimated.

The K coefficient is a factor of storage temperature. It can be related to shelf life using the predictive model in Equation (4.13).

$$shelf\ life(days) = 4.5 \times 10^{-15} \times e^{\left(\frac{9650}{K}\right)} \tag{4.13}$$

So, several empirical models can be developed for factors considering their effect on the type of food. Because empirical models are black box in nature, they are limited to a certain level of accuracy within a specific range.

4.5 CATEGORIES OF VARIOUS MODELS

4.5.1 Bacterial Growth Modeling

Growth models usually describe bacterial growth in food as a function of time. In view of this, some model requirements for bacterial growth can provide the prediction of the bacterial population in food at a given time. The prediction of these parameters usually describes the sigmoidal curve of bacteria in food with its different phases of growth. The sigmoidal growth curve has three different phases: lag, growth, and stationary. Kinetic growth models are widely used for predicting the microbial shelf life enhancement of foods. In the shelf life enhancement of food, predictive microbiological models are categorized into (1) primary (2) secondary, or (3) hybrid models. The first stage is called the primary growth phase and it defines the development of bacterial growth with respect to time. These models are given the main parameters of growth rate (μ_{max}) and lag time (*lag*). The primary kinetic models define the growth of bacteria in food as model constants. They usually include mathematical constants that can be obtained from inactivation or bacterial growth curve analysis and theoretical assumptions, not the properties correlated with the intrinsic or extrinsic properties of the food or environment of bacteria. Models that can correlate the primary model mathematical constants with those more attainable properties such as pH or water activity are referred to as secondary models. The secondary growth model explains the impact of different environmental variables on growth (μ_{max}, *lag*). It takes into consideration a_w, pH, temperature, and certain growth inhibitors such as organic acids, which are the main factors that affect bacterial behavior. The effect of all other minor factors is included in the food matrix. Finally, the tertiary models are available through a software application (Mahony and Seman, 2016).

4.5.2 Primary Models to Predict the Shelf Life of Foods

A series of mathematical functions were proposed that define and use sigmoid curves to model microbial growth development. Various primary and secondary models are studied and evaluated to find the best way to measure the shelf life of foods. The experimental data are fitted with the primary mathematical models like first-order kinetics, three-phase linear model, Gompertz, and the logistic model to predict microbial growth. The effect of temperature on microbial growth rate is predicted by using the secondary Arrhenius and square root models. Primary and secondary models or their combinations are models commonly applied to study the shelf life prediction of foods under dynamic conditions (Kreyenschmidt et al., 2009).

4.5.2.1 First-Order Kinetic Model

Temperature is one of the most important factors used to destroy microorganisms and to enhance the shelf life of foods. The rate of destruction increases with an increase in temperature, time of exposure, and storage period. The rate of their degradation can be determined from the first-order kinetic model, Equation (4.14):

$$ln\frac{N_t}{N_o} = -kt \tag{4.14}$$

where N_t is the surviving microorganisms at the time, $t(min)$ and N_o is the initial concentration of the microorganisms.

4.5.2.2 Three-Phase Linear Model

Buchanan et al. (1997) suggested three-phase linear equations for bacterial growth modeling, which define a complete sigmoidal growth curve. Three-phase sigmoidal growth rate with time is shown in Figure 4.1. The first inclination is the lag phase, the exponential phase (growth phase) is the phase in which the bacterial growth increases exponentially with time function, and in the stationary phase the growth remains constant. The advantage of this model is that it is relatively easy to implement and is more stable compared with other models involved. Appropriate three-step linear functions are used to model the growth and for predicting the parameters. The three stages of growth development are represented by a linear function and their steps are given in Equations (4.15)–(4.17) (Mahony and Seman, 2016; Dias and Piccoli, 2018).

$$\log N(t) = N_0 \quad \text{if } t \leq t_1 \tag{4.15}$$

$$\log N(t) = \log N_0 + k(t - t_1) \quad \text{if } t_1 < t < t_2 \tag{4.16}$$

$$\log N(t) = \log N_0 + k(t_2 - t_1) \quad \text{if } t \geq t_2 \tag{4.17}$$

where N_o is the initial bacterial concentration, $N(t)$ is the population at t time, t_1 represents the initial growth phase (*lag*), t_2 indicates the end of the third phase (stationary phase), and k is the slope of the exponential curve.

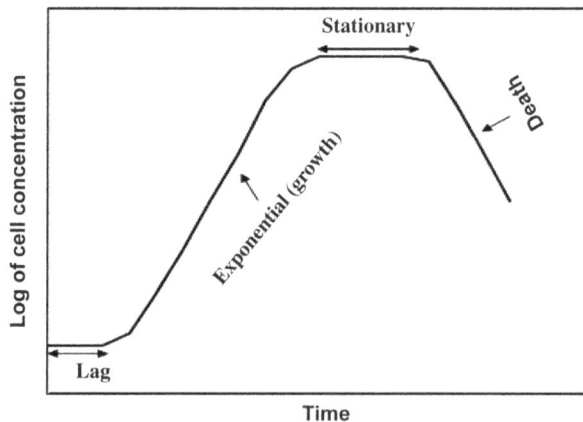

Figure 4.1   Phases of the microbial growth curve.

4.5.2.3 Gompertz Model

The Gompertz function is also a type of mathematical model commonly applied to examine bacterial growth as a function of time. The microbial growth count can be modeled using the Gompertz growth curve for each combination of salt, temperature, and pH. Among the various models, Gompertz is a highly versatile and widely used model for predicting growth parameters. Growth rate, generation time (GT), lag time (LT), and time to reach maximum bacterial growth are determined from the Gompertz model parameters (Hajmeer et al., 1997; Gibson et al., 1988) in

$$L(t) = A + C \exp\{-\exp[-B(t - M)]\} \tag{4.18}$$

where $L(t)$ is the log count for log10 at time t, and it is determined from the parameters A, B, C, and M, where C indicates the asymptotic log count, as t falls indefinitely, C is the asymptomatic sum of growth occurring as t grows indefinitely, and B is the relative development at time M, where the absolute growth rate is maximum. The term bacterial logarithmic count (LC) per hour is derived from the relation

$$\text{Growth rate}\,(LC/h) = BC/e \tag{4.19}$$

$$LT\,(h) = M - 1/B \tag{4.20}$$

$$GT\,(h) = \log10\,(2)\,e/B*C \tag{4.21}$$

4.5.2.4 Logistic Model

Decontamination of bacterial population and their effects can be modeled using the logistic growth model. Linear stability analysis is applied to different potential outcomes in which the bacterial population may find a positive balance or become extinct due to increased decontamination. The logistic model provides a good mathematical description for various microorganism biological populations (Kreyenschmidt et al., 2009).

$$N(t) = A + \frac{C}{1 + e^{-B(t-M)}} \tag{4.22}$$

where $N(t)$ is the log count for log10 at any time t, B is the relative development at time M where the absolute growth rate is maximum, and t is the time. C is the difference between the maximum growth and growth at zero time.

4.5.3 Secondary Models

4.5.3.1 Arrhenius Equation

Temperature is considered an important factor in the microbial growth of food as it changes with production and transportation while modifying the environment for bacterial growth. The impact of temperature on the microbial growth rate is evaluated by using the secondary models, Arrhenius and root square equations (Kreyenschmidt et al., 2009). The effect of temperature on growth rate is evaluated by using Equation (4.23):

$$\ln(B) = \ln F - \left(\frac{E_a}{R \cdot T}\right) \tag{4.23}$$

where B is the relative growth rate at time, M (h^{-1}); F is the pre-exponential factor, (h^{-1}); E_a is the activation energy; R is gas constant (8.314 Jmol^{-1}K^{-1}); and T is the temperature (K).

To determine the activation energy from Equation (4.23), the following steps need to be considered:

1. The rate constant "B" value is evaluated from the first-order kinetic model Equation (4.14) and the ln B values are calculated.
2. Convert temperature to absolute temperature (K) and determine the $1/T$.
3. Plot ln B versus $1/T$ and fit the linear equation $y = mx + c$.
4. Slope $(m) = -E_a/R$, and from the slope of the linear equation activation energy can be evaluated.

4.5.3.2 Square Root Equation

In several cases, the use of the Arrhenius equation is inadequate to explain the impact of suboptimal temperature on bacterial growth rates, since the apparent activation energy is dependent on temperature. Ratkowsky et al. (1982) proposed an empirical model, Equation (4.24), to solve this problem. The experimental growth rate data were fitted with the model and converted into the square root to stabilize its variance, and its various expansions are called square root type or Ratkowsky-type models (Ross and Dalgaard, 2004).

$$(C)^{0.5} = b(T - T_{min}) \qquad (4.24)$$

where C is the relative growth rate at time M (h^{-1}); b is the slope of the regression line; T is the temperature in (K); and T_{min} is the minimum temperature for growth, and its value can be range between 5 and 10°C.

Ratkowsky et al. (1983) modified Equation (4.24) into the expanded form to determine the optimal growth temperature given as

$$C = b(T - T_{min})^2 (1 - exp(d(T - T_{max}))) \qquad (4.25)$$

where b and d are constants, T is the temperature, and T_{min} and T_{max} are the theoretical minimum and maximum temperature. Growth cannot be achieved below the minimum temperature and beyond the maximum temperature.

Adarns et al. (1991) studied the growth of *Yersinia enterocolitica* as a response with varying pH and temperature. Adarns et al. (1991) and McMeekin et al. (1993) modified the previous Equation (4.25) and added two more parameters, pH and water activity (a_w), which could model the growth of microbiology of foods under suboptimal conditions, and their relationships are as follows:

$$(C)^{0.5} = b(T - T_{min})(pH - pH_{min})^{0.5} \qquad (4.26)$$

$$(C)^{0.5} = b(T - T_{min})(a_w - a_{wmin})^{0.5} \qquad (4.27)$$

4.6 INACTIVATION KINETICS

Enzymatic and non-enzymatic spoilage is one of the major factors in the deterioration of the quality of food products. Non-enzymatic spoilage, mainly Maillard reaction and lipid oxidation, degrades the quality of the product. The enzymatic reaction takes place when the enzymes present in the food product come in contact with the atmospheric air. Fruits and vegetables are mostly susceptible to enzymatic browning due to the presence of enzymes (polyphenol oxidase [PPO], peroxidase [POD], protease). Enzymatic browning leads to nutrition and color loss, and it may degrade the overall quality of the food product. To overcome the problems and to enhance shelf life, inactivation of enzymes has been carried out by the application of heat, pressure, and non-thermal technology (Shiferaw et al., 2010; Chakraborty et al., 2014). Various inactivation kinetic models are used to determine the rate of degradation of an enzyme. When the microorganisms come in contact with the external treatment (heat, pressure, and non-thermal technology), inactivation of microorganisms may take place. The rate of surviving microorganisms (S) after the treatment can be determined from the ratio of microorganisms at time $t = t(min)$ to the initial concentration at time, $t = 0$, as shown in Equation (4.28) (Huang et al., 2012).

$$S = \frac{N_{t=t}}{N_{t=0}} \qquad (4.28)$$

The residual activity of enzymes (R_A) is determined from the ratio of enzyme activity (A_t) after $t = t(min)$ to the initial activity (A_i) of enzymes and is measured using Equation (4.29).

$$R_A = \frac{A_t}{A_i} \qquad (4.29)$$

Residual activity of enzymes with respect to time is used to study the modeling kinetics. Inactivation kinetics is studied to predict and validate the degradation of enzymes. After understanding the process factors affecting the inactivation are evaluated, mathematical equations are used to fit the inactivation of enzymes, microbes, and degradation of nutritional compounds and to predict the model constants. Predictive parameters are used to evaluate the error that deviates from the complete fit of the model. A lower error value implies a better fit of the applied model. Various mathematical functions used to study the inactivation kinetics are discussed in the following sections.

4.6.1 Degradation Kinetics

The health-related nutrient loss is evaluated from the linear zero, first-order kinetic model equations (Nakilcioglu-taş and Otleş, 2018):

$$Zero\ order, C = C_0 - kt \qquad (4.30)$$

$$First\ order, C = C_0 \exp(-kt) \qquad (4.31)$$

where C_0 is the initial concentration (mg/mL of sample) at zero time, C is the concentration after time t (mg/mL of sample), t is the time, and k is the reaction rate constant. The rates of degradation of the nutrients are determined from the k value.

4.6.2 Weibull Model

The nonlinear model equation of nth order is used for the kinetic study and is given in Equation (4.32) (Illera et al., 2019):

$$log_{10}\frac{A}{A_o} = -bt^n \qquad (4.32)$$

where A in the activity of the untreated enzyme at time $t = 0$, A_o is the activity of treated enzyme after time t, b is the model constant, and n is the shape factor.

4.6.3 Two-Fraction Kinetic Model

When various isoenzymes exist in food products, their activity can be divide into liable and stable fractions. The rate of their inactivation is determined by using the first-order two-fraction kinetic models shown in Equation (4.33) (Klein et al., 2018):

$$A = A_L exp(-k_L) + A_s exp(-k_S t) \tag{4.33}$$

where A_L is the liability factor, A_s is the stability factor, and k_L and k_S are the first-order rate constants.

4.6.4 Fractional-Conversion Model

The non-zero residual activity of enzymes after the treatments (A_∞) (heat, pressure, electric voltage) are evaluated from the fractional conversion first-order kinetic model:

$$\ln(1 - f) = \ln\left[\frac{(A - A_\infty)}{(A_o - A_\infty)}\right] = -kt \tag{4.34}$$

4.7 ANALYSIS

Any model fitted data can be analyzed using three basic steps: (1) selection of a model, (2) model fitting, and (3) validation. The first step of modeling is to analyze the data through tabulations and graphs as an exploratory study. The concentration versus time data are used to fit the appropriate model and the best fit can be claimed depending on the model accuracy obtained. In the selection of the model, data plots, process information, and process assumptions are used to evaluate the shape of the model to fit the data. By using the model selected and possibly data information, a proper model-fitting technique is used to predict the unknown model parameters. The assumptions are then compared with the predicted parameters to validate the data.

Fujikawa et al. (2003) studied a new logistic growth modeling approach for the sigmoidal growth of bacteria under various initial conditions. The new logistic model has been compared with the Gompertz model, which is widely used for modeling of bacterial growth. Results showed that the new logistic model was more closed to the experimental values in the exponential period of growth than the Gompertz model. Predicted values are matched with the observed values, which signify the best fit of the sigmoidal growth modeling. However, a small deviation was observed in the fitted models. Among the models, a new logistic growth model fitted the experimental values better than the Gompertz function. Their studies also concluded that new logistic models can be used to predict the growth of bacteria under various temperature conditions.

Predictions of model parameters are important to study the degradation of compounds and to estimate the shelf life of foods. Due to the unstable composition of phenolic compounds, they are easily prone to degradation under different environmental conditions (light, pH, storage temperature and time). Phenolic compounds, antioxidants, and vitamins are highly sensitive to heat. When thermal treatment is used for the processing of foods, the increase in temperature may lead to the deterioration of the compounds. The storage of the food product for a period of time may also affect the nutritional value of the compounds. Hence, to retain the nutritional compounds, temperature and storage conditions are the main parameters to evaluate the quality. Kinetic model parameters are used to predict the degradation that occurs during the processing. The Arrhenius equation is usually used to determine the temperature effect on various chemical reactions. Kim et al. (2018) studied the kinetic models to predict the effect of temperature and storage on the quality change

(a)

(b)

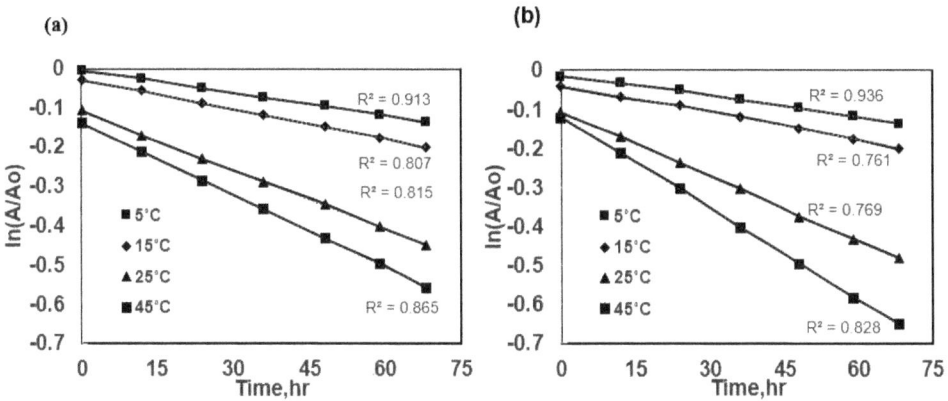

Figure 4.2 First-order kinetic plot for the degradation of total phenolic content (a) and DPPH radical scavenging activity (b) of hardy kiwi puree stored at different temperatures (5, 15, 25, and 45°C). (Adapted from Kim et al., 2018, with permission from Elsevier.)

(phenolic content and antioxidant activity) of hardy kiwifruit puree. The phenolic compounds and 2,2-diphenylpicrylhydrazyl (DPPH) radical scavenging activity against the storage time plot fitted with the first-order kinetic model and is shown in Figure 4.2. The plot of $\ln(A/A_o)$ versus storage time showed the degradation of phenolic compounds and antioxidant activity followed by first-order kinetics. However, the initial data with the highest R^2 best fits the first-order kinetic equation. The rate constant k value was determined from the slope of the curve. The studies reported that the higher k values obtained with increased storage temperature. The increase in k value indicates there is more degradation of nutritional quality in the higher temperature range. Similar results were also obtained with antioxidant activity.

4.8 VALIDATION

The role of model validation is to ensure that the performance of a statistical model is appropriate in relation to the output of the actual data-generating process to achieve the research objectives. The accuracy of a given model is determined by how similar the estimated values (LT, GT, maximum generation, model constants, residual activity of enzyme, etc.) are close to the experimental values.

To evaluate the precision of the primary models, the estimated and obtained parameters are determined. The performance of model fit can be determined as the coefficient of regression (R^2), which is also used for overall prediction achieved. The value close to $R^2 \leq 1$ signifies a better fit of the predicted model. Residual variability between the observed and predicted values are determined from the mean square error (MSE). Lower MSE tends to the best fitness of the model (Kreyenschmidt et al., 2009; Zhou et al., 2012).

$$R^2 = 1 - \frac{\Sigma_{i=1}^{n}(observed(obi) - estimated(esi))^2}{\Sigma_{i=1}^{n}(observed(obi) - mean)^2} \tag{4.35}$$

$$MSE = \sum \frac{(observed(obi) - estimated(esi))^2}{N} \tag{4.36}$$

In secondary modeling, estimated and observed parameters for accuracy are tested by measuring the bias factor (B_F) and accuracy factor (A_F). Ross (1996) suggested the use of expected ratios to observed values and developed the bias factor (B_F) and accuracy factor (A_F) to establish a simple method for model performance. A model is said to be in the "fail-safe" stage if the B_F obtained is (>1), whereas the value less than 1 signifies the "fail-harmful" model. The A_F = 1 indicates better unity between the estimated and the observed value, whereas greater than 1 represents how less reliable the model is (Angelo et al., 2013).The higher the accuracy factor, there is less acceptability of the estimated parameter.

$$bias\ factor\ (B_F) = 10 \sum_{i=1}^{n} \left(\frac{\left| \log\left(\frac{estimated(esi)}{observed(obi)} \right) \right|}{n} \right) \tag{4.37}$$

$$accuracy\ factor\ (A_F) = 10 \sum_{i=1}^{n} \left(\frac{\left| \log\left(\frac{estimated(esi)}{observed(obi)} \right) \right|}{n} \right) \tag{4.38}$$

Oscar (2005) evaluated the model performance by measuring the relative error of an individual measure and claiming the accuracy of the fit model. The estimation of B_F and A_F factors are the most commonly used methods to test the model accuracy of microbial foods. However, these two factors (B_F and A_F) shows certain limitations that may result in inaccurate model performance estimation. To overcome the limitations, the combined use of relative errors (R_e) is presented to enhance the acceptability of the model performance. Relative error of observed and predicted ratios are denoted as (λ) and (μ_{max}). The relative error (R_e) of each step is determined by using Equations (4.39) and (4.40)

$$R_e\ for\ \lambda = 1 - \frac{observed(obi)}{predicted(pdi)} \tag{4.39}$$

$$R_e\ for\ \mu_{max} = \frac{observed(obi)}{predicted(pdi)} - 1 \tag{4.40}$$

Fail-safe predictions are obtained in the zone where the R_e value is below zero, whereas the R_e value above zero indicates the fail-harmful prediction.

4.9 APPLICATION OF MATHEMATICAL MODELING FOR SHELF LIFE PREDICTION

There are enormous numbers of applications for mathematical modeling in processes. Any process optimization requires mathematical modeling. There were a number of mathematical models both theoretical and empirical in earlier sections of the chapter. The following are some of the examples of such applications:

- Mathematical models related to water activity as a factor are used for shelf life predictions of moisture-sensitive foods. Those models are mostly applied to packaged foods in co-occurrence with moisture transfer predictions (Azanha and Faria, 2005).
- Storage temperature is also a crucial factor. As mentioned earlier, the Arrhenius equation is used to study food spoilage with respect to temperature. It helps to predict reaction rates and shelf life at any temperature within the range, without actual testing (Tsironi et al., 2009).

- As mentioned in Equation (4.13), the shelf life of highly perishable food products can be calculated using models based on the concept of the exponential, Ratkowsky, and Arrhenius models.
- Food spoilage related to enzymatic reactions is predicted using the Michaelis-Menten equation. The equation is valid for a steady system (Pereira et al., 2017). Browning of fruit and vegetable tissue due to polyphenolase activity is an example of food spoilage due to enzyme action.
- Bacterial growth is better modeled using the three-phase equation as suggested in Section 4.4.

For most predicted models the applicability usually is limited to the particular food system studied. Most of the time the model is not the correlation of the actual mechanism of the reaction, so a substrate change in the food may have an effect on the rate of the quality parameter that cannot be predicted by it.

For specific cases, a detailed kinetic study of specific reactions crucial to food spoilage is required so that the effect of compositional changes can be studied. In these cases, the true mechanism of the reactions is supposed to be known. Those studies are modeled in experimental systems, rather than in actual foods, so that the composites are not likely to change under a controlled and monitored environment. Non-isothermal experiments are mostly considered for a robust set of data for prediction. Time and resource limitation leads us to follow the already predicted model more than the theoretical one to avoid much computation. Therefore, any extrapolation of kinetic results to apparently the same systems is to be done very cautiously.

4.10 CONCLUSION

This chapter focused solely on the modeling part of shelf life estimation. It gave a detailed overview of the types of models, categorized according to food type, and analysis and validation of the models and their applications. This chapter reviewed theoretical models based on physics-based laws and gave us a deeper understanding of how factors like processing conditions and environmental conditions interact with the food system. Empirical models, which are computationally easy to solve with less factors in consideration, can easily solve a processing problem. This may help food engineers working in industry to sort out a way to predict shelf life of their product and optimize the process.

REFERENCES

Adarns, M. R., Little, C. L., & Easter, M. C. (1991). Modelling the effect of pH, acidulant and temperature on the growth rate of *Yersinia enterocolitica*. *Journal of Applied Bacteriology*, *71*, 6571.

Angelo, D., Dalcanton, F., Maria, G., Aragão, F. De, Augusto, B., Carcio, M., & Laurindo, J. B. (2013). Assessing the prediction ability of different mathematical models for the growth of *Lactobacillus plantarum* under non-isothermal conditions. *Journal of Theoretical Biology*, *335*, 88–96. https://doi.org/10.1016/j.jtbi.2013.06.030

Azanha, A. B., & Faria, J. A. F. (2005). Use of mathematical models for estimating the shelf-life of cornflakes in flexible packaging. *Packaging Technology and Science*, *18*(4), 171–178.

Buchanan, R., Whiting, R., & Damert, W. (1997). When is simple good enough: a comparison of the Gompertz, Baranyi, and three-phase linear models for fitting bacterial growth curves. *Food Microbiology*, *14*(4), 313–326.

Chakraborty, S., Rao, P. S., & Mishra, H. N. (2014). Effect of pH on enzyme inactivation kinetics in high-pressure processed pineapple (*Ananas comosus* L.) puree using response surface methodology. *Food and Bioprocess Technology*, *7*, 3629–3645. https://doi.org/10.1007/s11947-014-1380-0

Corradini, M. G., & Peleg, M. (2007). Shelf-life estimation from accelerated storage data. *Trends in Food Science & Technology*, *18*(1), 37–47.

Dabadé, D. S., den Besten, H. M., Azokpota, P., Nout, M. R., Hounhouigan, D. J., & Zwietering, M. H. (2015). Spoilage evaluation, shelf-life prediction, and potential spoilage organisms of tropical brackish water shrimp (*Penaeusnotialis*) at different storage temperatures. *Food Microbiology*, *48*, 8–16.

Das, H. (2005). *Food Processing Operations Analysis*, Asian Books, Delhi, pp. 30–42.

Dias, L., & Piccoli, R. H. 2018. Primary and secondary modeling of *Brochothrix thermosphacta* growth under different temperature and pH values. *Journal of Food Science and Technology*, *2061*, 37–43.

Farhoosh, R., & Hoseini-Yazdi, S. Z. (2013). Shelf-life prediction of olive oils using empirical models developed at low and high temperatures. *Food Chemistry*, *141*(1), 557–565.

Fujikawa, H., Kai, A., & Mirozumi, S. (2003). A new logistic model for bacterial growth. *Journal of the Food Hygienic Society of Japan*, *44*, 155–160.

Galić, K., Ćurić, D., & Gabrić, D. (2009). Shelf life of packaged bakery goods—A review. *Critical Reviews in Food Science and Nutrition*, *49*(5), 405–426.

Garitta, L., Langohr, K., Elizagoyen, E., Ottaviano, F. G., Gómez, G., & Hough, G. (2018). Survival analysis model to estimate sensory shelf life with temperature and illumination as accelerating factors. *Food Quality and Preference*, *68*, 371–376.

Gibson, A. M., Bratchell, N., & Roberts, T. A. (1988). Predicting microbial growth : growth responses of salmonellae in a laboratory medium as affected by pH, sodium chloride and storage temperature. *International Journal of Food Microbiology*, *6*, 155–178.

Hajmeer, Maha N., Basheerb, Imad A., & Najjarb, Yacoub M. (1997). Computational neural networks for predictive microbiology II. Application to microbial growth. *International Journal of Food Microbiology*, *34*, 51–66.

Huang, K., Tian, H., Gai, L., & Wang, J. (2012). A review of kinetic models for inactivating microorganisms and enzymes by pulsed electric field processing. *Journal of Food Engineering*, *111*, 191–207. https://doi.org/10.1016/j.jfoodeng.2012.02.007

Illera, A. E., Chaple, S., Sanz, M. T., Ng, S., Lu, P., Jones, J., Carey, E., & Bourke, P. (2019). Effect of cold plasma on polyphenol oxidase inactivation in cloudy apple juice and on the quality parameters of the juice during storage. *Food Chemistry: X*, *3*, 100049. https://doi.org/10.1016/j.fochx.2019.100049

Kim, A., Kim, H., Chun, J., Jin, H., Kerr, W. L., & Choi, S. (2018). Degradation kinetics of phenolic content and antioxidant activity of hardy kiwifruit (*Actinidia arguta*) puree at different storage temperatures. *LWT – Food Science and Technology*, *89*, 535–541. https://doi.org/10.1016/j.lwt.2017.11.036

Klein, M. P., Ana, V. S., Hertz, P. F., Costa, R., & Ninow, J. L. (2018). Kinetics and thermodynamics of thermal inactivation of β-galactosidase from *Aspergillus oryzae*. *Brazilian Archives of Biology and Technology*, *61*, 1–12.

Kreyenschmidt, J., Hu, A., Beierle, E., Chonsch, L., Scherer, A., & Petersen, B. (2009). Determination of the shelf life of sliced cooked ham based on the growth of lactic acid bacteria in different steps of the chain. *Journal of Applied Microbiology*, *108*, 510–520. https://doi.org/10.1111/j.1365-2672.2009.04451.x

Lee, D. S. (2014). Comparison of two mechanistic microbial growth models to estimate shelf life of perishable food package under dynamic temperature conditions. *Mathematical Problems in Engineering*, *2014*(5), 392054.

Mahony, C. O., & Seman, D. L. (2016). *Modeling the Microbiological Shelf Life of Foods and Beverages, The Stability and Shelf Life of Food* (2nd ed.), Elsevier, pp. 253–289. https://doi.org/10.1016/B978-0-08-100435-7.00009-5

Martins, R. C., Lopes, V. V., Vicente, A. A., & Teixeira, J. A. (2008). Computational shelf-life dating: complex systems approaches to food quality and safety. *Food and Bioprocess Technology*, *1*(3), 207–222.

McMeekin, T. A., Olley, J., Ross, T., & Ratkowsky, D. A. (1993). Predictive Microbiology. *Theory and Application*, Research Studies Press, Taunton, UK.

Nakilcioglu-taş, E., & Otleş, S. (2018). Degradation kinetics of bioactive compounds and antioxidant capacity of Brussels sprouts during microwave processing. *International Journal of Food Properties*, *20*(Suppl. 3), 1–12. https://doi.org/10.1080/10942912.2017.1375944

Nicoli, M. C. (2012). An Introduction to Food Shelf Life: Definitions, Basic Concepts, and Regulatory Aspects. *Shelf Life Assessment of Food*, CRC Press, Boca Raton, FL, pp. 1–16.

Oscar, T. P. (2005). Validation of lag time and growth rate models for *Salmonella typhimurium*: acceptable prediction zone method. *Journal of Food Science*, *70*(2), M129–M137.

Pereira, M. J., Amaro, A. L., Pintado, M., & Poças, M. F. (2017). Modeling the effect of oxygen pressure and temperature on respiration rate of ready-to-eat rocket leaves. A probabilistic study of the Michaelis-Menten model. *Postharvest Biology and Technology*, *131*, 1–9.

Ratkowsky, D. A., Lowry, R. K., McMeekin, T. A., Stokes, A. N., & Chandler, R. E. (1983). Model for bacterial culture growth rate throughout the entire biokinetic temperature range. *Journal of Bacteriology*, *154*, 1222–1226.

Ratkowsky, D. A., Olley, J., McMeekin, T. A., & Ball, A. (1982). Relation between temperature and growth rate of bacterial cultures. *Journal of Bacteriology*, *149*, 1–5.

Ross, T. (1996). Indices for performance evaluation of predictive models in food microbiology. *Journal of Applied Bacteriology*, *81*, 501–508.

Ross, T., & Dalgaard, P. (2004). Secondary models. *Modeling Microbial Responses in Food*, CRC Press, New York.

Saguy, I., & Karel, M. (1980). Modeling of quality deterioration during food processing and storage. *Food Technology*, *34*(2), 78–85.

Shiferaw, N., Hong, Y., Knoerzer, K., Buckow, R., & Versteeg, C. (2010). High pressure and thermal inactivation kinetics of polyphenol oxidase and peroxidase in strawberry puree. *Innovative Food Science & Emerging Technologies*, *11*, 52–60. https://doi.org/10.1016/j.ifset.2009.08.009

Taoukis, P. S., Labuza, T. P., & Saguy, I. S. (1997). Kinetics of Food Deterioration and Shelf-Life Prediction. *Handbook of Food Engineering Practice*, CRC Press, Boca Raton, FL.

Tsironi, T., Dermesonlouoglou, E., Giannakourou, M., & Taoukis, P. (2009). Shelf life modelling of frozen shrimp at variable temperature conditions. *LWT – Food Science and Technology*, *42*(2), 664–671.

Zhou, K., Gui, M., Li, P., Xing, S., Cui, T., & Peng, Z. (2012). Effect of combined function of temperature and water activity on the growth of *Vibrio harveyi*. *Brazilian Journal of Microbiology*, *43*(4), 1365–1375.

CHAPTER **5**

Food Quality and Food Safety Assessment in the Supply Chain

Syed Insha Rafiq, Sourav Garg, Balaji G., Irshaan Syed,
Hikhmathunnisa Sultan Noordeen, and Syed Mansha Rafiq

CONTENTS

5.1 INTRODUCTION

Today, food industries are under immense pressure to improve food safety, maintain a risk management system, and sustain the quality of the product. Being more consumer-oriented, they need to ensure rapid response timing while dealing with any food safety and quality issues, or any other consumer grievances. An efficient traceability system throughout the entire food supply chain ensures production, supply, and distribution of safe and high-quality products, thereby reducing product recalls. According to the World Health Organization (2003), the responsibility of the management

Email: Syed Insha Rafiq, Syedinsha12@gmail.com

DOI: 10.1201/9781003091677-5

of food quality and food safety should be ensured by all sectors including governments, producers, processing industries, and consumers. With advancements in the global food trade, food products have to travel quite long distances from the producer to reach the consumer. Hence, the supply of safe and good-quality products has become a big hurdle for food handlers. In the past few years, various food crises like usage of genetically modified crops in food products, detection of high levels of dioxin in animal-based foods, mad cow disease, foodborne illness due to microbial contamination, and many more have largely challenged the credibility of food industries. As a consequence, consumers are becoming more concerned over the safety and quality aspects of food, and they demand transparency in the entire supply chain of foodstuffs (Beulens et al., 2005; Bertolini et al., 2006; Regattieri et al., 2007; Trienekens and Zuurbier, 2008).

5.1.1 Traceability

Traceability is a tool used to achieve numerous objectives, hence, its definition needs to be broad (Golan et al., 2004) and various organizations and legislation define it in their own way. International Organization for Standardization (ISO) 8402 defines traceability as "the ability to trace the history, application or location of an entity using recorded identification," whereas ISO 9000 defines it as "the ability to trace the history, application or location of that which is under consideration." As per the Codex Alimentarius Commission Procedural Manual, traceability is "the ability to follow the movement of a food through specified stage(s) of production, processing and distribution" (Olsen and Borit, 2013). Product traceability is an effective tool for the fulfilment of legislative compliances, improved food safety and quality standards, and a better connection between producers and consumers (Regattieri et al., 2007).

Food industries have developed several food safety and quality management strategies to combat the growing safety issues in the food supply chain. Quality assurance has become the base for the formation of their food safety policies for integration of food safety and quality management systems (Pinto et al., 2006). To boost consumer confidence and fulfill safety and quality requirements, food suppliers rely on two methodologies. The first method utilizes regulatory standards and certifications to manage the food supply chain, and the second one uses a traceability system for the entire production and logistical operations to achieve transparency in tracking forward and backward information (Hong et al., 2011).

5.1.2 Food Safety and Quality

As per the definition by the Food and Agriculture Organization of the United Nations(2003b), food safety is "an assurance that food will not cause harm to the consumer when it is prepared and/or eaten according to its intended use." Food safety refers to all food-related hazards that can potentially cause any illness to the consumer. These hazards are can happen at various steps of the food supply chain. Thus, it is the collective responsibility of all the members involved in the supply chain to ensure that the food is safe. The Hazard Analysis and Critical Control Point (HACCP) system is a vital preventive approach for tackling food safety hazards that can be used at all the steps in the food supply chain (Food and Agriculture Organization of the United Nations, 2003a).

As per the ISO, the definition of quality is "the totality of features and characteristics of a product that bear on its ability to satisfy stated or implied needs" (Van Reeuwijk and Houba, 1998). Alternate definitions of quality are "conformance to requirement," "fitness for use," or more suitably for foodstuffs, "fitness for consumption." Hence, quality is the necessities required to fulfill the requirements and expectations of consumers (Ho, 1994; Peri, 2006). Nevertheless, in the case of food quality, these expectations may vary largely with consumers. Quality also includes the parameters based on which the consumer decides the value of the product. Quality is not only governed by the attributes of the food product, it also includes the method used to achieve those attributes (Morris and Young, 2000).

5.1.3 Food Supply Chain

The food supply chain can be described as a series of activities that brings a product from the production site to the consumer's plate (Aramyan et al., 2006). It constitutes farmers (producers), cooperatives, chemical manufacturers, research centers, processing entities, food traders, logistics, transportation systems, storage, supermarkets, financial organizations, regulatory bodies, administration, and international organizations (Jaffee et al., 2010). The overall performance of the food supply chain depends on the individual performance of each stakeholder involved in the chain. It also depends on various characteristics such as shelf life of the food product, its availability, its sensory characteristics, food safety concerns, and environmental and storage factors. The interdependency of individual stakeholders and the factors listed previously make performance evaluation of the food supply chain even more complex (Sufiyan et al., 2019).

5.2 QUALITY AND SAFETY ASSESSMENT TECHNIQUES IN THE SUPPLY CHAIN

5.2.1 Radiofrequency Identification and Blockchain Technology

Radiofrequency identification (RFID) is an innovative technology initiated in the 20th century with applications in numerous fields. It has been used extensively in various industries like aeronautics, food, and medicine. RFID helps maintain traceability by controlling the flow of goods within the production chain, and is has several applications in supply chain processes across the globe (Kelepouris et al., 2007). In inventory tracking, there have been wide applications of RFID in supply chain management such as transportation, storage management, and the back-end quality control in the supply chain, theft control, stock monitoring, and automatic retail checkout (Nambiar, 2009). RFID technology involves the use of electromagnet or electrostatic coupling within the range of the RF portion (Yong-Dong et al., 2009) and the information related to the particular product is transmitted wirelessly through radio waves.

This wireless sensor works by detecting the electromagnetic signals and generally comprises three components: an RF tag (or transponder), an RF tag reader, and a computer (Nambiar, 2009). Basically, the transponder is a label or tag that has a memory chip and a coiled antenna and stores the important information related to the product like quantity, date of expiry, serial number, etc. The radio waves generated from an RF tag reader are detected by transponder, and the coil present in an RF tag powers the memory chip by creating a magnetic field. The data stored in this chip are then transmitted to the reader. The most common types of RFID transponders are active and passive tags. The active tags have small batteries for storing the required information, whereas the passive ones lack power systems and are relatively cheaper (Wamba et al., 2008). The antenna of the reader will supply energy to the transponder and receive data from the transponder. The reader transmits the received data to the computer system for further processing. The main advantages of RFID tags over optical barcode systems are that the data may be read automatically at a rate of several hundred tags per second and from a range of several meters (Weis et al., 2004).

An RFID system simplifies the supply chain process as several processes can be done automatically instead of using manual input and transferring data across the system. Hence, several companies have adopted RFID systems in their supply chains and are gaining better internal efficiencies (Attaran, 2012). In other words, it is an individual item-identification device that significantly improves efficacy of systems like quality control, accurate deliveries, customer services, product recalls, etc. (Beaumont et al., 2009). The RFID system offers a direct idea about the purchase habits of customers and increases accuracy and efficiency within the supply chain. It could drastically enhance performance of the supply chain by decreasing lead times, levels of inventories, shrinkage rates, and stock-outs. It can potentially improve the visibility of inventory, the throughput, order accuracy, customer services, product quality, inventory record accuracy, and cooperation among several units of the supply chain (Reyes and Frazier, 2007).

Blockchain technology (BT) is used for the design, operation, organization, and management of supply chains. It easily integrates various units in the supply chain; provides faster tracking, recording, and sharing of information; and helps in up-scaling the system (Underwood, 2016). It also helps in achieving better transparency and visibility of supply chain transactions by resolving trust-related issues. BT It helps organizations and all their key stakeholders to maintain a real-time digital record of all the transactions occurring throughout the supply network due to effective resource management, accurate demand forecasts, and reduction in inventory carrying costs. The main role of BT is to enhance the visibility and transparency of records and to improve the scale and scope of traceability and tracking system in the supply chains (Kamble et al., 2019).

The mode of transactions carried in supply chains is likely to bring an innovative paradigm shift (Tapscott and Tapscott, 2017). The transaction data is stored in blocks that are then put together sequentially to obtain a chain, which is then circulated among all the participating units. Tamper-proof nature and better traceability benefits further resolve the problems due to trust issues and ambiguities that occur in supply chain systems (Kshetri, 2018).

Transactions such as cash, material, and information flow take place at the interface of supply chain facilities like a plant, vendor, distribution center, etc., of every supply chain (e.g., between plant and distribution center, between vendor and plant). Thus, BT is expected to speed up the processes in supply chains. The ability of BT to pledge the reliability, traceability, authenticity of information, and resolving problems of contractual relationships for a trustless environment signifies its potential applications in supply chain management.

5.2.2 Artificial Intelligence and Machine Learning

Artificial intelligence (AI) refers to the utilization of computers for recognition of patterns, reasoning, comprehension of various behaviors (through collection, experience, and retention of knowledge), and the development of different means of problem-solving conclusions (Russell and Norvig, 1995). It understands the human intelligence phenomenon for the design of computers that imitate the patterns of human behavior and generate knowledge to find solutions to a particular problem. Hence, AI should possess the capability to learn and understand new concepts, impute meaning, perform reasoning, interpret symbols, and draw conclusions. Therefore, AI was introduced to develop and make "thinking machines," which can easily learn, imitate, and substitute human intelligence. It has been effectively applied in segments like semantic modeling, gaming, robotics, human performance modeling, data mining, machine learning (ML), genetic algorithms (GAs), neural networks, and expert systems (Luger, 2005). AI has shown an enormous guarantee of improved human decision-making processes and efficiency in diverse business activities due to the capability to seek information, learn business phenomena, business pattern recognition, and data analysis.

ML is "the scientific study of computational models and algorithms on computers using experience to improve the performance on a specific task or to make accurate forecasts" (Mohri et al., 2018). The term experience in this definition refers to "the historical data accessible to the learner for building a prediction model." ML was planned to make computers capable of learning without being explicitly programmed by the developer. Thus, the computer has to attain knowledge directly from the available data to solve a particular issue. Despite the learning task differences, ML techniques often challenge copying the knowledge and experiences of human beings. Furthermore, neurological studies of human brain function have motivated some of the ML techniques whereas others were based on the sociological hypothesis behind human collaborative actions.

ML can be classified into three broad learning methods, i.e., supervised, unsupervised, and reinforcement. In case of supervised learning, the labeled data with information about input variables and their desired output variables serves as a base for the development of a predictive model. On the other hand, supervised learning involves plotting the variables to the desired output variable (Zhu and Goldberg, 2009). These techniques include algorithms like decision trees, regression analysis, random forests, and Bayesian networks (Traore et al., 2017). The unsupervised learning algorithms use unlabeled data sets and do not have any idea about the input and output variables. They formulate hidden patterns based on the available unlabeled data set and are mainly used for reducing the dimensionality and exploratory data analysis (Jordan and Mitchell, 2015). Algorithms like GA, artificial neural networks (ANNs), deep learning, clustering, and instance-based learning models are included in unsupervised learning. In reinforcement learning the process involves a combination of the training and testing data sets and further interaction of the learner with the environment to collect information. The learner is rewarded for performing actions with respect to the environment and further creates a dilemma for exploration versus exploitation. The learner should work more on exploration of unknown actions to obtain more information instead of just exploiting the available information (Mohri et al., 2018). These algorithms are also used in real-time decision making, robot navigation, and machine skill acquisition.

5.2.3 Food Safety and Quality Risk Assessment Models

5.2.3.1 Optimization Model

In a study, Wang et al. (2009) formulated an integrated optimization model by simultaneous optimization of product batch size and batch dispersion policy in a cooked meat manufacturing case; however, this model also is widely applicable for any other perishable food manufacturing units. This model takes traceability, operational costs, and quality assurance into account for delivering a desirable product and reducing the effect of product recalls economically. Because the total cost function cannot be differentiated with respect to the batch size, other experiments were conducted simultaneously to analyze the performance of this model using the numerical method. Here, the risk associated with every raw material batch was found to affect the overall performance of the system considerably.

As the occurrences of product recall are quite low, the analytically optimized solution was obtained by ignoring the traceability factor. This model provides innovation in food production and supply chain management by demonstrating the advantages of integrated operational planning, considering food safety and quality parameters simultaneously. The study aimed to demonstrate the significance of traceability in the food supply chain operations to the food manufacturing and food handling enterprises. Although the model was proposed in the context of the perishable food production unit, the same integrated model applies to a wide variety of products involving the batch production process and assembling processes.

5.2.3.2 Aggregative Food Safety Risk Assessment Model

Fuzzy set theory has wide applications in evaluating the problems that are uncertain. It makes the decision-making process easier by representing the uncertainty and vagueness in mathematical form (Zadeh et al., 1996). Wang et al. (2012) proposed a collective food safety risk assessment model to analyze food safety risks in various steps of the food supply chain. This model was made using fuzzy set theory and the analytical hierarchy process. The linguistic terms for risk parameters are subjective, so they were represented in qualitative scale using triangular fuzzy numbers. Overall risk transformation was monitored along the entire supply chain by formulating a matrix model. For each product or process, hazards were identified using a hierarchical structure model to evaluate the aggregative food safety risk indicator. The applicability of this model was evaluated in a medium-sized food producer in the United Kingdom. Due to the evaluation of overall food safety in the entire supply chain, the model is expected to assist in determining critical control points and supporting operational decisions along various stages in the supply chain.

5.2.3.3 Hybrid Fresh Food Supply Chain Risk Assessment Model

The hybrid fresh food supply chain risk assessment model combines fuzzy logic with hierarchical holographic modeling techniques for risk identification and risk assessment, respectively. Initially, risk identification is done qualitatively using the hierarchical holographic modeling method, and then risk assessment is done using both a qualitatively (using a hierarchical holographic model) and a fuzzy logic-based approach to calculate the risk level and overall risk level. Here, the fuzzy logic-based approach is important as it is more useful in the evaluation of supply chain risks, which are imprecise and uncertain (Nakandala and Lau, 2012). The risk levels associated with the two methods are then compared to check the discrepancies, if any, and the overall risk level is evaluated using the root-mean-square of the two values. Arithmetically, the root-mean-square value of two numbers is greater than the mean value of these numbers; hence, this method is preferable for food manufacturing enterprises that are more conservative regarding food safety risk levels. The practical feasibility and validation of this approach was studied by taking the case of a fresh food supply chain company (Nakandala et al., 2017).

5.2.3.4 Fuzzy Comprehensive Evaluation Model

The fuzzy comprehensive evaluation model was developed for the evaluation of risk associated with food quality systems. Here, a quality risk indicator system was developed and then a fuzzy comprehensive evaluation model and failure mode, effects, and criticality analysis were applied to calculate the risk level associated with food quality. The resulting model can be used to emphasize the prominent objectives in evaluating food-quality risk levels as per the results of computational experimentation. However, this model can only be applied for the evaluation of risk associated with food quality, and the results obtained from the computational experiment cannot be generalized. There is a need for further research approaches to address these drawbacks (Bai et al., 2018).

5.3 REGULATION AND LEGISLATION

5.3.1 Codex Alimentarius Commission

The joint committee of the United Nations (UN) Food and Agricultural Organization (FAO) and World Health Organization (WHO) postulated an International Food Standards Program in 1963. This is the single reference point of contact for food standards, which is known as the Codex Alimentarius Commission (CAC) (Food and Agriculture Organization of the United Nations, 2003b). Codex is a collaborated and collated text that prescribes food standards that are accepted unanimously throughout the globe. The aim of this commission is to ensure safe food to consumers and protection of consumer health. It recommends the implementation of prerequisite programs such as Good Agricultural Practices (GAP), Good Hygiene Practices (GHP), Good Manufacturing Practices, Good Trade Practices, etc., as a base requirement for any concern along the food supply chain. Furthermore, these prerequisites lay a foundation for the implementation of HACCP, which is a risk-and-hazard-based preventive strategy for the effective control of food safety risk, thereby ensuring product safety. Also, ISO stipulates standards by enforcing HACCP, which is a good practice model along with a management commitment strategy to prevent the faults arising in the food supply chain. The detailed roles of ISO and HACCP are discussed in the following sections.

5.3.2 HACCP

HACCP is a coherent method of preventive food control. It analyses the probability of hazards at each stage of the process chain, thereby proactively measuring, analyzing, and preventing a range of hazards from affecting the consumer. It also facilitates the preventive quality assurance approach. The various hazards include physical, chemical, and biological types. HACCP ensures the safety of food to consumer, manufacturers, retailers, growers, food business operators (FBOs), etc. It involves interdisciplinary and inter-managerial personnel as everyone in the chain has a significant role to play for effective HACCP implementation and management. The seven principles of HACCP have been accepted across the globe, and its details have been published by the CAC (Food and Agriculture Organization of the United Nations, 2003b) and the National Advisory Committee on Microbiological Criteria for Foods.

5.3.2.1 Benefits of HACCP

Implementation of the HACCP model helps lower the risk of product failure and ensures product safety (Motarjemi and Lelieveld, 2013). Other benefits of HACCP include the following:

- Implementation and practice involve trained and experienced personnel, thereby reducing the bias and prioritization in making an informed judgment on food safety issues.
- Regulation of effective food safety management with documented evidence at each stage.
- May be expensive at the initial establishment, but it is extremely cost-effective after initial implementation.
- Implementation increases productivity as it reduces the non-conformities in the finished products by identifying the potential failures at early stages, which in turn implies a sound investment for a company to achieve high throughput.
- Fulfills the regulatory and statutory requirements for safe food production in a country.

5.3.2.2 HACCP in Supply Chain

The food supply chain is a complex interlinked chain with many interlinked processes and conditions (Dani, 2015). Apart from the manufacturing sector, HACCP could be applied to the packaging industry, retail industries, catering chains, storage and distribution chains, primary production sites, etc. However, a flexible approach needs to be developed that suits the practical system and postulates the necessities of each link in the supply chain model.

Although, there are many stakeholders in the simplified supply chain model, in reality, there will be more than hundreds of stakeholders for a single product. This includes producing farmers, raw material suppliers, distributors, logistic and storage operators, etc. Effective communication is crucial at all the stages of the supply chain, which is a great hindrance at present. However, each element of the model has a certain degree of control over others through supplier quality assurance (SQA). Alongside, SQA, implementation of the HACCP model at each stage and associated with each element ensures the vendors behind the chain have control over the critical parameters that influence the safety of material (Gehring and Kirkpatrick, 2020).

The major part of the mitigation of hazards comes from primary production. A greater challenge occurs when considering the application of the HACCP model in a forward chain. It is useful in identification and control of hazards that mitigate the food chain in the farm. For example, major foodborne illness happens by contamination of poultry and meat products with *Salmonella* sp., *Campylobacter* sp., and *Escherichia coli* OH 157. Contamination could be prevented by adopting GHP at the farm level by providing clean drinking water, proper vaccination, approved feed practices, fur cleaning before slaughter, etc. The GHP model, which is one of the prerequisites of HACCP implementations, ensures the effective implementation of HACCP in the system. Similarly, following the guidelines of PRPs (Prerequisite Program) will directly lead to the reduction in hazard migration into the food supply chain.

5.3.3 ISO

ISO is a non-profit organization that develops standards and guidelines that are accepted internationally, enabling fair trade practices across the globe. It is a worldwide federation of national standard bodies with each member body focusing on specific subjects. These standards apply to all fields and at various levels of management as they ensure integrity, a structured approach, and verified procedures, which dignify a company's image, etc. (Faergemand and Jespersen, 2017).

Our food supply chain is a complex interlinked system with numerous elements and stakeholders. Any discrepancy or failure in the food supply chain results in huge risk and can be potentially dangerous. ISO 22000 for Food Safety Management System (FSMS) ensures the absence of weak connections in the food supply chain system, thereby ensuring safe food. It applies to all types of firms involved in the food chain ranging from primary producers, processors, storage and logistic operators, distributors, dealers, and subcontractors to retail foodservice outlets (Panda and Priya, 2019).

5.3.3.1 ISO 22000

ISO 22000 provides the requirements for a FSMS across the supply chain where the organization has to plan and improvise all processes through effective planning, proactive actions, and frequent monitoring and verification of the system for effective control over food safety. This demonstrates the organization's capability to prevent and control the hazards related to food safety, thereby leading to a safe end product meeting both consumer and regulatory requirements.

ISO 22000 standards are based on risk-based thinking that incorporates a Plan-Do-Check-Act (PDCA) cycle. Thus, they provide an opportunity for an organization to plan and understand its various process and their interactions. The PDCA approach ensures that all processes carried out in an organization are managed and resourceful to act on improvement in the system.

The objectives of ISO 22000 standards include the following:

- Unified standard protocols that provide harmonized food safety management globally in all aspects of the food chain
- Integration of a management system such that it simplifies the usage of ISO standards
- Implementation and improvement of FSMS in an organization with a greater compactness of implementation
- Increases the level of customer satisfaction index by efficient control of food safety risks (Lokunarangodage et al., 2015)

5.3.4 European Standards

The General Food Law adopted by the European parliament in 2002 ensures a greater level of protection of human life and consumer's interests in relation to food while ensuring the effective functioning of the internal market. Under this law, it sets up the European Food Safety Authority and main procedures for the Rapid Alert System for Food and Feed (Bureau and Swinnen, 2018). This legislation has undergone many changes in the last few years. The new legislation encompasses a wide range of participants, from primary producers, distributors, retailers, processors, and consumers. This implies that new changes magnified each unit of the food chain to ensure safe and good quality food is being delivered on the plate (Panghal et al., 2018). The legislation demands product labeling, product quality and safety assurance, and product liability.

- *Product labeling:* Demand is created to make consumers aware of characteristics, origin, shelf life, claims, allergens, instructions for usage, ingredients, and instructions for use. It also involves cost, manufacturers and marketers, and importer details, if applicable.
- *Product quality and safety assurance:* This legislation focuses on the organizational measures to ensure the product safety and quality being delivered. This is achieved by HACCP, ISO, and BRC (British Retail Consortium) standards.
- *Product liability:* This legislation imposes liability on the company responsible for a food product in the market, if it causes any deficiency, disease, illness, etc. Even if there is no fault on the company's part, they are liable for any discrepancies arising out of their products.

5.3.5 GFSI

Food producers, retailers, distributors, food catering operators, industrial associations, technical experts, and regulatory and statutory bodies have a common understanding regarding food safety. They perceive that the consumer's faith level decreases with each food safety outbreak for any product in the last 10 years. In response to minimizing the public concern, manufacturers, retailers, etc., have begun to audit their suppliers so that they could trust safety assurance and have confidence in materials supplied by the supplier while estimating the supplier's ability to consistently deliver quality standards (Baines, 2010). This created repetitive audits and different standards for different products from different manufacturers. To address these issues and reduce the unnecessary audits, all industries came forward to create a unified framework covering the entire food supply chain known as the Global Food Safety Initiative (GFSI) (Sansawat and Muliyil, 2011).

GFSI aims to ensure the safety of food reaching end consumers throughout the world. This system also improves the efficiency of the process, promotes transparency, and paves the way for cost reduction while assuring food safety and creates a platform for continuous improvement. It harmonizes the global FSMS under a single roof and can act as the standard for benchmarking. GFSI aims to fulfill its mission to "provide continuous improvement in food safety management systems to ensure confidence in the delivery of safe food to consumers worldwide"(Crandall et al., 2012). GFSI recognizes schemes such as the Codex of general principles of food hygiene code of practice. These principals are adopted as essential and basic requirements of any food industry operating in this domain. These requirements amplify the robustness of the process and rigor to the fundamental requirement of food safety (Crandall et al., 2012).

5.3.6 Indian Regulations

The food processing industry with enormous processed food manufacturing and food chain systems is one of the massively growing industries in India. It is governed by several laws in zones such as 5S sanitation, licensing, and other necessary permits that are mandatory for a successful food business. The trade of food and food products (raw, minimally processed, primary and secondary processed, or various by-products) is a complicated, technical, and administrative operation involving the movement of substantial quantity and variety of food products. Therefore, to strengthen the food supply chain with essential standards, limits, and to ensure food safety and a safe marketing pattern, the government of India has enacted some crucial laws, regulations, and orders. The Prevention of Food Adulteration Act (PFA), 1954 managed the food safety of India initially but was replaced by the Food Safety and Standards Act, 2006 (FSSA) (FSSAI, 2020a). The FSSA was enacted to combine several laws existing in the country in relation to food safety and to formulate a single-point reference system (Vasanthi and Bhat, 2018). The FSSA, under the Ministry of Health and Family Welfare (MOHFW), integrated eight food laws that were formerly handled by various ministries and departments associated with food. These acts and orders regarding food are as follows (Jairath and Purohit, 2013; Kohli and Garg, 2015).

- The PFA, 1954, implemented by MOHFW
- Fruit Products Order (FPO), 1955, implemented by the Ministry of Food Processing Industries (MOFPI)
- Meat Food Products Order (MFPO), 1973, implemented by MOFPI
- Vegetable Oil Products (Control) Order, 1947, implemented by the Ministry of Consumer Affairs, Food, and Public Distribution (MCFPD)
- Edible Oils Packaging (Regulation) Order, 1998, implemented by MCFPD
- Solvent-Extracted Oil, De-Oiled Meal, and Edible Flour (Control) Order, 1967, implemented by MCFPD
- The Milk and Milk Products Order (MMPO), 1992, implemented by the Ministry of Agriculture (MOAGL)
- Any other order issued under the Essential Commodities Act, 1955, relating to food implemented by Inter-ministerial through issuance of control orders

5.3.6.1 FSSAI

In 2008, the Food Safety and Standards Authority of India (FSSAI) was established as a statutory body under the FSSA for creating science-based standards to cover all the sources of contamination that can possibly occur in the food chain. To provide the public with safe and wholesome food, every single process, such as manufacturing, processing, distribution, sale, and import of food, is completely controlled by this authoritative body. All sorts of organizations that are either associated with the food chain or involved in the evaluation of food chain characteristics are inspected by the guidelines developed by FSSAI. It deals with all the issues that would compromise the quality and safety of foods, for example, problems caused by food contamination, the presence of chemical residues and biological risks, food incidence, accidents caused in the food industry, and detection of emerging risks (Shukla et al., 2014). FSSAI also contributes to promoting awareness regarding food safety in India, progressing international technical standards in food, and disseminating information. It is also in charge of gathering and compiling data in connection with consumption and contamination of food, emerging risks, etc., and providing scientific advice and technical support to the central government (Trivedi et al., 2019).

In 2011, the Central Government implemented the Food Safety and Standards Rules and Act based on international legislation, instrumentalities, and the CAC, which provides core provisions to enhance safety in primary food throughout the supply chain right from the stage of manufacturing to consumption of the food product (Jairath and Purohit, 2013). On August 5, 2011, six regulations were implemented under the FSSAI that covered the interests of FBOs in the supply chain and consumers. These regulations from 2011 include the Food Safety Standards (FSS) (Licensing and Registration) Regulation; FSS (Food Product Standards and Food Additives) Regulation, Part I & Part II; FSS (Packaging and Labelling) Regulation; FSS (Prohibition and Restriction on Sales) Regulation; FSS (Contaminants, Toxins, and Residues) Regulation; and FSS (Laboratory and Sampling Analysis) Regulation (FSSAI, 2020a). A critical stipulation of this Act is that under any circumstances if a particular food product fails to satisfy the conditions of this Act, then the stakeholders like manufacturers, distributors, packers, wholesalers, and sellers associated with this product would be held accountable for the failure. The Indian food regulatory system includes the International Food Safety Objective (FSO) concept that allows the food businesses in India to handle risk hazards of food articles in the universal competitive market (Jairath and Purohit, 2013). Today there are about 21 regulations under the FSS Regulations implemented from 2016 to 2020 including the six regulations enacted on 2011 (FSSAI, 2020a).

5.3.6.2 Role of FSSAI in Food Safety and Security Supply Chain

As a regulating body for food safety by establishing the Indian food standards, FSSAI began several actions to execute novel statutes. While intending to attain effective implementation, extensive efforts were taken in various areas such as product labeling and nutritional claims, public consciousness, approval of a product, import regulations, and other relevant areas (Trivedi et al., 2019). To direct Food Safety Officers (FSOs) toward confirming that all the food laws were being accurately executed and applied in India with transparency, steadiness, and objectivity, the FSSAI provided the "Manual for Food Safety Officers" (FSSAI, 2017c).

Some of the critical regulations of FSSA promoting food safety and quality assessment in the food supply chain are explained in this section. For precise tracing and controlling of food quality, the Food Safety and Standards (Licensing and Registration of Food Businesses) Regulations, 2011 suggests all FBOs throughout the country get licenses for their businesses. This applies to all categories of food operations such as food processing units, hotels, restaurants, and even street food sold in portable stalls (Jairath and Purohit, 2013). The FSS (Food Product Standards and Food Additives) Regulations, 2011 sets rules and instructions for FBOs that are mandatory and must be followed while processing a standardized product to ensure food safety and quality to the manufactured products (FSSAI, 2011b). The FSS (Packaging and Labelling) Regulations, 2011 state that details about the (1) food article's name, (2) ingredients utilized in the product, (3) nutrition facts, (4) notification of veg and non-veg, (5) information about the food additives, (6) manufacturer's complete address and name, (7) net quantity, (8) lot/code/batch identification, (9) manufacturing or packing date, (10) best before and use by date, (11) name of the country in case of imported foods, and (12) directions for the usage of the food product have to be on the label of every packaged food. This facilitates easy traceability in the supply chain regarding the quality and safety of the processed food articles and educates customers about the product they consume (FSSAI, 2011b, 2020a; Jairath and Purohit, 2013). The FSS (Contaminants, Toxins and Residues) Regulation sets the tolerance limits for the presence of chemicals known as "crop contaminants." Any material that is unconsciously added to foods during any stage of its processing and handling operations is known as a crop contaminant (FSSAI, 2011a). The FSS (Laboratory and Sample Analysis) Regulations, 2011 carries out certifications for ensuring the quality and safety of food products by obtaining samples from manufacturing units and performing analysis through a "Notified laboratory" and "Referral laboratory," which refers to laboratories notified by the Food Authority and laboratories that are either set up or acknowledged by the Food Authority, respectively, under the Act (FSSAI, 2011c,d). The FSS (Food Recall Procedure) Regulations, 2017 verifies if all the foods mentioned under recall are eliminated throughout the food system in obedience to Section 28 of the Act; makes sure that the information is properly disseminated to concerned public people; and finally guarantees that all foods under recall are retrieved, destroyed, or reprocessed. A food recall is the action to remove food or food product that is determined or apparently taken into consideration as risky or as stated by the Authority of Food every now and then. These foods under recall are eliminated at any phase of the food system from the market (also comprising products that are already purchased by the customers) (FSSAI, 2017a). The FSS (Import) Regulation, 2017 stipulates guidelines for managing foods that are imported into the country. It deals with procedures that permit food articles into the country through international trade. Also, this regulation consists of many provisions associated with a food importer license; consent by the Authority of Food for food articles that are being imported; authorization of food import for particular reasons; storage, inspection, and sampling of food articles that are imported from other countries; analysis of imported food sample in laboratories; ban and constraints on food articles that are to be imported; controllable amenity for labeling of food in a manner that facilitates easy trade; and finally privilege for the importer to get their concerns, if any, addressed with respect of the clearance of their food products (FSSAI, 2017b).

In 1997, the Quality Council of India (QCI) was established by the Indian Industry and the Indian Government. Considering the intent of FSSAI for promoting food safety, the QCI introduced "IndiaGHP" and "IndiaHACCP." These are certifications that were established on the basis of universally acknowledged Codex Standards for utilization by operators in the food supply chain and by food producers. Worldwide HACCP certification is compulsorily required for areas that are highly at risk, e.g., dairy, meat, poultry, and seafood, whereas GHP is mandatory for all sectors of food. Because obtaining these international certifications are very much expensive and slow, the launch of these two schemes by QCI is expected to aid all Indian Food processing industries to show compliance with international standards (MOFPI, 2020). To enhance the Hygiene Rating in India through augmenting the number of authorized Hygiene Rating Audit Agencies, QCI, due to the initiative of FSSAI, launched a scheme for validation of Hygiene Rating Audit Agencies. The "Food Hygiene Rating Scheme" is a scheme that certifies all food-related businesses that directly deliver food to the public through online or dine-in modes. These businesses are certified based on cleanliness and safety circumstances detected during the auditing period. This scheme seems to be currently applicable for foodservice firms that provide cooked or prepared foods like catering services, small-scale dhabhas, restaurants, bakeries, etc., and also meat retail stores, thus, aiding the customers who are concerned with the hygiene and safety of their food products, particularly amid this pandemic (FSSAI, 2020b). The government could only conceptualize, pass rules and regulations, and punish if food-related businesses failed to act in accordance with these laws. It is the sole duty of all the stakeholders in the supply chain to successfully comply with these regulations for producing safe and quality food articles (Trivedi et al., 2019).

5.3.6.3 Other Food Standards in India

Quality standards are considered as specifications to manufacture high-quality products and are either established by the government or any regulatory body developed under the government. These standards set for various food products are also known as voluntary standards because they are not mandatory as legal standards. The Bureau of Indian Standards (BIS) and the Agricultural Marketing Standards (AGMARK) are the two types of Indian quality standards. These standards are to be met for the production of food articles that are to be exported internationally. Because they are not the lowest quality standards compulsorily expected for any food, BIS and AGMARK are taken into consideration as higher standards than those of the FSSAI. BIS is established for both processed foods and non-food products. Although BIS is a voluntary standard, it has been obligatory for certain products containing the Indian Standards Institute (ISI) mark, such as food colors, additives, etc., since the year 1987. BIS is considered to be a member of both the ISO and the International Electro-Technical Commission (IEC). BIS aims to exercise regulatory procedures throughout the food supply chain, which involves production process authentication, raw material testing, testing of the final product, storage, and final food article dispensation.

On the other hand, AGMARK applies to raw agricultural products like oilseeds, eggs, spices, cereals, pulses, butter, and ghee. One of the major purposes of AGMARK is to remove adulteration from the entire food supply chain (Gandhi et al., 2020). A traceability system within the Indian food supply chain is required to emphasize food quality and safety. Export in India has been augmented throughout the years. India exports farming products to different corners of the world under the Agricultural & Processed Food Products Export Development Authority (APEDA) regulations and AGMARK. Although the country lacks a compulsory traceability system, recently the Indian government has worked with private establishments and state and central governments, which comprise FSSAI, APEDA, GS1 India, and the National Bank for Agriculture and Rural Development (NABARD) for progressing the traceability system in the food industry and food supply chain of India (Dandage et al., 2017).

5.4 CONCLUSION

The food industries have developed several food safety and quality management strategies to combat the growing safety issues in the food supply chain. Quality assurance has become the basis for the formation of food safety policies of these industries for integration of food safety and quality management systems. An efficient traceability system in the entire food supply chain ensures production, supply, and distribution of safe and high-quality products, reducing the occurrences of product recalls. Traceability is the ability to trace the movement of a food through specified stages of production, processing, and distribution. It is an effective tool for the fulfillment of legislative compliances, improved food safety and quality standards, and better connection between producers and consumers. To boost consumer confidence and fulfill safety and quality requirements, food suppliers rely on two methodologies. The first method utilizes regulatory standards and certifications to manage the food supply chain, whereas the second one uses a traceability system for the entire production and logistical operations to achieve transparency in tracking forward and backward information. Regulations and legislation in the supply chain include the CAC, HACCP, ISO 22000, European standards, and GFSI. Regulations in India include those from the FSSAI, APEDA, BIS standards, AGMARK standards, and NABARD for developing the traceability system within the Indian food industry and food supply chain.

REFERENCES

Aramyan, L., Ondersteijn, C. J., Van Kooten, O., & Lansink, A. O. (2006). Chapter 5: Performance indicators in agri-food production chains. In Ondersteijn, C. J. M., Wijnands, J. H. M., Huirine, R. B. M., & van Kooten, O. (Eds.). *Quantifying the supply chain* (pp. 47–64). Springer, Heidelberg. Printed in the Netherlands.

Attaran, M. (2012). Critical success factors and challenges of implementing RFID in supply chain management. *Journal of Supply Chain and Operations Management*, 10(1), 144–167.

Bai, L., Shi, C., Guo, Y., Du, Q., & Huang, Y. (2018). Quality risk evaluation of the food supply chain using a fuzzy comprehensive evaluation model and failure mode, effects, and criticality analysis. *Journal of Food Quality*, (1), 1–19. https://doi.org/10.1155/2018/2637075

Baines, R. (2010). Quality and safety standards in food supply chains. In *Delivering performance in food supply chains* (pp. 303–323). Woodhead Publishing, Cambridge, UK.

Beaumont, N., Mo, J. P., Gajzer, S., Fane, M., Wind, G., Snioch, T., … & Wilson, F. (2009). Process integration for paperless delivery using EPC compliance technology. *Journal of Manufacturing Technology Management*, 20(6), 866–886.

Bertolini, M., Bevilacqua, M., & Massini, R. (2006). FMECA approach to product traceability in the food industry. *Food Control*, 17(2), 137–145.

Beulens, A. J., Broens, D. F., Folstar, P., & Hofstede, G. J. (2005). Food safety and transparency in food chains and networks Relationships and challenges. *Food Control*, 16(6), 481–486.

Bureau, J. C., & Swinnen, J. (2018). EU policies and global food security. *Global Food Security*, 16, 106–115.

Crandall, P., Van Loo, E. J., O'Bryan, C. A., Mauromoustakos, A., Yiannas, F., Dyenson, N., & Berdnik, I. (2012). Companies' opinions and acceptance of global food safety initiative benchmarks after implementation. *Journal of Food Protection*, 75(9), 1660–1672.

Dandage, K., Badia-Melis, R., & Ruiz-García, L. (2017). Indian perspective in food traceability: a review. *Food Control*, 71, 217–227.

Dani, S. (2015). *Food supply chain management and logistics: from farm to fork*. Kogan Page Publishers, London.

Faergemand, J., & Jespersen, D. (2017). ISO 22000 to ensure integrity of food supply chain. *ISO Insider*, September–October, 21–24.

Food and Agriculture Organization of the United Nations (FAO). (2003a). *Assuring food safety and quality. guidelines for strengthening national food control systems*. Food and Agriculture Organization of the United Nations, Rome, Italy.

Food and Agriculture Organization of the United Nations. (2003b). *Codex alimentarius: food hygiene, basic texts*. Food and Agriculture Organization of the United Nations, Rome, Italy.

Food and Agriculture Organization of the United Nations World Health Organization (FAO). (2003). *Assuring food safety and quality: guidelines for strengthening national food control systems*. FAO Food and Nutrition Paper 76, Food and Agriculture Organization of the United Nations World Health Organization, Rome, Italy.

FSSAI. (2011a). *Contaminants, toxins and residues*. Food Safety and Standards Authority of India, New Delhi, India.

FSSAI. (2011b). *Food products standards and food additives*. Food Safety and Standards Authority of India, New Delhi, India.

FSSAI. (2011c). *Laboratory and sample analysis*. Food Safety and Standards Authority of India, New Delhi, India.

FSSAI. (2011d). *Sampling and analysis*. Food Safety and Standards Authority of India, New Delhi, India.

FSSAI. (2017a). *Food recall procedure*. Food Safety and Standards Authority of India, New Delhi, India.

FSSAI. (2017b). *Food safety and standards import regulations*. Food Safety and Standards Authority of India, New Delhi, India.

FSSAI. (2017c). *Food safety officers manual*. Food Safety and Standards Authority of India, New Delhi, India.

FSSAI. (2020a). *Food safety and standards regulations*. Food Safety and Standards Authority of India, New Delhi, India.

FSSAI. (2020b). *QCI launches recognition scheme for hygiene rating audit agencies*. Food Safety and Standards Authority of India, New Delhi, India.

Gandhi, K., Sharma, R., Gautam, P. B., & Mann, B. (2020). *Chemical quality assurance of milk and milk products*. Springer, Singapore.

Gehring, K. B., & Kirkpatrick, R. (2020). Hazard Analysis and Critical Control Points (HACCP). In *Food safety engineering* (pp. 191–204). Springer, Cham.

Golan, E. H., Krissoff, B., Kuchler, F., Calvin, L., Nelson, K. E., & Price, G. K. (2004). Traceability in the U.S. food supply: economic theory and industry studies. Agricultural Economic Report No. AER-830. U.S. Department of Agriculture, Washington, DC.

Ho, S. K. (1994). Is the ISO 9000 series for total quality management?. *International Journal of Quality & Reliability Management*, *11*(9), 74–89.

Hong, I. H., Dang, J. F., Tsai, Y. H., Liu, C. S., Lee, W. T., Wang, M. L., & Chen, P. C. (2011). An RFID application in the food supply chain: a case study of convenience stores in Taiwan. *Journal of Food Engineering*, *106*(2), 119–126.

Jaffee, S., Siegel, P., & Andrews, C. (2010). Rapid agricultural supply chain risk assessment: a conceptual framework. Agriculture and Rural Development Discussion Paper 47, 1–64. International Bank for Reconstruction and Development/World Bank, Washington, DC.

Jairath, M., & Purohit, P. (2013). Food safety regulatory compliance in India: a challenge to enhance agri-businesses. *Indian Journal of Agricultural Economics*, *68*, 431–448.

Jordan, M. I., & Mitchell, T. M. (2015). Machine learning: trends, perspectives, and prospects. *Science*, *349*(6245), 255–260.

Kamble S., Gunasekaran A., & Arha H. (2019). Understanding the blockchain technology adoption in supply chains-Indian context, *International Journal of Production Research*, *57*, 2009–2033.

Kelepouris, T., Pramatari, K., & Doukidis, G. (2007). RFID-enabled traceability in the food supply chain. *Industrial Management & Data Systems*, *107*(2), 183–200. https://doi.org/10.1108/02635570710723804

Kohli, C., & Garg, S. (2015). Food safety in India: an unfinished agenda. *MAMC Journal of Medical Sciences*, *1*(3), 131.

Kshetri, N. (2018). Blockchain's roles in meeting key supply chain management objectives. *International Journal of Information Management*, *39*, 80–89.

Lokunarangodage, C. V. K., Wickramasinghe, I., & Ranaweera, K. K. D. S. (2015). Review of ISO 22000: 2005, Structural synchronization and ability to deliver food safety with suggestions for improvements. *Journal of Tea Science Research*, *5*(12), 1–12.

Luger, G. F. (2005). *Artificial intelligence: structures and strategies for complex problem solving.* Pearson Education, Boston, MA.

MOFPI. (2020). HACCP/ISO Standards/Food Safety/Quality Management Systems. Ministry of Food Processing Industries, New Delhi, India.

Mohri, M., Rostamizadeh, A., & Talwalkar, A. (2018). *Foundations of machine learning.* MIT Press, Cambridge, MA.

Morris, C., & Young, C. (2000). 'Seed to shelf', 'teat to table', 'barley to beer' and 'womb to tomb': discourses of food quality and quality assurance schemes in the UK. *Journal of rural studies, 16*(1), 103–115.

Motarjemi, Y., & Lelieveld, H. (Eds.). (2013). *Food safety management: a practical guide for the food industry.* Academic Press, Amsterdam.

Nakandala, D., & Lau, H. C. (2012). A novel approach to determining change of caloric intake requirement based on fuzzy logic methodology. *Knowledge-Based Systems, 36,* 51–58.

Nakandala, D., Lau, H., & Zhao, L. (2017). Development of a hybrid fresh food supply chain risk assessment model. *International Journal of Production Research, 55*(14), 4180–4195.

Nambiar, A. N. (2009). RFID technology: a review of its applications. Proceedings of the World Congress on Engineering and Computer Science 2009, WCECS 2009, vol. 2 (pp. 1–7). San Francisco, CA.

Olsen, P., & Borit, M. (2013). How to define traceability. *Trends In Food Science & Technology, 29*(2), 142–150.

Panda, S. K., & Priya, E. R. (2019). Overview of ISO 22000: 2018 FSMS. In Priya, E. R., Uchoi, D., Kumar, A., & K. Rejula (Eds.). *ISO-22000/HACCP for fish processing establishments* (pp. 40–48). Central Institute of Fisheries Technology, Cochin, India.

Panghal A., Chhikara N., Sindhu N., & Jaglan S. (2018). Role of food safety management systems in safe food production: a review, *Journal of Food Safety, 38,* e12464.

Peri, C. (2006). The universe of food quality. *Food Quality and Preference, 17*(1–2), 3–8.

Pinto, D. B., Castro, I., & Vicente, A. A. (2006). The use of TIC's as a managing tool for traceability in the food industry. *Food Research International, 39*(7), 772–781.

Reddy, A., Cadman, T., Jain, A., & Vajrala Sneha, A. (2017). *Food safety and standards in India.* ICAR-Indian Agricultural Research Institute, New Delhi, India.

Regattieri, A., Gamberi, M., & Manzini, R. (2007). Traceability of food products: general framework and experimental evidence. *Journal of Food Engineering, 81*(2), 347–356.

Reyes, P. M., & Frazier, G. V. (2007). Radio frequency identification: past, present and future business applications. *International Journal of Integrated Supply Management, 3*(2), 125–134.

Russell, S., & Norvig, P. (1995). Learning in neural and belief networks. In *Artificial intelligence: a modern approach.* Prentice–Hall, Upper Saddle River, NJ.

Sansawat, S., & Muliyil, V. (2011). Comparing Global Food Safety Initiative (GFSI) Recognised Standards. SGS, Geneva, Switzerland.

Shukla, S., Shankar, R., & Singh, S. P. (2014). Food safety regulatory model in India. *Food Control, 37,* 401–413.

Sufiyan, M., Haleem, A., Khan, S., & Khan, M. I. (2019). Evaluating food supply chain performance using hybrid fuzzy MCDM technique. *Sustainable Production and Consumption, 20,* 40–57.

Tapscott, D., & Tapscott, A. (2017). How blockchain will change organizations. *MIT Sloan Management Review, 58*(2), 10.

Traore, B. B., Kamsu-Foguem, B., & Tangara, F. (2017). Data mining techniques on satellite images for discovery of risk areas. *Expert Systems with Applications, 72,* 443–456.

Trienekens, J., & Zuurbier, P. (2008). Quality and safety standards in the food industry, developments and challenges. *International Journal of Production Economics, 113*(1), 107–122.

Trienekens, J. H., & Beulens, A. J, M. (2001). The implications of EU food safety legislation and consumer demands on supply chain information systems. 11th Annual World Food and Agribusiness Forum, Sydney, Australia.

Trivedi, S., Negi, S., & Anand, N. (2019). Role of food safety and quality in Indian food supply chain. *International Journal of Logistics Economics and Globalisation, 8*(1), 25–45.

Underwood, S. (2016). Blockchain beyond bitcoin. *Communications of the ACM, 59*(11), 15–17.

Van Reeuwijk, L. P., & Houba, V. J. G. (1998). *Guidelines for quality management in soil and plant laboratories (FAO Soils Bulletin-74).* Food and Agriculture Organization of the United Nations, Rome, Italy.

Vasanthi, S., & Bhat, R. V. (2018). Management of food safety risks in India. *Proceedings of the Indian National Academy of Science, 84,* 937–943.

Wamba, S. F., Lefebvre, L. A., Bendavid, Y., & Lefebvre, É. (2008). Exploring the impact of RFID technology and the EPC network on mobile B2B eCommerce: a case study in the retail industry. *International Journal of Production Economics*, *112*(2), 614–629.

Wang, X., Li, D., & O'Brien, C. (2009). Optimisation of traceability and operations planning: an integrated model for perishable food production. *International Journal of Production Research*, *47*(11), 2865–2886.

Wang, X., Li, D., & Shi, X. (2012). A fuzzy model for aggregative food safety risk assessment in food supply chains. *Production Planning & Control*, *23*(5), 377–395.

Weis S. A., Sarma S. E., Rivest R. L., & Engels D. W. (2004) Security and privacy aspects of low-cost radio frequency identification systems. In *Security in pervasive computing* (pp 201–212). Springer, Verlag Berlin Heidelberg.

World Health Organization. (2003) Diet, nutrition and the prevention of chronic diseases. World Health Organ Tech Rep Ser 916, World Health Organization, Geneva, Switzerland.

Yong-Dong, S., Yuan-Yuan, P., & Wei-Min, L. (2009). The RFID application in logistics and supply chain management. *Research Journal of Applied Sciences*, *4*(1), 57–61.

Zadeh, L. A., Klir, G. J., & Yuan, B. (1996). *Fuzzy sets, fuzzy logic, and fuzzy systems: selected papers* (pp. 394–432: Vol. 6). World Scientific, Singapore.

Zhu, X., & Goldberg, A. B. (2009). *Introduction to semi-supervised learning. In Synthesis Lectures on Artificial Intelligence and Machine Learning*, Morgan & Claypool Publishers, San Rafael, California.

CHAPTER 6

Food Product Development
Science, Shelf Life, and Quality

Michael V. Joseph and Chetan Sharma

CONTENTS

6.1 INTRODUCTION

To continue with growth and profitability, food companies continuously strive to create new products. Product development in the food industry is by no means different to similar developments in other fields, in it being a continuum. Continuous efforts are made to incorporate elements of change into an existing product or a new product to meet the demands of the changing market and consumer needs. It is a way to ensure that the company continues to grow and meet the expectations of the consumer as well as to enter newer markets in the same or different geographies. The imperativeness of product development is well understood; however, another critical aspect that cannot be overlooked is the risk involved. Modifications in existing products or new products may be rejected by the market for various reasons like not meeting consumer expectations, lack of promotion and product uniqueness, the product is so ahead of its time that consumers cannot appreciate the changes, etc.

New product development can be defined as the process of transforming a new market opportunity into a commercial product through a sequence of activities to achieve specific targets (Krishnan and Ulrich, 2001; Rudder et al., 2001; Ulrich and Eppinger, 2016). Developing new products requires an input of expensive resources such as substantial research, availability of personnel with appropriate skills, suitable physical facilities, and capital. With limited money available, product

Email: Michael Joseph, mvjoseph@ncsu.edu

DOI: 10.1201/9781003091677-6

development has to deal with internal competition along with the uncertainty of succeeding with a new product when introduced in the market.

What constitutes a new product? Would modifications in the existing product(s) make it to a new product category or would it be an improved product? For example, a comparison between corn flakes and frosted flakes immediately shows a major difference between the products: frosted flakes are corn flakes coated with sugar (minor ingredients like vitamin profile notwithstanding). A comparison between Kellogg's® Corn Flakes and Kellogg's Special K would reveal that both are made from different raw ingredients. Another example would be comparing different types of pasta like macaroni and fusilli or spaghetti and angel hair. In these pasta products, the raw materials used may be the same but different dies are used to create different shapes. Thus, changes in raw materials or the process can create a new product.

Another type of product that can be seen in the market is labeled "new and improved" or "improved." Are they new products? Yes, they are, and many times they have the same brand name. The word improved means "to better the quality," and without going into the definition of quality here, improvements can be achieved through additions or modifications of the existing product, process, product line, perception, etc. Would all of these be considered new products, or do only some types of changes qualify as creating new product? Thus, in an absolute sense, new products are those that have never been introduced in the market. These situations are rare as the product may already exist, but it may be unique for a specific company. However, what constitutes a new product in practical terms is discussed later in the chapter.

6.2 WHY PRODUCT DEVELOPMENT

The demand for product development arises because a business needs to grow and profit, improve the current product(s), and meet ever-changing consumer demands. With the changing landscape of taste, technology, and competition, manufacturers need to be proactive, as they cannot solely rely on existing products (Kotler and Armstrong, 1991). Companies understand these needs and develop products that are of value to customers and, in turn, increase their market share, revenues, and profitability. Thus, product development is a mutually beneficial activity for both the producer and consumer. The process, however, is expensive and fraught with risk and often fails. The extent of innovation and risk involved can be gauged by the fact that about 15,000 new food products are introduced each year and the failure rate can be as high as 90% (Aramouni and Deschenes, 2018). However, the risks do not act as a deterrent for many food processors as the motivation for them is the potential profits to be made from marketing and sales of successful products (Rudder et al., 2001). Graf and Saguy (1991) have reported that a great number of North American companies spend most of their internal research and development (R&D) money on product development and some on applied research in an approach commonly called "cook and look." Innovation and new product development are essential components in the company's strategy to gain market share and be successful. However, even with so much effort, many of the newly launched products end as failures (Buisson, 1993; Urban and Hauser, 1993; Hollingsworth, 1994; Rudolph, 1995). This shows that there is something inherently wrong in placing new products from labs to market shelves.

Overall, there are several driving factors that steer product development activities. It begins by understanding that all products have life cycles (introduction, growth, maturity, and decline). As the product navigates through its life cycle and sees a period of decline or nearing the end of its life cycle, it must be refreshed in some way to regain consumer attention, if the manufacturer wants to continue reaping profits. Introduction of successful new products also helps to achieve growth for the company. Demand for new types of foods (may include the need for fresh fruits and vegetables, sustainability and traceability, reduction in meat consumption, improvement in animal welfare, or need for organic products/produce) and introduction of new technologies also direct the company toward exploring new products. Changes in regulations, food legislation, agricultural policies, and other government rules also create the opportunity for new food product development. One or all of these factors would drive companies to pursue product development activities based on their capability and resources.

Product development is an activity that cannot be performed in isolation. It is a group activity that involves the active role of management and teams from marketing, operations, and finance.

Generally, companies have a business strategy, which is built with the vision of taking the company into the future. A section of this strategy is dedicated to innovation strategy, which focuses on innovation as a driver for creating value for potential customers and for itself. Within the innovation strategy is the product development strategy, which aims to mitigate the risk while developing a product concept with the idea of improving the fit of the product with market needs. This helps the company gain a competitive advantage by introducing products at the right time and in the right markets resulting in improved profits or revenues. Finally, the confluence of product development strategy and new product portfolio leads to a consolidated product development program (Earle et al., 2001). When designed in this way, the program fits in very well within the overall business strategy of the company.

For product development to be successfully pursued, all the relevant entities within the company must participate to create good strategies, clarify objectives and priorities, and help focus efforts around them (Pisano, 2015).

Before pursuing the path of new food product development, Aramouni and Deschenes (2018) have suggested doing a screening to get a clear picture of whether to proceed to the next steps (Figure 6.1). Their screening questions are as follows:

1. Who will use the product?
2. How will it be used?
3. What preparation is necessary?
4. How will the consumer benefit from it?

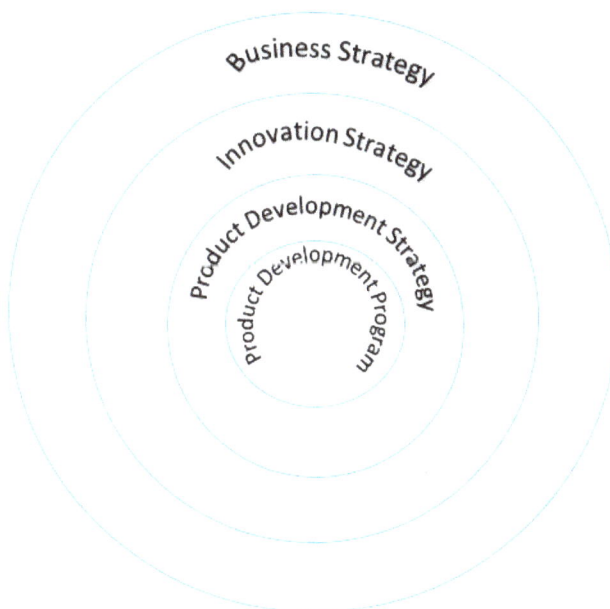

Figure 6.1 Schematic of the product development program.

5. Are there any other uses of the product?
6. What is the competition in the market and in what pack sizes and prices?
7. How is the product distinct from the competition?
8. Where will the product be available?
9. How will consumers know about the product?
10. At what price will the product(s) be offered?

Once there are answers and clarity for these questions, one can proceed to the next stages of feasibility (technology, regulations, and capital), test marketing (trial production run, promotion, and marketing while keeping the costs low), and finally commercialization (identify the place of manufacture, distribution, and other logistics).

6.3 TYPES OF PRODUCT DEVELOPMENT

According to Cooper (1993) and Crawford (1994), product development can be categorized into the following groups:

1. *New-to-the-world:* These products comprise innovations that have never existed and are different from pre-existing products and have never been in the market. These types of products are less common than other types of newly developed products. Usually, they are a result of the application of newfound ingredients and/or technological advancements, e.g., use of traditional ingredients like ancient grains, new protein derivatives (pea proteins, legume proteins) from plants and animals (insect proteins), and genetically modified (GM) crops, and use of new technologies like extrusion and biotechnology (for GM crops). Some successful products are the probiotic drink Yakult®; Pringles®, the saddle-shaped stackable potato-based chips; and Beyond Meat®, which produces plant-based meat substitutes.

2. *New product lines:* These are existing products but are new for the company. It occurs when the company diversifies its product range into an area in which it has not previously operated. This approach can be called "me-too," which can lead to the issue of product imitation. An example would be if a beverage manufacturing company forays into new categories like snack foods, baked products, or frozen foods. These types of products already exist in the market but are new for the company.

3. *Product line extensions:* This category appears when the company introduces additional products that are similar to the existing product range. Generally, this approach is taken when the company identifies a new market segment, sees a decline in sales, sees new consumer demand, etc. The changes may not be drastic and may include some changes like the introduction of new flavors, package size, package features, and health-related variants. For example, beverage manufacturer Coca-Cola® has several lines of its popular soda drink Coke with variants like Coca-Cola Classic, Coca-Cola Zero, Coca-Cola Life, and Coca-Cola Light.

4. *Product improvements:* These are simply current products made better. These changes may be minor or significant depending on the need and may generally involve changing one or two of the product features. Corn soy blend (CSB), a fortified blended food (FBF), has long been used by many international and multilateral agencies like U.S. Agency for International Development (USAID) and World Food Programme (WFP) to combat undernutrition in developing countries. As CSB was ineffective in addressing moderate acute malnutrition, a newer variant of CSB was developed called CSB+, which had a better micronutrient profile. Later another variant, CSB++, was designed with a better micronutrient profile, better protein quality, reduced fiber, higher fat content, and tighter microbiological specifications.

5. *Product repositioning:* This involves either substantially modifying the image/application of an existing product or assigning it to a new market segment. This approach is not very common and is utilized when the product performance is poor or has stalled, even though the product has potential benefits for the consumer. While repositioning, the product package and appearance may be

changed to give it a "new look." This helps in refreshing the customers' perception of the brand and almost gives a new start to the brand in the market. An example would be oat-based products positioned as a product that helps reduce cholesterol.

6. *Cost reduction:* Companies replace existing products with new products and provide a similar performance level to the customer at a lower cost.

All of these groups are different forms of product development, but the highest risk for the company is for new-to-the-world, and new product lines as these are considered uncharted territory for the company.

6.4 PRODUCT DEVELOPMENT STAGES

New product development, from conceptualization to marketing, has to follow a cycle for it to reach its intended users. (Is it being successfully launched and marketed?) Earle et al. (2001) divided the product development process into four stages:

Stage 1—Product strategy development: This constitutes the initial phase where the product development program is integrated into the company's business strategy combined with a comprehensible explanation of the market and consumers and identifying market and consumer needs.

Stage 2—Product design and process development: This stage encompasses the use of new techniques in addition to quantitative design specifications, feasibility analysis, and multidiscipline integration.

Stage 3—Product commercialization: Ideas that were generated, designed, tested, and validated shape into a viable product in this stage and have the potential to obtain consumer acceptance in the path toward mainstream adoption. The product would be ready to generate a financial return on the investment. However, some precursors to commercialization are multifunctional integration, planning, and scheduling (commercialization blueprint). The blueprint would consist of a mix of strategic and tactical concepts, which include timing, targeting, positioning, partnerships and alliances, pricing, communication, product, and distribution. Finally, a business analysis of the product based on the acceptance of the product in the designated market(s) completes the loop.

Stage 4—Product launch and evaluation: This is achieved when stages 1–3 have been completed, and the company is ready to launch the newly developed product. During this phase, the company wants to maximize the exposure of their business and the product. The when, how, and where questions get answered with respect to the product being made available to consumers. Some of the elements considered during this stage for maximizing exposure includes the design of logos, visuals, developing advertising and marketing materials, identifying optimal marketing and distribution channels, and evaluation of the outcomes.

Many other researchers have described the process of product development in multiple steps and are different from each other in terms of the number of steps involved. Kotler and Armstrong (1991) divided the new product development process into eight major steps: (1) idea generation, (2) idea screening, (3) concept development and testing, (4) marketing strategy development, (5) business analysis, (6) product development, (7) test marketing, and (8) commercialization. Urban and Hauser (1993) identified a five-step decision process that could be applied to new product development activities: (1) opportunity identification, (2) design, (3) testing, (4) introduction, and (5) life cycle management. Graf and Saguy (1991) divided a project into five phases: (1) screening, (2) feasibility, (3) development, (4) commercialization, and (5) maintenance. Though the names of stages and activities may differ between companies, they all essentially undertake the same steps in some form or another. While undertaking this process, it is essential to involve people with different skills and knowledge from other departments. The key factor, however, is the need for continuous assessment through feasibility studies, business analysis, and post-launch studies while the project is being worked on and implemented later.

6.5 ROLE OF SCIENCE

New product development is not all creativity; instead, it is creativity backed by social and scientific knowledge (creativity will be discussed later in the chapter) that forms a basis for product development. It is a combined investigation of a blend of natural sciences, social sciences, dispensation, and consumer sciences for creating a new food product (Sharif et al., 2018). These different constituents must seamlessly blend to come together harmoniously to successfully create a new product. The first step is to understand the consumer insight (discussed in detail in the Concept of Quality section), followed by product innovation ideas. A cross-functional team is put in place comprising experts from culinary, nutrition, regulatory, marketing, product development, packaging, engineering, process safety, and others to create and test product ideas. This team is essential to develop the formulation, process, and guardrails related to ingredient specifications, nutrition, and packaging claims. While the product development process is underway, it may be possible *not* to meet some of the aspects of the target characteristics or features due to lack of individual skill within a company or due to limitations in technology. While working on a new product, it is critical never to lose focus on the needs of the consumer. The major factors that affect new food product development are discussed in the following sections.

6.5.1 Generation of New Ideas

Generating ideas, let alone useful ones, can sometimes feel challenging and overwhelming for several reasons. A kind of "tunnel vision" can develop in people who have been working on a new product idea for a long time. Due to their long experience, they consider themselves experts in that category and can sometimes stifle their creativity. Such a situation may improve a product's performance or optimize its technology relatively quickly or more effortlessly than evaluating the product's relevance to the user. Nonetheless, when a company enters a new product category, the team may face different challenges in ideation. The team may have limited exposure to the product category they are entering into, and they may not be familiar with users of that product. This lack of familiarity may restrict their capability to develop an innovative and competitive product relevant and purposeful to the consumer. Sometimes, the proposition of just coming up with new ideas may be challenging because it may be challenging to be creative. This "chokehold" on ideas can occur even under the backdrop of having substantial research and findings on the target consumers because the transition from research findings to meaningful concept generation is indeed challenging. In such situations, often the solutions offered from concept generation may lack significance back into the user research (L. Weigel, 2015, Chapter 5).

    Creativity, a key in product development, has been defined in various ways, especially considering its deep roots in culture (Shao, 2019). Nevertheless, creativity is a multifaceted and complex construct, but Leslie Skarra of Merlin Development Inc. proposed creativity in an equation form to fill the linguistic gap between scientists and so-called product developers,

$$Creativity = \frac{knowledge^{(1+imagination)}}{evaluation} \tag{6.1}$$

Based on Leslie's equation, creativity does not come from *ex nihilo*, but it starts from a base of knowledge, and one should certainly have knowledge of the domain to be creative in that domain. Creativity facilitator's view knowledge more broadly than those directly related to the problem at hand because they use knowledge as a source of stimulating analogies to find solutions. Creativity flows from the knowledge raised by the power of imagination, and since imagination does not ever diminish knowledge, "1+" in the exponent was used. In case if there is no imagination, one still can have knowledge but no enhancement. Because imagination is an exponent in Equation (6.1), creativity significantly increases as imagination increases. Certain cultural elements noted to enhance creativity are no fear, encouragement, facilitation, and collaboration. Evaluation, as a necessary part of the creativity equation, focuses on the possibilities germane to the problem. Simultaneously, evaluation as a denominator shows the impact of negative evaluation. Good creativity facilitators always separate the evaluation from creative activities, and Equation (6.1) nicely explains why this is appropriate.

Interestingly, a large amount of negative evaluation was found to not be as devastating as a tiny amount. If the evaluation is blatantly negative, it is quickly discarded for its obvious hostile intent. However, if the evaluation is negative but not blatant, it has a significant negative effect and no easy means of negation. So, the question remains: How can one know if the evaluation is positive or negative? Skarra's idea is to recognize the direction of the evaluation by its effect on the organization (Skarra, 2006). If an organization views evaluation negatively or is delivered negatively, it will eventually diminish creativity. Creativity is an essential tool, but simultaneously, it can be a boon or bane since it cultivates at the cost of experience.

In the ongoing state of affairs, innovation is a process that produces the desired ultimate business result, and creativity is only one factor in the innovation equation, as per Equation (6.2). Despite assuming that the job is done once creativity occurs, product developers should understand that creative ideas produce nothing tangible and are, on their own, inadequate to spur innovation. Once we have a creative idea, Equation (6.2) guides us to the realization. Action drives innovation, and by taking positive actions, one can convert creative ideas into real innovations. More extensive positive actions help more than small positive actions. Emotional intelligence, also known as emotional quotient or EQ, is defined as "the capacity for recognizing, understanding, using and managing our feelings and those of others, for motivating ourselves and overcoming challenges and conflicts." Skarra echoed that a strong EQ goes a long way toward helping innovations become a reality. Persistence is the characteristic that boosts the base emotional intelligence skill already present. Assuming zero persistence, which may not be possible in the popular case, one will still have the base of emotional intelligence, reflected by 1+ in the equation. Like action, a large amount of persistence may help a great deal but persistence alone can be problematic and the addition of EQ in the equation recognizes this fact. It is possible to persistently move creative ideas forward by counterbalancing a suboptimal evaluation process using good emotional intelligence. Finally, Equation (6.2) shows that the product of action, persistence, and EQ can be compounded to increase innovation. This equation also highlights the importance of other factors in addition to creativity.

$$Innovation = creativity \times action \times emotional\ intelligence \times (1 + persistence) \qquad (6.2)$$

6.5.2 Shelf Life

One of the critical features of a newly developed product is to effectively communicate the shelf life of the product. Various ingredients in a product, when subjected to process conditions, lead to alterations in the physicochemical properties due to changes occurring at physical and molecular levels. That may alter the path of spoilage and thus affect the product's shelf life. Whatever the mix of ingredients and the process adopted to manufacture it, the quality, safety, and consumer trust are greatly affected by shelf life. The aim of shelf life is to help consumers make safe and informed use of foods, and here it will be useful to define shelf life. Many definitions have been proposed by governments and different organizations; the Institute of Food Technologists (1974) defines it as "the period between manufacturer and retail purchase of a food product during which the product is of satisfactory quality." The European Food Information Council (2013) defines shelf life as "length of time a food can be kept under stated conditions while maintaining its optimum safety and quality." Whereas the Institute of Food Science and Technology (1993) defines it as "the time during which the product will a) remain safe, b) be certain to retain the desired physical, chemical, microbiological and sensory characteristics, and c) comply with any label declaration of nutrition data, when stored under recommended conditions." Words like "satisfactory quality," "desired…characteristics," "optimum safety," and "stored under recommended conditions" may often be ambiguous when used in the practical sense.

Both intrinsic and extrinsic factors influence the overall shelf life of a food product. The initial nature of the food (fresh or dry), quality of ingredients used (high quality, low microbial load), and product formulation (presence or absence of preservatives or antioxidants) constitute the intrinsic factors. Extrinsic factors include process conditions, barrier and other properties of packaging material, transportation and storage conditions, and consumer handling (Zweep, 2015).

During shelf life testing, chemical, microbial, and sensorial attributes should be tested. Readers are encouraged to review Chapter 3, "Sensory Methods for Shelf Life Assessment of Foods." After the test attributes are finalized, analytical methods should be identified. This should be followed by selecting the storage conditions (temperature, relative humidity, and lighting) to represent either optimal, typical, or worst-case conditions. A target end point and testing frequency should be determined. Real-time lengthy studies or accelerated shelf life studies can be performed based on the needs of the testing. An accelerated shelf life study should be performed if there is a relationship between storage behavior in normal conditions and accelerated conditions. The accelerated shelf life studies predict the deterioration in a product over a period of time, but they cannot replace real-time shelf life testing. However, accelerated studies can be used as a useful tool when decisions have to be made within time limitations for product launch. Thus, careful consideration of the previously mentioned factors (primarily experimental design and test parameters) would help ensure a safe, quality food product that meets customers' expectations.

Once the product has been tested for its shelf life, it is time for an important component, labeling the dates on the package. There is little government regulation when it comes to expiration dates stamped on packages (IFT, 2016). According to the United States Department of Agriculture (USDA) Food Safety and Inspection Service (FSIS), a few different types of dates include the following:

- *Sell-by:* This guides stores on how long they can display a product before its quality starts to decline. Although the item may still be edible after this date, it may not be of the highest quality. It is not a safe date.
- *Best if used by/best before:* This does not refer to purchase or safety date; instead, it is a guide to the quality or flavor profile of the food.
- *Use-by:* Using a product after this date is not recommended as the food may lose nutrients, or potentially develop harmful bacteria, which can lower the quality. It is not a safety date except for when used in infant formula.
- *Freeze-by:* This indicates when a product should be frozen to maintain peak quality. It is not a purchase or safety date.

Food manufacturers must display dates such that the product can be useful to the maximum extent possible. This is because the USDA estimates that 30% of the food supply is lost or wasted at the retail and consumer levels (USDA-ERS, 2020). One source of food waste is the confusion among retailers and consumers about the meaning of dates displayed on the label. Therefore, FSIS recommends that food manufacturers and retailers use a "best if used by" date. Research has shown that this phrase conveys to the consumers that the product quality will be of the best quality if used by the calendar date shown.

6.5.3 Concept of Quality

Just like creativity, "quality" incredibly has many definitions as there are specialized interests in food (Booth, 1995; Grunert, 1995; Pierson et al., 1995). Harry Lawless, a renowned sensory scientist, in fact, stated that no single definition of quality is entirely satisfactory (Lawless, 1995). The importance of this topic can be apprehended from the fact that the *Food Quality and Preference* journal devoted an entire issue to understand the meaning of quality under the title of "The Definition and Measurement of Quality" in 1995. To date, this topic is as relevant as it was in the recent past and probably will remain so until humans stop comparing things. From a product development standpoint, the notion of quality perceived by its users should be recognized first to implement it in the production for better product fortune.

First and foremost, the dialogue of quality as *perceived quality* means a perception process that may have different content for various persons, products, and places (Ophuis and van Trijp, 1995). Indeed, it is not enough that a product has quality features; instead, they must be perceived by the consumers (Silvestri and La Sala, 2018), such as perceived safety or perceived nutrition (Cardello 1995). Perception, as a mediator, plays an essential role between the quality offered and the quality demanded by consumers. Quality has been long recognized as an unmeasurable philosophical phenomenon, identified as transcendent (Garvin, 1984) or metaphysical (Steenkamp and van Trijp, 1989), often a synonym with *innate excellence* (Ophuis and van Trijp, 1995), and described metaphorically, such as *mercy* by Lawless. Against this backdrop *perceived quality* lies in-between the transcendent and objective quality continuum (Garvin, 1984; Zeithaml, 1988). This assumption opens the scope of *measurement*, which appreciates into dimensions of quality measurement (Garvin, 1984; Grunert, 1995; Lawless, 1995). Of all dimensions, the objective and subjective dimensions of measurements have been debated extensively and will be touched on briefly here for relevance.

Objective quality refers to the physical characteristics built into the product and is typically dealt with by engineers and food technologists. On the other hand, subjective quality is the quality perceived by consumers (Grunert, 2005). On the conceptual front, few scholars (Lawless, 1995; Lancaster, 1971; and Molnar, 1995, through his proposed model, and not by his definition of food quality) echoed the importance of the objective dimension of quality, relating to physical and chemical characteristics, whereas others (Cardello, 1995) emphasized the subjective dimension. Instead of visualizing the two schools of thought in separate spaces, it is a relationship between them that has economic importance, and which makes the concept of quality quantifiable by using the principles and approaches derived through science. It is the same relationship that has been widely adopted by both practitioners and academic marketers and is of importance to product developers. It is essential to understand that the relationship between these two dimensions provides an economic value to the quality and thereby product development (Silvestri and La Sala, 2018). Klaus Grunert (2005) stated that both objective and subjective quality dimensions can remain at the core of economic importance only when producers can translate consumer wishes and only when consumers can infer desired qualities from the way the product has been built (Figure 6.2). A necessary provision of *abstractness* and *concreteness* observed in Grunert's assertion connects formulators or product developers (abstract → concrete) to the consumers (concrete → abstract). Steenkamp and van Trijp (1989) developed the concept of perceived quality as an attempt to mediate between the product characteristics and consumer preferences.

According to Ophuis and van Trijp (1995), perceived quality is generally considered as an overall, global concept, like an attitude or *person* factor. Another definition provided by Aaker (1991) is "the customer's perception of the overall quality or superiority of a product or service with respect to its intended purpose, relative to alternatives." A crucial element in discussing perceived quality from a product development standpoint is the concept of quality indicators. Quality is considered as a multifaceted concept that is based on several dimensions that cannot all be evaluated by a

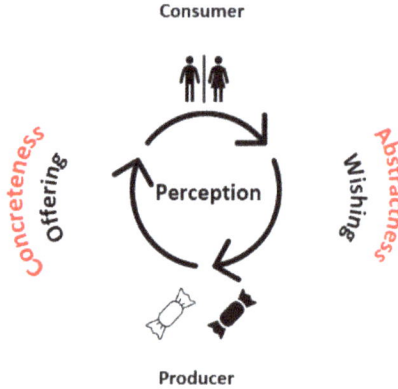

Figure 6.2 Quality circle.

consumer. Therefore, consumers use surrogate or indirect indicators of quality, in the form of *quality cues* and *quality attributes*, both of which can be further classified into subcategories, to make a judgment of perceived product quality. This type of approach has been termed the "multidimensional approach," where the quality depends on the combination of attributes (Silvestri and La Sala, 2018). These attributes can and de facto were categorized into different other major categories, such as *search*, *experience*, and *credence* (Nelson, 1970, 1974; Darby and Karni, 1973). Likewise, the multi-attribute attitude model classifies attributes into extrinsic and intrinsic cues (Olson and Jacoby, 1972). Steenkamp and van Trijp (1989) extended the work of Olson by developing a more complex model in which a distinction is made between cues and attributes. Garvin (1984) offered one of the most known multidimensional models of quality, consisting of eight dimensions of quality. In contrast to the multidimensional approach, the "hierarchical approach" (Zeithaml model) does not focus on the attributes but on the association between these attributes. Readers are suggested to refer otherwise for in-depth insights on the previously mentioned approaches (Silvestri and La Sala, 2018; Garvin, 1984; Zeithaml, 1988).

Garvin (1984) defined quality from five approaches: (1) the *transcendent* approach, (2) the *product*-based approach, (3) the *user*-based approach, (4) the *manufacturing*-based approach, and (5) the *value*-based approach. Quality is *innate excellence* in the transcendental approach, irrespective of geographical limitations, and is an absolute property in a platonic sense. Like other such terms that philosophers consider to be "logically primitive," beauty and perhaps quality as well, can be understood only through experience. Under the *product*-based approach, quality is a precise and measurable variable in the form of some ingredients or attributes. This approach offers a vertical or hierarchical dimension to quality, and goods can be ranked according to the amount of the desired features they possess. Building on the economic aspect of research, this approach emphasized first that the higher quality can be obtained only at a higher cost because quality reflects the quantity of attributes, such as the amount of wool in a suit. Secondly, that quality is an inherent characteristic that is measurable in an objective manner. Aligning with this approach, Joseph (2016) studied the physicochemical and nutritional attributes of a newly developed extruded FBF.

The structure and function testing of food can be categorized as:

1. *The functionality of physical attributes:* These would include tests for particle size and its distribution, moisture content, bulk density, flowability, and more.
2. *Product quality analysis:* This includes tests like bulk density, Brix, acidity, moisture content, water activity, microbial status, texture, viscosity, anti-nutritional factors, allergen tests, and so on.
3. *Nutritional values:* These help in maintaining the quality of the product by meeting label claims and informing consumers about the nutrient status of the food.

The *user*-based approach contrasts with the *product*-based approach by premising quality "in the eyes of the beholder." An idea of subjective quality poses a practical problem of aggregation, i.e., how to aggregate widely varying individual preferences to lead to a meaningful definition of quality at the market level. This problem is solved assuming that the high-quality products are those able to meet the needs of most consumers, even if in this way it ignores the relative weight that each associates with each product attribute (Silvestri and La Sala, 2018). This approach also differentiates quality from consumer satisfaction by echoing that a product with maximizing satisfaction would undoubtedly be preferable to one that meets fewer consumer needs, but it does not mean it is better.

The *manufacturing*-based approach is typical of those involved on the supply side and defines quality as "conformance to requirements." Accordingly, any deviation implies a reduction in quality once a specification has been established. The focus of this approach is internal and cost reduction and can include process variability studies. These are tested to measure process fluctuations like flow rate, temperature, and pressure profiles to produce a consistent quality product and ensure cost-efficiency.

Finally, the *value*-based approach defines quality in terms of costs and prices. In this approach, a quality product is one that provides performance at an acceptable cost. All approaches are interrelated to some extent and these interrelationships are by no means straightforward and easy. After unfolding the five approaches Garvin (1984) proposed a theoretical framework that includes eight dimensions of quality (Table 6.1).

Each of these eight, or nine in the case of Plsek (1987), dimensions are important in defining the quality, but only some of these are emphasized in approaches, such as the *product*-based approach emphasizes *performance*, *features,* and *durability.* The central proposition of this multi-dimensional approach is that consumers perceive *value* in a product to the extent that they believe that the consumption of this product will lead to *self-relevant consequences*, which is discussed in detail in the forthcoming section. Most definitions of existing quality may fall into one or more of the approaches described by Garvin, such as Joseph M. Juran, a specialist of the quality management system, who stated that the quality of a manufactured product should be defined in pragmatic and functional terms, related to "aptitude at use" (Juran, 1983; Daget, 1987). This definition seems to lean towards a *user*-based approach instead of a *transcendent*-based approach. The existence of different approaches is useful to understand the concept of quality. The quality should be first identified through market research, i.e., *user*-based approach, then converted into product attributes, i.e., through *product*-based approach, and finally the process of production should be organized to conformance to specifications, i.e., through the *manufacturing*-based approach. Instead of focusing on quality, this should be done to create value for the whole economic system, i.e., the *value*-based approach.

Table 6.1   Dimensions of Quality

1	Performance
2	Features
3	Reliability
4	Conformance
5	Durability
6	Serviceability
7	Aesthetics
8	Perceived quality

Source: Garvin (1984).

A crucial element in discussing perceived quality is the concept of quality indicators, and the work of Steenkamp and van Trijp (1989) deserves to be discussed here. Quality cues are concrete product characteristics that can be observed by a consumer without actual consumption, whereas quality attributes are abstract product benefits that can only be experienced because of the consumption of the product. Quality cues are categorized as intrinsic or extrinsic. Intrinsic quality cues cannot be changed or experimentally manipulated unless changing the physical characteristics of the product, for example, appearance, color, shape, size, structure, etc. In contrast, extrinsic quality cues can be manipulated without the need to modify the physical characteristics of the product, such as price, brand name, and country of origin, just to name a few; these are probably the best-known extrinsic indicators of quality. Likewise, quality attributes are categorized as experience and credence quality attributes. Quality attributes are the functional and psychosocial benefits provided by the product. Within the context of foods, taste, convenience, or freshness are some of the most essential experience quality attributes. Credence quality attributes include healthfulness, wholesomeness, naturalness, ethicalness, etc., of food products, and these attributes are gaining immense importance these days (Sharma, 2019). Personal values will affect the degree to which one attaches importance to specific credence quality attributes. Together, these surrogates may eventually lead to *self-relevant consequences* (Grunert, 1995). Once the notion of perceived quality has been recognized as an essential concept, the next question is how to implement this in product development, production, and subsequent marketing. Hereafter, two methods will be discussed to comprehend the physical characteristics of perceived quality via the quality guidance model of Steenkamp and van Trijp (1989) and the means-end chain (MEC) model of Grunert (1995).

A *quality guidance* model was suggested to relate *perceived quality* judgments with physical product characteristics (Steenkamp and van Trijp, 1989; Ophuis and van Trijp, 1995). Quality guidance is a consumer-based philosophy, the concept in which consumer perceptions are taken as the point to start rather than end, and later these perceptions are made meaningful to product developers. The goal of quality guidance is the formulation of technical product specifications that are related to consumers' quality perceptions. According to this approach, identify the perceived quality judgments first, disentangle them into constituents (quality cues and attributes), and later translate them into physical product characteristics. A lot about *perceived quality* has already been discussed in the preceding sections, and Steenkamp's approach is no different from others (Steenkamp and van Trijp, 1989). An impression of Garvin's *user*-based approach is highlighted in the quality guidance model's first step in which perceived quality is viewed as an overall unidimensional evaluative judgment, i.e., "poor" or "good". However, as a process, this model assumes that quality perception is formed at two different points in time: before purchase, which was called "quality expectation" and after consumption, which was called "quality experience." The quality guidance model is concerned with intrinsic cues and quality attributes only. Perceptions can be made actionable by finding their appropriate physical counterparts. A team of sensory evaluators, food scientists, and marketers may be of huge importance in this search. Finally, the translation of the consumer perceptions into physical product characteristics was comprehended by the psychophysical principle. The only exception was that consumer perceptions were considered the starting point rather than the physical variable in this principle. The concepts that are relevant to the quality guidance model have been discussed in brief, but simultaneously readers are encouraged to refer to original work for more details. Based on Brunswik's lens model (Brunswik, 1952), Steenkamp and van Trijp (1989) proposed an integrated model to accommodate the previously mentioned concepts into a model that relates perceived quality judgments to physical product characteristics.

Figure 6.3 shows an integrated model of the linkage of physical characteristics to perceived quality judgments both before purchase and on consumption. Readers can refer to the original work (Steenkamp and van Trijp, 1989) for a sample-specific (beefsteak) overview of the model. The upper part of the model represents the relation between physical characteristics, perceptions about the intrinsic cues, and quality expectation. Likewise, the lower part represents how physical

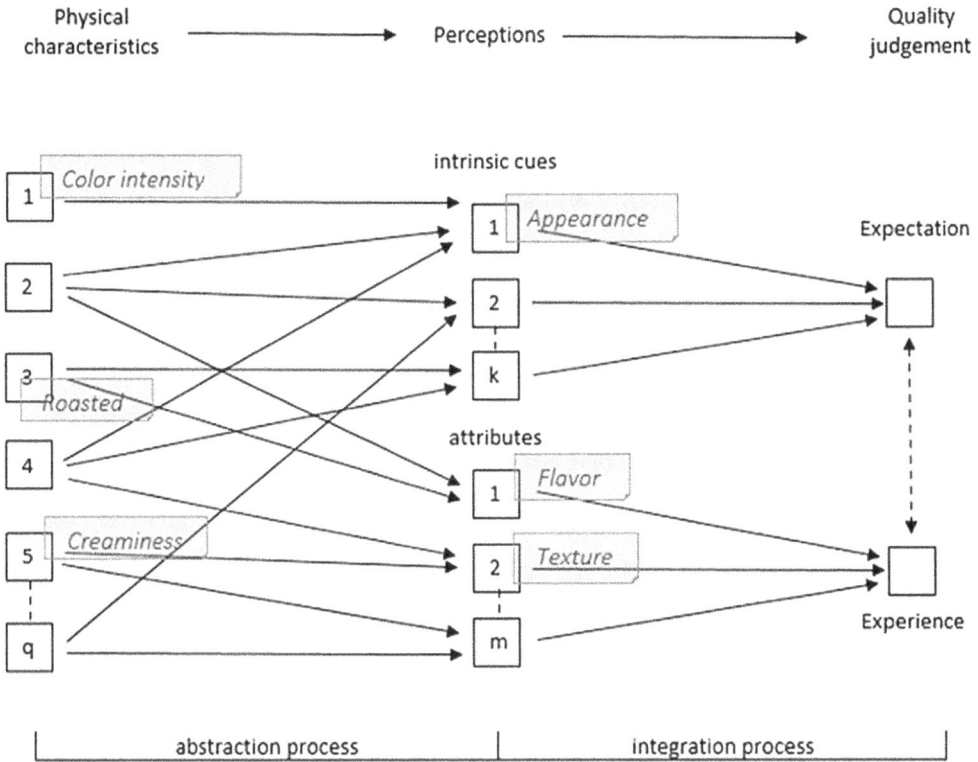

Figure 6.3 Integrated model. Labels were added for clarity. (Reproduced from *Food Quality and Preference*, 6(3), Peter A.M. Oude Ophuis and Hans C.M. Van Trijp, Perceived quality: A market driven and consumer oriented approach, 177–183, Copyright (1995) with permission from Elsevier.)

characteristics are related to product experience through quality attribute perceptions. The quality cues and attributes are mediating variables toward the physical attributes of a product. Once the relation between physical characteristics and consumer perceptions is known, it will be possible to manipulate physical product parameters to optimize overall perceived quality. The mutual influence of quality expectation and quality experience is represented by the dashed two-headed arrow between them. Based on Brunswik's lens model, which models the effect of one or more distal variables on an individual's inferential response through the perception of intervening proximal cues, this model also consists of two phases, abstraction and integration. The abstraction process of the integrated model comprehends the relationship between physical product attributes and intrinsic cue/attribute perceptions based on psychophysical relationship. The integration phase models how intrinsic cue perceptions and quality attribute perceptions are integrated into a judgment about quality expectation and quality experience, respectively (Steenkamp and van Trijp, 1989). Quality expectations and quality experience are overall quality judgments that are supposed to be formed (integrated) based on quality cues (expectation) and quality attributes (experience). By comparing this model to other models of consumers' judgment formation, such as those of Lancaster (1971) and traditional multi-attribute types, the existence of perceptions was ignored in the Lancaster model, whereas determinants of consumer perceptions were not covered in the multi-attribute model type, respectively. Later, Steenkamp and van Trijp tested this model with blade steak and found a modest but significant correlation between quality expectation and quality experience. About 30–40% of the variance in the quality cues could be explained by the technical physical measurements. The abstraction of technical characteristics toward quality attributes was less clear, which opens it up for further research, especially from a psychologic and psychophysics point of view.

In this ongoing state of affairs, the comprehension of quality by Gail V Civille (1991) follows the *manufacturing-* and *product-*based rut of the multidimensional approach. A two-phase program focusing on *conformance* and *reliability* dimensions was devised to assess food quality via (1) determining the product requirements and (2) measuring conformance. Diverging from the meaning of *perceived quality*, more emphasis was given on the simple, objective aspect of quality. This approach may fit into the "multi-attribute approach." This interpretation of quality aligns with the traditional view, which emphasizes determining a level of quality that is economically acceptable to the company, and then inspecting the products to ensure that nothing below that level gets shipped (Wolff, 1986).

To incorporate consumers' quality perception insights into the production and product development, the MEC model is discussed here to ensure that products introduced in the market fit in with consumers' demand for quality. Grunert (1995) demonstrated a method for the identification of the perceived quality judgments and later investigated the link between concrete product characteristics and self-relevant consequences in the form of an MEC model. Grunert (1995) politely explained the attractiveness of concrete product characteristics among consumer researchers in the 1970s and 1980s instead of understanding the link between these characteristics and self-relevant consequences. Self-relevant consequences can also be termed *buying motives*, because these consequences, not the product itself, are what the consumer is interested in (Grunert, 1995). The idea of linking the characteristics of the product to deep purchase motivations via the so-called MEC model is one of the most popular "hierarchical approaches" (Silvestri and La Sala, 2018; Olson, 1989). MEC theory is also important from product development side because it restricts treating products as a mere collection of physical attributes. For instance, Crofton and Scannell (2020) found that the rich protein content of brewers spent grain (BSG) was not seen relevant in the context of a cereal snack product category and its use as a sustainable ingredient was not viewed as a means to attain a desired end state (environment friendly) in the consumers' mind. Instead, these attributes must be considered, developed, and promoted concerning the benefits they offer to consumers (Chooi, 2020). In contrast to multidimensional approaches, "hierarchical approaches" do not focus on attributes but on the association between these attributes and abstractness such as values, which can create interest in certain products. The MEC model explicates the mechanism, to some extent, of achievement of *values*, a more common domain of psychology than sociology. The basic assumption of MEC theory is that consumers organize the information at different levels of abstraction (Silvestri and La Sala, 2018) and are not interested in products per se, but in what the product is doing for them, in the form of *self-relevant consequences*, in the way the product helps them attain their life values (Grunert, 2005; Olson and Jacoby, 1972). This model infers that consumers' subjective product perception is established by associations between product attributes and more abstract, more major cognitive categories such as values, which can motivate behavior and create interest in product attributes. Grunert (1995) proposed calling these abstract product characteristics *quality aspects*. This model depicts how concrete product characteristics are linked to self-relevant consequences, irrespective of whether they are functional or psychosocial in nature. This model consists of three steps: (1) elicit product attributes relevant to the subjects in question, (2) reveal how subjects in question relate product attributes to consequences and values by an in-depth interview process or so-called *laddering*, and (3) derive hierarchical value maps (HVM) from depicting the aggregate MECs of the subjects in question (Zanoli and Naspetti, 2002). Initially, product attributes can be elicited by focus groups, surveys, open-ended questions, or other techniques, such as free sorting, ranking, projective mapping, etc. Grunert (1995) asked the respondents to generate concrete product characteristics by presenting triads of products, whereas Zanoli and Naspetti (2002) grounded findings by providing no specific product at the outset. Practitioners are cautioned not to rely on instrumental techniques, such as trained assessors, for product attribute aggregation. Following

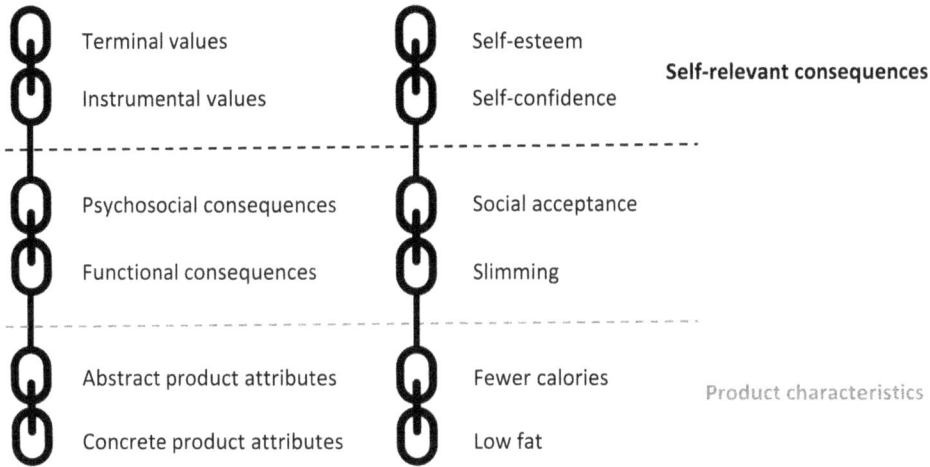

Figure 6.4 Means-end chain model for low-fat milk. (From Grunert 1995.)

elicitation, a certain number of the most important product attributes are retained for laddering. Consumers are driven to build up their own sequence of *attributes-consequences-values* and in this way, they naturally reveal their personal reasons. Figure 6.4 shows an example of the concrete characteristics of low-fat milk linked to the abstract product characteristics and self-relevant consequences. Different kinds of laddering techniques, such as soft, or hard, have been reported (Zanoli and Naspetti, 2002). The natural flow of speech is restricted as little as possible in soft laddering, whereas the respondent is forced to produce ladders one by one and give answers at an increasing level of abstraction in hard laddering. The hard laddering approach is preferred in budgetary and timing restriction situations, whereas soft laddering is preferred more in probing scenarios (Crofton and Scannell, 2020), especially where the interviewee wants more subtle details of either low or highly involved respondents. Subsequently, consumer-relevant motivations are decoded in chunks of meaning and then coded in categories of attributes-consequences-values (content analysis). An *implication matrix* based on codes is computed, and from this an HVM can be derived. An example of HVM is shown in Figure 6.5 for regular and non-regular buyers of organic food. Software, such as MECanalyst PLUS, LADDERMAP or LadderUX are also available to derive the implication matrix and the relevant HVMs. Self-relevant consequences, as shown in Figure 6.4 can find application in product development, segmentation (Sharma, 2020), and marketing. As a model of consumers' cognitive structures, MEC models attempt to show how consumers structure experience with and information about products so that this information can be used in the future to develop and determine *quality* in product development.

Grunert (1995) exhibited the applicability of the MEC model to measure the *perceived values* and recognized that consumers do not derive *value* from the product characteristics per se but from the *self-relevant consequences* that they see associated with these characteristics. This model has the potential of providing insight into the reason why consumers value certain consequences by linking the self-relevant consequences to the consumer's internalized *value* structure. These insights can be used in product development and production to ensure that products introduced in the market fit in with consumers' demand for quality. Without any doubt, the means-end theory is unfortunately not problem free, and the same is highlighted by Ophuis and van Trijp (1995) in their commentary.

Later, Grunert (1995) added a contextual dimension into the quality perception, in the form of *procedural knowledge*, which refers to skills that are usually motoric or perceptual. Previously,

(a)   **(b)**

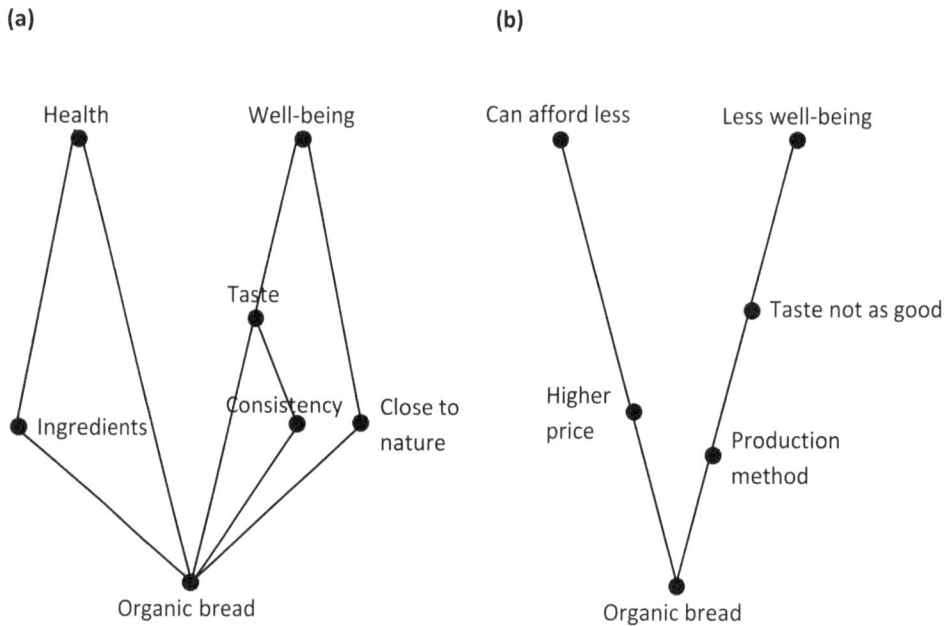

Figure 6.5   Hierarchical value map. (a) Regular buyer of organic foods. (b) Non-regular buyer of organic food. (Reproduced from *Food Quality and Preference*, 6(3), Klaus G Grunert, Food Quality: A means-end perspective, 171–176, Copyright (1995) with permission from Elsevier.)

Daget (1987) echoed the importance of the consumption environment in the *utility value* characteristics of consumers. Likewise, Cardello (1995) stated quality is not relative to not only who is doing the evaluation but to a wide range of situational and contextual factors. By doing so, Grunert (1995) implicitly added the concept of *quality experience*, a standard quality aspect in the foodservice industry or restaurant context. In food services, the product is regarded as an experience only part of which involves food (Pierson et al., 1995). MEC models and HVMs are models of cognitive structures, and, in general, contain two types of knowledge, *declarative* and *procedural*. Both MEC and HVM are examples of declarative knowledge. The attainment of self-relevant consequences will not depend on the product's characteristics only but also on how and where the product will be used. This extension of the means-end perspective by Grunert (1995) to integrate consumers' procedural knowledge by explicitly incorporating *scripts*, such as meal preparation scripts and shopping scripts, raises questions on the stable relationship between preferences to perceived concrete product characteristics that is taken for granted. For example, Grunert (1995) illustrated the operation of the previously mentioned model and found the major concrete product characteristics, quality aspects, meal preparation scripts, shopping scripts, and buying motives in connection with beef employing focus groups. Later, the relationship between these constructs was quantified through an extended form of conjoint analysis. It was found that preferences for concrete product characteristics may be unstable when the salience of buying motives varies with different types of consumption situations.

     In the aftermath, the author(s) of this chapter certainly agree on the fact that food quality should be viewed and measured from the consumer's perspective (Cardello, 1995; Molnar, 1995), and those objective product characteristics should be related to subjective perception. To provide a reliable and acceptable product, objective evaluation techniques must be associated with sensory evaluation. A chef's continuous search for the elusive ultimate level of sensory excellence often perplexes the food scientists attempting to replicate a stated level of excellence. Such divisions can be better understood by focusing on the dimensions of quality emphasizing a particular trade. The concept of *quality experience* may produce more valid answers to public house consumer demands. If a product is to fulfill consumer expectations, then systematic consumer research is necessary to establish what the features of the product should be ideally. Grunert (1995) wisely said that consumer behavior will become more difficult to predict in the future as collectivism is depleting; general lifestyle typologies are shrinking; and as a corollary of this, a preference inferred based on concrete product characteristics will only become less stable and therefore less useful.

6.6 CONCLUSION

For a product to be successful, management has to be proactive and must include new food product development in its long-term goals. This is because products have a finite life cycle and new impetus is needed to regain the consumer attention. Accompanying this, declining kinship or prosocial structures, and thereby collective values, are paving the way for new challenges in product development by realizing the fact that acceptance is an individual's decision. It should be propelled by facts and "gut feeling" (unless validated) does not have a place, if the exercise is to be successful. *In toto*, due to changes in consumer preferences, availability of new ingredients and technologies, and to keep pace with changing regulations, food product development becomes an important tool in the growth of a company.

REFERENCES

Aaker, D.A. (1991). *Managing Brand Equity: Capitalizing on the Value of a Brand Name*. New York: Free Press.

Aramouni, F. & Deschenes, K. (2018). *Methods for Developing New Food Products – An Instructional Guide* (2nd ed.). Lancaster, PA: DEStech Publications.

Booth, D.A. (1995). The cognitive basis of quality. *Food Quality and Preference*, 6(3), 210–207.

Brunswik, E. (1952). The conceptual framework of psychology. *Psychology Bulletin*, 49(6), 654–656.

Buisson, D.H. (1993). Marketing interaction and independency in product development process. Food Technology in New Zealand. https://scholar.google.com/scholar_lookup?hl=en&publication_year=1993&author=+Buisson%2C+D.H.&title=%E2%80%98Marketing+Interaction+and+Independancy+in+Product+Development+Process%E2%80%99

Cardello, A.V. (1995). Food quality: relativity, context and consumer expectations. *Food Quality and Preference*, 6, 163–170. https://doi.org/10.1016/0950-3293(94)00039-X

Chooi, B.-J. (2020). Cultural priming conditions and decision making on food consumption: means-end evidence for everyday consumer goods. *Journal of Marketing Theory and Practice*, 28(4), 403–417. https://doi.org/10.1080/10696679.2020.1772673

Civille, G.V. (1991). Food quality: consumer acceptance and sensory attributes. *Journal of Food Quality*, 14, 1–8.

Cooper, R.G. (1993). *Winning at New Products*. Reading, MA: Addison-Wesley.

Crawford, C.M. (1994). *New Products Management* (4th ed.). Burr Ridge, IL: Irwin.

Crofton, E.C. & Scannell A.G.M. (2020). Snack foods from brewing waste: consumer-led approach to developing sustainable snack options. *British Food Journal*, 122(12), 3899–3916. https://doi.org/10.1108/BFJ-08-2018-0527.

Daget, N. (1987). Commercial sensory quality of food products. In M.H. Thompson (Ed.), *Food Acceptability* (pp. 55–66). New York: Elsevier Applied Science.

Darby, M.R. & Karni, E. (1973). Free competition and the optimal amount of fraud. *Journal of Law and Economics*, 16(1), 67–88. https://www.jstor.org/stable/724826

Earle, M., Earle, R., & Anderson, A. (2001). *Food Product Development* (1st ed.). Boca Raton, FL: CRC Press LLC.

European Food Information Council (2013). Food shelf-life and its importance for consumers. https://www.eufic.org/en/healthy-living/article/food-shelf-life-and-its-importance-for-consumers

Garvin, D.A. (1984). What does "product quality" really mean? MITSloan Management Review. https://sloanreview.mit.edu/article/what-does-product-quality-really-mean/

Graf, E. & Saguy, S.I. (1991). *Food Product Development from Concept to the Market Place*. Chapman and Hall, New York: Van Nostrand Reinhold.

Grunert, K.G. (1995). Food quality: a means-end perspective. *Food Quality and Preference*, 6, 171–176. https://doi.org/10.1016/0950-3293(95)00011-W

Grunert, K.G. (2005). Food quality and safety: consumer perception and demand. *European Review of Agricultural Economics*, 32(3), 369–391. https://doi.org/10.1093/eurrag/jbi011

Hollingsworth, P. (1994). The perils of product development. *Food Technology*, 48(6), 80–88.

IFT (2016). Food Storage and shelf-life. *Food Facts*. https://www.ift.org/career-development/learn-about-food-science/food-facts/food-storage-and-shelf-life

Institute of Food Science and Technology (1993). *Shelf-life of Foods – Guidelines for Its Determination and Prediction*. London: Institute of Food Science and Technology.

Institute of Food Technologists (1974). Shelf-life of foods. *Journal of Food Science*, 39, 1–4.

Joseph, M.V. (2016). Extrusion, physico-chemical characterization and nutritional evaluation of sorghum-based high protein, micronutrient fortified blended foods. Ph.D. dissertation, Kansas State University, Manhattan, Kansas. http://hdl.handle.net/2097/32907

Juran, J.M. (1983). *Gestion de la Qualité*. Normes Techniques Series, Paris: AFNOR.

Kotler, P. & Armstrong, G. (1991). *Principles of Marketing* (5th ed.). Englewood Cliffs, NJ: Prentice-Hall.

Krishnan, V. & Ulrich, K.T. (2001). Product development decisions: a review of the literature. *Management Science*, 47, 1–21.

Lancaster, K. (1971). *Consumer demand: a new approach*. New York: Columbia University Press.

Lawless, H. (1995). Dimensions of sensory quality: a critique. *Food Quality and Preference*, 6(3), 191–195. https://doi.org/10.1016/0950-3293(94)00023-O

Molnar, P.J. (1995). A model for overall description of food quality. *Food Quality and Preference*, 6(3), 185–190. https://doi.org/10.1016/0950-3293(94)00037-V

Nelson, P. (1970). Information and consumer behavior. *Journal of Political Economy*, 78(2), 311–329.

Nelson, P. (1974). Advertising as information. *Journal of Political Economy*, 82(4), 729–754. https://www.jstor.org/stable/1837143

Olson, J.C. (1989). Theoretical foundations of means-end chains. *Werbeforschung & Praxis*, 5, 174–178.

Olson, J.C. & Jacoby, J. (1972). Cue utilization in the quality perception process. In M. Venkatesan (Ed.) *SV-Proceedings of the Third Annual Conference of the Association for Consumer Research* (pp. 167–179). Chicago, IL: Association for Consumer Research.

Ophuis, P.A.M. O. & van Trijp, H.C.M. (1995). Perceived quality: a market driven and consumer oriented approach. *Food Quality and Preference*, 6, 177–183. https://doi.org/10.1016/0950-3293(94)00028-T

Pierson, B.J., Reeve, W.G., & Creed, P.G. (1995). "The quality experience" in the food service industry. *Food Quality and Preference*, 6, 209–213. https://doi.org/10.1016/0950-3293(94)00042-T

Pisano, G.P. (2015). You need an innovation strategy. *Harvard Business Review*. https://hbr.org/2015/06/you-need-an-innovation-strategy

Plsek, P. E. (1987). Defining quality at the marketing development interface. *Quality Progress*, 20(6), 29–36

Rudder, A., Ainsworth, P., & Holgate, D. (2001). New food product development: strategies for success? *British Food Journal*, 103, 657–671. https://doi.org/10.1108/00070700110407012

Rudolph, M. (1995). The food production development process. *British Food Journal*, 97(3), 3. https://doi.org/10.1108/00070709510081408

Sharma, C. (2019). Sensory and consumer profiling of potatoes grown in the USA. Doctoral dissertation, Kansas State University, Manhattan, Kansas. http://hdl.handle.net/2097/40044

Sharma, C. (2020). Segmentation of potato consumers based on sensory and attitudinal aspects. *Foods*, 9(2), 161. https://doi.org/10.3390/foods9020161

Shao, Y., Zhang, C., Zhou, J., Gu, T., & Yuan, Y. (2019). How does culture shape creativity? A mini-review. *Frontiers in Psychology*, 10, Article No. 1219. https://doi.org/10.3389/fpsyg.2019.01219

Sharif, M.K., Zahid, A., & Shah, F.H. (2018). Role of food product development in increased food consumption and value addition. In *Food Processing for Increased Quality and Consumption* (pp. 455–479). San Diego: Elsevier Inc. https://doi.org/10.1016/B978-0-12-811447-6.00015-1

Silvestri, R. & La Sala, P. (2018). Food Quality perception. In F. Conto, M.A.D. Nobile, M. Faccia, A.V. Zambrini, & A. Conte (Eds.), *Advances in Dairy Products* (pp. 355–366). Hoboken, NJ: John Wiley & Sons Ltd.

Skarra, L. (2006). A left-brained approach to creativity, a very right brained subject. *Cereal Foods World*, 51(5), 277–278, 286. DOI: 10.1094/CFW-51-0277

Steenkamp, J.E.M. & van Trijp, H.C.M. (1989). Quality guidance: a consumer-based approach for product quality improvement. In G.J. Avlonitis (Ed.) *Marketing Thoughts and Practice in the 1990s: Proceedings of the 18th Annual Meeting of the European Marketing Academy* (Vol. II), Athens: EMAC, 1191–1217.

Ulrich, K.T. & Eppinger, S.D. (2016). *Product Design and Development*. New York: McGraw-Hill Education.

Urban, G. & Hauser, J. (1993). *Design and Marketing of New Products* (2nd ed.). Englewood Cliffs, NJ: Prentice-Hall.

USDA-ERS (2020). Food Loss. https://www.ers.usda.gov/data-products/food-availability-per-capita-data-system/food-loss/

Weigel, L. (2015). Design thinking to bridge research and concept design. In M.G. Luchs, K.S. Swan, & A. Griffin (Eds.), *Design Thinking: New Product Development Essentials from the PDMA* (Ch. 5, pp. 59–69). Hoboken, NJ: John Wiley and Sons, Inc.

Wolff, M.F. (1986). Quality/process control: What R&D can do. *Research Management*, 29(1), 9–11.

Zanoli, R. & Naspetti, S. (2002). Consumer motivations in the purchase of organic food: a means-end approach. *British Food Journal*, 104(8), 643–653. https://doi.org/10.1108/00070700210425930

Zeithaml, V.A. (1988). Consumer perceptions of price, quality, and value: A means-end model and synthesis of evidence. *Journal of Marketing*, 53(3), 2–22. https://doi.org/10.2307/1251446

Zweep, C. (2015). Shelf-life determination: the manufacturer's challenge. New Food. https://www.newfoodmagazine.com/article/21317/shelf-life-determination/

CHAPTER 7

Fresh and Refrigerated Foods
Science, Shelf Life, and Quality

Thoithoi Tongbram, Jinku Bora, and H.A. Makroo

CONTENTS

Email: Jinku Bora, jinkubora@gmail.com

DOI: 10.1201/9781003091677-7

7.1 INTRODUCTION

Our fast-paced modern world is based on safety, quality, convenience, and the growing realization of the health benefits of the food that we consume. The umbrella of fresh foods encompasses food that is free from preservation and spoilage, food that is freshly harvested, or food that is just butchered or slaughtered, and food that is treated properly (e.g., cooling after harvest). Traditional cooling techniques reinforced by modern methods are generally used to maintain the shelf life of fresh foods. However, for long-haul storage, long transit times, or foods requiring immediate cold preservation, refrigeration and cold storage techniques are the preferred form of food safety practice, thereby keeping them shelf stable for later consumption. The ability to store and save food right after harvest for transport, processing, or future use, without decay and spoilage, can be mostly attributed to the invention, and thereafter, the evolution of refrigeration and cold storage chains around the world. Without this vital part (a means to keep our harvest and food "cold," below a certain critical temperature) modern food production would collapse under the pressure of compromised quality and safety.

Fresh foods include fruits, vegetables, eggs, meat, poultry, and seafood that have not been cooked, canned, dried, frozen, or preserved by any of the known processing methods. They contain nutrients in their natural quality and quantity, unaltered by processing, hence, retaining their original flavors, colors, and texture. In absence of food processing, fresh foods are also free from artificial additives and preservatives. Therefore, they are relatively cheaper and preferred by consumers who are aware of their nutritional and healthy benefits. These days, as consumers become increasingly conscious of processes and food additives, the demand for fresh foods is on the rise. This being said, almost all fresh foods have short shelf lives and their quality rapidly deteriorates in response to internal physiological reactions and environmental factors. Generally, fruits and vegetables are seasonal, whereas meat, poultry, and seafood require proper storage and handling to remain free from contamination. To curve such losses and ensure maximum consumption of fresh produce, precooling techniques and special packaging (e.g., modified atmosphere) are employed in conjunction to refrigeration.

Refrigeration is the foundation underneath that holds the entire picture together right from harvest, primary processing, cold chain storage through retail markets, and finally to end-consumer domestic storage. To realize how important a refrigerator (or refrigeration) is in our lives, let us start at the perspective of our domestic living spaces. For example, it is peak summertime, and you are just home from some work outside in the blazing heat. Almost instantly, what do you desire: How about a cold glass of water, a bottle of a refreshing drink, some fresh fruit, or perhaps even ice cream or your favorite dessert from yesterday? It is true that a refrigerator is indispensable in our homes; rather, it has become a part of our daily lives. We may even fail to grasp its basic importance when everything is going alright, but once in a while, we are reminded duly, to say, when the power goes off or an electrical circuit blows up inside the unit, thereby putting all of our stored food in harm's way. A refrigerator is undoubtedly one of the crucial electronic items for keeping food safe in the kitchen inside our homes. The electronic wonder that we see today was once little more than a rudimentary box, the utility of which was to supply cold temperature air sourced from a block of ice placed in it. Figure 7.1 demonstrates how refrigeration helps us store and preserve our harvest.

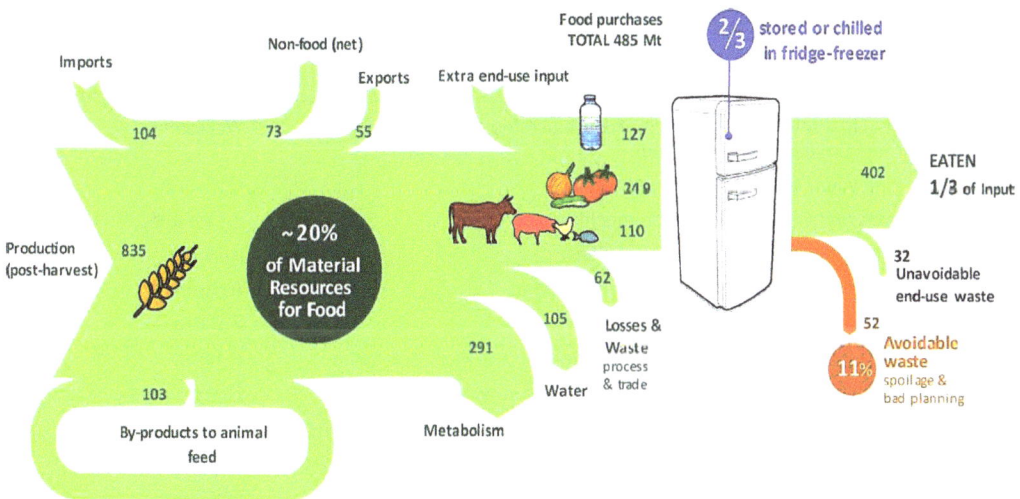

Figure 7.1 Simplified European Union flow diagram for material resource input, food products, and waste from 2011 to 2012 (forage expressed in dry hay equivalent). (Adapted from van Holsteijn and Kemna, 2018, with permission.)

7.2 HISTORY OF REFRIGERATION

The journey of refrigeration started with a discovery of the benefits of storing foods in springs or packed in snow and ice. So, cold nights, cool caves, and natural ice and snow were used from prehistoric times for most of history, until people started harvesting ice from lakes and rivers in winters. Frozen water bodies became the source of ice, and there were icehouses to store them until summer. Where there was no ice, cold cellars (insulated tube wells dug into the ground) and spring houses built on top of springs were used to keep food cold. In some parts, an intermediate achievement in cooling foods was made in the form of pushing water temperature down by adding chemicals like sodium nitrate or potassium nitrate for the storage of wine; this was recorded around 1550.

It wasn't until the advent of mechanical refrigeration, with a compressor and refrigerant, in the last quarter of the 19th century that commercial icehouses that manufactured blocks of ice for use in iceboxes blossomed. A long, tedious process of development ensued, from thereon, to elevate cold preservation as a large-scale, booming industrial process around the globe. Environmental concerns regarding the use of "old refrigerant," commercially known by its trade name, "freons" grew in the world of mechanical refrigeration. It was replaced in accordance with the rules laid down by the Regulatory Clean Air Act, Title 6 by a new refrigerant, HFC 134a, which is less injurious to the ozone layer and just as effective in refrigeration applications (U.S. Department of Agriculture, Food Safety and Inspection Service, 2015). It is interesting to note that frozen foods made a short appearance before World War II. Some milestones in the history of mechanical refrigeration are briefly represented in Table 7.1 (Powitz, 2005; Berk, 2018).

7.3 IMPORTANCE OF REFRIGERATION AS A FOOD PRESERVATION TECHNIQUE

Microorganisms grow when favorable conditions such as temperature, moisture, and nutrients (from food) are freely available. They could be food, air, water, or soilborne pathogens capable of inducing illnesses and food poisoning in our systems if consumed. Some microorganism such as bacteria thrive well in the temperature range of 4–60°C (40–140°F), marked as the "danger zone." In this

Table 7.1   Milestones in the History of Mechanical Refrigeration – A Brief Timeline

Year	Milestone	By
1748	First recorded demonstration of artificial refrigeration by using vacuum evaporation of ether.	W. Cullen
1805	First vapor compression system	O. Evans
1834	Improved vapor compression machine	J. Perkins
1844	First refrigerator using O. Evans design	J. Gorrie
1856	First commercial application of refrigeration	A. Twinning
1859	First ammonia machine	F. Carré
1868	Refrigerated transatlantic maritime transport of meat attempted	P. Tellier
1873	First industrial refrigeration systems in brewery	C. von Linde
1876	Modern refrigeration technology patented	C. von Linde
1918	First household refrigerators	—
1920	Commercial air conditioning	W. Carrier
1938	Frozen food industry	C. Birdseye
1974	Suspected refrigeration gases destroying ozone layer	S. Rowland and M. Molino

Sources: Powitz (2005) and Berk (2018).

Table 7.2 Storage Life of Common Food Items without Refrigeration

Food Product	Average Useful Storage Life (Days)		
	0°C	22°C	38°C
Dry fruits	>1000	>350 and <1000	>100 and <350
Dry meats and fish	>1000	>350 and <1000	>100 and <350
Dry seeds	>1000	>350 and <1000	>100 and <350
Fish	2–7	1	<1
Fruits	2–180	1–20	1–7
Leafy vegetables	3–20	1–7	1–3
Meat	6–10	1	<1
Poultry	5–18	1	<1
Root crops	90–300	7–50	2–20

Source: https://nptel.ac.in/content/storage2/courses/112105129/pdf/RAC%20Lecture%203.pdf. NPTEL (2008).

temperature range, bacteria multiply rapidly, even doubling their numbers in a matter of 20 minutes, if conditions persist to favor their growth. Along with their exponential multiplication, the food gets spoiled, making it unsafe and unhealthy for human consumption. In the confinement of our homes, they are everywhere in the air, water, or food. So, refrigeration is a must to curb their spoilage activity and rapid multiplication. Technically, storing foods below 4°C is considered safe, as the low temperature will protect most foods (U.S. Department of Agriculture, Food Safety and Inspection Service, 2015). Some fruits and vegetables along with their shelf lives under refrigeration and normal storage are shown in Table 7.2.

7.3.1 Preservation Action of Refrigeration

The principle behind refrigeration as a food preservation technique is based on one of the concepts of physical chemistry, depression of molecular mobility at low temperatures. First, when food is stored at refrigeration temperatures (below 4°C), due to decreased molecular mobility, enzymatic and chemical reactions are hampered; consequently, biological processes are also slowed down. This, in turn, halts microorganisms from inflicting further damages to the food matrix. Second, this depression also affects the food's mechanism of physical and biochemical reactions that lead to natural maturity and decay. Even after harvest for fruits and vegetables and primary processing for meat and other perishables, the very chemical reactions that sustained their lives also cause them to age and decay. Refrigeration is helping these severed tissues save themselves from such deterioration by arresting metabolic activities. In contrast to heat treatment that destroys cellular enzymes and wipes out microbes, cold temperature merely depresses their activity. Some points to note while using cold storage preservation techniques are as follows:

1. The safest way to hold a quality food product in refrigeration is to ensure its initial high microbial quality before preservation because refrigeration does not kill or remove microorganisms once present. The same applies to fresh products and their maturity levels.
2. Refrigeration is not permanent preservation. Even under the influence of cold temperatures, food items still have a definite shelf life, which again depends on the storage temperature.
3. The action of preservation is valid as long as the cold temperature exists. This means there is no room for temperature fluctuations or gaps in cold treatment along with the entire cold storage and transport chain.
4. The hurdle principle states that to ensure food safety, refrigeration has to be employed along with other techniques of food preservation (Berk, 2018).

7.4 REFRIGERATION SYSTEMS USED FOR THE STORAGE OF FOODS

For use in food preservation, two kinds of refrigeration systems exist. The first type is known as a closed mechanical refrigeration system. This is the more common system and has a recirculating refrigerant. The second type is known as an open cryogenic system, where the cooling action is provided by specific cryogens such as LIN or carbon dioxide in the solid/liquid phase (CO_2).

7.4.1 Closed Mechanical Refrigeration System

As illustrated in Figure 7.2, inside a closed mechanical system four components work continuously and in cycles to keep the cooling action alive. Refrigerants such as hydrochlorofluorocarbon (HCFC) and ammonia (NH_3) (Barbosa-Cánovas et al., 2005) flow through the condenser, compressor, evaporator, and expansion valve (the four parts) by cyclic phase changes with respect to changes in pressure and temperature. This is the basic working mechanism inside a closed mechanical system. The cycle starts when the refrigerant absorbs heat and phase changes from liquid to gas, thereby lowering the temperature of the food. Controlled by a thermostat, the compressor sucks in the vaporized refrigerant, pressurizes it, and passes it on to the condenser, where it releases heat and phase changes again. The refrigerant, now in liquid form, passes into the expansion valve, where it expands to become a liquid-gas mixture at reduced pressure. This mixture is recirculated through the coils of the refrigeration chamber (evaporator), drawing out heat again and vaporizes. The vapor is passed back into the compressor for another round of freezing and the cycle thus repeats. The next section discusses this in detail.

7.4.2 Open Cryogenic System

Instead of refrigerants, the open cryogenic system uses cryogens such as LIN or solid/liquid CO_2. One of the other differences between the two systems is the way in which the cooling agent is consumed. Unlike in modern closed mechanical refrigeration systems, which recirculate refrigerants,

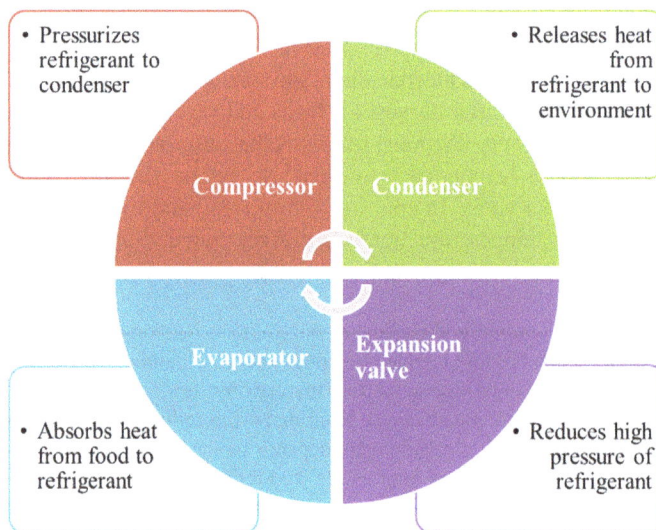

Figure 7.2 Closed mechanical refrigeration system (one stage).

these traditional open systems (as the name suggests) expel the spent cryogen into the atmosphere and are not collected for reuse.

7.4.3 Refrigerants

Following phasing out of the use of chlorofluorocarbons (CFCs) as an old refrigerant, HCFCs, and hydrofluorocarbons (HFCs) are superseded as the less harmful, low ozone-depleting, alternative refrigerants. Because they are active greenhouse gases, they are under fluorinated gas (F-gas) regulations and monitored through such legislation. Choosing a refrigerant is based on the physical, thermodynamic, and chemical properties of the fluid, and taking into account the economic, safety, and hazardous concerns it may yield. Most halocarbons, due to their potentially hazardous nature, have been banned. Ammonia, even though a thermodynamically very efficient fluid was often left out due to safety concerns during the performance. New refrigeration technologies make these old refrigerants perform better. Now, ammonia, chlorofluoromethane, and tetrafluoroethane are the most commonly used refrigerants for food industries. Additionally, carbon dioxide, air, and hydrocarbons have been successfully utilized as refrigerants (Stoecker and Jones, 1982; Persson and Londahl, 1993; Barbosa-Cánovas et al., 2005; Shannon, 2011).

7.5 COMPONENTS OF REFRIGERATION SYSTEMS

Mechanical refrigerators are based on the principle of vapor compression, hence, the name vapor compression refrigeration (VCR) systems. A suitable gas of fixed quantity, acting as the refrigerant, is filled in a closed VCR system. The gas changes phase with changes in temperature and pressure and moves in a cycle through tubes connected among four basic elements: evaporator, compressor, condenser, and expansion valve (refrigeration flow control). A block diagram of a typical VCR system is shown in Figure 7.3.

7.5.1 Compressor

The most important part and the workhorse that accounts for 80–100% of the total energy consumption of a VCR system is the compressor. The energy spent by a compressor is a factor of the

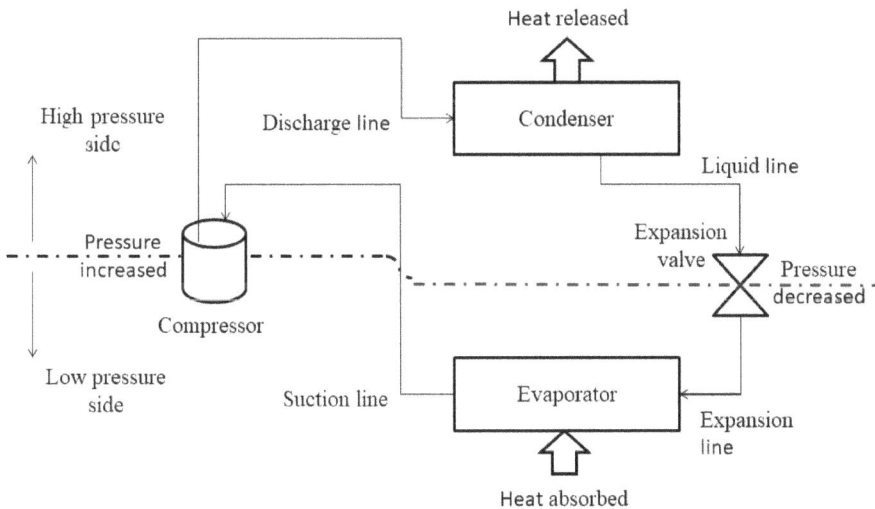

Figure 7.3 Block diagram of a reciprocal VCR system.

compressor type, load, and temperature difference (degrees the system is required to cool). It can be of many types, but let us discuss how a reciprocating compressor works its cooling action. A motor, a shaft, and a piston are housed inside the compressor unit. When the motor rotates the main shaft, it pushes the piston up and down inside the cylinder. The downward motion of the piston creates a suction pressure that pumps in the low-pressure–high-temperature refrigerant (LP-HT) from the evaporation chamber in gaseous form. The LP-TH is the result of heat picked up from the evaporation chamber (food chamber). As the piston moves upward, it compresses LP-TH with a high pressure that converts it to a high-pressure–high-temperature refrigerant (HP-HT) and discharges it into the condenser via a discharge valve.

7.5.2 Condenser

The HP-HT is introduced into a section with a large area for heat exchange called the condenser. This are is called a condenser because it "condenses" the gaseous HP-HT into a liquid form, ejecting its latent heat into water or air, as per the system's design and later into the atmosphere. Here, HP-HT is converted to a high-pressure–low-temperature refrigerant (HP-LT) in a gas-liquid mix form. The temperature of HP-LT is such that it is below the required temperature to be maintained inside the evaporator. HP-LT is stored in the receiver before it is passed on to the evaporator via an expansion valve.

7.5.3 Pressure Reduction or Expansion Valve

The main function of the expansion valve is to throttle pressure between two components, the condenser and the evaporator. The condenser functions at high pressure and stores HP-LT in the receiver, whereas the evaporator requires low pressure to function. The means to mediate pressure between these two components is facilitated by the pressure reduction valve. It also regulates the flow of HP-LT from the receiver according to the load on the evaporator. Some devices that are used to perform such functions include the following:

1. Capillary tube
2. Pressure control or automatic expansion valve
3. Thermostatic expansion valve
4. High-side float valve
5. Low-side float valve
6. Solenoid valve

7.5.4 Evaporator

The evaporator is the chamber that houses the food to be refrigerated. HP-LT is pressure throttled by the expansion valve and comes into the evaporator as a low-pressure–low-temperature refrigerant (LP-LT) ready to pick up heat from the food to be cooled. At low pressure, LP-LT flash vaporizes into a low-pressure–high-temperature refrigerant (LP-HT) in vapor form withdrawing heat from the chamber. The suction pressure created by the piston moving down sucks out the LP-HT from the evaporator and into the compressor for another cycle of refrigeration. Different categories of evaporators used for food refrigeration include the following:

1. Flood evaporators
2. Dry expansion evaporators
   a.   Natural convection evaporators
   b.   Forced convection evaporators

3. Shell and tube evaporators
4. Double pipe evaporators
5. Plate evaporators
6. Baudelot cooler

7.5.5 Primary Refrigerant

The fluid that undergoes the four cycles (evaporation, compression, condensation, and expansion) of a VCR system is called the primary refrigerant. In short, it transfers heat from food in the evaporator to air or water in the atmosphere. Fluids that qualify as a primary refrigerant have very low boiling points (e.g., ammonia boils at −33°C under atmospheric pressure) (e-Krishi Shiksha, 2013; Yahia, 2019).

7.6 FACTORS AFFECTING QUALITY OF FRESH FOODS

Fresh produce growers and exporters are constantly faced with the limitations of reduced shelf lives and quality of fresh foods due to the variability of a number of factors throughout the supply chain. These challenges may be presented as internal or external factors that play crucial roles in determining the shelf life and quality of fresh produce.

7.6.1 Inherent Attributes

Fresh produce continues to maintain two metabolic processes even after harvest: respiration and transpiration. Respiration consumes oxygen from the environment but releases heat (post-harvest, it is known as residual heat), carbon dioxide, and water. This is necessary for the tissues to continue to "live." Transpiration in fresh fruits and vegetables allows gaseous exchange. Due to this continued metabolic activity, there is a buildup of moisture, residual heat, and a significant amount of ethylene (in certain fruits and vegetables) that together can cause ripening, premature aging, fermentation, discoloration, and spoilage (presence of mold and other undesirable microcontaminants). Produce with a higher level of water activity as well as those that can release high concentrations of ethylene will have higher perishability presented as lower shelf life and fresh food quality after harvest. Arresting these internal factors is one of the keys to regulating shelf life and the quality of fresh foods.

7.6.2 Initial Microbial Load

The microbial contamination harnessed by fresh produce at harvest due to bad handling practices and incompetence in maintaining hygiene during packaging is bound to amplify during storage and transport. Introduction of sanitary and absolute handling practices can maximize shelf life and quality that would otherwise be compromised due to microbial growth and contamination arising out of poor arrangements during handling and storage.

7.6.3 Macrobiological Elements

When the moisture content surrounding produce is more than 11% and humidity falls between 10 and 35°C, insect and pest infestation can take place during storage. The infestation of rodents and mice in warehouses poses another risk to safety and quality of stored produce. Proper packaging and sanitation in storage areas are the most effective measures against infestation by macrobiological elements.

7.6.4 Time, Temperature, and Packaging

The shelf life and quality of fresh produce go hand in hand, as it is packed for transport after harvest. Temperature variations; humidity changes; spacing; number of pieces per primary, secondary, and tertiary level of packaging; and management of produce once received at destination warehouses each have a role in protecting the shelf life of fresh produce. Consultation with specialists and following expert advice on optimum temperature, range of storage humidity, packing spaces, and packaging materials during the supply chain is a must to reduce post-harvest losses and ensure fresh foods are maintained at their longest shelf life.

7.6.5 Consumer Handling

Displacement of fresh produce from market to our home storage cabinets and refrigerators exposes food to another set of temperature and humidity variables until it is consumed. Quality deterioration and shortening of shelf life can still happen during this interval. A higher than optimal temperature can aggravate physiological oxidation processes during transportation to home, and chances are that it can stimulate rapid microbial spoilage while it is stored in domestic spaces (Dauthy, 1995; Zweep, 2018).

7.7 FACTORS AFFECTING QUALITY OF REFRIGERATED FOODS

The primary objective of food preservation by refrigeration is to slow down processes that cause deterioration and spoilage and other unacceptable changes that lead to a decline in food quality and safety. These processes could be the following:

- *Physical processes:* Involve loss of texture or water (by evaporation).
- *Microbial processes:* Comprise growth and multiplication of bacteria, mold, and pathogens.
- *Chemical and biochemical process:* Cover enzymatic degradation or side reactions (Bøgh-Sørensen, 2006).

The question of how well food is preserved depends on certain factors such as pretreatment, packaging, refrigeration rates, final temperature, etc. The changes that food undergoes during refrigeration are the underlying factors that affect refrigeration and final product quality.

7.7.1 Cold Temperature Inducing Moisture Migration

Moisture loss is a common occurrence that plays a part in determining how effective refrigeration is for the quality preservation of food. When fresh products like meat or vegetables are left exposed directly to the cold air inside the refrigeration chamber (evaporator), moisture movement occurs from food to the cold air. This is due to greater vapor pressure of water on the food surface than that of the drier cold air contacting it. Texture, appearance, and juiciness may be lost in fruits and vegetables. Affected frozen meat products have a glassy appearance due to ice sublimation, which is also known as freezer burn. The loss of water from refrigerated food has the same economic value in mass as the food product itself, amounting to typical losses of about 1–2%, 1%, and 0.5% during chilling, freezing, and storage and transport per month, respectively, for meat processing (Pham and Willix, 1984; Pham, 1987). Not all is lost due to moisture migration; instead, certain degrees of loss of moisture are beneficial to refrigeration. In meat processing, a layer of sublimed ice on meat surfaces retards or prevents microbial spoilage due to a layer of low-level water activity protecting the meat. To minimize moisture migration, wrapping food with an impervious layer of packaging material, fast chilling, and freezing operations will keep losses in check. Alternately, introducing cold air with a high level of relative humidity (RH) (as close to 100% as possible), low temperature, and low velocity for unwrapped products will minimize losses. Also, keeping away from direct light, warmth, or any type of irradiating sources is a good practice.

7.7.2 Ice Crystal Formation, Ice Crystal Growth, and Thawing

For freezing, under moderate operations, the formation of ice crystals causes distortion and damage of cellular structures like cell membranes. Due to extracellular ice formation, not all is reabsorbed during thawing and generally results in drip losses. On the other hand, during fast freezing or instantaneous freezing such as with liquid nitrogen (LIN), smaller crystals form and are known as intracellular ice crystals. Due to these intracellular smaller crystals, lesser damage occurs, in principle, with some exceptions to this rule (Añón and Calvelo, 1980). The benefits of fast freezing are undone when large pieces of food such as meat chunks are frozen in packed cartons or boxes. During freezing operations in such cases, water migrates from smaller ice crystals, leading to their disappearance, to bigger ones, increasing the average size of ice crystals eventually (Pham and Mawson, 1997). This loss is amplified with fluctuating temperature through the entire cold chain between and during storage, transport, retail, and domestic refrigeration. Another factor with large chunks of food (e.g., frozen tuna) is cracking during freezing, as the water expands by about 9% in volume when turning into ice (Hung and Kim, 1996).

7.7.3 Physiological Forces

Chilling operations, on some level, inflict damage on the tissue structure of food under refrigeration. For fruits and vegetables, which rely heavily on visual appearance, color, texture, and other quality factors, these changes may be unwelcome. As these products respire, the design of chilling operations must take into account the heat released during respiration. Even though respiration slows down at low temperatures, and finally stops on freezing, there are chances of a chilling injury waiting to happen at temperatures before freezing can occur. Therefore, uniform air circulation supplemented with slots or holes in walls of cartons and containers used for packaging of respiring products must be employed in designs. To add to this, gaps in between products or product stacks will help carry the respiration heat away. For non-respiring products, proper flow in channels between and around the product stack must happen to keep away heat gain.

7.7.4 Biochemical Forces

Most biochemical reactions such as ripening, maturity, senescence, vitamin degradation, starch-sugar coupled reactions, oxidation, and other metabolic reactions are slowed down by refrigeration, and even stops at freezing (most reactions). Some biochemical reactions, however, continue to occur near or even beyond freezing points. Due to the concentration effect of solutes in the food matrix, enzymatic reactions are favored, such as protein denaturation proceeding rapidly just below freezing points. Lipid auto-oxidation also occurs even beyond freezing is established, producing rancidity in fish and meat products. Some of these reactions are desirable before carrying out chilling or freezing operations; for example, beef meat after slaughter, if put immediately to chilling below 15°C, develops cold shortening and the meat becomes very hard. To avoid this, it is electrically stimulated and proceeds to aging. Other undesirable cases of enzymatic reactions can be eliminated by blanching pretreatment before freezing.

7.7.5 Microbial Activity

Pathogenic and spoilage microorganisms are detrimental to stored food. They not only deteriorate the food quality but are also sources of ailments and food poisoning. There is no doubt that they destroy the taste, appearance, and smell of food, but a food devoid of such unpleasant qualities may still be a containment zone for dangerous microorganisms. Refrigeration aims to stop such accidents and food losses from happening. Several factors contribute to the growth of microbes in food: temperature, pH, water activity, gaseous composition, preservatives, etc. Under all other optimal conditions for microbial activity, bacteria can grow in temperatures as low as $-7°C$, whereas some exceptional yeast and molds survive at even $-10°C$ (Gill, 2012). Because microorganisms are surface dwellers, the risk for cut, ground, or minced food is usually high, with minced food having the highest risk. Thawing also presents another risk because food items are left untouched above $0°C$ for a considerably long time, which may be long enough for microorganisms to kick start their activities. A practice of precaution would be to treat refrigeration as nothing more than a preservation technique and not a sterilization method (Tuan Pham, 2014).

7.8 SAFETY AND ENVIRONMENTAL CONSIDERATIONS

7.8.1 Safety

Keeping and preserving food inside the refrigerator to keep it safe also comes with certain safety checks and practice manuals as discussed in the following sections.

7.8.2 Internal Temperature

For safety, refrigerators must work to keep the inside temperature at $4°C$ or below. Some refrigeration units have inbuilt thermometers to keep track of internal temperatures. In case the unit is without one, a thermometer should be safely placed inside it for monitoring. In the event of power failures, refrigerators should be able to hold the internal temperature for some time. The same must be checked when the power comes back on. All consumables should be discarded if food is found to be held at temperatures above $4°C$ for more than 2 hours without power. Refrigeration doors should be closed tight, operated only when necessary, and closed in the shortest time possible.

7.8.3 Handling before Storage

Before food goes into a refrigeration unit it should be prepared for the process first. Hot foods can be chilled in ice or given a cold bath to take off some of the load (heat) before it goes into the refrigerator. Food in open containers can be moisture sealed to prevent water loss and odor gain. Larger portions of food should be divided into smaller ones for faster and uniform chilling and placed in shallow containers before refrigeration.

7.8.4 Shelves and Compartments

A refrigerator's internal environment should be $4°C$ or less throughout the entire unit so that any shelf or compartment is safe for storage. Some domestic units are equipped with adjustable shelves, chiller compartments, door bins, crispers, and drawers to optimize food storage space and convenience. Table 7.3 demonstrates the effect temperature-wise compartmentalization can have on the shelf life of refrigerated foods. Chiller compartments are designed for meat, poultry, or cheese storage, whereas crisper drawers are for fresh fruits and vegetables. Additionally, chiller compartments

Table 7.3 Recommended Temperatures for Storage of Common Food Items into Six Temperature Sections

Food Items	Freezer −18°C	Meat Chiller −1°C	Salad Chiller 2°C	Fresh 4°C	Cellar 12–14°C	Pantry 17°C
Bread						
Meat						
Fish and shellfish						
Eggs						
Cheese			When ripe			
Strawberries						
Broccoli, cauliflower, chicory						
Carrots, lettuce, spinach						
Leek, green onions, celery						
Mushrooms						
Apples, pears, kiwi			Max storage time			Ripening
Peaches, nectarines			Max storage time			Ripening
Milk						
Yogurt						
Butter/margarine						
Beans						
Citrus fruits						
Avocados						
Bell peppers, eggplant						
Courgette, cucumber						
Onions						
Potatoes						
Bananas, pineapple, mango, melon						Ripening
Tomatoes						Ripening

Note: Shaded cells show recommended temperature for storage of corresponding items.
Source: Adapted from van Holsteijn and Kemna (2018), with permission.

come equipped with an adjustable temperature setup and unique cool air vents to keep food chilled. Also, crisper drawers come with customizable settings to allow the user to control humidity inside the drawers: low for fruits and high for vegetables. Figure 7.4 represents the extended shelf life of some refrigerated items with such improved and organized compartmentalization refrigeration.

7.8.5 Environmental Considerations

The use of refrigeration equipment is known to contribute to global warming: directly when the greenhouse gases (refrigerants) are released from leakages through gradual use or breakdowns and indirectly from the power stations burning fossil fuels to supply power for consumption. To combat such a scenario, alternative systems and system designs are being proposed worldwide in a push to end greenhouse gas emissions. In this attempt, the total equivalent warming impact (TEWI) factor is being increasingly used to

Extended shelf lives with improved refrigerator

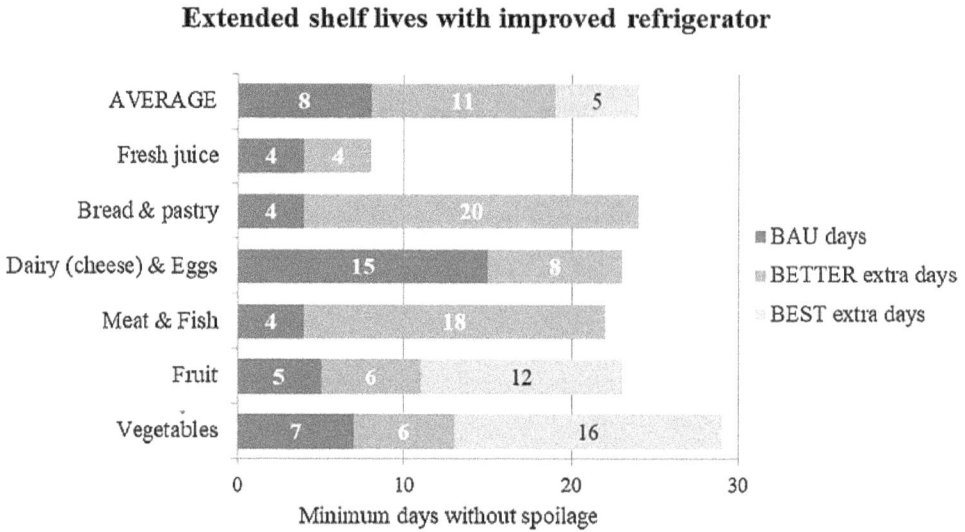

Figure 7.4   Effect of improved refrigeration on the shelf life of common food items. *Abbreviations:* BAU, business as usual refrigeration (conventional); BETTER and BEST, organized refrigeration (improved). (Adapted from van Holsteijn and Kemna, 2018, with permission.)

- Account for both direct and indirect contributions to global warming and compare alternative feasible systems and
- Measure and compare the impact of alternative system designs and refrigerants in a life cycle on global warming.

Such calculations are relevant only when measuring and comparing systems that are designed to address the same application requirements. TEWI has two basic components (BRA, 2006):

1. The amount of refrigerant released in a lifetime and that which is unrecovered during scrapping of the unit converted into an equivalent mass of CO_2
2. The impact of supplying energy for the equipment's operation by burning fossil fuel throughout its lifetime in kWh units converted into an equivalent mass of CO_2 (Klemes et al., 2008).

7.9 TYPES OF SPOILAGE MICROORGANISMS IN FRESH AND REFRIGERATED FOODS

Microorganisms are present everywhere including food, soil, water, air, or contact surfaces. Bacteria, molds, yeasts, and other microbes contaminate our living spaces, and grow on nutrient-rich food, making it unsuitable for human consumption. When the environment is dry and acidic, yeast and molds thrive. Molds are also found contaminating moisture-rich environments and on the surface of spoiled foods. However, bacteria are known to be the leading cause of food spoilage and food poisoning cases, especially in foods high in moisture (Dhil Engineering, 2014). The types of bacteria that contaminate food can be pathogenic or spoilage. Pathogenic bacteria, as the name implies, are carriers and cause foodborne illnesses, whereas spoilage bacteria, as the name suggests, cause spoilage and deterioration prematurely. Contamination by anyone or both types of bacteria in food can easily defeat the purpose of refrigeration.

Pathogenic bacteria are most commonly present in the temperature range of 4–60°C, which is marked as a danger zone. Unlike spoilage bacteria, which offsets a warning signal (smell or taste) to the consumer, pathogenic bacteria is undetectable visually, which make them a special concern when refrigerating. They generally do not produce any changes to the quality of food like odor, taste, and texture.

Spoilage bacteria are dwellers even in low-temperature conditions of refrigeration. Their spoilage activities cause the food to produce a foul smell and/or taste. Such food is deemed unsafe to eat and if consumed, results in sickness and illness. Food that has been left out over the counter for a long period could seem fine but may be dangerous to consume. Food that has been in refrigerated storage for too long could have deteriorated in quality but not enough to cause detectable changes. It is a question of safety versus quality (U.S. Department of Agriculture, Food Safety and Inspection Service, 2015).

7.10 SHELF LIFE AND HANDLING OF FRESH AND REFRIGERATED FOODS

Food storage at proper temperatures maintains safety and shelf life, thereby keeping favorable attributes of food such as color, flavor, texture, nutrients. and other qualities intact. All perishable foods such as fresh fruits, vegetables, meat, poultry, and other items need ambient storage and refrigeration, if not treated with other preservation operations. To make the best of one's fresh and refrigerated produce buy from the market, knowledge about their shelf life and proper handling (storage at the right temperature) is a must. Some common fresh and refrigerated foods and their shelf lives under ambient storage temperatures are given in Table 7.4. Even when processed, say cooking, the cooked leftover food will need safe cold temperature storage to prevent it from spoilage. When stored in cold or freezing temperatures, almost all of the factors that work to make food undergo spoilage are suppressed and food stays fresh and safe for longer periods than usual. The shelf life of these food items are thus increased with refrigeration by reducing the rate at which food deteriorates. Therefore, it is of utmost importance to utilize refrigeration operations effectively to get useful results at best. Some commonly refrigerated foods and their storage life data are given in Table 7.5.

7.10.1 Handling of Fresh Foods

Perishable food items such as fruits and vegetables are associated with residual heat due to the continuation of respiration and other metabolic activities even after harvest. This heat is an enemy to the shelf life of fresh produce and must be dealt with immediately. A cost-effective and economical method to do this is to use precooling techniques prior to induction into the supply chain. Precooling works by removing some of the residual heat and arresting metabolic activities that may cause produce ripening and spoilage.

7.10.2 Precooling

Different techniques are employed to achieve precooling in freshly harvested produce. Although different, these techniques justify the principle and objective of precooling all the same; their utilization depends on nature and type of produce, design and type of packaging, cooling capacity, economic efficiency, and resource availability. A comparison of some of the preferred methods of precooling is briefly illustrated in Table 7.6.

7.10.3 Supplemental Atmosphere

In vital supplementation to precooling, controlled atmospheric and modified atmospheric conditions are used to maintain produce shelf life. Controlled atmosphere creates a dynamic environment

Table 7.4   Common Fresh Produce and Their Shelf Life under Room Temperature and Refrigerated Storage

Produce	How Long It Lasts
Apples	4–8 weeks in the fridge
Avocado	4–7 days at room temperature
Bananas	2–5 days at room temperature
Blueberries	1–2 weeks in the fridge
Broccoli	7–14 days in the fridge
Carrots	3–4 weeks in the fridge
Cucumbers	1 week in the fridge
Garlic	3–6 months at room temperature
Iceberg and romaine lettuce	7–10 days in the fridge
Lemons	3–4 weeks in the fridge
Onions	2–3 months at room temperature
Oranges	3–4 weeks in the fridge
Peaches	1–3 days at room temperature
Potatoes	3–5 weeks in the pantry
Strawberries	3–7 days in the fridge
String beans	3–5 days in the fridge
Tomatoes	1 week at room temperature
Watermelon	7–10 days at room temperature
Whole mushrooms	7–10 days in the fridge
Zucchini	4–5 days in the fridge
Chicken or turkey (whole/pieces)	1–2 days in the fridge
Eggs, raw in shell	3–5 weeks in the fridge
Fresh beef, veal, lamb or pork	3–5 days in the fridge
Fresh fish or shellfish	1–2 days in the fridge
Raw egg yolks, whites	2–4 days in the fridge

Sources: Doster (2020; https://www.tasteofhome.com/article/heres-how-long-your-fresh-produce-will-really-last/); Institute of Agriculture and Natural Resources (2021; https://food.unl.edu/free-resource/food-storage#meatpoultryfisheggs).

in which there is a balance of gaseous constituents (carbon dioxide, oxygen, ethylene, etc.) through the use of gas generators and quenchers. This tunes down the produce respiration to a level that is safe for storage. Modified atmosphere makes use of a saturated environment that is achieved over time due to produce respiration in a sealed package. It may be passive (produce gradually creates its own saturated environment gradually) or active (package is flushed with variable quantities of gases that modulate metabolism). Some combinations of storage temperature and gases for fresh produce are represented in Table 7.7.

7.10.4 Handling of Refrigerated Foods

Storing food at refrigeration temperatures not only reduces incidences of fool spoilage and poisoning but also maintains and/or preserves certain nutritional qualities and organoleptic properties of food to an extent. Fluctuations in cold temperature due to transit periods during handling throughout the cold chain can lead to catastrophic failures in refrigeration. A list of common food items are given in Table 7.8 along with the refrigeration actions to be taken.

Table 7.5   Cold Storage Chart for Meat, Poultry, and Egg Products

Storage Times for Refrigerated Foods	
Ground Meat, Ground Poultry, and Stew Meat	
Ground beef, turkey, veal, pork, lamb	1–2 days
Stew meats	1–2 days
Fresh meat (beef, veal, lamb, and pork)	
Steaks, chops, roasts	3–5 days
Variety meats (tongue, kidneys, liver, heart, chitterlings)	1–2 days
Fresh poultry	
Chicken or turkey, whole	1–2 days
Chicken or turkey, parts	1–2 days
Giblets	1–2 days
Bacon and sausage	
Bacon	7 days
Sausage, raw from meat or poultry	1–2 days
Smoked breakfast links, patties	7 days
Summer sausage labeled "Keep Refrigerated"	Unopened, 3 months; Opened, 3 weeks
Hard sausage (such as pepperoni)	2–3 weeks
Ham, corned beef	
Ham, canned, labeled "Keep Refrigerated"	Unopened, 6–9 months; Opened, 3–5 days
Ham, fully cooked, whole	7 days
Ham, fully cooked, half	3–5 days
Ham, fully cooked, slices	3–4 days
Corned beef in pouch with pickling juices	5–7 days
Hot Dogs and Luncheon Meats	
Hot dogs	Unopened package, 2 weeks; Opened package, 1 week
Luncheon meats	Unopened package, 2 weeks; Opened package, 3–5 days
Deli and vacuum-packed products	
Store-prepared (or homemade) egg, chicken, tuna, ham, and macaroni salads	3–5 days
Pre-stuffed pork, lamb chops, and chicken breasts	1 day
Store-cooked dinners and entrees	3–4 days
Commercial brand vacuum-packed dinners with/USDA seal, unopened	2 weeks
Cooked meat, poultry, and fish leftovers	
Pieces and cooked casseroles	3–4 days
Gravy and broth, patties, and nuggets	3–4 days
Soups and stews	3–4 days
Fresh fish and shellfish	
Fresh fish and shellfish	1–2 days
Eggs	
Fresh, in shell	3–5 weeks
Raw yolks, whites	2–4 days
Hard-cooked	1 week
Liquid pasteurized eggs, egg substitutes	Unopened, 10 days; Opened, 3 days
Cooked egg dishes	3–4 days

Source: U.S. Department of Agriculture (2015).

Table 7.6   Comparative Analysis of Different Operating Parameters of Various Precooling Techniques

Variable	Cooling Method				
	Ice Cooling	Hydro Cooling	Vacuum Cooling	Forced-Air Cooling	Room Cooling
Cooling times (h)	0.1–0.3	0.1–1.0	0.3–2.0	1.0–10.0	20–100
Water contact with the product	Yes	Yes	No	No	No
Product moisture loss (%)	0–0.5	0–0.5	2.0–4.0	0.1–2.0	0.1–2.0
Capital cost	High	Low	Medium	Low	Low
Energy efficiency	Low	High	High	Low	Low

Source: Kader and Rolle (2004).

Table 7.7   Combination of Gases and Temperature for Use in Supplemented Atmospheric Storage of Fresh Produce

Vegetables	Temperature[a]	Atmosphere[b]	
	(°C)	O_2	CO_2
Broccoli	5–10	1–2	1–10
Brussels	0–5	1–2	5–7
Cantaloupe	2–7	3–5	1–20
Cauliflower	0–5	2–3	3–4
Cucumber	8–12	1–4	0
Lettuce (crispy)	4–5	1–3	0
Spinach	4–5	7–10	5–10
Tomatoes			
Mature green	12–14	3–5	2–3
Ripe	10–14	3–5	3–5

[a] Optimum and range of usual and/or recommended temperatures. RH of 90–95% is usually recommended.
[b] Specific controlled atmosphere recommendations depend on the cultivar, temperature, and duration of storage.
Source: Kader and Saltveit (2003).

Table 7.8   Refrigeration Actions for Some Common Food Items

Food Item	Action: Refrigerate
Butter	In the icebox
Cheese	Soft cheeses (Brie and Mozzarella) or follow maker's instructions
Corn on the cob	With or without husk for up to 2 days
Eggs	Recommended by USFDA
Frosted cupcakes	Immediately
Jelly	Once opened
Ketchup	To maintain flavor and freshness
Maple syrup	Once opened, up to 1 year
Melon	Once sliced
Natural peanut butter	For use longer than 1 month
Nuts	In an airtight container for use longer than 1 month
Pies	Within 2 hours
Sliced tomatoes	Immediately
Tortillas	Once opened

Source: Food Network (2020).

To ensure continuity of this cold chain, here are some recommendations, right from bringing food home from retailers to storing them in our domestic units and finally consuming it.

7.10.5 While Buying Food

a. Save the most fragile and perishable frozen products from the freezer for the last.
b. Use shopping bags with thermal insulation for perishable food items.
c. Ensure your means of mobility is tuned to a temperature fit to carry perishables. Use shades or screens when you leave it in the parking lot.
d. The less transit time between your shopping center and your home, the better.
e. Pick out the most temperature-sensitive items and place them so that they are the first ones to go into the refrigerator once you reach home.

7.10.6 While Storing Food

a. Place raw meat, fish, domestic preparations, and leftovers in the coldest part of your refrigeration unit (below 4°C). For leftover items, airtight containers or plastic food wraps should be used.
b. Remove cardboard or any secondary packaging materials that come with food to keep contamination at bay.
c. Hot food must be precooled outside to avoid internal condensation and overloading the fridge.
d. Manually check internal temperatures regularly and make sure not to overload the refrigeration unit.

7.10.7 While Consuming Food

a. Minimize withdrawal time of sensitive food items. Their exposure to room temperature must be minimal and returned to storage immediately.
b. Rapidly consume sensitive food items once their primary packaging is pried open. Leftovers make the top of the priority list, which when left for too long at ambient temperatures should be thrown away.
c. Strictly follow the user before or use by date (ANSES, 2016a,b).

7.11 TEMPERATURE AND RELATIVE HUMIDITY REQUIREMENTS OF FRESH AND REFRIGERATED FOODS

Uniform and unrestricted air circulation inside storage chambers are necessary to avoid development of warm and cold spots (can cause produce to sweat), decay, and deterioration. Thermometers can be installed in warehouses to check for temperature fluctuations. RH can be recorded with a hygrometer and kept in check using a humidistat. It is a well-known fact that refrigeration can arrest the speed of food degradation and various biochemical reactions that cause ripening and maturity in fruits and vegetables. This significantly affects the storage life of food items. Storage life depends on water activity, which in turn is a function of available water and its temperature. The cold temperature required to achieve preservation status refers to the temperature of this moisture inside food. To achieve this, refrigeration, which may be preceded by a precooling operation, starts right after harvest to get rid of post-harvest heat and continues throughout the cold chain until the domestic refrigerator for a short time of storage until consumption. The shelf life of most fruits and vegetables improves when storage is controlled at a temperature between 0 and 10°C.

Along with cold temperature requirements, food items under refrigeration need a RH balance for preservation to be effective. It is a common recommendation to increase RH closer to saturation to prevent condensation inside the unit and loss of moisture from the food items. In stores, water condenses when the air temperature is above refrigerant temperature while frosting on cooling coils when the refrigerant temperature is less than the freezing point of water. This reduces the cooling efficiency of the unit. When moisture is removed from stored air or there is a vapor pressure gradient toward the store air from the surface of the food item, the food loses moisture (vapor sub-limation). Food temperature is to be maintained by balancing all other factors such as removal of respiratory heat by the refrigerant, heat from fans or leakage through doors, cracks, and insulation within the store (Bodbodak and Moshfeghifar, 2016). Temperature and RH requirements of common food items are shown in Tables 7.9–7.11.

Table 7.9 Practical Shelf Lives of Various Food Items with Temperature and Humidity Range Requirements: Part I

Products	Optimum Temperature (°C)	Relative Humidity (%)	Storage Life
Fresh products			
Apple	−1 to 4.5	90–95	1–12 months
Asparagus	2	95–100	2–3 weeks
Banana	13–15	90–95	1–4 weeks
Blueberries	0.5–1	90–95	2–3 weeks
Broccoli	0	95–100	2 weeks
Cabbage	0	95	1 week
Cucumber	7–10	95	1 week
Eggplant	8–12	90–95	2–3 weeks
Ginger	13	65	5–10 weeks
Guava	5–10	90	1–2 weeks
Green beans and field peas	3–7	95	2–3 months
Leafy vegetables	0	95	1–3 weeks
Lime	9–10	85–90	3–5 weeks
Mango	13	90–95	2–4 weeks
Onion	0	70	2–3 weeks
Papaya	7–13	8590	1–3 weeks
Passion fruit	7–10	85–90	3–5 weeks
Peach	0	95–98	2–4 weeks
Peppers	7–10	90–95	3–5 weeks
Potato	3–4.5	90–95	5–8 months
Sweet corn	0	90–95	5–7 days
Sweet potato	13	90	6–12 months
Tomato, pink	9–10	85–95	7–14 days
Turnip	0	95	4–5 months
Watermelon	10–15.5	90	2–3 weeks
Fish	0–3	90	2–3 weeks
Milk	4	—	3–5 days
Cooked food			
Vegetables	0–4	—	2–4 days
Fish	0–4	—	2–3 days
Meat	0–4	—	3–5 days
Soup	0–4	—	2–3 days

Source: Cooling India (2018).

Table 7.10   Practical Shelf Lives of Various Food Items with Temperature and Humidity Range
Requirements: Part II

Food Item	Storage Temperature (°C)	Relative Humidity (%)	Maximum Recommended Storage Times	Storage Time in Cold Storage for Vegetables in Tropical Countries
Apples	0–4	90–95	2–6 months	
Beetroot	0	95–99	4–6 months	
Cabbage	0	95–99	5–6 months	2 months
Carrots	0	98–100	5–9 months	2 months
Cauliflower	0	95	3–4 weeks	
Cucumber	10–13	90–95	10–14 days	
Eggplant	8–12	90–95	7 days	
Lettuce	0	95–100	2–3 weeks	
Melons	7–10	90–95	2 weeks	
Mushrooms	0–4	95	2–5 days	1 day
Onions	0	65–70	6–8 months	
Oranges	4	85–90	3–4 months	
Peas, green	0	90–95	1–2 weeks	
Pears	0	90–95	2–5 months	
Potatoes	4–16	90–95	2–8 months	
Pumpkin	10–13	70–75	6–8 months	
Spinach	0	95	1–2 weeks	1 week
Tomatoes	13–21	85–90	1–2 weeks	1 week

Source: https://nptel.ac.in/content/storage2/courses/112105129/pdf/RAC%20Lecture%203.pdf (NPTEL, 2008).

7.12 QUALITY OF FRESH AND REFRIGERATED FOODS

7.12.1 Quality of Fresh Foods

The sensitivity of fresh produce requires post-harvest fast shipping and storage under proper temperature-humidity conditions. Disruptions in the supply chain or impairment in storage conditions can seriously affect the quality of fresh foods. Generally, the quality of fresh foods are based on the following.

7.12.2 Appearance, Texture, and Flavor

This set (appearance, texture, and flavor) forms one of the most important aspects of fresh produce, a quality that a consumer strongly invests in. Before buying, one makes a quick visual assessment of the size, shape, color, freshness, and/or presence or absence of defects (e.g., blemishes). A trained hand may be able to assess desirable texture characteristics of fresh food by the slightest application of pressure from fingers or by a gentle tapping. Taste and smell are another superficial way to assess fresh food quality.

7.12.3 Nutrition

The nutrition profile of fresh foods is another of its attractions. Because it is free from processing and additives, its contents remain intact and as they are at the time of harvest. With careful handling and storage practices, consumers are able to reap the full health benefits provided by fresh foods.

Table 7.11   Practical Shelf Lives of Various Food Items with Temperature and Humidity Range Requirements: Part III

Commodity	Temperature (°C)	Relative Humidity (%)	Storage Time
Apple	−4	90–98	2–8 months
Apricots	−0.5	90–95	1–3 weeks
Asparagus	0–2	95–97	2–3 weeks
Avocado	7–13	85–90	4–5 weeks
Beans, green	4–7	90–95	2–3 weeks
Beet root	0–2	95–97	3–5 months
Blackberry	−0.5	95–97	1–2 weeks
Brinjal	0–2	90–95	1–2 weeks
Broccoli	0–2	90–95	10–14 days
Cabbage	0–2	90–95	1–2 months
Carrots	0–2	90–95	4–6 months
Cauliflower	0–2	90–95	2–4 weeks
Cherries	−1 −0.5	90–95	2–4 weeks
Cucumber	7–10	90–95	10–14 days
Grapes	−2	85–90	2–4 weeks
Lemons	4–15	86–88	2–6 weeks
Lettuce	0–1	95–98	2–3 weeks
Lime	3–10	85–90	2–4 weeks
Mango	11–18	85–90	5–7 weeks
Melon water	2–4	85–90	2–3 weeks
Onions	0–2	65–70	6–7 months
Orange	0–10	85–90	2–4 weeks
Peach	−2	88–92	4–6 weeks
Potato	1.5–4	90–94	4–9 months
Pears	0–2	90–95	2–7 months
Peas	0–2	90–95	1–3 weeks
Spinach	0–2	90–95	10–14 days

Source: Camelo (2004).

7.12.4 Maturity and Harvesting

The stage at which food is harvested, the time of day, and the methods used therein are crucial to its quality. Climacteric fruits and vegetables should be harvested at mature but green stage, whereas non-climacteric crops should be harvested at full maturity. Harvest should be made and completed during the coolest time of the day using the method that produces minimum damage to harvested crops.

7.12.5 Handling and Transport

Post-harvest handling and transportation are the final crucial stages to ensure a yield of high-quality fresh foods. Precleaning, precooling, and sorting may be done either at the field or at packaging warehouses.

7.12.6 Packaging and Storage

Packaging is a multipurpose enclosure that prevents produce from physical injuries, retains internal optimal environment for maintaining shelf life, reduces external contamination, and eases handling and transportation. All of these contribute to the final quality of produce. Temperature and humidity requirements are set by storage conditions to ensure produce obtains its longest shelf life.

7.12.7 Miscellaneous

Sanitary, safe, and good handling practices during harvest, transport, packaging, marketing, and final preparation are also necessary to deliver a safe, fresh food with uncompromised quality (Singh et al., 2019).

7.12.8 Quality of Refrigerated Foods

Factors that affect refrigeration operations, such as physical, biological, physiological, chemical, and microbiological processes, are the same ones that influence the quality of refrigerated foods. As such, the extent of changes set forth by these processes, even under the influence of cold temperature preservation, are directly responsible for the outcome of the final quality of these foods. Quality is often affected by these four parameters of refrigeration.

7.12.9 Cooling and Freezing Rate

For freezing, food texture is influenced by the kind of ice crystals that are formed during freezing. The smaller the size, the greater the number, and the faster the ice crystals are formed, the better. In short, rapid freezing or instantaneous freezing favors an acceptable food quality. The rate of cooling and freezing depends on the following:

a. *Composition:* Presence of food components that hinder heat transfer, e.g., proteins and fats acting as insulators
b. *Temperature gradient:* The higher the gradient, the faster the rate
c. *Product dimension and rate of heat transfer:* The shorter the distance of the minor axis from surface to center, the greater the heat transfer and, hence, the faster the rate
d. *Cold air circulation:* Higher air velocity leads to faster heat exchange and faster rate
e. *Contact surface:* The larger the contact area, the faster the rate

7.12.10 Final Temperature

Whether you have to cool or freeze food depends on several factors: texture, perishability, period of storage, chemical changes, etc. So, food could be chilled at 4°C, frozen at −10°C, or deep frozen at −18°C. However, the rate of deterioration is highly governed by the food's internal changes due to chemical processes and its physical structure. Table 7.12 gives a list of some frozen foods and the effect of temperature on storage life.

7.12.11 Storage Temperature Stability

Temperature fluctuations, no matter how minor, in refrigeration can release immobilized water in the form of ice crystals during freezing. As the temperature rises, not all water is reabsorbed but

Table 7.12   Some Foods and Effect of Temperature on Storage Life

Product	Storage Temperature		
	−18°C	−12°C	−6.7°C
Beef (raw)	13–14 days	5 days	<2 days
Fat fish (raw)	2 days	1.5 days	24 days
Lean fish (raw)	3 days	<2 days	<1.5 days
Peaches	12 days	<2 days	6 days
Pork (raw)	10 days	<4 days	<1.5 days
Raw chicken (well packed)	27 days	15.5 days	<8 days
Spinach	6–7 days	<3 days	1 days
Strawberries	12 days	2.4 days	10 days
Turkey pies	>30 days	9.5 days	2.5 days

Source: UBC Wiki (2020).

expelled as drip, which refreezes when the temperature goes back to normal. This results in the movement of water from smaller ice crystals to bigger ones. Ultimately, fewer and bigger ice crystals are formed, thereby damaging the tissue structure of the food.

7.12.12 Thawing Rate

Refrigerated foods, particularly frozen ones, are at their highest susceptibility when being thawed. Ice is a good heat conductor and better than water. When food is thawed, the outermost ice melts forming a layer of liquid water. This layer acts as a barrier to heat transfer from the outside environment to the next subsequent layers of ice. The reverse is true for freezing as the outermost layer of ice helps freezing proceed faster to the core. Thus, thawing takes longer than freezing, so the fate of the food is left at the activities of microorganisms and other internal processes set to resume as soon as cold temperature preservation is withdrawn. So, rapid thawing methods help retain the maximum quality of food (Lorentzen, 1978; UBC Wiki, 2020).

7.13 ADVANTAGES AND DISADVANTAGES OF REFRIGERATION

7.13.1 Advantages

a. Refrigeration is the backbone of cold chain storage, transport, and distribution. It helps maintain cold chain continuity and hamper microbial developments.
b. End consumers/customers are benefited the most, as they can start further processing in the best quality conditions.
c. Refrigeration depresses molecular mobility and enhances the shelf life of all stored foods.
d. There exists different equipment that fulfils different application requirements. With proper application methods, refrigeration is one of the easiest, most flexible preservation operations with minimal changes in the initial quality of food items.
e. Operating (energy consumption) cost could be subsidized, as per government regulation, in developing countries. Lowering percentages or reducing the unit price for energy consumption can enhance production, making it economically convenient.

7.13.2 Disadvantages

a. Some foodborne pathogens such as *Clostridium botulinum* type E, *Yersinia enterocolitica*, entero-toxigenic *Escherichia coli*, *Listeria monocytogenes,* and *Aeromonas hydrophilia* can easily survive refrigeration temperatures (around 4°C).

b. Another group of pathogens such as *Salmonella* sp., *Staphylococcus aureus*, *Vibrio parahaemolyticus*, and *Bacillus cereus* can grow for longer periods, even around 4°C. Contamination with any of these pathogens could prove to be harmful.

c. Temperature abuse and fluctuations not only increase hazards from these types of microorganisms but also create losses in the form of weight (due to dripping) and texture (due to large ice crystals).

d. Thawing, if not handled properly and rapidly carried out, is another source of safety concern from a microbial point of view.

e. High initial capital investment and high running costs may be an economic burden to developing countries in terms of feasibility of the process for longer periods.

f. Operating refrigeration equipment contributes to greenhouse gases directly through emissions, leakage, or release of refrigerants and indirectly through supplying energy by running fossil fuel-based power plants throughout the lifetime of a unit (Palumbo, 1986; Barbosa-Cánovas et al., 2005).

7.14 CONCLUSION

A large quantity of food produced is spoiled because there are no systems in place for storage and preservation. According to the Food and Agriculture Organization (FAO), high-income countries around the world lose or waste about 670 million tons of food and middle- and low-income countries account for 630 million tons, together amounting to a whopping 1.3 billion tons of food wasted every year. In its 2017 report, "The Future of Food and Agriculture – Trends and Challenges," FAO quotes 800 million as the number of hungry people in the world and 2 billion people suffering from nutrient deficiency on a global scale (FAO, 2017). Agricultural lands are becoming limited, and the world needs to feed more mouths. The world population stands at almost 8 billion and to feed them reducing losses and saving is the key. Refrigeration can help achieve this and certainly has a great deal to contribute to global food security.

REFERENCES

Añón, M. C., & Calvelo, A. (1980). Freezing rate effects on the drip loss of frozen beef. *Meat Science*, *4*(1), 1–14.

ANSES. (2016a). The importance of cold-chain continuity. Accessed 10 August 2020. Available at: https://www.anses.fr/en/content/importance-cold-chain-continuity

ANSES. (2016b). The refrigerator and food hygiene. Accessed 10 August 2020. Available at: https://www.anses.fr/en/content/refrigerator-and-food-hygiene

Barbosa-Cánovas, G. V., Altunakar, B., & Mejía-Lorío, D. J. (2005). *Freezing of Fruits and Vegetables: an Agribusiness Alternative for Rural and Semi-rural Areas* (Vol. 158). Food & Agriculture Organization. Accessed 10 August 2020. Available at: https://www.fao.org/3/y5979e/y5979e00.htm#Contents

Berk, Z. (2018). *Food Process Engineering and Technology* (pp. 439–460). Academic Press.

Bodbodak, S., & Moshfeghifar, M. (2016). Advances in controlled atmosphere storage of fruits and vegetables. In *Eco-friendly Technology for Postharvest Produce Quality* (pp. 39–76). Academic Press.

Bøgh-Sørensen, L. (2006). *Recommendations for the Processing and Handling of Frozen Foods*. International Institute of Refrigeration.

BRA. (2006). *Guideline Methods of Calculating TEWI, Issue 2*. British Refrigeration Association.

Camelo, A.F.L. (2004). *Manual for the Preparation and Sale of Fruits and Vegetables: From Field to Market* (Vol. 151). Food & Agriculture Organization.

Cooling India. (2018). *Refrigeration in Food Processing & Cold Chain.* Accessed 10 August 2020. Available at: https://www.coolingindia.in/refrigeration-in-food-processing-cold-chain/

Dauthy, M. E. (1995). *Fruit and Vegetable Processing* (pp. 1–6). Food & Agriculture Organization.

Dhil Engineering. (2014). *Refrigeration and Freezing of Foods.* Accessed 10 August 2020. Available at: http://dhilreefer.blogspot.com/2014/11/refrigeration-and-freezing-of-foods.html#:%7E:text=Refrigeration%20and%20freezing%20of%20perishable,and%20the%20loss%20of%20quality

Doster, N. (2020). *Here's How Long Your Fresh Produce Will Really Last.* Taste of Home. Accessed 30 June 2021. Available at: https://www.tasteofhome.com/article/heres-how-long-your-fresh-produce-will-really-last/

e-Krishi Shiksha. (2013). *Agricultural Engineering (Version 2.0). (2013) Lesson 31: Refrigeration Systems.* Accessed 10 August 2020. Available at: http://ecoursesonline.iasri.res.in/mod/page/view.php?id=124105

Food & Agriculture Organization (FAO). (2017). The future of food and agriculture – trends and challenges. *Annual Report.* FAO: Rome.

Food Network. (2020). Foods that must be refrigerated. Accessed 10 August 2020. Available at: https://www.foodnetwork.com/healthy/packages/healthy-every-week/healthy-tips/foods-that-must-be-refrigerated

Gill, C.O. (2012). *Handbook of Frozen Food Processing and Packaging* (pp. 83–100). Marcel Dekker.

Hung, Y. C., & Kim, N. K. (1996). Fundamental aspects of freeze-cracking. *Food Technology (Chicago), 50*(12), 59–61.

Institute of Agriculture and Natural Resources. (2021) *Home Food Storage.* Accessed 30 June 2021. Available at: https://food.unl.edu/free-resource/food-storage#meatpoultryfisheggs

Kader, A. A., & Rolle, R. S. (2004). *The Role of Post-harvest Management in Assuring the Quality and Safety of Horticultural Produce* (Vol. 152). Food & Agriculture Organization.

Kader, A. A., & Saltveit, M. E. (2003). Respiration and gas exchange. In Bartz, J., & Brecht, J., editors. *Post-harvest Physiology and Pathology of Vegetables* (2nd ed.). Marcel Dekker, Inc.

Klemes, J., Smith, R., & Kim, J. K. (Eds.). (2008). *Handbook of Water and Energy Management in Food Processing.* Elsevier.

Lorentzen, G. (1978). Food preservation by refrigeration, a general introduction. *International Journal of Refrigeration, 1*(1), 9–12.

NPTEL. (2008). *Applications of Refrigeration and Air Conditioning.* (Version 1 ME, IIT Kharagpur). Accessed 10 August 2020. Available at: https://nptel.ac.in/content/storage2/courses/112105129/pdf/RAC%20Lecture%203.pdf

Palumbo, S. A. (1986). Is refrigeration enough to restrain foodborne pathogens?. *Journal of Food Protection, 49*(12), 1003–1009.

Persson, P. O., & Londahl, G. (1993). Freezing technology. In Mallett, C. P., editor. *Frozen Food Technology* (pp. 20–58). Blackie Academic and Professional.

Pham, Q. T. (1987). Moisture transfer due to temperature changes or fluctuations. *Journal of Food Engineering, 6*(1), 33–49.

Pham, Q. T., & Mawson, R. F. (1997). Moisture migration and ice recrystallization in frozen foods. In *Quality in Frozen Food* (pp. 67–91). Springer.

Pham, Q. T., & Willix, J. (1984). A model for food desiccation in frozen storage. *Journal of Food Science, 49*(5), 1275–1281.

Powitz, R. W. (2005). Cold, hard facts about refrigeration equipment. *Foodsafety Magazine.* Accessed 10 August 2020. Available at: https://www.food-safety.com/articles/4684-the-cold-hard-facts-about-refrigeration-equipment

Shannon, L. (2011). *Refrigeration and Freezing for Food Preservation.* US Cooler. Accessed 10 August 2020. Available at: https://www.uscooler.com/blog/refrigeration-and-freezing-for-food-preservation/

Singh, A., Chaurasiya, A., & Mitra, S. (2019). Quality of fresh and processed products. In *Trends & Prospects in Post Harvest Management of Horticultural Crops* (pp. 135–155). Today and Tomorrows Printers and Publishers. Available at: https://www.researchgate.net/publication/336071435_QUALITY_OF_FRESH_AND_PROCESSED_PRODUCTS

Stoecker, W. F., & Jones, J. W. (1982). Compressors. *Refrigeration and Air Conditioning* (2nd ed.). McGraw-Hill.

Tuan Pham, Q. (2014). Refrigeration in food preservation and processing. In Bhattacharya, S., editor. *Conventional and Advanced Food Processing Technologies* (pp. 357–386). John Wiley & Sons.

UBC Wiki. (2020). *Preservation of Foods by Low Temperature.* (Course:FNH200/Lessons/Lesson 07). Accessed 10 August 2020. Available at: https://wiki.ubc.ca/Course:FNH200/Lessons/Lesson_07

U.S. Department of Agriculture, Food Safety and Inspection Service. (2015). *Safe Food Handling. Refrigeration and Food Safety.* Accessed 10 August 2020. Available at: https://www.fsis.usda.gov/ wps/portal/fsis/topics/food-safety-education/get-answers/food-safety-fact-sheets/safe-food-handling/ refrigeration-and-food-safety/ct_index

van Holsteijn, F., & Kemna, R. (2018). Minimizing food waste by improving storage conditions in household refrigeration. *Resources, Conservation and Recycling, 128,* 25–31.

Yahia, E. M. (Ed.). (2019). *Postharvest Technology of Perishable Horticultural Commodities* (pp. 209–270). Woodhead Publishing.

Zweep, C. (2018). Determining product shelf life. Accessed 28 June 2021. Available at: https://www. foodqualityandsafety.com/article/determining-product-shelf-life/

CHAPTER 8

Dried Foods
Science, Shelf Life, and Quality

Darakshan Majid, Sajad Ahmad Sofi, Aabida Jabeen, Farhana Mehraj Allai,
H.A. Makroo, and Shahnaz Parveen Wani

CONTENTS

8.1 INTRODUCTION

Drying is the most primitive method of preservation, but the technique has come a long way. Drying of foods implies the removal of water and is accomplished by vaporizing the water from the food by the applied latent heat of vaporization. There are two important process-controlling factors that enter into the unit operation of drying: (1) transfer of heat to provide the necessary latent heat of vaporization and (2) movement of water or water vapor through the food material and then away from it to effect separation of water from food. The drying process is a preservation method while involving heat and mass transfer. The various definitions related to drying and

Email: Darakshan Majid, syed.darakshan@gmail.com

DOI: 10.1201/9781003091677-8

Table 8.1   Definitions of Commonly Encountered Terms Used in Drying

S. No.	Terms	Definition
1	Adiabatic saturation temperature	Equilibrium gas temperature reached by unsaturated gas and vaporizing liquid under adiabatic conditions
2	Constant rate drying period	When evaporation rate per unit drying area is constant under constant drying conditions
3	Dry bulb temperature	Temperature of air recorded by an ordinary thermometer with clean and dry sensing elements (td or tdb)
4	Wet bulb temperature	Temperature of air recorded by a thermometer when its bulb is covered with a wet cloth (tw or twb)
5	Equilibrium moisture content	When the vapor pressure of the water held by a product is equal to the water vapor pressure of the surrounding air, the moisture content of the product is the equilibrium moisture content (emc)
6	Critical moisture content	Critical moisture content is the average material moisture content at which the drying rate begins to decline
7	Free moisture	Moisture content in excess of the equilibrium moisture content
8	Bound moisture	Monolayer water: bound in food; it is restricted in its movement due to charges, hydrogen bond, and physical entrapment; hard to remove from food; and never be able to remove water completely
9	Unbound moisture	Multilayer water: additional layer of water around food particle; not as hard to remove as the monolayer
10	Absolute humidity	Mass of water vapor present in a unit mass of dry air
11	Relative humidity	Ratio of the amount of moisture the air holds (mv) to the maximum amount of moisture the air can hold at the same temperature (mg)
12	Water activity	Availability of water for biological reactions and to support the growth of microorganisms

drying mechanisms are mentioned in the Table 8.1. The majority of the water present inside the food is removed by the application of heat under controlled conditions through evaporation by the drying process (Fellows, 2000). An increase in the shelf life of the product is noted due to the decrease in water activity, which limits the growth of microorganisms. Preservation is the fundamental rationale behind drying, but it also decreases the weight and bulk of the product resulting in the reduction of overall transportation costs, including packaging. The dried product showed a significant change in the basic quality attributes such as texture, color, flavor, and nutritional value with better acceptability and longer shelf life. Water activity is low in the dried product, which essentially signifies that most of the reactions as well as microbial growth will be limited. The drying process showed changes in quality in terms of nutritional composition, appearance, structure, lipid oxidation, and other chemical reactions such as ascorbic acid browning, caramelization, and Maillard's reaction (Perera and Baldwin, 2001; Gabas et al., 2002). The various drying pretreatments of fruits and vegetables, such as blanching and preservative dipping, also reduced the food's nutrient loss (Sablani, 2006). The use of advancement in the use of optimized drying techniques for dried food reduced the altered quality changes in the product and improved consumer acceptability along with shelf stability.

8.2 MECHANISM OF DRYING PROCESS

Drying is a complex process that involves momentary transfer of heat and mass transfer with the occurrence of a physical change like crystallization, glass transition, shrinkage, and puffing. Some desirable and undesirable changes also occur that ultimately result in changes in color, texture, and other properties. Drying involves removal of moisture from food, and to accomplish this moisture content the food needs to be vaporized by supplying latent heat of vaporization. Thus, the two important factors responsible for drying are

1. Heat transfer to accomplish the latent heat of vaporization, and
2. Separation and vaporizations of moisture from food.

Thus the process of drying involves three stages:

1. Removal of water from food with the help of air or any hot contact surface under pressure;
2. Removal of water from food by means of conduction or sometimes by radiation, but under vacuum; and
3. Removal of water by the process of sublimation of frozen foods.

In Figure 8.1, the first stage of drying is known as the initial period. When the temperature is applied to food, it absorbs heat and results in an increase in temperature. The evaporation of the moisture takes place simultaneously and tends to cool down the drying surface. After some time the temperature stabilizes (heating and cooling at equilibrium) and the temperature equals the wet bulb temperature. After this stage the constant rate period reaches a point where the temperature remains constant with a constant rate of drying. The moisture is evaporated and replaced by water diffusing from the interior surface of food. At the end of this stage the rate of diffusion is equal to the rate of evaporation, and this point is also known as the critical moisture content (CMC). After this stage is over, the next stage, which is called as falling rate period, begins. In this stage dry spots begin to appear and the rate of drying begins to fall off because the surface water is no longer replaced at a rate fast enough to maintain a film on the surface. Then comes the second falling rate period in which the rate of drying falls even more rapidly than the first time and the rate of drying is dependent on the rate of diffusion. This point is called as the equilibrium moisture content, and beyond this point drying rate is equal to zero.

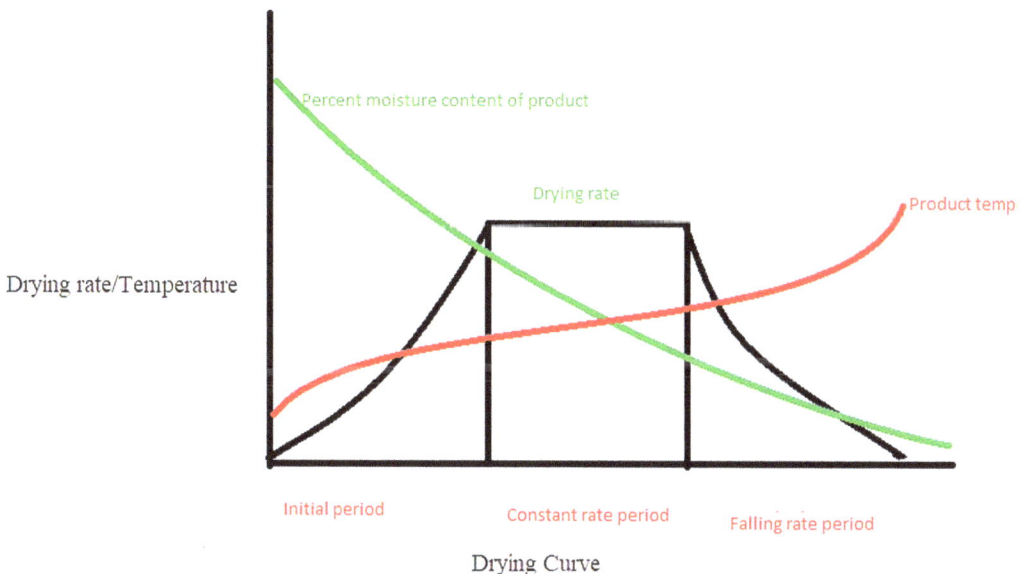

Figure 8.1 Drying curve of food materials.

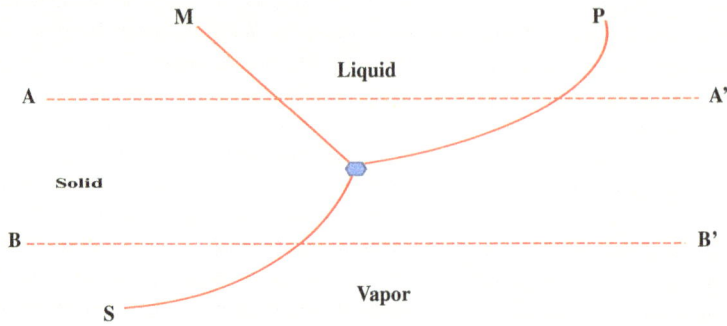

Figure 8.2 Phase diagram of water.

8.3 PHASE DIAGRAM FOR WATER

Water in its purest form exists in three states: liquid, solid, and gaseous. If we want to know about the state of water we can choose any condition regarding temperature and pressure; the corresponding point on the phase diagram will lie in any of the three specified regions (i.e., liquid, solid or gas) (Figure 8.2). In certain conditions there are the chances of two states existing side by side along the lines of the diagram. Similarly, under one condition, there is the chance of the existence of all three states occurring simultaneously at one point represented by center blue dot on the diagram, and this point is called the triple point. If heat is applied in any state of water at a constant pressure, there is an increase in temperature and the condition moves horizontally along the diagram, resulting in a change in the state of the water. Figure 8.2 shows that when heat is applied at point A the ice starts melting and is converted into liquid. If the heat is continued the liquid starts boiling and finally evaporates and reaches condition A′. Similarly, if heat is applied to ice at point B under vacuum then the ice is converted to a gaseous state bypassing the liquid state. Phase change in water occurs at 0.0098°C and 0.64 kPa. Liquid and gas can coexist only along the line from blue dot to P, which is then called the vapor pressure/temperature line.

8.4 DRYING KINETICS

Drying kinetics is considered to be a way to choose the appropriate drying methods and as a means to control the drying process. This is a very important process for optimization of various parameters and is an important way to express the removal of moisture and its relation to various variables in a process. Thus, for developing a drying model the drying rate of the product should be known. Various models have been developed by various scientists and a few of them are listed in Table 8.2.

8.5 DRYING METHODS

Drying involves the application of heat energy to reduce a product's volume and weight, wastage, and transportation costs, and increase shelf life. The drying of food products involves a heat and mass transfer mechanism. The removal of moisture content to an optimal level for designing and

Table 8.2   Drying Kinetic Using Kinetic Models

Models for Drying	Formulas	Reference
Two term	$MR = a \exp(-k_0 t) + b \exp(-k_1 t)$	Henderson (1974)
Fick's diffusion	$MR = a \exp[-c(t/L^2)]$	Kemp (2011)
Henderson and Pabis	$MR = a \exp(-kt)$	Henderson and Pabis (1961)
Modified page II	$MR = \exp(\exp(-c(t/L^2)^n)$	Diamante and Munro (1991)
Lewis model	$MR = \exp(-kt)$	Bruce (1985)
Simplified Fick's diffusion	$MR = a \exp(-c(t/L^2))$	Diamante and Munro (1991)
Page model	$MR = \exp(-kt^n)$	Tanaka et al. (2015)
Modified page	$MR = \exp[(-kt)^n]$	Overhults et al. (1973)

developing quality dried food products depends on the methods and techniques of drying involved. The following sections describe various drying methods that improve the quality of the drying process and food products.

8.6 SOLAR-DRYING

Sun-drying is one the cheapest methods used for drying food products such as fruits, vegetables, meat, and fish (Guiné, 2010). Solar-drying uses the energy from the sun as its source of heat. Sun-drying, along with its easy availability, also showed some drawbacks in quality maintenance of the final dried products. The sun-drying of food products is susceptible to atmospheric conditions, and contaminating agents such as insects, rodents, birds, and microorganisms. Solar-drying with a greenhouse for efficient drying of food products was used to overcome the drawbacks of sun-drying and protect food products from contaminated sources (Ziaforoughi & Esfahani, 2016). The solar method uses direct and indirect methods of drying for food products. The direct method involves the use of solar radiation and the indirect method collects solar energy and heats the surrounding air around the food products (Guiné, 2010).

8.7 HOT AIR CONVECTIVE DRYING

The convective mode of drying in food industries plays a significant role in removing moisture content. For food products with high water content and saturation capacity of moisture, the use of air as a drying medium is the most suitable and efficient method (Dobre et al., 2016). The hot air used as a drying medium is versatile and sufficient to remove moisture from the food product without much change in the product quality. The use of hot air as a convective method is widely applied for the preservation of food products. This method is used in various types of drying equipment, such as tray, tunnel, rotatory drum, and fluidized bed dryers, depending on the type of food products to be dried (Guiné, 2010). The use of hot air convective drying is generally considered a destructive operation in terms of the quality of final dried food products. The quality parameters that are highly affected by hot air convective drying are shrinkage, case hardening, color change, nutrient loss, and low consumer acceptability (Guiné, 2010). Food products that are sensitive to heat are not suitable for drying in the convective hot air drying method. This drying method is associated with advanced methods of drying as a hybrid method such as ultrasound, microwave, and radiofrequency, which were used to reduce the detrimental effects of convective drying and improve the quality of the final dried food products (Szadzińska et al., 2017).

8.8 FREEZE-DRYING METHOD

Freeze-drying is one of the drying methods used for dehydration of foods with high value, for sensitive materials, and for the preservation of microbial cultures. Freeze-drying involves freezing of moisture content and then sublimation under special conditions in temperature and pressure. Freeze-drying is a slower method of drying and involves higher costs, but it maintains the structure of food products (Wang et al., 2017). It is mostly preferred for heat-sensitive foods and for preserving their native quality characteristics (Guiné, 2010). In the food industry this method is operated in batch mode, requires more processing time, has low output, and may cause chill injury. These drawbacks of freeze-drying limit their use in the food processing sector (Haque & Adhikari, 2014).

8.9 INFRARED-DRYING

The infrared-drying uses electromagnetic radiation of wavelengths ranging from 0.8 to 1000 µm. This method is used to heat the surface of food products without heating the surrounding air (Guiné, 2010). Infrared-drying is a quick, uniform, and efficient heat transfer that does not compromise the organoleptic and nutritional deterioration of food products (Haque & Adhikari, 2014). Using infrared radiation in drying removes more water content from food materials and reduces the drying time up to half compared with conventional methods. Infrared-drying with a combination of hot air as a preheating treatment in food materials increased the efficiency of drying rate and reduced the energy of drying by 20% (Adak et al., 2017).

8.10 MICROWAVE-DRYING

Microwave-drying is a volumetric heating method with the use of electromagnetic radiations with wavelengths ranging from 1 mm to 1 m. In microwave-drying, when the electromagnetic radiations pass through the food materials they produce molecular oscillations that generate heat and remove moisture content from the materials (Calín-Sánchez et al., 2011). The most common frequencies used in microwave-drying are 915 and 2450 MHz. Microwave-drying is used to produce products of high quality as it efficiently removes moisture content and has a higher heating rate and uniform heating (Pu et al., 2016). Microwave-drying in post-harvest processing in the food industry is used as a final stage of dehydration and to lower water content in food products (Zielinska & Michalska, 2016). The use of microwave heating causes overheating of food products and degradation of sensitive compounds. The use of microwave associated with a vacuum system, however, minimizes the overheating and degradation of bioactive compounds in dried food products (Zielinska & Michalska, 2016).

8.11 RADIOFREQUENCY-DRYING

Radiofrequency-drying is an alternative to convective-drying methods without affecting textural and sensory qualities of the dried food product (Marra et al., 2009). It involves heating food products by electromagnetic radiations of radiofrequency and molecular entities present in the food. The heat is generated due to molecular friction between polar molecules and alternative electric fields. The water content in food has bipolar molecules and is evaporated due to the heat generated by frictional heat (Babu et al., 2018). Radiofrequency-drying heats the food product directly without overheating the surface and is a suitable method of drying at low humidity conditions. It is a novel method of drying with less processing time, high efficiency, and it maintains the quality of dried food products compared with the conventional drying method (Michael et al., 2014; Zhu et al., 2012).

8.12 OSMOTIC DEHYDRATION

Osmotic dehydration refers to the partial removal of moisture content from fruits and vegetables placed in a solution of high osmotic pressure. The solutions with high osmotic potential used in osmotic drying are sugar at 40–70% for fruits, and salts (sodium chloride) at 5–20% for vegetables (Barman & Badwaik, 2017). Osmotic dehydration removes 50% of the moisture content from fruits and vegetables for 4–6 hours. The loss of moisture is high initially in osmotic dehydration of fruits and vegetables, but after 1–2 hours the moisture loss is reduced and it takes several days to reach equilibrium stages (Guiné, 2010). Osmotic dehydration needs the addition of other drying methods such as microwaves, air, ultrasound, and ohmic heating drying (Da Costa Ribeiro et al., 2016).

8.13 REFRACTANCE WINDOW DRYING

Refractance window (RW) drying is a novel method of drying used for liquid and semisolid foods for the production of dried products (Bolland, 2000). RW drying uses convection, conduction, and radiation modes of heat transfer. It is similar to drum dryers in terms of the formation of a dried thin film, except the drying process occurs at low temperature (Abonyi et al., 2002). The RW drying equipment consists of infrared transparent material placed on the heated circulatory system over which the drying of food materials occurs (Guiné, 2010). The method is energy efficient and the drying rate of the food product is very short (Sabarez, 2016). RW drying is a low-cost method of drying that is suitable for heat-sensitive materials operated under atmospheric pressure and low temperature (Nindo & Tang, 2007).

8.14 SHELF LIFE OF DRIED FOODS

Shelf life is an important quality parameter of food products and techniques associated with processing food products. Shelf life is a time period of food products from production to the point of unacceptable conditions. The food products during the period of shelf life stages such as processing, handling operations, and transport and storage conditions are prone to physicochemical, enzymatic, and microbial spoilage. Decreasing shelf life of food products by different types of spoilage impacts the overall quality, consumer acceptability, and health issues of end-use consumers. Different drying methods for food products are being adopted in the food industry to enhance their shelf life and consumer qualities.

Post-harvest preservation of dried foods against the microbial spoilage is done by performing physical and chemical pretreatment to improve food qualities and minimize post-harvest losses (Schmidt et al., 2018). A decrease in the shelf life of dried foods is related to spoilage by microorganisms. The physical treatment involves use of thermal, ionizing, and non-ionizing radiations to inhibit the growth of fungi and bacteria; extend the shelf life; and increase the stability of dried food products (Schmidt et al., 2018). The heat pretreatments such as hot water dips, hot dry air, or superheated steam used for dried foods can possibly reduce post-harvest losses (Klein & Lurie, 1991). Heat treatment by blanching fruits and vegetables for the drying process inactivates enzyme activities, enhances their shelf life, and prevents deterioration against enzymatic spoilage (Ben Zid et al., 2015). The uses of heat treatment prior to storage reduced the microbial load and decreased the water activity of food products (Nielsen et al., 2004). Ionizing irradiations are electromagnetic waves used as a post-harvest treatment for drying food products; they are a rapid, efficient, and safe way to reduce spoilage from microorganisms and increase shelf life (Nemţanu et al., 2014). Shelf life of dehydrated vegetables was increased by the use of gamma irradiation as a decontamination process (Zhou et al., 1996). Wang and Chao (2003) also reported quality improvements and increased shelf life of dried potatoes using gamma irradiation as a quality pretreatment. Wei et al. (2014) reported shelf stability and nutrient retention of gamma-irradiated dried apricots. Decontamination as pretreatment in dried mushrooms using irradiation was used to increase their postharvest shelf life (Culleré et al., 2012; Fernandes et al., 2012). Ultraviolet (UV) light on dried food products after drying extended their shelf life and reduced microbial load; the range used was from 100 to 400 nm in the electromagnetic spectrum (Gomez-Lopez et al., 2012).

The use of chemical pretreatment before drying foods reduced the enzymatic activity, delaying lipid oxidation, color preservation, and reduction of microbial load. The use of sulfur dioxide and sodium metabisulfite dips as one of the pretreatment chemicals used for increasing and reduction of microbial load results in enhancement of the shelf stability of dried food products (Andrews and Godshall, 2002). Sulfur dioxide treatment used in dried pomegranate aril preserves the color quality and increases their shelf stability (Thakur et al., 2010). Ascorbic acid pretreatment was also used in drying foods with a retention of nutrients during drying and shelf stability in dried gac aril powder and mangoes (Chen et al., 2007; Tuyen et al., 2011; Chuyen et al., 2017). Tomato slices treated with 1% (w/v) calcium chloride ($CaCl_2$), 2% potassium metabisulfite, and 2% sodium chloride (NaCl) solutions before oven-drying reduced carotenoid loss and improved color, nutrient quality, and shelf life for up to 3 months (Baloch et al., 1997). The use of olive oil, potassium carbonate, and fatty acid ethyl esters used as pretreatment in drying whole fruits reduced nutrient loss and stability during stored conditions (Vásquez-Parra et al., 2013). Doymaz (2006) investigated the use of ethyl-oleate and potassium carbonate as chemical treatments on berries and found reduced drying time and improved quality characteristic and shelf stability. During the storage period dried fruits are sensitive to oxygen, humidity, and direct sunlight, which affect their moisture content and shelf stability (Ranđelović et al., 2014). Dried foods and their sensitivity to surrounding environmental conditions are protected by packaging materials. Properties like gas permeability, film thickness, packaging material type, and gas composition are important to consider while designing the packaging materials for dried foods. Packing dried foods in high-oxygen permeability packaging reduced the shelf life and quality of dried foods (Simal et al., 1999). Microwave dried potato cubes stored in biaxially oriented polypropylene (BOPP) laminate after flushing oxygen resulted in nutrient loss and reduced their stability period (Shakouri et al., 2015). Prolonged storage period and better quality of dried food can be increased by using packaging materials thick enough to be permeable to air and moisture (Ornelas-Paz et al. (2012). Similar results related to packaging thickness and shelf life were reported by Dak et al. (2014) with dried pomegranate arils packed in aluminum-laminated polyethylene or high-density polypropylene (HDPP) pouches. Dried pomegranate arils packed in 90-μm thickness extended shelf life from 3 months to 6 months compared with a 40-μm thickness. The use of different packaging materials also showed a significant impact on the shelf life of dried foods. Dried apricots stored in polyester/aluminum/polyethylene packaging materials showed lower water activity compared with paper/aluminum/polyethylene for 12 months (Ranđelović et al., 2014). Packaged dried apricots in polyester/aluminum/polyethylene packaging materials with modified atmospheric conditions (28.48% carbon dioxide, 69.93% nitrogen, 1.23% oxygen) reportedly had lower water activity and better shelf life qualities compared with non-modified atmospheric packaging conditions (Ranđelović et al., 2014). The shelf life of dried foods was extended by using modified atmosphere packaging and replacing normal atmospheric air with a combination of protective inert gases (Ranđelović et al., 2014). Modified atmospheric conditions in dried food materials maintain nutrient and hygienic qualities (Sergio et al., 2016). Modified atmospheric packaging in dried foods under low oxygen concentration reduced the oxidative effects and maintained their nutrient values and shelf life (Ramesh et al., 1999).

Edible coatings as pretreatment for minimally processed and dried foods improved nutritional quality and storage stability due to decreased moisture loss, gas exchange, respiration, and oxidative reaction (Díaz-Mula et al., 2012; Ferrari et al., 2013). The texture, appearance, and nutrients of dried foods were maintained by applications of polysaccharides, proteins, and lipid-based edible coatings

(Lago-Vanzela et al., 2013). Edible coatings used as pretreatments before drying act as a barrier to oxygen and water, resulting in reduced moisture loss from the products and acting as a sacrificial agent (Ferrari et al., 2013). The use of edible coatings on food products reduced the time of exposure to oxygen during the drying process, hence, resulting in a reduction in oxidative damage during drying time. The edible coatings before drying retained nutrients and kept the qualities of dried foods during the convective drying process (Garcia et al., 2014). Edible coatings with active ingredients such as ascorbic acids improve the color, flavor, texture, and nutritional quality and control microbial growth (Ayranci and Tunc, 2004). The use of pectin-based coating enhances the quality and stability of dried papaya during the air-drying process (Canizares and Mauro, 2015). Edible coatings used as pretreatments during drying protects dried food compounds against oxidative changes, prevents nutrient loss, acts as a carrier for additives, and maintains physical and sensory properties and shelf life of food products.

8.15 QUALITY OF DRIED FOODS

Drying is one of the unit operations used in the food industry for preserving and maintaining the quality of food products. The quality of dried products is the main focus of the processing industry with good acceptability among consumers. Drying affects the physicochemical, nutritional, textural and sensory qualities of food products and their effect is dependent on the amount of moisture to be removed, types of food product, amount of products, and drying methods (Bonazzi and Bimbenet, 2008). Texture, flavor, color, safety, and shelf stability are the main quality parameters of dried foods, along with convenience, which is very important to consumers. All these components are dependent on one another. Methods of processing and preservation, the composition of food, and storage and packaging are mainly responsible for attaining the food quality (Bonazzi & Dumoulin, 2011).

There are various quality parameters involved in the food industry that monitor and control the quality of dried food products and their acceptance by consumers, including physicochemical, biological, and nutritional parameters. The physical parameters in determining the quality of dried foods are color, appearance, porosity, texture, and rehydration (Perera, 2005). Color is the main physical quality parameter and most affected parameter when drying food products due to chemical and biochemical reactions during drying periods (Chua et al., 2000). The enzymatic browning, Maillard reactions, caramelization, and ascorbic acid browning reactions that occurred in the drying process have an adverse effect on the quality of dried foods (Perera & Baldwin, 2001). The appearance of dried foods also has a significant effect on acceptability among consumers. The changes in color, size, shape, and chemical composition of food products during drying have a direct effect on the visual appearance and appeal of dried foods (Perera, 2005). Porosity determines the shrinkage of dried foods during drying, affects their final shape and size (Ayrosa and Nogueira-de-Moraes-Pitombo, 2003), and determines the quality of finished products. The rehydration property of dried food also depends on the porosity of the product and the quality of the drying method (Perera, 2005). The texture of dried food products is one of the main physical parameters in determining consumer acceptance and quality of the product. The quality of the dried product with acceptable texture shows that freeze- and vacuum-drying methods have been applied to the product. Freeze- and vacuum-drying showed that products with high porosity, low density, high rehydration, and good organoleptic qualities were air-dried with a collapsed texture (Gabas et al., 2002). The chemical parameter, which alters the quality in terms of taste, consists of concentrations of oxygen and carbon dioxide (Jensen et al., 2003). The presence of oxygen concentration below 0.5% and carbon dioxide above 80% reduced the oxidative chemical reactions and quality of the dried products (Perera, 2002). The use of advanced packaging technology such as vacuum and nitrogen flushing for dried foods controls their deterioration and unpleasant taste and extends its quality and storage stability for a long time. The water activities of dried foods are very low and auto-oxidation reaction starts at a very low water activity for dried foods, resulting in taint and off-flavor, which significant effect the quality of dried foods (Perera, 2002). Lowering water activity by proper drying methods and preservatives overcomes the effects of off-flavor, taste, and low stability in dried products. Dried foods also develop unacceptable taste and flavor from the microbial action, chemical reactions, and absorbing odors from the surrounding atmosphere (Mottram, 1998). The presence of microbial load and pests and contaminants also determined the quality of dried foods. The microbial load in dried foods should be at an acceptable limit for the end-use consumer for safety purposes. Maintaining the quality of dried foods is related to acceptable limits of microbial load and controlling the humidity temperature and handling (Rockland & Beuchat, 1987). Using packaging, proper storage, and fumigation techniques improve the quality in terms of dried food free from pests and contaminants (Perera, 2002). Drying foods using different methods has a significant effect on the nutritional quality of the final product. During drying, protein undergoes denaturation, which alters the nutritive value and functionality, which alters the texture and nutritional quality of the dried products. Also, lipids undergo oxidation, which results in off-flavor carbohydrates due to Maillard compounds and vitamin loss caused by the drying temperature and improper storage conditions (Damoderan, 1996; Larsen et al., 2003). The retention of these nutrients can be achieved by using optimized drying conditions and methods to improve the nutritional and acceptable organoleptic quality in dried foods.

8.15.1 Quality Change in Dried Foods

In food industries, processing and preservation of food products using the drying method controls various properties of the final product. The processing varies from raw food to the final products, type and quantity of products, and desired quality of products (Bonazzi & Dumoulin, 2011). During this process, biochemical, physical, structural, and storage qualities change, which affects the overall qualities of the final products. The biochemical changes that occur during drying are color change, vitamin degradation, fat oxidation, protein denaturation, and enzyme activity. The change in biochemical properties of dried foods during drying may be desirable or undesirable depending on the end-product specifications. During the drying of food products, nutrients compositions were significantly affected. Among the various nutrients, vitamin C is the most affected by drying temperature, and its retention is dependent on the time and temperature combinations of drying methods (Santos & Silva, 2008). Vitamin C in dried foods can be retained by adopting low-temperature drying methods, such as freeze- and vacuum-drying, with better nutritional quality compared with conventional drying methods (Bonazzi & Dumoulin, 2011). The color of dried foods is the most important quality attribute and it determines consumer acceptance and methods of drying used. The presence of pigments such as carotenoids, chlorophylls, anthocyanins, and betalains is responsible for the color of food products. During the drying process, these pigments undergo degradation by enzymatic and non-enzymatic reactions resulting in color change of the final dried food products (Marty-Audouin et al., 1992). Proteins showed a change in biological value and functional properties during the drying process. The change in properties was dependent on the time and range of temperature used for drying (Bonazzi & Dumoulin, 2011). Lipid oxidation during the drying period initiated by-products of moisture and heat, which resulted in a loss of fat-soluble vitamins and off-flavor of the final dried products (Perera, 2005). The non-enzymatic browning reactions common in dried food products during drying were caramelization and Maillard reactions, which shows the significant color and nutrient change in final dried food products. These browning reactions were high at low moisture and at high temperature, showing that these browning reactions are more prone in dried food products (Bonazzi & Dumoulin, 2011). The control of these biochemical reactions on dried food products depends on the drying methods and their related processing parameters, such as time and temperature and proper packaging techniques, to maintain the nutritional and functional qualities of the final dried foods. The physical changes during the drying period include decreased water activity, glass transition, crystallization, and melting of fat. Migration or retention of volatile components also was observed in dried foods (Rocha Mier, 1993). The rubbery or crystallized structure of final dried food products depended on the nature of the product and methods of drying used (Alves-Filho & Roos, 2006). The cracks or fissures on the dried foods were due to non-uniform drying, mechanical stresses, and shrinkage produced during the drying process (Kamst et al., 2002). The phenomenon of surface cracking was observed in dried food during the drying process, which significantly affects the overall quality and commercial value of final products. The handling, storage, and post-operation conditions maintained may also have an important effect on the quality of the final dried product. The physicochemical, biochemical, functional, and mechanical properties that changed during these conditions alter the quality parameters of dried foods (Perera & Baldwin, 2001).

8.16 CONCLUSION

Food preservation involves different food processing steps to maintain food quality at the desired level so that maximum benefits and nutrition values can be achieved. Today, consumers demand novel and healthier ready-to-eat products with long shelf life and dehydrated products meet these criteria. Food products are dried to enhance their storage stability, minimize packaging requirements, and reduce transport weight. Preservation of products through optimum drying techniques for the production of quality dried products is a cost-effective approach that shortens drying time, reduces energy utilization, and minimizes damage to dried food products.

REFERENCES

Abonyi, B. I., Feng, H., Tang, J., Edwards, C. G., Chew, B. P., Mattinson, D. S., & Fellman, J. K. (2002). Quality retention in strawberry and carrot purees dried with Refractance Window™ system. *Journal of Food Science, 67*(3), 1051–1056.

Adak, N., Heybeli, N., & Ertekin, C. (2017). Infrared drying of strawberry. *Food Chemistry, 219*, 109–116.

Alves-Filho, O., & Roos, Y. H., (2006). Advances in multi-purpose drying operations with phase and state transitions. *Drying Technology, 24*(3), 383–396.

Andrews, L. S., & Godshall, M. A. (2002). Comparing the effects of sulphur dioxide on model sucrose and cane juice systems. *Journal of American Society of Sugarcane Technologists, 22*, 90–101.

Ayranci, E., & Tunc, S. (2004). The effect of edible coatings on water and vitamin C loss of apricots (*Armeniaca vulgaris* Lam.) and green peppers (*Capsicum annuum* L.). *Food Chemistry, 87*, 339–342.

Ayrosa, A. M. I. B., & Nogueira-de-Moraes-Pitombo, R. I (2003). Influence of plate temperature and mode of rehydration on textural parameters of precooked freeze-dried beef. *Journal of Food Processing and Preservation, 27*(3), 173–180.

Babu, A. K., Kumaresan, G., Raj, V. A. A., & Velraj, R. (2018). Review of leaf drying: mechanism and influencing parameters, drying methods, nutrient preservation, and mathematical models. *Renewable and Sustainable Energy Reviews, 90*, 536–556.

Baloch, W. A., Khan, S., & Baloch, A. K. (1997). Influence of chemical additives on the stability of dried tomato powder. *International Journal of Food Science & Technology, 32*(2), 117–120.

Barman, N., & Badwaik, L. S. (2017). Effect of ultrasound and centrifugal force on carambola (*Averrhoa carambola* L.) slices during osmotic dehydration. *Ultrasonics Sonochemistry, 34*, 37–44.

Ben Zid, M., Dhuique-Mayer, C., Bellagha, S., Sanier, C., Collignan, A., Servent, A., & Dornier, M. (2015). Effects of blanching on flavanones and microstructure of Citrus aurantium peels. *Food and Bioprocess Technology, 8*(11), 2246–2255.

Bolland, K. M. (2000). A new low-temperature/short-time drying process. *Cereal Foods World, 45*(7), 293–296.

Bonazzi, C., & Bimbenet, J. J., (2008). Sechage des produits alimentaires – Materiels et applications. *Techniques de l.ingenieur-Traité Agroalimentaire F3. Editions T. I.*, Paris, France, pp. 1–17.

Bonazzi, C., & Dumoulin, E. (2011). Quality Changes in Food Materials as Influenced by Drying Processes. Edited by Tsotsas, E. & Mujumdar, A.S. *Modern Drying Technology: Product Quality and Formulation*, Wiley-VCH Verlag GmbH & Co. KGaA, pp. 1–20.

Bruce, D. M. (1985). Exposed layer barley during, three models fitted to new data up to 150°C. *Journal of Agricultural Engineering Research, 32*, 337–347.

Calín-Sánchez, Á., Szumny, A., Figiel, A., Jałoszyński, K., Adamski, M., & Carbonell-Barrachina, Á. A. (2011). Effects of vacuum level and microwave power on rosemary volatile composition during vacuum–microwave drying. *Journal of Food Engineering, 103*(2), 219–227.

Canizares, D., & Mauro, M. A. (2015). Enhancement of quality and stability of dried papaya by pectin-based coatings as air-drying pretreatment. *Food and Bioprocess Technology, 8*(6), 1187–1197.

Chen, J. P., Tai, C. Y., & Chen, B. H. (2007). Effects of different drying treatments on the stability of carotenoids in Taiwanese mango (*Mangifera indica* L.). *Food Chemistry, 100*(3), 1005–1010.

Chua, K. J., Mujumdar, A. S., Chou, S. K., Hawlader, M. N. A., & Ho, J. C. (2000). Convective drying of banana, guava and potato pieces: effect of cyclical variations of air temperature on drying kinetics and color change. *Drying Technology, 18*(45), 907–936.

Chuyen, H. V., Roach, P. D., Golding, J. B., Parks, S. E., & Nguyen, M. H. (2017). Effects of pretreatments and air drying temperatures on the carotenoid composition and antioxidant capacity of dried gac peel. *Journal of Food Processing and Preservation, 41*(6), e13226.

Culleré, L., Ferreira, V., Venturini, M. E., Marco, P., & Blanco, D. (2012). Evaluation of gamma and electron-beam irradiation on the aromatic profile of black truffle (*Tuber melanosporum*) and summer truffle (*Tuber aestivum*). *Innovative Food Science & Emerging Technologies, 13*, 151–157.

Da Costa Ribeiro, A. S., Aguiar-Oliveira, E., & Maldonado, R. R. (2016). Optimization of osmotic dehydration of pear followed by conventional drying and their sensory quality. *LWT – Food Science and Technology, 72*, 407–415.

Dak, M., Sagar, V. R., & Jha, S. K. (2014). Shelf-life and kinetics of quality change of dried pomegranate arils in flexible packaging. *Food Packaging and Shelf Life, 2*(1), 1–6.

Damoderan, S. (1996). Amino Acids, Peptides and Proteins. Edited by Fennema, O.R. *Food Chemistry* (3rd ed.). Marcel Dekker: New York, pp. 321–429.

Diamante, L. M., & Munro, P. A. (1991). Mathematical modeling of hot air drying of sweet potato slice. *International Journal of Food Science & Technology, 26*, 99–109.

Díaz-Mula, H. M., Serrano, M., & Valero, D. (2012). Alginate coatings preserve fruits quality and bioactive compounds during storage of sweet cherry fruit. *Food and Bioprocess Technology, 5*, 2990–2997.

Dobre, T., Pârvulescu, O. C., Stoica-Guzun, A., Stroescu, M., Jipa, I., & Al Janabi, A. A. (2016). Heat and mass transfer in fixed bed drying of non-deformable porous particles. *International Journal of Heat and Mass Transfer, 103*, 478–485.

Doymaz, I. (2006). Drying kinetics of black grapes treated with different solutions. *Journal of Food Engineering, 76*(2), 212–217.

Fellows, P. (2000). *Food Processing Technology – Principles and Practice*. Woodland Publishing Limited, CRC Press, pp. 369–380.

Fernandes, Â., Antonio, A. L., Oliveira, M. P. P., Martins, A., & Ferreira, I. C. F. R. (2012). Effect of gamma and electron beam irradiation on the physico-chemical and nutritional properties of mushrooms: a review. *Food Chemistry, 135*, 641–650.

Ferrari, C. C., Sarantópoulos, C. I. G. L., Carmello-Guerreiro, S. M., & Hubinger, M. D. (2013). Effect of osmotic dehydration and pectin edible coatings on quality and shelf life of fresh-cut melon. *Food and Bioprocess Technology, 6*, 80–91.

Gabas, A. L., Menegalli, F. C., Ferrari, F., & Telis-Romero, J. (2002). Influence of drying conditions on the rheological properties of prunes. *Drying Technology, 20*(7), 1485–1502.

Garcia, C. C., Caetano, L. C., Silva, K. S., & Mauro, M. A. (2014). Influence of edible coating on the drying and quality of papaya (*Carica papaya*). *Food Bioprocess Technology, 7*, 2828–2839.

Gomez-Lopez, V. M., Koutchma, T., & Linden, K. (2012). Ultraviolet and Pulsed Light Processing of Fluid Foods. Edited by Cullen, P.J., Tiwari, B. K. & Valdramidis, V. *Novel Thermal and Non-Thermal Technologies for Fluid Foods*, Academic Press: New York, pp. 185–223.

Guiné, R. (2010). Analysis of the drying kinetics of S. Bartolomeu pears for different drying systems. *Journal of Environmental, Agricultural and Food Chemistry*, *9*(11), 1772–1783.

Haque, M., & Adhikari, B. (2014). Drying and Denaturation of Proteins in Spray Drying Process. *Handbook of Industrial Drying* (4th ed.). CRC Press, Taylor & Francis Group, NW, USA, pp. 971–983.

Henderson, S. M. (1974). Progress in developing the thin layer drying equation. *Transactions of the American Society of Agricultural Engineers*, *17*, 1167–1172.

Henderson, S. M., & Pabis, S. (1961). Grain drying theory I: Temperature effect on drying coefficient. *Journal of Agricultural Research Engineering*, *7*, 85–89.

Jensen, P. N., Sorensen, G., Brockhoff, P., & Bertelsen, G. (2003). Investigation of packaging systems for shelled walnuts based on oxygen absorbers. *Journal of Agricultural & Food Chemistry*, *51*(17), 4941–4947.

Kamst, G. F., Bonazzi, C., Vasseur, J., & Bimbenet, J. J. (2002). Effect of deformation rate and moisture content on the mechanical properties of rice grains. *Transactions of the American Society of Agricultural Engineers*, *45*(1), 145–151.

Kemp, I. C. (2011). Drying models, myths and misconceptions. *Chemical Engineering & Technology*, *34*, 1057–1068.

Klein, J. D., & Lurie, S. (1991). Postharvest heat treatment and fruit quality. *Postharvest News and Information*, *2*(1), 15–19.

Lago-Vanzela, E. S., Nascimento, P., Fontes, E. A. F., Mauro, M. A., & Kimura, M. (2013). Edible coating from native and modified starches retain carotenoids in pumpkin during drying. *LWT – Food Science and Technology*, *50*, 420–425.

Larsen, H., Magnus, E. M., & Wicklund, T. (2003). Effect of oxygen transmission rate of the packages, light, and storage temperature on the oxidative stability of extruded oat packaged in nitrogen atmosphere. *Journal of Food Science*, *68*(3), 1100–1107.

Marra, F., Zhang, L., & Lyng, J. G. (2009). Radio frequency treatment of foods: review of recent advances. *Journal of Food Engineering*, *91*(4), 497–508.

Marty-Audouin, C., Lebert, A., & Rocha-Mier, T. (1992). Influence of drying on the color of plant products. Edited by Mujumdar, A.S. *Drying of Solids*, Oxford & IBH Publishing Co.: New Delhi, India, pp. 326–346.

Michael, M., Phebus, R. K., Thippareddi, H., Subbiah, J., Birla, S. L., & Schmidt, K. A. (2014). Validation of radio-frequency dielectric heating system for destruction of *Cronobacter sakazakii* and *Salmonella* species in nonfat dry milk. *Journal of Dairy Science*, *97*(12), 7316–7324.

Mottram, D. S. (1998). Chemical tainting of foods. *International Journal Food Science and Technology*, *33*(1), 19–29.

Nemţanu, M. R., Braşoveanu, M., Karaca, G., & Erper, İ. (2014). Inactivation effect of electron beam irradiation on fungal load of naturally contaminated maize seeds. *Journal of the Science of Food and Agriculture*, *94*(13), 2668–2673.

Nielsen, K. F., Holm, G., Uttrup, L. P., & Nielsen, P. A. (2004). Mould growth on building materials under low water activities. Influence of humidity and temperature on fungal growth and secondary metabolism. *International Biodeterioration & Biodegradation*, *54*(4), 325–336.

Nindo, C. I., & Tang, J. (2007). Refractance window dehydration technology: a novel contact drying method. *Drying Technology*, *25*(1), 37–48.

Ornelas-Paz, J. D. J., Zamudio-Flores, P. B., Torres-Cisneros, C. G., Holguín-Soto, R., Ramos-Aguilar, O. P., Ruiz-Cruz, S., Guevara-Arauza, J. C., González-Aguilar, G. A., & Santana-Rodríguez, V. (2012). The barrier properties and potential use of recycled-LDPE films as a packaging material to preserve the quality of jalapeño peppers by modified atmospheres. *Science Horticulturea (Amsterdam) 135*, 210–218.

Overhults, D. D., White, G. M., & Hamilton, M. E. (1973). Drying soybeans with heated air. *Transactions of the American Society of Agricultural Engineers*, *16*, 195–200.

Perera, C. O. (2005). Selected quality attributes of dried foods. *Drying Technology*, *23*(4), 717–730.

Perera, C. O., & Baldwin, E. A. (2001). Biochemistry of Fruits and Its Implications on Processing. Edited by Arthey, D. & Ashurst, P.R. *Fruit Processing; Nutrition, Products and Quality Management* (2nd ed.). Aspen Publishers: Gaithersburg, MD, pp. 19–36.

Pu, H., Li, Z., Hui, J., & Raghavan, G. V. (2016). Effect of relative humidity on microwave drying of carrot. *Journal of Food Engineering*, *190*, 167–175.

Ramesh, M. N., Wolf, W., Tevini, D., & Jung, G. (1999). Studies on inert gas processing of vegetables. *Journal of Food Engineering*, *40*(3), 199–205.

Ranđelović, D., Lazić, V., Tepić, A., & Mošić, I. (2014). The influence of packaging materials protective properties and applying modified atmosphere on packed dried apricot quality changes. *Hemijska Industrija*, *68*(3), 289–295.

Rocha Mier, T. (1993). Influence des pretraitements et des conditions desechage sur la couleur et l.arome de la menthe (Mentha spicata Huds.) et du basilic (Ocimum basilicum). PhD thesis, ENSIA, Massy, France.

Rockland, L. B., & Beuchat, L. R. (1987). *Water Activity: Theory and Applications to Food*, Marcel Dekker: New York, p. 340.

Sabarez, H. (2016). Drying of Food Materials. *Reference Module in Food Science*, Elsevier BV: Amsterdam, the Netherlands, pp. 102–112.

Sablani, S. S. (2006). Drying of fruits and vegetables: retention of nutritional/functional quality. *Drying Technology*, *24*(2), 123–135.

Santos, P. H. S., & Silva, M. A., 2008. Retention of vitamin C in drying processes of fruits and vegetables – a review. *Drying Technology*, *26*(12), 1421–1437.

Schmidt, M., Zannini, E., & Arendt, E. K. (2018). Recent advances in physical post-harvest treatments for shelf-life extension of cereal crops. *Foods*, *7*(4), 45.

Sergio, L., Gatto, M. A., Spremulli, L., Pieralice, M., Linsalata, V., & Di Venere, D. (2016). Packaging and storage conditions to extend the shelf life of semi-dried artichoke hearts. *LWT – Food Science and Technology*, *72*, 277–284.

Shakouri, S., Ziaolhagh, H. R., Sharifi-Rad, J., Heydari-Majd, M., Tajali, R., Nezarat, S., & Teixeira da Silva, J. A. (2015). The effect of packaging material and storage period on microwave-dried potato (*Solanum tuberosum* L.) cubes. *Journal of Food Science & Technology*, *52*(6), 3899–3910.

Simal, S., Sanchez, E. S., Benedito, J., & Rossello, C. (1999). Effect of temperature and gas composition on the shelf-life of dehydrated apricots. *Food Science & Technology International*, *5*(5), 377–383.

Szadzińska, J., Łechtańska, J., Kowalski, S. J., & Stasiak, M. (2017). The effect of high power airborne ultrasound and microwaves on convective drying effectiveness and quality of green pepper. *Ultrasonics Sonochemistry*, *34*, 531–539.

Tanaka, F., Tanaka, F., & Tanaka A (2015). Mathematical modeling of thin layer drying according to particle size distribution in crushed feed rice. *Journal of Biosystems Engineering*, *136*, 87–91.

Thakur, N. S., Bhat, M. M., Rana, N., & Joshi, V. K. (2010). Standardization of pre-treatments for the preparation of dried arils from wild pomegranate. *Journal of Food Science and Technology*, *47*(6), 620–625.

Tuyen, C. K., Nguyen, M. H., & Roach, P. D. (2011). Effects of pre-treatments and air drying temperatures on colour and antioxidant properties of gac fruit powder. *International Journal of Food Engineering*, *7*(3).

Vásquez-Parra, J. E., Ochoa-Martínez, C. I., & Bustos-Parra, M. (2013). Effect of chemical and physical pretreatments on the convective drying of cape gooseberry fruits (*Physalis peruviana*). *Journal of Food Engineering*, *119*(3), 648–654.

Wang, B., Timilsena, Y. P., Blanch, E., & Adhikari, B. (2017). Characteristics of bovine lactoferrin powders produced through spray and freeze drying processes. *International Journal of Biological Macromolecules*, *95*, 985–994.

Wang, J., & Chao, Y. (2003). Effect of gamma irradiation on quality of dried potato. *Radiation Physics and Chemistry*, *66*(4), 293–297.

Wei, M., Zhou, L., Song, H., Yi, J., Wu, B., Li, Y., ... & Li, S. (2014). Electron beam irradiation of sun-dried apricots for quality maintenance. *Radiation Physics and Chemistry*, *97*, 126–133.

Zhou, Q. C., Jin, R. H., Wei, J. Y., Fu, J. K., & Xiong, L. D. (1996). Irradiation preservation and its dose control for dehydrated vegetables. *Acta Agriculture Zhejiangensis (China)*, *8*, 255–256.

Zhu, X., Guo, W., Wu, X., & Wang, S. (2012). Dielectric properties of chestnut flour relevant to drying with radio-frequency and microwave energy. *Journal of Food Engineering*, *113*(1), 143–150.

Ziaforoughi, A., & Esfahani, J. A. (2016). A salient reduction of energy consumption and drying time in a novel PV-solar collector-assisted intermittent infrared dryer. *Solar Energy*, *136*, 428–436.

Zielinska, M., & Michalska, A. (2016). Microwave-assisted drying of blueberry (*Vaccinium corymbosum* L.) fruits: drying kinetics, polyphenols, anthocyanins, antioxidant capacity, colour and texture. *Food Chemistry*, *212*, 671–680.

CHAPTER 9

Frozen Foods
Science, Shelf Life, and Quality

H.A. Makroo, Preetisagar Talukdar, Baby Z. Hmar, and Pranjal Pratim Das

CONTENTS

9.1 INTRODUCTION

Freezing of food is one of the foremost techniques of food conservation. As early as 1755 the first artificial ice was produced without using natural cold. By the late 19th century fish and meat were preserved using freezing technology. Meanwhile, scientists and researchers continued their studies to improve and develop the freezing process. By the early 20th century, in the United States, fruit and vegetables were frozen, but this process was met with the hurdle of deterioration in texture and taste. In 1929, it was concluded that by blanching for enzyme inactivation the degradation in texture and taste could be minimized. From 1928, modern freezing began with the advancement of double-belt contact freezers and led to a revolution in quick-freezing equipment and methods. This led to the successful freezing of packaged foods like fruits and vegetables. In the modern world, an increase in population growth has led to an increase in demand for food products. By freezing, the quality of agricultural produce can be retained over a long duration and fruits and vegetables can

Email: H.A. Makroo, hilalmakroo@gmail.com

DOI: 10.1201/9781003091677-9

be made available throughout the year. Freezing also has high retention in terms of nutritive and sensory properties compared with canning and dehydration, without causing changes to the food. During the freezing process, the temperature of the product is decreased beneath the freezing point, i.e., $-18°C$ or below. This extreme cold stops microbial growth and arrests the chemical changes in food. In the freezing process, the preservation of food products is obtained by a combined factor of low temperature, which leads to a reduction in biological, physical, chemical, and microbial activity. In some cases, blanching plays an important factor in decreasing enzymatic activity (Fennema et al. 1973; Fennema 1977; George 1993).

For food product freezing, a low-temperature medium is used for an adequate duration to eliminate the latent heat of fusion and sensible heat. Thus, with the removal of both parameters, the temperature of the food product decreases, leading to the conversion of water to ice. While freezing meat, fish, fruits, and vegetables the freezing rate and formation of ice crystals are critical to preserving the quality of the product. Due to the formation of ice crystals during freezing, there are physical changes in the structure of the products. Although the formation of small ice crystal is critical as it minimizes the deterioration of tissue in produce, the formation of large ice crystals is unwanted as it maximizes tissue breakage in the product (Fellows 2009; Singh and Heldman 2001).

Freezing methods are carried out by an indirect or direct contact system. For direct contact freezers, the food product to be frozen is completely enclosed by the freezing medium, i.e., the refrigerant. This maximizes the heat transfer efficacy, whereas, for indirect contact freezers, the food product is in contact with a belt or plate, which in turn interacts with the freezing medium, leading to indirect exposure of the food product to the refrigerant. To select a method of freezing, the following points are considered: the dimension of the product, shape, thickness, specific heat, rate of freezing, and packaging.

9.2 FREEZING PRINCIPLE

The freezing point of a food material is defined as the temperature in which small ice crystals initially appear in the food material and are in equilibrium with the surrounding water in the food material. The process of freezing consists of kinetics and thermodynamics, which dominate each other during different stages of the process of freezing. The material heat content reduction is a major thermal event during the freezing process, which leads to the cooling of the food material to be frozen. This cools the temperature of the material to a temperature at which the process of nucleation starts. During the nucleation process, the water molecules begin to gather into small clusters, arranging in a way that defines the crystal structure of the solid frozen water or liquid material. This growth of ice crystals in the solution initiates the process of phase change from liquid to solid in the food material. These are the first steps of the freezing process that cumulates in a complete and thorough phase change (Sahagian and Goff 1996). There are two types of nucleation: heterogeneous (starts nucleation sites on the surface) and homogenous (starts nucleation sites away from the surface). Because food material constitutes components like lipids, proteins, carbohydrates, and minerals, it is likely to follow heterogeneous nucleation (Fellows 2009; Singh and Heldman 2001). For food material the freezing curve and its components are as follows:

- *A–S:* The food material is cooled below its freezing point (θ_f). At S supercooling occurs, and this is the point where the water is in liquid form even though the temperature of the surroundings is below freezing point. This is where the process of nucleation starts. The duration of supercooling depends on the food compositions and the rate of freezing.
- *S–B:* The rapid rise of temperature to the freezing point since the formation of ice crystals begins and the latent heat of crystallization is released.
- *B–C:* The rate of heat removal from the food material is the same as in S–B. Here, due to ice formation, the latent heat is removed while maintaining the temperature at the freezing point. With the increase of solute concentration in the food material, the freezing point is depressed. The

temperature falls as the formation of ice increases. A major part of ice formation takes place in this stage.

- *C–D:* Among many solutes in the food material, one attains supersaturation and is completely crystallized. At this stage the release of the latent heat of crystallization takes place and leads to a rise in temperature to eutectic temperature for that particular solute.
- *D–E:* At this stage, the solute and water crystallization continues. The total time (t_f) taken for the growth of ice crystals is dependent on the rate of heat removal and mass transfer of water to the nuclei from the liquid phase. The temperature is the combination of water and ice, which is in equilibrium with that of the freezer.
- *E–F:* At this stage, the formation of ice and concentration of solute takes place until there is no water to be frozen. With the removal of sensible heat from ice the temperature falls. At point F (θ_d) the temperature is known as the glass transition temperature.

9.3 FREEZING SYSTEM

Different types of freezing equipment are available for most food materials. Same food material can be frozen by using more than one type of freezer. Important factors like functionality, economics, and feasibility must be considered while selecting the type of freezing systems (Mallett 1993; Hung and Kim 1996; Singh and Heldman 2001; Fellows 2009).

9.3.1 Direct Contact

There is direct contact between the refrigerant and the food material to be frozen. This system is more efficient as there is no barrier between the refrigerant and the food material for heat transfer.

9.3.1.1 Air-Blast Freezers

Air-blast freezers are the oldest and most commonly used freezing systems. In this system, air is the freezing medium as forced or still air. The air is allowed to circulate around the food material for a particular time to freeze. It is, however, the slowest method of freezing.

Different types of air-blast freezers include the following:

- *Tunnel freezer:* The air is circulated in a tunnel and food materials are loaded on trays that are placed in trolleys or racks and passed through the tunnel.
- *Belt freezer:* They are set up in blast rooms with a wire mesh conveyor. Belt freezers provide a continuous product flow and a vertical airflow to pass air through the food material layer.
- *Fluidized bed freezer:* This freezer consists of a perforated bottom bed, and through this bed cold air is passed vertically upward. Only particular food material can be frozen using this freezer.

9.3.1.2 Immersion Freezer

This type of freezer uses a liquid refrigerant for the freezing product, such as carbon dioxide and nitrogen. The food materials are passed through a bath containing the liquid refrigerant. As the product is passed through the refrigerant, it transforms from an aqueous to a gaseous state, thereby absorbing the product heat.

9.3.2 Indirect Contact Systems

In this system, the refrigerant is separated from food materials during the freezing process with the help of a barrier. Packaged food cannot be frozen as the package provides resistance to heat transfer.

9.3.2.1 Plate Freezer

In the plate freezer the food materials are pressed between two refrigerant plates. The plate and package mostly act as a barrier for the refrigerant. Heat transfer for cooling can be enhanced by applying pressure.

9.3.2.2 Contact Belt Freezer

The contact belt freezer is straightforward or drums with a single band or double band. Mostly thin layers of food materials are frozen using this method.

9.4 FREEZING TIME

Freezing time is defined as the time required to decrease the temperature of the food material from its initial temperature to the required temperature, at its thermal core. The freezing time of a food material depends on various factors like dimensions, product shape, heat transfer coefficient, thickness, temperature of the refrigerant medium, initial and final temperature, variation in product enthalpy, and thermal conductivity (Nagaoka et al. 1955; Planck 1980; Singh and Heldman 2001; Fellows 2009).

The foremost equation used to calculate the time of freezing is Planck's equation. Due to assumptions included in the calculation, such as the freezing and initial temperature of the food material are in equilibrium, this equation is merely helpful to provide an estimation of the freezing time. The freezing time obtained by Equation (9.1) gives the freezing period.

$$t_f = \frac{\rho_f L_f}{T_F - T_a}\left(\frac{P'a}{h} + \frac{R'a^2}{k_f}\right) \tag{9.1}$$

where ρ_f is the density of the frozen material (kg/m^3); L_f is variation in latent heat of the food (kJ/kg); T_F is the temperature of freezing (°C); h is the convective heat transfer coefficient at the material surface (W/m°C); a is the object thickness (m); k is the thermal conductivity of the frozen material (W/m°C); and the constant P' and R' are considered for the shape of the product with $P' = \frac{1}{2}$ and $R' = \frac{1}{8}$ for infinite plate, $P' = \frac{1}{4}$ and $R' = \frac{1}{16}$ for infinite cylinder, and $P' = \frac{1}{6}$ and $R' = \frac{1}{24}$ for sphere.

Because Planck's equation only gives the freezing period, it is necessary to calculate the real freezing time. The extent of heat elimination needed for a reduction in product temperature from its initial temperature to the freezing temperature along with the extent of heat released (during the phase change) and heat eliminated (achieving the freezing temperature), respectively, can be determined from the Nagaoka equation:

$$t_f = \frac{\rho \Delta H}{T_F - T_e}\left(\frac{Pl}{h} + \frac{Re^2}{k}\right)[1 + 0.008(T_i - T_F)] \tag{9.2}$$

where T_i is the food temperature at the start of freezing, ΔH is the alteration among the food enthalpy at initial and final freezing temperatures, Pl and Re represent the dimensionless numbers, and h and k are the heat transfer coefficient and thermal conductivity, respectively. A dimensionless feature has been extensively utilized in equations to determine the time of freezing for products with asymmetrical shapes, particularly fruits and vegetables (Cleland et al. 1987a,b).

9.5 EFFECT OF FREEZING ON THE QUALITY OF DIFFERENT FOOD PRODUCTS

Freezing is a method that preserves quality that is feasible in ranges of food products; thus, studies of freezing and its effect on food quality are important (Dawson et al. 2018). It is one of the commonly used long-term preservation methods for foods that allows food to maintain attributes associated with freshness much better than other conventional preservation methods like canning and drying. Yet, the quality and texture of cellular foods can be strongly impacted by the freezing process in various food products such as meat, vegetables, and fruits (Bulut et al. 2018; Van der Sman 2020).

The process of freezing slows down the physical deterioration of foods as well as their biological and chemical reactions. It also causes degradation of food quality such as its changes in sensory properties, enzymatic activity, lipid oxidation, and damage in the ice crystal structure occurs.

The changes in freezing process parameters such as freezing rate, freezing temperature, and time cause a major effect for its resulting reaction on the quality of different food products. Many researchers have also reported the effect of fast and slow freezing. Fast freezing resulted in rapid ice nucleation within the intracellular areas of food products that create uniform and small-sized ice crystals, thus, causing less effect to the food structure (Silva et al. 2008). Fast freezing also changes physical properties such as a reduction in weight, an increase water-holding capacity, discoloration, and textural changes, i.e., intracellular/extracellular ice crystal growth effects on the structure of different seafood products (Dawson et al. 2018). This does not occur for all food products as fast freezing and small ice crystal formation sometimes produces better quality. This is due to the existence of a limit-freezing rate after which a further rise in freezing rate cannot impact overall product quality during storage; whereas slow freezing causes denaturation of the protein as the process depends on temperature (Johnston 1994). Low freezing rates are also desirable in reducing energy consumption, increasing throughput, and improving yield. Therefore, further research is required to find a balance between maintaining food quality and maximizing the efficiency of energy from mechanical freezers (Reid 1997; Remington 2017).

This shows that there is no sufficient knowledge on how freezing impacts the matrix of tissue-based food materials including fruits, vegetables, and meat, using freeze concentration and the effect of mechanical stress imparted by the growing ice crystals (Gómez and Sjoholm 2004). For example, drip loss occurring after thawing is one of the major impacts of freezing on vegetables and fruits that occurs due to changes in the water-holding capacity of the food imparted by freezing. Samples frozen at faster freezing rates had lower mass loss than samples frozen at higher freezing rates. These changes and their physical causes are hardly discussed in literature, with a few early exceptions that discussed freezing of starch-rich foods and meat (Sikorski 1978; Leygonie et al. 2012).

Research in food freezing is advancing at a very rapid pace to enhance shelf life, maintain product quality as close to its original unfrozen product, reduce the cost of freezing, and meet changing consumer needs and perceptions.

9.6 SHELF LIFE EXTENSION OF FOOD BY FREEZING

Frozen food can be safely kept at a temperature of -1 to $-18°C$ for 3–12 months with no loss of its quality. The freezing time may depend on the composition of food. To further increase the shelf life of frozen products, freezers should be kept at or below $-18°C$, but this may cause undesirable changes. Unlike refrigerators, frozen products must be packed in airtight containers to prevent freeze burn (Li and Sun 2002). By slowing down the enzyme activity that causes food spoilage before freezing, the process of deterioration also slows down and extends the shelf life of food by preventing microorganism growth. With water forming into ice crystals during freezing, there is no available water for the proper growth of microorganisms; however, most microorganisms are not destroyed but only inactivated, which requires frozen foods to be handled carefully once defrosted (Van der Sman 2020).

To further extend the shelf life of food, the process of blanching, i.e., immersion of food into boiling water for a short period, is practiced before freezing to inactivate enzymes and yeasts that would continue to cause food spoilage even during storage. In other cases, because harvested produce can take days to be sorted, transported, and distributed to stores, freezing may be performed to improve the condition of vegetables and fruits soon after harvesting, thus reducing field losses (Sriket and La-ongnual 2018; Qian et al. 2021).

For losses during storage, frozen fresh fruits and green vegetables can lose as much as 15% ± 5% of vitamins and minerals; whereas in frozen meats, fish, and poultry there are almost no vitamin A and D and mineral losses because of the type of amino acid proteins contained in these foods. There could be a loss of liquid containing water-soluble vitamins and mineral salts during defrosting during the cooking process. (Cheng et al. 2015). During a longer storage period, while freezing, the physical characteristics of food products slowly continue to deteriorate. However, chemical characteristics like lipid oxidation enzymatic activity and microbial growth are important factors affecting food quality. Due to the difference in quality and the effects slow- and fast-term freezing methods have on food products, studying the effects of both phases of freezing is necessary (Dawson et al. 2018).

According to studies, there is less effect of lipid oxidation and discoloration in the muscle of meat, whereas on other foods, vitamins and minerals are lost during storage at −20°C for 14 days (Sriket and La-ongnual 2018). Maintaining the muscle structure of meat takes place due to the interaction between water and proteins. Long-term frozen storage leads to the formation of ice crystals and volumetric expansion, leading to a series of changes to myofibrillar protein (50%–55%) of the muscle proteins and having a significant effect on the physicochemical properties such as appearance, sensory, tenderness, and water-holding capacity of meat (Chan et al. 2011).

9.7 CHANGES IN FROZEN PRODUCTS DURING STORAGE

Freezing is a commonly used method for food preservation to maintain nutritional quality and sensory characteristics. Although freezing can be applied to food as a preservation technology, it is not suitable for all foods because of physical and chemical changes it produces in some foods. Freezing also reduces the quality of the thawed material or the final product. Food enterprises are also looking to improve food quality through the production of smaller ice crystals (to reduce damage caused by large ice crystals) and at the same time lower the cost of freezing. Measurements used to evaluate the changes caused by freezing were the freezing time, color, firmness, drip loss, vitamin C, and microstructure of the frozen products (James et al. 2015).

During long-term frozen storage, due to the storage temperature, various biological and chemical reactions occur. Significant impact was found on its enzymatic activity and lipid oxidation during long-term frozen storage and due to the storage temperature. These factors lead to slow deterioration in the quality of food. Lipid oxidation during storage leads to a decrease in the sensory quality of food products. These changes can also be influenced by the handling and processing of food. Oxidation and denaturation of proteins that occur during long-term frozen storage increase discoloration and induce tough texture on meat muscle. Based on research, frozen storage at −20°C after 8 weeks of storage affected the qualities of fish. The formation of ice crystals during frozen storage further causes physical changes to the food structure. In high-moisture foods, the formation of large ice crystals destroys the structure of the food resulting in an increased drip on thawing (Préstamo et al. 1998). During frozen storage, freeze-induced protein denaturation tends to be less affected with lower freezing temperatures. However, studies also showed that the difference in frozen storage temperature (−12 and −18°C) did not show significant changes in the protein properties during a 6-month storage period. In this regard, lowering the freezing temperature might not minimize myofibrillar protein denaturation during a limited storage duration, which was also confirmed by the standard properties of beef samples after frozen storage (Li and Sun 2002).

In addition, understanding components of cell structure change during freezing is important to develop controlled freezing processes to minimize damages. Freezing damage to cell structures is a complex issue due to the composition of food materials and its dynamics, and thermodynamics. Freezing also includes depolymerization of the cell wall, the rupture of the cell membrane, the alteration of osmotic pressure, etc. These changes in permeability or rupture of cell membrane result in the loss of intercellular water, spillage of cellular contents, and loss of nutrition contents and sensory properties. It may also cause loss of cell viability and osmotic pressure. In the freezing water transfer process, crystallization of local water in cells or outside cells can lead to the formation of some local high concentrations. This will result in strong absorption of adjacent water due to the difference in osmotic pressure (Ngapo et al. 1999; Ceballos et al. 2012).

The phenomenon of food freezing involves a thermodynamic process of heat and mass transfer. In most of the studies on heat and mass transfer during food freezing, the food materials are assumed as cellular capillary-porous materials (Datta 2007), resulting in a complex way to transport water and heat during the process. All of these phenomena occur due to the complex matrix of food tissue, and as the cell structure of food encompasses most water into the matrix, it becomes an obstacle in the transfer of water and heat (Delgado and Sun 2001). This causes undesirable changes, which can be further studied for different food products. In a number of studies, the process of pretreatment, such as blanching samples, decreased firmness to a great extent compared with unbalanced frozen products. This is due to the formation of a gel from the effect of the heat on the pectic substances. Evaluation of changes during freezing and during storage can be measured by color, firmness, drip loss, vitamin and mineral composition as well as microstructure of the frozen products (Delgado and Sun 2001; Datta 2007).

9.8 ADVANTAGES AND DISADVANTAGES OF FREEZING

9.8.1 Advantages

Freezing takes less time than canning or drying, and, unlike drying and canning, freezing retains the natural color, flavor, and nutritive value of the food. Freezing food materials allows the preservation of a vast amount of seasonal fruit and vegetables, making them available throughout the year. Frozen products are nutritionally high, as freezing prevents the loss of sensitive bioactive compounds during transportation and storage. Also, by freezing fruits and vegetables they can be stored without the help of preservatives at their natural state. Economically, frozen foods are cheaper when compared with fresh food as they are preserved quickly and can be stored for a long time. Fresh and frozen fruits and vegetables have equivalent nutritional components. Due to long storage time and ease of transportation to a different part of the world, frozen foods decrease the amount of food waste (Fennema et al. 1973; Li and Sun 2002; Fellows 2009; James et al. 2015).

9.8.2 Disadvantages

Frozen foods suffer from the nutritional loss of vitamins C and B. Also, there are fewer bioactive compounds in frozen foods compared with that in fresh foods. Some frozen foods like pizza, turkey, sausages, salami, etc., contain more sodium than recommended for daily intake. During the freezing process, the formation of ice crystals can cause tissue and structural damage to the food material. Freeze burn (dried out, gray-brown edges of meat or dull coloring in fruits and vegetables) affects the texture, color, and flavor of the food material. Improper storage of frozen foods can expose the frozen material to air, leading to the formation of damaging ice crystals, which degrade the quality of food material by loss of moisture, texture, and flavor. Due to freezing the natural crispness of vegetables and fruits are lost, even after blanching the food material. Improperly packed food material has an increased chance of picking up the flavor and smell of other foods around it. In some frozen food, freezing changes its texture so that it does not feel or taste the same after thawing (Li and Sun 2002; Fellows 2009; James et al. 2015).

9.9 CONCLUSION

Freezing is a promising and well-established process of food preservation. The restriction of microbial growth and slowing down the enzyme activity due to low temperature is the principle reason for the shelf life extension of food during freezing. The freezing process depends on various factors including the type of product, product dimensions, freezing medium, temperature difference, etc. Planck's equation is one of the most commonly adapted approaches used to calculate the freezing time of a product. The freezing process has many advantages including maintaining the quality of the product and is a less time-consuming process. Although freezing can be applied to food as a preservation technology, it is not suitable for all foods because of the physical and chemical changes it introduces to some foods. It is also reduces the quality of the thawed material or the final product. Food enterprises are also looking to improve food quality through the production of smaller ice crystals (to reduce damage caused by large ice crystals) and lower the cost of freezing. On the other side, the freezing process has many limitations such as the requirement of a special supply chain and transportation and packaging and storage conditions when compared with other conventional food preservation methods.

REFERENCES

Bulut, M., Bayer, Ö., Kırtıl, E., & Bayındırlı, A. (2018). Effect of freezing rate and storage on the texture and quality parameters of strawberry and green bean frozen in home type freezer. *International Journal of Refrigeration*, *88*, 360–369.

Ceballos, A. M., Giraldo, G. I., & Orrego, C. E. (2012). Effect of freezing rate on quality parameters of freeze dried soursop fruit pulp. *Journal of Food Engineering*, *111*(2), 360–365.

Chan, J. T., Omana, D. A., & Betti, M. (2011). Effect of ultimate pH and freezing on the biochemical properties of proteins in turkey breast meat. *Food Chemistry*, *127*(1), 109–117.

Cheng, X., Zhang, M., Xu, B., Adhikari, B., & Sun, J. (2015). The principles of ultrasound and its application in freezing related processes of food materials: A review. *Ultrasonics Sonochemistry*, 27, 576–585.

Cleland, D. J., Cleland, A. C., & Earle, R. L. (1987a). Prediction of freezing and thawing times for multi-dimensional shapes by simple formulae. Part 1: Regular shapes. *International Journal of Refrigeration*, *10*(3), 156–164.

Cleland, D. J., Cleland, A. C., & Earle, R. L. (1987b). Prediction of freezing and thawing times for multi-dimensional shapes by simple formulae. Part 2: Irregular shapes. *International Journal of Refrigeration*, *10*(4), 234–240.

Datta, A. K. (2007). Porous media approaches to studying simultaneous heat and mass transfer in food processes. I: Problem formulations. *Journal of Food Engineering*, *80*(1), 80–95.

Dawson, P., Al-Jeddawi, W., & Remington, N. (2018). Effect of freezing on the shelf life of salmon. *International Journal of Food Science*, *2018*, 1686121.

Delgado, A. E., & Sun, D. W. (2001). Heat and mass transfer models for predicting freezing processes–a review. *Journal of Food Engineering*, *47*(3), 157–174.

Fellows, P. J. (2009). *Food Processing Technology: Principles and Practice*. Elsevier.

Fennema, O. (1977). Loss of vitamins in fresh and frozen foods. *Food Technology*, *12*, 32–38.

Fennema, O. R., Powrie, W. D., & Marth, E. H. (1973). Low-Temperature Preservation of Foods and Living Matter. Marcel Dekker.

George, R. M. (1993). Freezing processes used in the food industry. *Trends in Food Science & Technology*, *4*(5), 134–138.

Gómez, F., & Sjoholm, I. (2004). Applying biochemical and physiological principles in the industrial freezing of vegetables: A case study on carrots. *Trends in Food Science & Technology*, *15*(1), 39–43.

Hung, Y. C., & Kim, N. K. (1996). Fundamental aspects of freeze-cracking. *Food Technology (Chicago)*, *50*(12), 59–61.

James, C., Purnell, G., & James, S. J. (2015). A review of novel and innovative food freezing technologies. *Food and Bioprocess Technology*, *8*(8), 1616–1634.

Johnston, W. A. (1994). *Freezing and Refrigerated Storage in Fisheries* (Vol. 340). Food & Agriculture Organization of the united Nations.

Leygonie, C., Britz, T. J., & Hoffman, L. C. (2012). Impact of freezing and thawing on the quality of meat. *Meat Science*, *91*(2), 93–98.

Li, B., & Sun, D. W. (2002). Novel methods for rapid freezing and thawing of foods – a review. *Journal of Food Engineering*, *54*(3), 175–182.

Mallett, C. P. 1993. *Frozen Food Technology*. Chapman and Hall.

Nagaoka, J., Takaji, S., & Hohani, S. (1955). Experiments on the freezing of fish in an air-blast freezer. *Journal of the Tokyo University of Fisheries*, *42*(1), 65–73.

Ngapo, T. M., Babare, I. H., Reynolds, J., & Mawson, R. F. (1999). Freezing and thawing rate effects on drip loss from samples of pork. *Meat Science*, *53*(3), 149–158.

Planck, R. (1980). *El Empleo del Frío en la Industria de la Alimentación*, Reverte.

Préstamo, G., Fuster, C., & Risueño, M. C. (1998). Effects of blanching and freezing on the structure of carrots cells and their implications for food processing. *Journal of the Science of Food and Agriculture*, *77*(2), 223–229.

Qian, S., Li, X., Wang, H., Mehmood, W., Zhang, C., & Blecker, C. (2021). Effects of frozen storage temperature and duration on changes in physicochemical properties of beef myofibrillar protein. *Journal of Food Quality*, *2021*, 8836749.

Reid, D. S. (1997). Overview of physical/chemical aspects of freezing. In *Quality in Frozen Food* (pp. 10–28). Springer.

Remington, M. C. A. (2017). The effect of freezing and refrigeration on food quality. MS thesis, Department of Food, Nutrition and Packaging Sciences, Clemson, University, Clemson, South Carolina.

Sahagian, M. E., and Goff, H. D. (1996). Effect of freezing rate on thermal, mechanical and physical aging properties of the glassy state in frozen sucrose solutions. *Thermochimica Acta*, *246*(2), 271–283.

Sikorski, Z. E. (1978). Protein changes in muscle foods due to freezing and frozen storage. *International Journal of Refrigeration*, *1*(3), 173–180.

Silva, C. L., Gonçalves, E. M., & Brandao, T. R. (2008). Freezing of fruits and vegetables. In *Frozen Food Science and Technology* (pp. 165–183). Blackwell Publishing.

Singh, R. P., & Heldman, D. R. (2001). *Introduction to Food Engineering*. Gulf Professional Publishing.

Sriket, P., & La-ongnual, T. (2018). Quality changes and discoloration of Basa (*Pangasius bocourti*) fillet during frozen storage. *Journal of Chemistry*, *2018*, 5159080.

Tu, J., Zhang, M., Xu, B., & Liu, H. (2015). Effects of different freezing methods on the quality and microstructure of lotus (*Nelumbo nucifera*) root. *International Journal of Refrigeration*, 52, 59–65.

Van der Sman, R. G. M. (2020). Impact of processing factors on quality of frozen vegetables and fruits. *Food Engineering Reviews*, *12*, 399–420.

CHAPTER 10

Thermal Treatment of Foods
Science, Shelf Life, and Quality

**Arshied Manzoor, Bisma Jan, Insha Zahoor, Nadira Anjum,
Aarifa Nabi, Farhana Mehraj Allai, Qurat Ul Eain Hyder Rizvi,
Rayees Ahmad Shiekh, Mohd Aaqib Sheikh, and Saghir Ahmad**

CONTENTS

10.1 INTRODUCTION

Unlike the past during which new technologies were focused exclusively on cost reduction and yield improvement, the present era makes preservation and the distribution of foods a priority due to the ample amount of market opportunities. Since the last decade, researchers, consumers, and food industries have been miserably challenged by food allergies witnessed by an upsurge in the severity of clinical manifestations. Food allergens have caused a public health threat that has created allergic reactions from a wide variety of foods. Also, consumers opt for thermally processed foods that have the least damage to nutritive compounds in addition to the cooking and nutrient loss, color intensity reduction and texture deterioration, and shrinkage (Kong et al., 2007). Hence, researchers have been working on the processes causing food spoilage and have come up with several remedial techniques. Among them, thermal processing is considered a major and primitive

Email: Arshied Manzoor, arshidfe12@gmail.com

DOI: 10.1201/9781003091677-10

processing technology in the food industry. Of the extrinsic factors affecting growth, quality attributes, and those leading to microbial inactivation, the temperature is the potent candidate. Thermal processing provides safety and shelf life enhancement of food products by inactivating pathogens. Microbial growth and metabolism causing undesirable changes to the food are also inhibited.

Humans were unaware of the reduction in allergenicity by heat they started eating cooked foods through heat treatments centuries before. Moreover, thermal treatments furnish some desirable changes, such as protein coagulation, texture softening, and formation of aromatic components. The purpose of application along with the severity of heat treatment determines the thermal process such as pasteurization and sterilization (Karel and Lund, 2003). Food products also contain some nutrients that will be lost with limited destruction due to heat treatments coupled with the alteration of natural taste and flavor, which is favored through thermal processing. The commonly used thermal treatment methods include pasteurization and sterilization, which are often determined as critical control points (CCPs). The degree to which a foodstuff is preserved and rendered safe for consumption depends on the effect of microbial destruction. Various heat treatment processes certainly increase quality, safety, and storage through inactivation of toxins and microbes. Nevertheless, enhancement in palatability, digestibility, and nutritional quality coupled with improvements in appearance, texture, flavor, and taste cannot be ignored. Thermal processing among the food processing techniques has resulted in the structural alteration of proteins (the allergenic component in foods), thereby reducing the allergenicity of the entire food. Hence, thermal processing is considered a promising method to cater to the food allergic consumer. In the United States, a $5–6 billion loss is recorded on an annual basis due to foodborne diseases. Also, 6.5–33 million cases of diarrheal disease are reported and about 9000 persons die due to pathogenic bacteria such as *Campylobacter*, *Escherichia coli* O157:H7, *Listeria monocytogenes*, *Salmonella*, and *Staphylococcus aureus* every year.

Broadly, pasteurization and sterilization comprise two main categories of thermal processing based on temperature causing a reduction or destruction in microbial activity and enzyme activity. Food industries have utilized various heat processing methods in food and food product processing. Based on the heat severity, heat treatment methods used in food processing are blanching and pasteurization (mild processes) and canning, baking, frying, and roasting (more severe processes). The temperature range to inactivate microbes and enzymes is 50–150°C. An increase in temperature during the thermal processes exhibits a higher rate of microbial destruction compared with the rate of destruction of the nutrients and sensory components and enhances the antioxidant activity of flavonoids in particular (Manzoor et al., 2020). Among the thermal processes, such as pasteurization, commercial sterilization, operations for food tenderization, and blanching, the process type used is a function of pH, microbial load, and desired shelf life (Tijskens et al., 2001; Aamir et al., 2013). Conventionally the thermal processes are categorized as pasteurization, sterilization, and ultrahigh temperature (UHT) processes with a temperature range of 65–85°C, 110–121°C, and 140–160°C, respectively. Up to a certain limit, both microbial growth and enzyme activity increase proportionally with temperature after which both start to decrease and inactivation begins. Most of the vegetative microorganisms responsible for foodborne illness such as *Salmonella* are destroyed in foods by thermal processing, which is a traditional method still in use. It has proved to be inexpensive, preservative free, efficient, and environmentally friendly.

10.2 BLANCHING

In various unit operations applied in the processing of fruits and vegetables, blanching represents a very important process. It is defined as a mild heat treatment process used as a pretreatment of raw materials (vegetables) after preparation but preferably before further processing (canning, freezing, dehydrating, etc.) (Arroqui et al., 2003). The inactivation of enzymes (responsible for the quality deterioration) in vegetable products is the prime function of blanching. The other functions of blanching include the release of entrapped air from intercellular spaces within fresh foods before canning to reduce internal stresses in the can, surface microbial contamination reduction, to ease filling in the containers through softening of vegetable tissues, and material heating before filling for the vacuum creation after the boiling process. In fruits and vegetables, catalase and peroxidase (POD) enzymes are very heat resistant. Enzyme inactivation is insufficient in foods through preservation techniques such as freezing, dehydration, and canning. These faults allow blanching to be employed in the inactivation of enzymes such as lipoxygenase, polygalacturonase, polyphenoloxidase, and chlorophyllase. Blanching, also called scalding, involves the use of heat in the form of either steam or boiling water. The time-temperature combination in the process is decided based on the nature and composition of the material, and the processing method incorporated after blanching.

The POD enzyme acts as an indicator enzyme to confirm the adequacy of the blanching process as POD exists in the vegetable systems as the most thermally stable enzyme (Akyol et al., 2004). In addition, the degree of inactivation of POD and polyphenol oxidase (PPO) enzymes, loss of ascorbic acid (vitamin C), and color and texture change represent indicators for assessing the efficacy of the blanching process, as presented in Figure 10.1. Blanching stabilizes food and prevents the degradation of vegetables by inactivating the enzymes that catalyze the respective reactions during storage. Blanching is also applied to furnish uniformity in color after the frying process. For example, the texture of potatoes is somewhat improved in addition to enzyme inactivation after blanching. Moreover, it provides a compact surface through starch gelatinization on the potato surface, reducing the oil uptake and surface pores/air cells (Krokida et al., 2001). However, food products undergoing a severe blanching process may lead to a substantial loss in nutrients, product taste, color, and texture. Generally, blanching causes significant loss of vitamin C (found in higher content in green leafy vegetables) as it is unstable at higher temperatures, which causing its inactivation. Vitamin C shows fair solubility in water so it is washed away during blanching. The effect of blanching on various green leafy vegetables of tropical origin regarding vitamin C (%), total phenolic content,

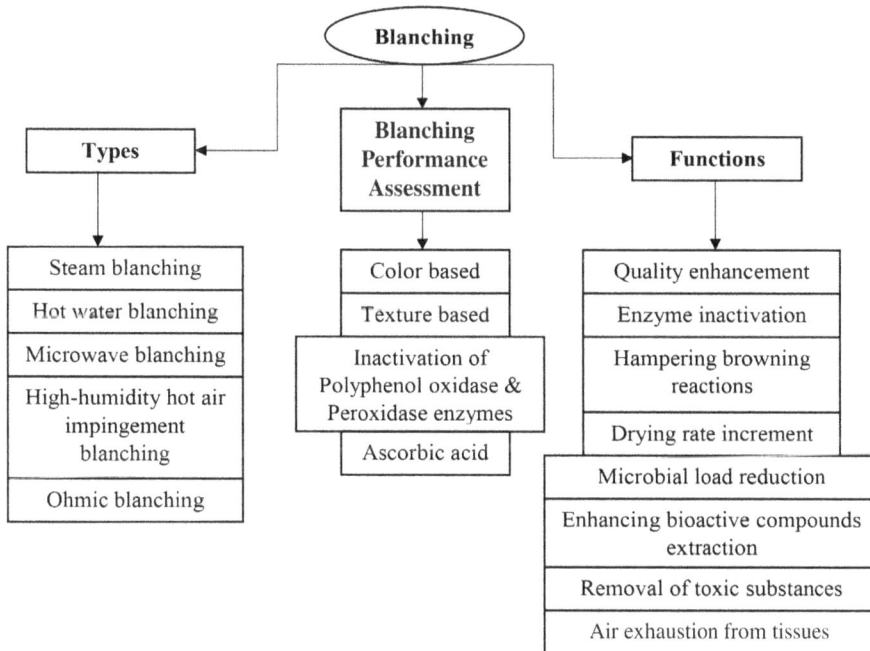

Figure 10.1 Blanching: Types, assessment, and functions.

Table 10.1   Loss/Gain of Vitamin C, Total Phenolic Content, and the Effect on Properties (Reducing and Free Radical Scavenging Activity) during Blanching (Hot Water for 5 Minutes) in Some Tropical Green Leafy Vegetables

Tropical Green Leafy Vegetable	Loss in Vitamin C (%)	Gain in Total Phenolic Content (%)	Loss in Reducing Property (% Absorbance at 700 nm)	Loss in Free Radical Scavenging Activity
Telfairia occidentalis	82.4	33.3	66.7	5.8
Structium sparejanophora	67.6	200.0	0.0	0.0
Corchorus olitorius	63.7	33.3	60.0	18.0
Solanum macrocarpon	60.9	100.0	20.0	6.8

reducing property, and free radical scavenging activity is presented in Table 10.1 (Oboh, 2005). In blanching, foods (vegetable or fruit) are put into water (boiling), scalded, removed after a while (usually 1–10 minutes), and then eventually submerged into iced water (shocking or refreshing) to stop the cooking process. Inactivation time of the POD and polyphenoloxidase enzymes determines the blanching required for a particular fruit or vegetable (Xiao et al., 2017). The disadvantages of contamination of raw fruits and vegetables that starts just after harvesting through transportation and finally during storage make the blanching process necessary as it kills the responsible microorganisms (Cruz et al., 2006; Mukherjee and Chattopadhyay, 2007).

10.2.1 Methods of Blanching

10.2.1.1 Hot Water Blanching

This is the most commonly used blanching method because of its simple establishing nature and easy operation (Mukherjee and Chattopadhyay, 2007). In this method hot water at a temperature of 70–100°C is used, in which fruits or vegetables are dipped for some time and drained, and finally cooled in running cold or iced water before proceeding to the next process. Sodium sulfite or sodium metabisulfite are commonly added to the blanching water to preserve the product color that could otherwise be leached into the bleaching water. The water becomes laden with the nutrients leached out of the products necessitating the substitution of the used water with fresh water after some calculated time. To date, this method has been applied for various products. For example, high-quality paprika or chili powder is produced through endogenous enzyme inactivation caused by blanching of the chili in hot water and steam with different time-temperature combinations (80°C for 10 minutes, 90°C for 5 and 10 minutes, or 100°C for 5 and 10 minutes) (Schweiggert et al., 2005). These combinations resulted in an almost 98% reduction in POD activity in chili and paprika. Moreover, they reported the inactivation of lipid oxidase enzymes at blanching conditions of 90°C for 5 minutes and 100°C for 5 minutes. In another study, hot water blanching of Brussels sprouts at 96–98°C resulted in a decrease of total amino acids from 2783 mg/100 g in fresh samples to 2345 mg/100 g in blanched samples (Lisiewska et al., 2009). Blanching also reduced the total amino acids in cassava leaves and broccoli in much greater amounts compared with Brussels sprouts (Ngudi et al., 2003). Hot water blanching (at a temperature of 60, 70, 80, and 88°C for 12 minutes) helps to render almond kernels free of pellicles (Harris et al., 2012) and in combination with freezing restricts the oil uptake and preserves the color and texture desired (Pimpaporn et al., 2007). Combined with citric acid blanching, hot water blanching can also lead to the inactivation of microorganisms such as *Salmonella* in fresh fruits and vegetables such as carrots (Dipersio et al.,

2007). An increased amount of total ascorbic acid loss was determined in vegetables such as beans (green), peas, slices of leek, zucchini and carrots, broccoli, spinach branches, hashed spinach, French beans (yellow), cauliflower, and mushrooms by using hot water blanching among various treatments (Bureau et al., 2015).

However, this traditional method of blanching is prone to disadvantages and/or limitations. The leaching going side by side during this method leads to nutrient loss (Mukherjee and Chattopadhyay, 2007). Moreover, substances such as ascorbic acid, aroma, and flavor compounds are sensitive to the hot water blanching that causes their degradation in addition to the leaching of vitamins, flavors, minerals, carbohydrates, sugars, and proteins. A study reported the reduction of ascorbic acid content from 94.6 to 69.7 mg/100 g dry matter in French fries (Haase and Weber, 2003). Hot water blanching also leads to total phenolic content loss as reported by Ismail et al. (2004). The water left after hot water blanching of the food products is a potent threat to the environment due to high concentrations of biochemical elements, soluble solids, and chemical oxygen demand from leaching during the process causing eutrophication (Liu and Yang, 2012). To cater to these problems, new energy-efficient, environmentally friendly methods with reduced nutrient loss have been developed among which high-humidity hot air impingement blanching (HHAIB) and microwave, ohmic, and infrared methods combined with hot air blanching are worth mentioning.

10.2.1.2 Steam Blanching and High-Humidity Hot Air Impingement Blanching (HHAIB)

HHAIB works on the principle that the superheated steam with high enthalpy contents transfers its latent heat to the product through condensation. The temperature of the product increases, leading to the inactivation of the microorganisms and enzymes. This method is used to inactivate the deteriorating enzymes resulting in the shelf life enhancement. A study has reported the negligible leaching effects leading to retention of most minerals and water-soluble components compared with traditional water blanching and it is also less expensive than blanching alone (Roy et al., 2009). However, steam blanching has some disadvantages, such as tissue softening and quality changes that reduce heat transfer and increase the heating time. Steam blanching coupled with microwave blanching affected chlorophylls a and b the most in spinach. Moreover, the formation of pyropheophytins and the development of pyrochlorophylls a and b were reported using steam blanching (Teng and Chen, 1999). Steam blanching was also applied to kiwifruit, and the study reported a reduction in fruit firmness and microstructure alteration caused by steam blanching, which was witnessed through transmission electron microscopy (TEM) and fluorescence microscopy (FM). A color change was also seen in the fruit due to chlorophyll degradation (Llano et al., 2003).

Superheated steam blanching reduced the water loss and prevented the quality deterioration in potatoes (Sotome et al., 2009). Steam blanching is used to extend the shelf life of fruit products such as mango slices (94°C for 0–10 minutes) and peeled garlic (100°C for 4 minutes) by inactivating PPO, inulinase, and POD enzymes (Ndiaye et al., 2009; Fante and Noreña, 2012). However, prolonged treatment of hot water blanching could cause color degradation as well as nutrient and texture loss in food products. Steam blanching leads to non-uniform blanching effects when carried out in thick layers over moving belts. It is also more time-consuming, less economical, and less effective against POD inactivation than hot water blanching.

HHAIB technology is a recently developed technique with advantages such as uniformity, faster rate, and efficient processing (Xiao et al., 2014). In some food products, such as yam slices, the prime focus of this technique is to preserve color coupled with the prevention of the browning reaction (Xiao et al., 2010). In addition, the HHAIB technique has various beneficial impacts on shelf life enhancement, such as the inactivation of PPO enzyme with an increased rate of drying in grapes, optimization of color and texture in sweet potatoes, and reduction in microbial load in poultry and fresh-cut lettuce (Kondjoyan and Portanguen, 2008; Rico et al., 2008; Bai et al., 2013).

10.2.1.3 Microwave Blanching

Microwave blanching is far superior to the conventional blanching methods due to high heating rates, reduced nutrient leaching, and short processing times in addition to energy efficiency and volumetric heating. Heat generation is created with direct interaction between the electromagnetic field and food materials used for processing. This method uses microwaves (electromagnetic waves) with wavelengths measuring 1 mm to 1 m, and 300 MHz to 300 GHz (frequency) ranges. In heating applications concerned with industrial, scientific, and medical (ISM) applications, the U.S. Federal Communications Commission (FCC) recommends only 915 and 2450 MHz microwaves. The mechanism of microwave heating lies in the conversion of the absorbed microwave energy into heat through the dielectric heating effect caused by the molecular dipole rotation and agitation of charged ions within a high-frequency alternating electric field (Chandrasekaran et al., 2013). Moreover, volumetric heating in microwave heating is caused by the collision of rotating atoms or molecules, and the heating takes place within the food product in addition to the food surface.

10.2.1.4 Ohmic Blanching

Ohmic blanching is another method of blanching in which food products are sandwiched between two electrodes. Electrical resistance is posed by the food material, causing heat generation and a rapid temperature rise in the product (Assiry et al., 2003). Ohmic heating, also called Joule heating, electrical resistance heating, or electro-heating, generates heat based on product conductivity and current induced in the product (Reznick, 1996). The potential advantages, such as fast and uniform heating and high energy conversion efficiencies, of ohmic blanching make it a potential replacement for the conventional blanching methods. Moreover, volumetric heating in ohmic blanching minimizes the processing time and yields better quality products due to color retention and reduction in solids and nutrient leaching (Leizerson and Shimoni, 2005). Ohmic blanching is used for blanching fruits and vegetables in larger volumes, thereby retaining their quality due to a higher convection heat transfer rate. Ohmic blanching leads to enzyme inactivation in less time than hot water blanching and can preserve the total protein and polyphenolic contents up to 3 months of canning storage of artichoke heads (Guida et al., 2013). Ohmic heating leads to the enhancement in mass transfer kinetics but the loss of firmness at higher temperatures (50°C) during osmotic dehydration of strawberries. However, an increase in voltage leads to increased degradation of vitamin C in acerola pulp during ohmic blanching. Juices and milk can also be preserved for longer periods by ohmic blanching, which helps to inactivate enzymes such as alkaline phosphatase, pectin methylesterase, and POD at a greater enhanced rate than conventional methods (Jakób et al., 2010). Ohmic heating coupled with osmotic dehydration and vacuum impregnation in apples leads to the least amount of change in color and firmness and PPO inactivation, which is the biggest problem in the food industry, and a shelf life extension by more than 28 days at 5°C (Moreno et al., 2013). Ohmic blanching (heating) causes a minimum change in food structure, thereby keeping the nutritional value intact coupled with quality enhancement depending on the electrical field strength (Jan et al., 2021).

Among the disadvantages of ohmic blanching are difficulty in controlling the blanching temperature, a decrease in heating rate with increase in frequency, and nutrient degradation due to oxygen and hydrogen generation during electrolyzation of water. The catalytic property of oxygen generated may also lead to color changes, anthocyanin oxidation, and ascorbic acid degradation in acerola pulp if a low electric field frequency (10 Hz) is used (Mercali et al., 2014). It is necessary for food processing technologies in today's advanced world to boost nutrient retention, process sustainability, and be energy efficient to produce better quality products coupled with the reduction in nutrient loss, environment load, and production cost. Hence, selecting a particular blanching technology with the previously mentioned qualities is an essential parameter for food products.

However, it is impossible to apply a single blanching technology with the same outcome for all food products that have different properties.

10.3 PASTEURIZATION

Among the processing technologies in the food industry, pasteurization represents one of the oldest and most commonly used ones. Pasteurization is still considered an effective method of processing even if many upgraded and new technologies are available. The process of pasteurization got its name from the invention of mild heat treatment (55°C) of some liquids (wine and beer) by famous scientist Louis Pasteur (Wilbey, 2014). Later, pasteurization was used in the destruction of pathogenic microbes and to reduce the number of spoilage microorganisms in milk. Pasteurization is a unit operation widely used as a CCP in Hazard Analysis and Critical Control Point (HACCP) in the food industry and extends the shelf life of food products by using mild heat treatment (<100°C) to kill pathogens (harmful bacteria). Pasteurization leads to the destruction of heat-sensitive non-spore-forming bacteria, yeasts, molds, and some enzymes. These properties make the pasteurization process a better unit operation typically for the extension of shelf life by several days (e.g., milk) or months (e.g., bottled fruit). Pasteurization can also be used for various food products such as juices, cider, eggs, cheeses, butter, vinegar, sauerkraut, almonds, beer, and acidic canned foods. Studies have shown a reduction in tuberculosis cases in the United States because of milk pasteurization. The heat treatment level to be used in the pasteurization from a particular food product is decided by the D value (decimal reduction time or time to reduce numbers by a factor of 10% or 90% of the initial load). Moreover, the pH of food also determines the extent of the heat treatment and the shelf life enhancement. In juices, particularly apple juice, thermal pasteurization eliminates microbial spoilage; however, the applied heat may affect the overall quality of the final product through undesirable biochemical and nutritional changes. Pasteurization increases the shelf life of food with minimal effect on nutrition by days or even some weeks by killing the harmful pathogens responsible for food spoilage and can prevent the alteration of food's sensory characteristics by limiting the enzymatic activity such as browning or other undesirable color changes.

Pasteurization is a mild heat treatment of packaged and non-packaged foods (such as milk and fruit juice), usually to less than 100°C (212°F), to extend shelf life through the destruction of pathogens. Pasteurization can be applied before packaging or after packaging of food products, for example, pasteurization of juice or raw milk before packaging in containers and that of solid foods after packaging in jars. Pasteurization can reduce an initial load of 10,000 microorganisms to mere zero, i.e., 5 log cycle microbial reduction. However, refrigeration is required after pasteurization for safety and quality maintenance. Many pasteurization technologies have been used, but the two most commonly used techniques are low temperature, long time (LTLT) and high temperature, short time (HTST). In HTST, a higher temperature is used for a shorter period and varies according to the product; for example, a temperature of 161°F for 15 seconds is used in HTST for milk. However, a lower temperature for an extended period is used in the LTLT process; for example, milk treated at a temperature of 145°F for 30 minutes. Moreover, factors such as target pathogens and food variety decide the time-temperature combination needed to pasteurize the food product. Because the temperature used in pasteurization is below 100°C, all the microorganisms present are not eliminated (Silva and Gibbs, 2010). Pasteurization helps to preserve the freshness of vegetables by inactivating the endogenous enzymes, thereby keeping the quality intact throughout processing and storage. Spoilage organisms are either reduced in number or even killed in foods like vinegar through pasteurization, whereas disease-causing microorganisms are destroyed (e.g., pasteurization of milk). The alkaline phosphatase enzyme present in raw milk has the same D value as the heat-resistant pathogens, making this enzyme the indicator of pasteurization effectiveness. Pasteurization is effective when it ensures that phosphatase activity is eliminated. Pasteurization is also defined as a process, or treatment, or a combination of both in which the microorganisms responsible for public health are reduced to an amount that does not pose a threat to public health in normal conditions (National Advisory Committee on Microbiological Criteria for Foods, 2006).

Pasteurization can be performed using thermal and non-thermal techniques in different foods. Among the thermal techniques, steam and hot water heating, ohmic heating, microwave heating, and infrared processing are prominent ones. High-pressure processing, ultraviolet radiation, irradiation, pulsed electric field (PEF), chemical treatments, ultrasound, filtration, and high voltage arc discharge are the non-thermal technologies of pasteurization. Studies have suggested considering these points during the development of the pasteurization process: determining the most resistant microorganism concerning public health that could survive the selected pasteurization method; to assess and make sure that the target microorganisms are inactivated to the level at which there is no public health risk; and to evaluate the right temperature for storage, distribution, and shelf life. Moreover, the operating conditions, type of equipment, and the effect of the food matrix on the pathogen should also be included. The properties of raw food, especially the sensory ones, are largely retained by mild temperature (95°C) treatment for a specified time duration. Moreover, most vegetative pathogens such as *Salmonella* are inactivated, whereas spore-forming bacteria survive such mild heat treatments. Hence, refrigerated conditions (temperature below 7°C) are preferred for the storage, transportation, and sale of low-acid, raw, and pasteurized foods (Silva and Gibbs, 2009).

Various time-temperature combinations applied during milk pasteurization are 63°C for 30 minutes for 1 second, 90°C for 0.5 seconds, 94°C for 0.1 seconds, and 100°C for 0.01 seconds. These combinations, along with eliminating yeasts, molds, and bacteria (gram-negative and gram-positive), destroy almost all the heat-sensitive non-spore-forming pathogenic organisms. However, pasteurization cannot kill all the microorganisms, for example, both thermophilic (*Streptococcus thermophilus*) and thermoduric microorganisms (*Streptococcus* and *Lactobacillus*) can survive higher temperatures. Nevertheless, thermoduric ones do not grow at higher temperatures, whereas the growth of thermophilic microorganisms is facilitated at these higher temperatures. Inactivation time of *Mycobacterium tuberculosis* determines the pasteurization required for a particular fruit or vegetable and its destruction confirms the elimination of other more heat-resistant non-spore-forming, disease-causing microorganisms. In the food industry, several pasteurization methods are used. In one type of pasteurization called the batch (holding) method, milk is heated to at least 63°C followed by holding for at least 30 minutes. In HTST, milk is heated throughout to at least 72°C followed by holding for at least 15 seconds, which is carried out as a continuous process.

10.3.1 Pasteurization of Packaged Foods

Packaging material varies according to the food product to be packed and pasteurized. If the food is packaged in glass, hot water is used for pasteurization to minimize the chances of glass breakage due to thermal shock. In liquid foods, pasteurization is carried out after packing into containers such as beer and fruit juices with a maximum temperature gradient of 20°C between liquid and the container and 10°C for cooling. Liquid foods are pasteurized in either a batch or continuous manner. Food crates placed in a water bath are heated to a preset temperature and then cooled by draining and adding cold water in a simple batch type method, whereas in the continuous type, food crates are passed through a hot water bath followed by a cold water bath. Acid products such as fruit or acidified vegetables like beetroot can be pasteurized in a retort.

10.3.2 Effects of Pasteurization on Food Products

Consumers today prefer the fresh-like qualities of food products (vegetables), which is another factor along with microbial inactivation to consider when selecting the pasteurization method (Dadali et al., 2007; Nisha et al., 2011). Thermal pasteurization in foods, especially vegetables, does not provide percentage inactivation of the endogenous enzymes that cause the quality deterioration in processed vegetables during storage. Various enzymes affect vegetable quality and cause quality deterioration. POD results in texture alteration through oxidative cross-linking of cell wall polymers and catalyzing browning through phenolic oxidation. Hence, the product's shelf life is reduced by the combined action of spoilage bacteria and the endogenous enzymes. Color texture and flavor constitute the quality attributes of vegetables that are affected by various enzymes.

Color: Prior to eating, a consumer accepts a food product based on color furnished by pigments such as chlorophylls, anthocyanins, and carotenoids including lycopene, which also provides health and nutritional benefits. Hence, color is an important characteristic and plays a vital role in a product's acceptance. Thermal pasteurization results in vegetable color degradation and the extent of the degradation is dependent on factors such as heat intensity, duration, media, compounds responsible for color, and storage time as reported by some studies. Color change in vegetables is caused through carotenoid oxidation by lipoxygenase (LOX), which also results in off-flavor development through oxidation of polyunsaturated fatty acids. A recent study revealed that the color change of sweet potatoes has more resistance over broccoli florets during thermal processing (Koskiniemi et al., 2013). Thermal pasteurization results in color degradation in vegetables depending on the heat intensity, duration, media, compounds responsible for color, and storage time. For example, a temperature range of 70–98°C and 80–100°C reduces the chlorophyll content followed by color degradation in asparagus, mixed vegetable juice, butterhead lettuce, celery, parsley, apple concentrate, and kalamansi lime. Color change due to thermal pasteurization is also reported in spinach leaves.

Carotenoids: Lycopene (red/orange), xanthophyll (yellow), lutein, α- and β-carotenes, and zeaxanthin (green/yellow) constitute the carotenoids in vegetables. Vegetables comprise carotenoids as the most prominent organic pigment and act as antioxidants to prevent diseases along with having a role as bioactive compounds, particularly in carrots. Orange color in vegetables such as carrots and sweet potatoes is provided by vitamin A precursors, and α- and β-carotenes are essential for vision. In tomatoes, red color is furnished by lycopene and α- and β-carotenes that also help in cancer prevention due to antioxidant activity. Unintentionally, thermal pasteurization results in oxidation and other chemical changes of lycopene and carotene (α-, β-); however, total carotenoids are resistant to mild pasteurization (Shi et al., 2003). No prominent change was seen in carrots after pasteurization ranging from mild ($F_{70°C}$ D 2 minutes) to severe type ($F_{90°C}$ D 10 minutes) and the credit was given to the protective food matrix preventing their degradation (Lemmens et al., 2013). Pasteurization has also been found to reduce the number of carotenoids in red sweet pepper and carrot juice after pasteurization at 70 and 100°C for 10 minutes(Rayman and Baysal, 2011). Contrary to this, an increase in total carotenoid content, lycopene, and β-carotene was observed in tomato juice after pasteurization at 90°C for 30 s or 60 s (Odriozola-Serrano et al., 2009). In vegetables, pasteurization also leads to the effect on the bioaccessibility of carotenoids during digestion.

Texture: Pasteurization also has an impact on the texture, which consumers consider as an attribute for in fruits and vegetables. The impact includes breakage of cellular membranes and cell walls causing turgor loss and disassembly resulting from pectin transformations by enzymatic and non-enzymatic processes (Sila et al., 2008; Peng et al., 2014). In contrast, pasteurization helps to make the tissues of vegetables firmer due to the pectin de-esterification. Pasteurization results in a reduction in texture to half in red bell pepper soon after the processing, and almost complete texture reduction was noted after storing for 60 days at 30°C. Moreover, a recent study showed that a continuous microwave system (3.5 kW) for 4 minutes at 75°C resulted in texture reduction during storage (Koskiniemi et al., 2013). The addition of NaCl and citric acid also favors tissue softening after pasteurization, particularly during storage.

Phenolics and antioxidant activity: Phenolics are always associated with the antioxidant activity of foods. Moreover, they are important constituents of phytochemicals in vegetables that act as bioactive compounds. Thermal pasteurization effects on phenolics in vegetables is a function of the properties of the food material, package, and storage conditions. During storage of pumpkin and carrot juice at a temperature of 4°C up to 3–4 months, a reduction in the phenolic content was reported in pasteurization and post-pasteurization (Rayman and Baysal, 2011; Zhou et al., 2014). However, pasteurization may also result in the inactivation of enzymes that cause phenolic degradation and prevent their decrease in foods. This is supported by the study reporting no significant change in tomato juice phenolic content through pasteurization at 90°C for 30–60 seconds, and the phenolics content was maintained constant during storage. However, the same authors reported that thermal pasteurization decreases the phenolic content of tomato juice (Odriozola-Serrano et al., 2009) explaining it through the role of POD. Pasteurization also results in the reduction of flavonoids such as quercetin due to their degradation by hydroxyl and ketone groups in vegetables (Odriozola-Serrano et al., 2009). For example, pasteurization causes total quercetin reduction in onion by-products and its frozen products.

10.4 STERILIZATION

Food preservation, higher quality, and the freshness of products are the aspects that have attracted researchers and companies today to develop food preservation methods. Among these methods, sterilization represents a widely used and most effective method. The techniques used in thermal sterilization are well advanced and economical. The objective of sterilization is that the product after processing must be devoid of toxin-producing and spoilage microorganisms and be safe during shelf life up to the consumer level for consumption. However, in practice nutrient retention and product quality deterioration are caused by complete sterilization. If a product in normal conditions neither spoils nor puts consumer's health under threat, it is declared to be commercially sterile.

The heating process, which provides nearly complete elimination of microorganisms (including spores) from food products at a high-temperature range of 121–140°C, is defined as sterilization (Deak, 2014). Today, researchers are more inclined toward the non-conventional methods due to their potential to provide better quality, fresh-like taste, and be economical. In the food industry, the word sterilization refers to commercial sterilization and is achieved through a combination of relatively mild heat treatment, other processing parameters, and storage conditions (Teixeira, 2015). Heinz and Hautzinger (2007) suggested that the inhibition of microbial growth instead of microbial presence or absence should be the criterion to determine the adequacy of the process of sterilization. In sterilization, heating is done through the use of retorts resulting in food products heating at temperatures higher than 100°C. The packaging process during sterilization may lead to recontamination of foods (Wani et al., 2014). UHT processing with aseptic filling and in container processing (retorting) are the most common methods of thermal sterilization. The elimination or reduction in the number of heat-resistant spore-forming spoilage microbes shows how efficient the process of sterilization is for food products (Deak, 2014). Sterilization extends the shelf life of food products extraordinarily compared with pasteurization and leads to the destruction of all the otherwise heat-resistant spores and bacteria. However, it leads to color change and nutritional loss along with changing the flavor and texture of the product (Featherstone, 2015; Teixeira, 2015).

Sterilization can be performed by the use of heat, radiation, or chemicals, or by physical removal of cells. For complete sterilization, a food product first needs to be heated to a temperature of 110-125°C and equilibrated for some time at this temperature so that the heat reaches the cold point within the product. Next, the temperature is maintained to reach a predetermined F_o value of sterilization. At last, cooling of the product is done to avoid container bursting if the temperature is further increased. The severity of sterilization is determined by the pH of the product.

10.4.1 Methods of Sterilization

Broadly speaking, sterilization comprises two methods, namely thermal (using heat) and non-thermal (without heat). Of the two categories, non-thermal methods are effective because they are devoid of any drawbacks on quality and nutrition. The two methods of thermal sterilization are aseptic processing in which the food product is sterilized before packaging and canning and then packed and sterilized (Barbosa-Cánovas et al., 2017). Thermal processing causes quality and nutritive loss in foods and does not affect some bacteria types; still, it is largely practiced in the present times.

10.4.2 Conventional Sterilization Methods

The extreme conditions applied to food products during the sterilization process ensure safety in packaged low-acid foods, which leads to the quality reduction. This form of the process achieves a goal sterilization temperature, e.g., 121°C, where durability for a given time comes from various heating media, including water immersion, steam/air, and water/vapor spray, among many others. Studies have shown that the limitation of ensuring proper quality by this process lies in the alteration of only two variables, namely time and temperature (Ibarz and Barbosa-Cánovas, 2003). The time used in sterilization to reduce the initial load of *Clostridium botulinum* by 12 log cycles is termed as 12D, and D represents the time required to reduce the number of microorganisms to the tenth. Sterilization is done to ensure 12D at every point including the cold points (worst-case processing scenario). Recent studies are focused on overcoming such limitations of sterilization through new technologies.

Thermal processing is carried out through sterilization of food products in containers (in-container sterilization) or before filling in the container (aseptic processing), and the principle remains the same in both cases. In the former case, heart transfer is slow, hence, the sterilization process takes a longer time compared with the latter case. A product is sterilized by heating the product rapidly up to a temperature of 130–145°C then holding for a while at this temperature followed by rapid cooling. However, the kind of heat (dry or moist) and time of application coupled with temperature (organism specific) should be kept in mind during the process to ensure the destruction of all microorganisms. Sterilization is considered complete when the food product is free of bacterial endospores. Lethal time depends on the species, initial microbial count, product's nature, pH, and temperature. Sterilization is carried out in sterile packaging processes such as canning and bottling where it kills all the microorganisms by heat.

Retorting, also called a heat processing method of sterilization, is applied by using high-pressure retorts or sterilizers provided by agitators. Steam or pressurized hot water is used as a heating medium in batch-type high-pressure retorts. High pressure in the process is maintained with steam in the heating cycle and compressed air in the cooling cycle. During the process, heat transfer is improved through the movement of the product in the container, thereby improving the quality of the final product. Moreover, in sterilization, saturated steam is condensed on the outer surface of the container, thereby transferring the latent heat to the food.

10.4.3 Non-Conventional Methods

Sterilization in combination with high pressure (pressure = 100 MPa and treatment time = 14 minutes), regardless of the temperature, in conjugated linoleic acid (CLA)-enriched milk has proven to retain more than 80% of the CLA, which is both health promoting and effective against diseases. Moreover, the addition of the antioxidant catechin (1 g/kg) increased CLA retention up to >90% in milk anhydrous milk fat (Martinez-Monteagudo et al., 2012). Hence, the shelf life of milk is enhanced through high-pressure thermal sterilization (HPTS) and preservation of CLA activity. Moreover, HPTS helps to produce better quality food products due to shorter thermal treatments compared with thermal retorting. Sevenich et al. (2014) reported that temperatures of 90 and 105°C, with a pressure of 600 MPa in baby food puree resulted in up to 81–96% reduction of *Geobacillus stearothermophilus* furan compared with the retorting.

Another appealing technology used in food sterilization is microwave-assisted thermal sterilization (MATS). This technique has avoided many problems encountered by various conventional methods or by using microwaves. It helps retain nutrients better through in-container sterilization at 915 MHz than conventional methods. In this method, water is used as an intermediate step to ensure uniform heating and prevent edge effects. The PEF method uses short pulses of electricity that result in the inactivation of microorganisms through the process of cell wall electroporation. In this method, a high electric field (20–80 kV/cm) for 1–100 μs is used in the processing of foods (liquid, solid, and semi-liquid), thereby preventing quality loss in foods and maintaining their freshness. Thermal pasteurization can easily be carried out by this technique. PEF efficiency is determined through various parameters such as microorganism parameters (types, growth phase, and size and shape of microbes), process parameters (such as electric field intensity, power, and treatment time), and medium parameters (Raso et al., 2014). PEF combined with heat has great potential to be used as a sterilization technique (Jaeger et al., 2014). There is no clear justification for the spore inactivation in this method, nevertheless, some researchers believe in its similarity to vegetative cell inactivation. The elimination of microorganisms by this method depends on the type of bacteria present; for example, gram-negative bacteria proved to be more sensitive to PEF compared with gram-positive bacteria and yeasts, and spores are more resistant to PEF than vegetative cells.

10.4.4 Ultrahigh Temperature Process

UHT sterilization ensures minimum loss in quality coupled with the retention of natural nutritional and sensory food attributes. This is credited to the application of higher temperatures and then pasteurized for a short period. Several studies have used the UHT method for the production of beer, juices, coffee, Chinese rice wine, and milk (Singh et al., 2009; Sopelana et al., 2013; Chen et al., 2016; Yin et al., 2017; Yang et al., 2019). UHT is a means of thermal sterilization that is mostly applied for the production of sterile milk with a shelf life of 6 months (Tran et al., 2008). UHT is a better method to treat food products compared with pasteurization; however, UHT results in an alteration of milk characteristics such as flavor, physical stability, and nutritional profile. In some cases, the long-term exposure of food products to high temperatures in UHT causes chemical changes depending on the length of time of exposure (Lewis and Heppell, 2000). A UHT process described as145°C for 3 seconds means the exposure of a product to that temperature and time in the holding tube. UHT reduces the total phenolic content of juices as witnessed in ginger juice in a recent study undertaken by Chen et al. (2016) for 91 days at 4 and 25°C. However, a 14.18% increase in gingerol and no change was reported in monoterpenoids.

10.5 CONCLUSION

Blanching, pasteurization, and sterilization have shown great potential for sterilization combined with thermal treatment. They are capable of reducing the microbial threat to food products causing shelf life extension and improving nutritional quality. Other quality parameters such as texture, flavor, aroma, and appearance of foods are improved. Moreover, thermal processing is a very important unit operation particularly in fruits and vegetables leading to the inactivation of enzymes such as PPO and POD, thereby preventing deteriorative reactions. According to current knowledge, thermal processing should be promoted for food processing and preservation. However, the side effects of thermal processing when used at higher temperatures present lacunae that need to be explored; hence, their usage should be analyzed with utmost concern.

REFERENCES

Aamir, M., Ovissipour, M., Sablani, S. S., & Rasco, B. (2013). Predicting the quality of pasteurized vegetables using kinetic models: a review. *International Journal of Food Science*, 2013. DOI: 10115/2013/271271.

Akyol, C., Bayindirli, A., & Alpas, H. (2004). Effect of combined treatment of high hydrostatic pressure and mild heat on peroxidase inactivation in green beans, peas and carrots. Presented at IFT Annual Meeting, July 12–16, Las Vegas, NV.

Arroqui, C., Lopez, A., Esnoz, A., & Virseda, P. (2003). Mathematical model of heat transfer and enzyme inactivation in an integrated blancher cooler. *Journal of Food Engineering*, *58*(3), 215–225.

Assiry, A., Sastry, S. K., & Samaranayake, C. (2003). Degradation kinetics of ascorbic acid during ohmic heating with stainless steel electrodes. *Journal of Applied Electrochemistry*, *33*(2), 187–196.

Barbosa-Cánovas, G. V., Ma, L., & Barletta, B. (2017). *Food Engineering Laboratory Manual*. CRC Press, Boca Raton.

Bai, J. W., Sun, D. W., Xiao, H. W., Mujumdar, A. S., & Gao, Z. J. (2013). Novel high-humidity hot air impingement blanching (HHAIB) pretreatment enhances drying kinetics and color attributes of seedless grapes. *Innovative Food Science & Emerging Technologies*, *20*, 230–237.

Bureau, S., Mouhoubi, S., Touloumet, L., Garcia, C., Moreau, F., Bédouet, V., & Renard, C. M. (2015). Are folates, carotenoids and vitamin C affected by cooking? Four domestic procedures are compared on a large diversity of frozen vegetables. *LWT-Food Science and Technology*, *64*(2), 735–741.

Chandrasekaran, S., Ramanathan, S., & Basak, T. (2013). Microwave food processing—A review. *Food Research International*, *52*(1), 243–261.

Chen, D., Pan, S., Chen, J., Pang, X., Guo, X., Gao, L., … & Wu, J. (2016). Comparing the effects of high hydrostatic pressure and ultrahigh temperature on quality and shelf life of cloudy ginger juice. *Food and Bioprocess Technology*, *9*(10), 1779–1793.

Cruz, R. M., Vieira, M. C., & Silva, C. L. (2006). Effect of heat and thermosonication treatments on peroxidase inactivation kinetics in watercress (*Nasturtium officinale*). *Journal of Food Engineering*, *72*(1), 8–15.

Dadali, G., Demirhan, E., & Özbek, B. (2007). Color change kinetics of spinach undergoing microwave drying. *Drying Technology*, *25*(10), 1713–1723.

Deak, T., 2014. Food Technology: Sterilization. In: *Encyclopedia of Food Safety*, Volume 3. Elsevier, Amsterdam, pp. 245–252.

DiPersio, P. A., Kendall, P. A., Yoon, Y., & Sofos, J. N. (2007). Influence of modified blanching treatments on inactivation of Salmonella during drying and storage of carrot slices. *Food Microbiology*, *24*(5), 500–507.

Fante, L., & Noreña, C. P. Z. (2012). Enzyme inactivation kinetics and colour changes in Garlic (Allium sativum L.) blanched under different conditions. *Journal of Food Engineering*, *108*(3), 436–443.

Featherstone, S. (2015). 12-Sterilization Systems. In: *Fundamental Information on Canning. A Complete Course in Canning and Related Processes* (14th ed.). Series in Food Science, Technology and Nutrition. Woodhead Publishing, pp. 239–267.

Guida, V., Ferrari, G., Pataro, G., Chambery, A., Di Maro, A., & Parente, A. (2013). The effects of ohmic and conventional blanching on the nutritional, bioactive compounds and quality parameters of artichoke heads. *LWT-Food Science and Technology*, *53*(2), 569–579.

Haase, N. U., & Weber, L. (2003). Ascorbic acid losses during processing of French fries and potato chips. *Journal of Food Engineering*, *56*(2–3), 207–209.

Harris, L. J., Uesugi, A. R., Abd, S. J., & McCarthy, K. L. (2012). Survival of *Salmonella enteritidis* PT 30 on inoculated almond kernels in hot water treatments. *Food Research International*, *45*(2), 1093–1098.

Heinz, G., & Hautzinger, P. (2007). Canning/Sterilization of Meat Products. In: *Meat Processing Technology for Small- to Medium-Scale Producers*. Food and Agriculture Organization of the United Nations, Regional Office for Asia and the Pacific, Bangkok.

Ibarz, A., & Barbosa-Cánovas, G. V. (2003). *Unit Operations in Food Engineering*. CRC Press, Boca Raton, FL.

Ismail, A., Marjan, Z. M., & Foong, C. W. (2004). Total antioxidant activity and phenolic content in selected vegetables. *Food Chemistry*, *87*(4), 581–586.

Jaeger, H., Meneses, N., and Knorr, D. (2014). Food Technologies: Pulsed Electric Field Technology. In: *Foods, Materials, Technologies and Risks, Encyclopedia of Food Safety*, Volume 3. Elsevier, Switzerland, pp. 239–244.

Jakób, A., Bryjak, J., Wójtowicz, H., Illeová, V., Annus, J., & Polakovič, M. (2010). Inactivation kinetics of food enzymes during ohmic heating. *Food Chemistry*, *123*(2), 369–376.

Jan, B., Shams, R., Rizvi, Q. E. H., & Manzoor, A. (2021). Ohmic heating technology for food processing: a review of recent developments. *Journal of Postharvest Technology*, *9*(1), 20–34.

Karel, M., and Lund, D. B. (2003). *Physical Principles of Food Preservation* (2nd ed.). Marcel Dekker, New York, Chap. 6.

Kondjoyan, A., & Portanguen, S. (2008). Effect of superheated steam on the inactivation of *Listeria innocua* surface-inoculated onto chicken skin. *Journal of Food Engineering*, 87(2), 162–171.

Kong, F., Tang, J., Rasco, B., & Crapo, C. (2007). Kinetics of salmon quality changes during thermal processing. *Journal of Food Engineering*, 83(4), 510–520.

Koskiniemi, G. B., Truong, V. D., McFeeters, R. F., & Simunovic, J. (2013). Quality evaluation of packaged acidified vegetables subjected to continuous microwave pasteurization. *LWT-Food Science and Technology*, 54, 157–164.

Krokida, M. K., Oreopoulou, V., Maroulis, Z. B., & Marinos-Kouris, D. (2001). Deep fat frying of potato strips—quality issues. *Drying Technology*, 19(5), 879–935.

Leizerson, S., & Shimoni, E. (2005). Effect of ultrahigh-temperature continuous ohmic heating treatment on fresh orange juice. *Journal of Agricultural and Food Chemistry*, 53(9), 3519–3524.

Lemmens, L., Colle, I., Knockaert, G., Van Buggenhout, S., Van Loey, A., & Hendrickx, M. (2013). Influence of pilot scale in pack pasteurization and sterilization treatments on nutritional and textural characteristics of carrot pieces. *Food Research International*, 50(2), 526–533.

Lewis, M., & Heppell, N. (2000). *Continuous Thermal Processing of Foods – Pasteurization and UHT sterilization*. Aspen Publishers, Gaithersburg, MD.

Lisiewska, Z., Słupski, J., Skoczeń-Słupska, R., & Kmiecik, W. (2009). Content of amino acids and the quality of protein in Brussels sprouts, both raw and prepared for consumption. *International Journal of Refrigeration*, 32(2), 272–278.

Liu, J., & Yang, W. (2012). Water sustainability for China and beyond. *Science*, 337(6095), 649–650.

Llano, K. M., Haedo, A. S., Gerschenson, L. N., & Rojas, A. M. (2003). Mechanical and biochemical response of kiwifruit tissue to steam blanching. *Food Research International*, 36(8), 767–775.

Manzoor, A., Dar, I. H., Bhat, S. A., & Ahmad, S. (2020). Flavonoids: Health Benefits and Their Potential Use in Food Systems. In *Functional Food Products and Sustainable Health*. Springer, Singapore, pp. 235–256.

Martinez-Monteagudo, S. I., Saldana, M. D.A., Torres, J. A., & Kennelly, J. J. (2012). Effect of pressure-assisted thermal sterilization on conjugated linoleic acid (CLA) content in CLA-enriched milk. *Innovative Food Science and Emerging Technologies*, 16, 291–297.

Mercali, G. D., Schwartz, S., Marczak, L. D. F., Tessaro, I. C., & Sastry, S. (2014). Ascorbic acid degradation and color changes in acerola pulp during ohmic heating: effect of electric field frequency. *Journal of Food Engineering*, 123, 1–7

Moreno, J., Simpson, R., Pizarro, N., Pavez, C., Dorvil, F., Petzold, G., & Bugueño, G. (2013). Influence of ohmic heating/osmotic dehydration treatments on polyphenoloxidase inactivation, physical properties and microbial stability of apples (cv. Granny Smith). *Innovative Food Science & Emerging Technologies*, 20, 198–207.

Mukherjee, S., & Chattopadhyay, P. K. (2007). Whirling bed blanching of potato cubes and its effects on product quality. *Journal of Food Engineering*, 78(1), 52–60.

National Advisory Committee on Microbiological Criteria for Foods. (2006). Requisite scientific parameters for establishing the equivalence of alternative methods of pasteurization. *Journal of Food Protection*, 69, 1190–1216.

Ndiaye, C., Xu, S. Y., & Wang, Z. (2009). Steam blanching effect on polyphenoloxidase, peroxidase and colour of mango (*Mangifera indica* L.) slices. *Food Chemistry*, 113(1), 92–95.

Ngudi, D., Kuo, Y. H., & Lambein, F. (2003). Amino acid profiles and protein quality of cooked cassava leaves or "saka-saka." *Journal of the Science of Food and Agriculture*, 83(6), 529–534.

Nisha, P., Singhal, R. S., & Pandit, A. B. (2011). Kinetic modelling of colour degradation in tomato puree (*Lycopersicon esculentum* L.). *Food and Bioprocess Technology*, 4(5), 781–787.

Oboh, G. (2005). Effect of blanching on the antioxidant properties of some tropical green leafy vegetables. *LWT-Food Science and Technology*, 38(5), 513–517.

Odriozola-Serrano, I., Soliva-Fortuny, R., Hernández-Jover, T., & Martín-Belloso, O. (2009). Carotenoid and phenolic profile of tomato juices processed by high intensity pulsed electric fields compared with conventional thermal treatments. *Food Chemistry*, 112(1), 258–266.

Peng, J., Tang, J., Barrett, D. M., Sablani, S. S., & Powers, J. R. (2014). Kinetics of carrot texture degradation under pasteurization conditions. *Journal of Food Engineering*, 122, 84–91.

Pimpaporn, P., Devahastin, S., & Chiewchan, N. (2007). Effects of combined pretreatments on drying kinetics and quality of potato chips undergoing low-pressure superheated steam drying. *Journal of Food Engineering*, 81(2), 318–329.

Raso, J., Condon, S., & Alvarez, I. (2014). Non-Thermal Processing | Pulsed Electric Field. In: Batt, C.A. (Ed.), *Encyclopedia of Food Microbiology* (2nd ed.). Elsevier, pp. 966–973.

Rayman, A., & Baysal, T. (2011). Yield and quality effects of electroplasmolysis and microwave applications on carrot juice production and storage. *Journal of Food Science*, 76(4), C598–C605.

Reznick, D. (1996). Ohmic heating of fluid foods: ohmic heating for thermal processing of foods: government, industry, and academic perspectives. *Food Technology (Chicago)*, 50(5), 250–251.

Rico, D., Martín-Diana, A. B., Barry-Ryan, C., Frías, J. M., Henehan, G. T., & Barat, J. M. (2008). Optimisation of steamer jet-injection to extend the shelflife of fresh-cut lettuce. *Postharvest Biology and Technology*, 48(3), 431–442.

Roy, M. K., Juneja, L. R., Isobe, S., & Tsushida, T. (2009). Steam processed broccoli (*Brassica oleracea*) has higher antioxidant activity in chemical and cellular assay systems. *Food Chemistry*, 114(1), 263–269.

Schweiggert, U., Schieber, A., & Carle, R. (2005). Inactivation of peroxidase, polyphenoloxidase, and lipoxygenase in paprika and chili powder after immediate thermal treatment of the plant material. *Innovative Food Science & Emerging Technologies*, 6(4), 403–411.

Sevenich, R., Kleinstueck, E., Crews, C., Anderson, W., Pye, C., Riddellova, K., … & Knorr, D. (2014). High-pressure thermal sterilization: food safety and food quality of baby food puree. *Journal of Food Science*, 79(2), M230–M237.

Shi, J., Le Maguer, M. A. R. C., Bryan, M., & Kakuda, Y. (2003). Kinetics of lycopene degradation in tomato puree by heat and light irradiation. *Journal of Food Process Engineering*, 25(6), 485–498.

Sila, D. N., Duvetter, T., De Roeck, A., Verlent, I., Smout, C., Moates, G. K., … & Van Loey, A. (2008). Texture changes of processed fruits and vegetables: potential use of high-pressure processing. *Trends in Food Science & Technology*, 19(6), 309–319.

Silva, F. V., & Gibbs, P. A. (2010). Non-proteolytic *Clostridium botulinum* spores in low-acid cold distributed foods and design of pasteurization processes. *Trends in Food Science & Technology*, 21(2), 95–105.

Silva, F. V. M., & Gibbs, P. A. (2009). Principles of Thermal Processing: Pasteurisation. Chapter 2. In: Simpson, R. (Ed.), *Engineering Aspects of Thermal Food Processing, Contemporary Food Engineering Series*. CRC Press, Boca Raton, FL.

Singh, R. R. B., Ruhil, A. P., Jain, D. K., Patel, A. A., & Patil, G. R. (2009). Prediction of sensory quality of UHT milk–a comparison of kinetic and neural network approaches. *Journal of Food Engineering*, 92(2), 146–151.

Sopelana, P., Pérez-Martínez, M., López-Galilea, I., de Peña, M. P., & Cid, C. (2013). Effect of ultra high temperature (UHT) treatment on coffee brew stability. *Food Research International*, 50(2), 682–690.

Sotome, I., Takenaka, M., Koseki, S., Ogasawara, Y., Nadachi, Y., Okadome, H., & Isobe, S. (2009). Blanching of potato with superheated steam and hot water spray. *LWT-Food Science and Technology*, 42(6), 1035–1040.

Teixeira, A. A. (2015). Chapter 6, Thermal Food Preservation Techniques (Pasteurization, Sterilization, Canning and Blanching). In: Bhattacharya, S. (Ed.), *Conventional and Advanced Food Processing Technologies*. Wiley Blackwell, Hoboken, NJ, pp. 115–128.

Teng, S. S., & Chen, B. H. (1999). Formation of pyrochlorophylls and their derivatives in spinach leaves during heating. *Food Chemistry*, 65(3), 367–373.

Tijskens, L. M. M., Schijvens, E. P. H. M., & Biekman, E. S. A. (2001). Modelling the change in colour of broccoli and green beans during blanching. *Innovative Food Science and Emerging Technologies*, 2(4), 303–313.

Tran, H., Datta, N., Lewis, M. J., & Deeth, H. C. (2008). Predictions of some product parameters based on the processing conditions of ultra-high-temperature milk plants. *International Dairy Journal*, 18(9), 939–944.

Wani, A. A., Gotz, A., Langowski, H.-C., & Wunderlich, J. (2014). Food Technologies: Aseptic Packaging. In: Motarjemi, Y. (Ed.), *Encyclopedia of Food Safety*, Volume 3. Elsevier Science, Burlington, MA, pp. 124–134.

Wilbey, R. A. (2014). Heat Treatment of Foods: Principles of Pasteurization. *Encyclopedia of Food Microbiology* (2nd ed.). Academic Press, San Diego, pp. 169–174.

Xiao, H. W., Bai, J. W., Sun, D. W., & Gao, Z. J. (2014). The application of superheated steam impingement blanching (SSIB) in agricultural products processing–A review. *Journal of Food Engineering*, 132, 39–47.

Xiao, H. W., Pan, Z., Deng, L. Z., El-Mashad, H. M., Yang, X. H., Mujumdar, A. S., … & Zhang, Q. (2017). Recent developments and trends in thermal blanching – A comprehensive review. *Information Processing in Agriculture*, 4(2), 101–127.

Xiao, H. W., Pang, C. L., Wang, L. H., Bai, J. W., Yang, W. X., & Gao, Z. J. (2010). Drying kinetics and quality of Monukka seedless grapes dried in an air-impingement jet dryer. *Biosystems Engineering*, *105*(2),
233–240.

Yang, Y., Xia, Y., Wang, G., Tao, L., Yu, J., & Ai, L. (2019). Effects of boiling, ultra-high temperature and
high hydrostatic pressure on free amino acids, flavor characteristics and sensory profiles in Chinese rice
wine. *Food Chemistry*, *275*, 407–416.

Yin, H., Deng, Y., He, Y., Dong, J., Lu, J., & Chang, Z. (2017). A preliminary study of the quality attributes of
a cloudy wheat beer treated by flash pasteurization. *Journal of the Institute of Brewing*, *123*(3), 366–372.

Zhou, C. L., Liu, W., Zhao, J., Yuan, C., Song, Y., Chen, D., ... & Li, Q. H. (2014). The effect of high hydrostatic
pressure on the microbiological quality and physical–chemical characteristics of Pumpkin (*Cucurbita
maxima* Duch.) during refrigerated storage. *Innovative Food Science & Emerging Technologies*, *21*,
24–34.

CHAPTER 11

Non-Thermal Processing of Foods
Science, Shelf Life, and Quality

Nitamani Choudhury, Farheena Iftikhar, and H.A. Makroo

CONTENTS

Email: H.A. Makroo, hilalmakroo@gmail.com

DOI: 10.1201/9781003091677-11

11.1 INTRODUCTION

Food preservation deals with the destruction or inactivation of the spoilage and disease-causing microorganisms and enzymes present in food. To achieve this purpose, the preservation treatments that are used largely rely on conventional food processing methods, which have several disadvantages including loss of nutrients, desired flavor, and desired functionality. For over a decade there has been a tremendous increase in consumer demands for better sensory and nutritional qualities in processed foods while keeping the intended shelf life. This is possible with a shift from conventional methods of processing to novel ones. Conventional food processing methods like drying, chemical preservation, and those involving intense heat treatments such as pasteurization and sterilization produce microbiologically safe products, but at the same time tend to destroy nutrients like vitamins, minerals, and bioactive compounds like carotenoids, polyphenols, etc., present in food (Raso and Barbosa-Cánovas, 2003). Conventional food preservation methods have evolved from time to time with better applicability, but the limitations and the disadvantages have not subsided as per market customization. The inherent nutrients present in foods, using conventional processing, lose their biological functionality when exposed to extreme processing conditions, rendering the food devoid of essential nutrients. These limitations have fostered the development of innovative techniques to enhance the shelf life of foods while retaining the nutritional and qualitative benefits. Thus, non-thermal methods of food processing are evolving as potential alternatives to the thermal and chemical preservation methods to achieve the concept of "minimal processing," which describes food safety approaches designed to retain natural characteristics of the foods (Allende et al., 2006). These methods not only produce minimally processed foods with better organoleptic properties but are also considered energy efficient when compared with some of the conventional methods. Non-thermal processing includes all preservation methods that work effectively at ambient or near-ambient temperatures, such as adjustment of pH, and providing modified atmospheres, thus eliminating the need to increase the temperature for the inactivation of microbes and enzymes, preventing the deleterious effect of heat. This is the reason food technologists and scientists are shifting their focus toward such alternative technologies. These technologies can preserve food without much alteration in the sensory and nutritional properties and are in line with the competitive global market so they can be used for bulk quantities of food in industries (Sarika and Bindu, 2018). Numerous non-thermal technologics, such as high-pressure processing, pulsed electric field, pulsed white light technology, ultrasonication, irradiation, ultraviolet (UV) light, cold plasma technology, etc., have been developed. A few of the most extensively researched non-thermal processes are high-pressure processing, ultrasonication, pulsed white light, and pulsed electric field and these appear to be promising in terms of application in food industries. Although the non-thermal technologies can inactivate microorganisms and enzymes, the high intensities used may have some adverse effects. For example, using high pressure can alter the structure of carbohydrates and proteins, resulting in textural changes and loss of functionality. High-intensity ultrasonication can denature the proteins present in food and generate free radicals that can have an adverse effect on the flavor and aroma of high-fat foods. Therefore, studies on the combination of different non-thermal technologies to produce synergistic antimicrobial effects with reduction in the intensities of treatment required and energy input have been performed in recent years. Each of these technologies and their potential use in treating different foods and limitations will be discussed further in this chapter.

11.2 HIGH-PRESSURE PROCESSING

11.2.1 Principle

High-pressure processing is a non-thermal processing method implemented in foods, wherein high pressures are used to destroy microorganisms and endogenous food enzymes. When high pressures are applied to food packages, the pressure distributes uniformly throughout the food and destroys the microorganisms and enzymes. The effectiveness of high-pressure processing is governed by a combination of factors, eventually leading to the damage of cell walls and cytoplasmic membranes of the microorganisms by disrupting their intracellular vacuoles. The process has been found to be more effective when applied on the microbes during their log phase of growth curve compared with the stationary phase and death phase. Some studies have also shown that high pressures have a disrupting effect on the metabolic processes of the enzymes (Earnshaw, 1996; Knorr, 1999; Patterson et al., 2007; Yordanov and Angelova, 2010). Enzymes show differences in the sensitivities to high pressure. Some can withstand pressure up to 1000 MPa and some others become inactivated at pressures of a few hundred MPa (Serment-Moreno et al., 2014). Temperature, pH, and the substrate composition are additional factors responsible for the effectiveness of high-pressure processing in foods.

11.2.2 Effect on the Quality of Foods and Their Storage Life

Until now, it is known that high-pressure processing affects only the non-covalent chemical bonds, leaving the covalent chemical bonds integral, thereby allowing destruction of microorganisms and enzymes without affecting the food molecules responsible for the flavor, texture, and nutritional properties of food. Because high-pressure processing can be performed at or near ambient temperatures, there is no significant damage to the nutrients, natural colors, and flavors of the food, hence producing high quality products. Houška et al. (2006) investigated the effect of high-pressure processing on apple-broccoli juice (550 MPa for 10 minutes) and reported that inactivation of microbial population was more than 5 logs without any deleterious effects on antidiabetic and antimicrobial properties of the broccoli. Moreover, there is uniformity in the treatment as the pressure is distributed evenly throughout the food, thus overcoming the problem of lack of uniformity in other processing methods such as conductive or convective heating. High-pressure processing can affect biopolymers like carbohydrates and proteins present in the food. It leads to unfolding and aggregation in some proteins and gel formation in the others. However, such gels tend to maintain their natural flavor and color when compared with heat-induced gels.

11.2.3 Application in Foods

High-pressure processing is mainly applied to fruit and vegetable products. It can be used in the pasteurization and sterilization of products like salad dressings, pickles, sauces, etc. It is also used for enzyme inactivation in fruits and vegetables, and different enzymes in different products show different barosensitivities. For example, endogenous food enzymes, i.e., polyphenol oxidase, peroxidase, and pectin methyl esterase, are extremely resistant to high pressures and are partially inactivated at conditions that are commercially feasible when compared with enzymes, like lipoxygenase and polygalacturonase, which are sensitive to high pressures and are substantially inactivated (Terefe et al., 2014). It is also used in unfolding low-quality proteins for improving functional properties like emulsification. Puppo et al. (2005) reported an improvement in the emulsifying properties of soybean proteins when subjected to high-pressure processing. Other applications include tenderization of meat, tempering of chocolate, and preservation of dairy products and seafoods.

11.2.4 Limitations

- Approximately 40% free water should be present in foods for antimicrobial effects, when subjected to high pressures.
- The packaging options are limited. As foods tend to reduce volume under high pressures, there is a chance that the package may get distorted, hence, packaging material should be carefully selected. Conventional plastic and foil pouches are somewhat suitable.

11.3 PULSED ELECTRIC FIELD PROCESSING

11.3.1 Principle

Pulsed electric field processing is a non-thermal technique based on the utilization of short pulses of electricity discharged through food for destruction of microorganisms and enzymes. In this process, energy is obtained from a high-voltage power supply (10–80 kV/cm) and discharged through foods that are either static or flowing through the treatment chamber, for less than 1 second (Jan et al., 2017). The mechanisms involved in producing lethal effect microorganisms include the following. (1) Formation of pores in the cell membrane once the electric potential of the membrane exceeds the normal potential of 1 V is due to the applied electric field. These pores lead to rupturing of the cells, as shown in Figure 11.1. (2) Disruption in the metabolic processes is due to oxidation and reduction reactions within the cells. (3) Free radicals are generated from the food components due to the applied electric field. (4) There is transformation of induced electric energy to heat energy. Although the accurate mechanism for enzyme inactivation is still ambiguous, available literature suggests that thermal or electrochemical effects of pulsed electric field or both may be responsible for inactivation by changing the conformation of enzymes (Terefe et al., 2015). Critical factors responsible for microbial inactivation in foods include electric field intensity, pulse duration, temperature, electrical conductivity, and pH of foods.

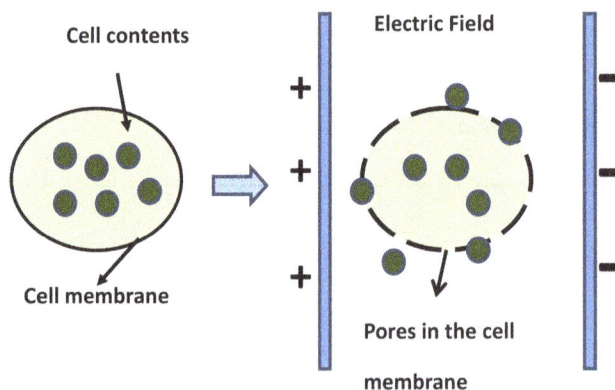

Figure 11.1   Pore formation in the cell membrane due to the pulsed electric field.

11.3.2 Effect on the Quality of Foods and Their Storage Life

Pulsed electric field, along with producing microbiologically safe products, causes minimal or no detrimental effects on the nutritional and organoleptic properties of foods and helps retain the freshness due to low processing temperatures (Min et al., 2007). It has a lethal effect on the microbes and a few enzymes, which are of prime concern in the juice industry. Altuntas et al. (2010) reported inactivation of *Listeria monocytogenes* by 3 log colony-forming units (CFU)/mL in sour cherry juice on exposure to pulsed electric field treatment of 27 kV/cm for 131 µs at approximately 20°C. These results were in concordance with the study conducted by Timmermans et al. (2011) on orange juice with pulsed electric field of 23 kV/cm for 36 µs. The storage life of the orange juice was increased up to 2 months when stored at 4°C, with microbial counts of <10 CFU/mL for Enterobacteriaceae, <1000 CFU/mL for total plate count, and <10 CFU/mL for lactic acid bacteria. Endogenous enzymes present in fruit juices can be destroyed by the pulsed electric field (Giner et al., 2005; Espachs-Barroso et al., 2006; Min et al., 2007; Buckow et al., 2012). Sentandreu et al. (2006) treated orange juice with a pulsed electric field of 25 kV/cm for 330 µs and reported that more than 90% of the pectin methyl esterase was inactivated. Studies have reported that pulsed electric field minimally affects the nutritional and sensorial attributes of the foods while increasing the storage life. Pulsed electric field treatment to tomato juice at <68°C minimally degraded vitamin C (Min et al., 2003). The results were in concordance with Torregrosa et al. (2006) in orange-carrot juice. Yeom et al. (2000) performed a comparative study on the color of the pulsed electric field-treated and heat-pasteurized orange juice and reported that pulsed electric field-treated juice had less browning and more whiteness compared with heat-pasteurized juice.

11.3.3 Application in Foods

Pulsed electric field is mostly suitable for liquid foods and not for solid food products. This technology can be successfully used in pasteurization of juices, yogurt, milk, and soups. Foods with low electrical conductivity and no air bubbles are preferred for treatment with pulsed electric field. Processing by this method has been found to be successful in less viscous juices with low electrical conductivity like apple and orange juices. Apart from microbial destruction, it can also be used as an aid in filtration methods and for extracting sugars and starches from root vegetables (Jan et al., 2017).

11.3.4 Limitations

- Pulsed electric field cannot be extensively used for solid food products and cannot be used with highly conductive materials.
- Electrolysis products due to the applied electric field may adversely affect foods.

11.4 PULSED LIGHT PROCESSING

11.4.1 Principle

Pulsed white light is a novel technology used mainly for surface sterilization of foods and packaging materials (Bank et al., 1990). It consists of a spectrum of white light, ranging from UV wavelengths of 200 nm to infrared wavelengths of 1000 nm, i.e., the non-ionizing part of the electromagnetic spectrum. The food or the packaging material to be treated is exposed to high-intensity pulses for a duration of a few hundred microseconds. The lethal effect of pulsed white light is attributed to the photochemical and photothermal effects (Anderson et al., 2000; Takeshita, 2003). The UV component of the light has a photochemical effect. In the process of photochemical inactivation of microbes, the carbon-carbon double bonds of their proteins and nucleic acids absorb UV light, leading to the structural changes in DNA and RNA, thereby yielding an antimicrobial effect. (Chang et al., 1985; Miller et al., 1999). Whereas the photothermal effect involves heating up a thin layer of food or packaging material when it is exposed to pulsed white light, hence, destroying the vegetative cells.

11.4.2 Effect on the Quality of Foods and Their Storage Life

Pulsed white light has found its applicability in the surface decontamination of foods and packaging materials, without affecting the product quality. As the intensity of the light is 20,000 times brighter than sunlight and it lasts for only a second, the thermal effect is not very prominent, hence, it minimally affects nutritional and sensorial attributes of the product. Thus, such treatment when applied on foods increases their storage life by microbial destruction, while retaining food quality (Krishnamurthy et al., 2004). In an experiment conducted by Aguilo-Aguayo et al. (2014) for evaluation of pulsed light on the color of avocadoes after storing for 15 days at 4°C, it was found that the color was maintained as it was. Dunn et al. (1995) observed complete destruction of *Staphylococcus aureus* in milk when treated with pulsed light in a continuous system. Artíguez and de Marañón (2015) investigated the impact of pulsed light on *Listeria innocua* inactivation in whey, skimmed whey, diluted whey, and distilled water. Microbial inactivation was found to increase with increased number of pulses. Further, the inactivation was found to depend on the amount of light transmitted through the fluid, suggesting highest inactivation in the diluted whey samples. Xu and Wu (2016) conducted a study on the inactivation of *Escherichia coli* O157:H7 and *Salmonella* spp. using pulsed light. The study was conducted on raspberries stored at 4°C for 10 days and results concluded that the samples treated with pulsed light had lower pathogen survival than the untreated samples.

11.4.3 Application in Foods

Pulsed white light is applied in the treatment of baked products like cakes, bread, and pizza, increasing the shelf life of such products for a few days. It is also used for surface decontamination of fresh fruits and vegetables, seafood, meats, and milk products. Other applications include decontamination of the equipment surfaces, packaging material, and water.

11.4.4 Limitations

- One of the major limitations of this preservation method is that it only produces surface effects, and its use is difficult on complex surfaces. Folds or fissures present in the foods might possibly shield the microbes from pulsed light exposure, thereby decreasing its efficiency (Koh et al., 2016). Dunn et al. (1995) reported efficiency of pulsed white light treatment on simple surfaces higher than relatively complex surfaces, such as meat.
- Some microbial strains such as *L. monocytogenes* might show resistance to pulsed white light (Jan et al., 2017).

11.5 ULTRASOUND PROCESSING

11.5.1 Principle

Ultrasound waves are sound waves with a frequency of 20,000 Hz or more. The application of ultrasound technology in food processing is categorized into low-intensity ultrasound (<1 Wcm^{-2}), a non-destructive analytical method for determining the composition and structure of food, and

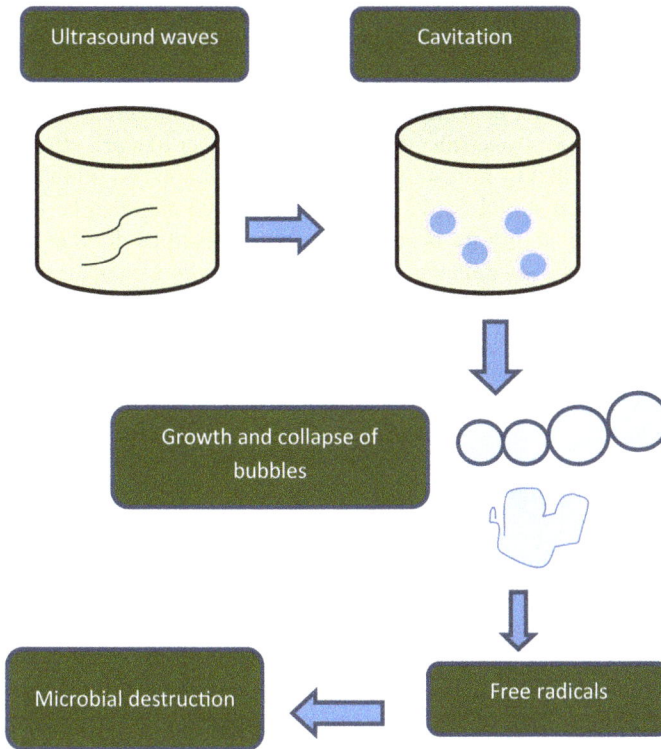

Figure 11.2 Cavitation in liquids due to ultrasound.

high-intensity ultrasound (10–1000 Wcm^{-2}), which is used for microbial destruction, cleaning equipment, and creating emulsions. When a food surface is exposed to ultrasound waves, a force is generated. The generated force leads to the production of compression and shear waves within the food. The underlying mechanism of ultrasound involves thinning of the cell membranes of microbes, localized heating, and generation of free radicals, pertaining to the microbial destruction. (Piyasena et al., 2003). Ultrasound can produce localized changes in pressure and temperature that lead to occurrence of cavitation, i.e., formation of bubbles in the liquid food, as shown in Figure 11.2. These bubbles produce a very high pressure and temperature when they immediately burst by expansion. Such high pressures are mainly responsible for its lethal effect on microorganisms. The hot zones also kill some microbes, but the effect is localized and does not cover a large area.

11.5.2 Effect on the Quality of Foods and Their Storage Life

The compression and shearing effect of ultrasound is responsible for the denaturation of proteins and reducing the enzyme activity in foods. However, ultrasonication is unlikely to be used as a method of preservation by itself, as the resistance of microorganisms and enzymes to ultrasound is so high that higher intensities of ultrasound treatment would result in undesirable changes in sensory properties. Hence, this method can be used in amalgamation with other available and applied preservation methods. Ultrasound reduces the resistance of the microorganisms to heat by disruption in their cell structures, which makes them more sensitive to heat denaturation. Hence, ultrasound can assist in reducing the intensities of conventional heat treatments resulting in improved sensory and nutritional properties. Milani and Silva (2017) did a comparative study on inactivation of *Saccharomyces cerevisiae* ascospore by using ultrasound-assisted thermal treatment and only thermal pasteurization of beer. The study concluded that ultrasound-assisted thermal treatment led to higher ascospore inactivation in beer than thermal treatment alone. Additionally, Deng et al. (2018) studied the impact of thermosonication on physicochemical and sensory properties of beer and revealed that the physicochemical properties of beer like pH, ethanol content, and bitterness were not very affected by ultrasonication. The treatment could inhibit the growth of yeast and spoilage bacteria when stored for 12 months. Ultrasound can also be used in combination with drying, creating a possibility for heat-sensitive materials to be dried at a lower temperature and rapidly, without causing any damage. Tekin et al. (2017) did a comparative study on the drying behavior and quality parameters of green beans that were dried using ultrasound-assisted vacuum drying and vacuum drying alone. Their results revealed that the drying time was reduced by 1 hour when ultrasound-assisted vacuum drying was used. Furthermore, in their study, ultrasound-assisted vacuum drying resulted in more phenolic compounds in the samples compared with unsonicated ones.

11.5.3 Application in Foods

Ultrasonication finds application in preservation of foods when combined with other preservation methods, as discussed previously. It can be used in the processing meat and meat products. For meat and its products, the prolonged exposure to ultrasound tenderizes the meat tissues, aiding the release of myofibrillar proteins in meat and its products, eventually resulting in better water binding capacity, tenderness, and cohesiveness (McClements, 1995). It can extensively be used in combination with a drying method for drying products like papaya, peppers, and certain fruits so that the loss of bioactive components in such products is minimized. Other applications in the food industry include filtration of juices, degassing or deaeration in carbonated beverages and cutting of soft products like cheese by using cavitation phenomenon.

11.5.4 Limitations

- Ultrasonication cannot be used by itself and needs to be combined with other processes.
- The air present in the product affects the penetration depth of ultrasonic waves.
- Some unwanted textural and structural modifications in foods might occur.

11.6 IRRADIATION

11.6.1 Principle

Irradiation or ionizing radiation is a non-thermal process applied for the processing of foods. The technology is based on the destruction of the spoilage and pathogenic microorganisms in foods by targeting their DNA. Ionizing radiation take the form of gamma rays or electrons, which can break the chemical bonds on absorption by different food materials. On exposure, the ionization products are generated including ions or free radicals, which due to enhanced activity, react further in the process pertaining to the radiolysis in the material exposed to irradiation. These reactions are responsible for destruction of microorganisms, enzymes, and pests present in foods. The process of microbial inactivation by irradiation involves two mechanisms, namely direct interaction of the radiation with the microbial cells and the indirect interaction of the radiolysis products with the cells. Radiolysis products, though short-lived, are sufficient to destroy the microbes. The reactive ions produced by irradiation affect both the RNA and DNA in the cells, which are of primal importance in microbial replication. Due to such an effect, the DNA double-helix does not unwind, hence stopping the replication of microorganisms by cell division. Destruction of microorganisms also takes place by change in structure of the cell membranes, disrupting the metabolic enzyme activity.

11.6.2 Effect on the Quality of Foods and Their Storage Life

The effect of irradiation on foods directly depends on the exposure dose. When exposed to low-dose levels, digestibility of proteins or amino acid composition are not affected by irradiation, but at higher dose levels, sulfhydryl groups from sulfur amino acids break down, causing a change in the taste and aroma of the food. Fatty foods are unsuitable for irradiation as the free radicals formed in the process lead to the oxidation of fats, hence producing rancid flavors. However, as there is no or very little heating in the process, the nutritional properties are not affected. Jan et al. (2017) reported that microbiological safety of green onions is enhanced with an irradiation dose as low as 1 kGy, increasing the shelf life up to 14 days, retaining the sensory properties at the same time. The effect of gamma irradiation on the storage life and quality of beef was studied by Haque et al. (2017).The study concluded with an improvement in the storage life of beef due to reduction in the total plate count, total yeast count, total mold count, and total coliform count. Results also revealed an increase in the color, tenderness, and juiciness of irradiated samples. The irradiation dose and the physical state of foods affects its irradiation efficiency on vitamins. The effect of irradiation dose on the physicochemical parameters and bioactives of jujube fruit was studied by Najafabadi et al. (2017). The results revealed augmented polyphenol content and total monomeric anthocyanins in the irradiated samples up to a certain dose, with darker color and slight decrease in the water-soluble vitamins at higher doses, i.e., >2.5 kGy. Thus, suggesting that up to a certain dose, irradiation can successfully be used without significant loss in bioactive compounds.

11.6.3 Application in Foods

Irradiation at different doses finds numerous applications in disinfestation, sterilization, sprout inhibition, etc. Spices and herbs are often contaminated by heat-resistant, spore-forming bacteria, and irradiation helps in the sterilization (radappertization) of spices and herbs under the exposure dose of 8–10 kGy to reduce the microbial load to a safe and desired level without hampering the quality parameters. Comparatively low doses of irradiation like 3–10 kGy finds application in the destruction of disease-causing bacteria (radicidation) like *Salmonella typhimurium.* Irradiation is useful in increasing the shelf life by reducing the vegetative cells (radurization) in the food products. Irradiation minimizes post-harvest losses. An extension of storage life by two to three times has been observed in some fruits and vegetables by irradiating them and controlling the ripening process. Irradiation also inhibits sprout formation in potatoes and onions. Disinfestation in grains, legumes, and spices by using doses below 1 kGy is a common application of irradiation. One main advantage of using irradiation over chemicals like ethylene dioxide is that it does not leave any chemical residue, which may have harmful effects on health. Irradiation is also widely used in the sterilization of packaging materials.

11.6.4 Limitations

- Irradiation is limited to a few foods only and cannot be applied on all the foods, particularly high fatty foods, which on exposure yield undesirable odors and flavors Some fruits may also become discolored.

- If the spoilage microorganisms in the food are destroyed but the pathogenic ones remain, consumers may not get any indication of the unwholesomeness of the products.
- Harmful radiolytic products may be formed in the foods when the ions and free radicals produced by irradiation react with the food components.

11.7 DENSE-PHASE CARBON DIOXIDE

11.7.1 Principle

Carbon dioxide has been used in non-thermal food applications due to its purity, low cost, and safety. Dense-phase carbon dioxide (DPCD) is used in the non-thermal treatment of liquid foods, as it can inactivate the microorganisms without deterioration in the quality parameters. The pressurized carbon dioxide exhibits bacteriostatic action through different steps that occur simultaneously in a complex manner including compressed carbon dioxide dissolution in the external liquid phase; modification of the cell membrane of the microbes, decrease in intracellular pH due to which cellular metabolism becomes inhibited, inhibition of cellular metabolism or denaturation of enzymes required for metabolic processes by direct effects of molecular carbon dioxide and HCO_3, and vital constituents may be extracted from the cell membranes, leading to physical disruption of the membranes.

11.7.2 Effect on Food Quality and Shelf Life

Treatment with DPCD has protective effects on some nutrients like vitamins, antioxidants, and phytochemicals due to the use of comparatively lower processing temperatures, vacuum of oxygen in the processing environment, and decreasing pH. Chen et al. (2009) reported that beta-carotene was better retained in Hami melon juice when treated with DPCD as compared with pasteurization. The effect of high-pressure carbon dioxide on the activity of the endogenous enzyme polyphenol oxidase in apples and the stability of corresponding apple juice was investigated by Manzocco et al. (2017). The study revealed that such treatment could partially inactivate polyphenol oxidase in apple juice and increase the microbial safety when compared to untreated sample, without impairing the freshness and sensory characteristics of the juice.

11.7.3 Application in Foods

DPCD is mainly used with liquid foods and solid-liquid mixtures. The technology presents great potential for the inactivation of microbes and food enzymes in the dairy industry, and encapsulation of certain bioactives of fruit and vegetable origin. Tang et al. (2021) investigated the effectiveness of DPCD (20 MPa, 60°C, 30 minutes) on microbial and enzyme inactivation in mango in syrup. The results showed that the treatment led to complete destruction of *S. aureus* and *E. coli* with initial concentration of >8 log CFU/mL. Furthermore, peroxidase and polyphenol oxidase inherently present in many foods were completely deactivated while retaining the desirable color and nutritional properties compared with thermal pasteurization. Damar et al. (2009) reported that a DPCD-treated (34.5 MPa, 25°C, 6 minutes) coconut water beverage had more likeability compared with a heat-treated sample in terms of sensorial quality.

11.7.4 Limitations

- Due to the higher cost of operation, DPCD is not commercially viable.
- Environmental regulations are stringent when it comes to release of carbon dioxide into the atmosphere. New systems need to be designed for total capture and recycling of the gas.

11.8 COLD PLASMA

11.8.1 Principle

Cold plasma is yet another innovative, novel, and emerging method from the category of non-thermal technologies. Recent studies suggest the use of this emerging method in microbial inactivation as well as food enzyme deactivations. By subjecting a gas to an electric field of constant or alternating amplitude, plasma can be generated. Cold plasma is a gas that is partially ionized and comprises ions, electrons, free radicals, and radiation ranging from UV to visible (Lebrun, 2015; Mir et al., 2016). The roles of different constituents in the deactivation of microbes and food enzymes varies according to the gas, operating pressure, and the available moisture in the organism. The process effectiveness was higher for moist organisms compared to dry ones, as reported by Dobrynin et al. (2010). The reactive species of plasma inactivates the microbes by damaging the DNA, as shown in Figure 11.3. The reactive species interact with water to form hydroxyl radicals (Zou et al., 2004), which then react with the organic molecules, thus damaging the DNA, cell membranes, and other cellular components by causing chain oxidation. The lipid bilayer of the cells is disrupted by the bombardment of the reactive oxygen species, which form surface lesions that cannot be rapidly repaired by living cells. This process is called etching (Misra et al., 2011; Thirumdas et al., 2015; Dey et al., 2016). The effectiveness in enzyme deactivation is supported by the peptide bond breakage and uncoiling of the three-dimensional structure of proteins in enzymes.

11.8.2 Effect on the Quality of Foods and Their Storage Life

The innate food enzymes, i.e., polyphenol oxidase and peroxidase, are responsible for browning in fruits and vegetables during post-harvest handling (Surowsky et al., 2013). The studies conducted by Pankaj et al. (2013) suggested that "plasma" can inactivate such enzymes, thereby preventing enzymatic browning. The studies were supported by a decreased peroxidase activity in tomato when treated with dielectric barrier discharge plasma produced using atmospheric air. Cold plasma can effectively be applied to fresh products to enhance their storage life by microbial destruction. Misra et al. (2014) studied the total mesophilic count in strawberries subjected to cold plasma. The team reported a total decrease in mesophilic count by 12–85%. Furthermore, a reduction percentage of 54 was noted for *E. coli* in raw milk when treated with cold plasma for 3 minutes (Gurol et al.,

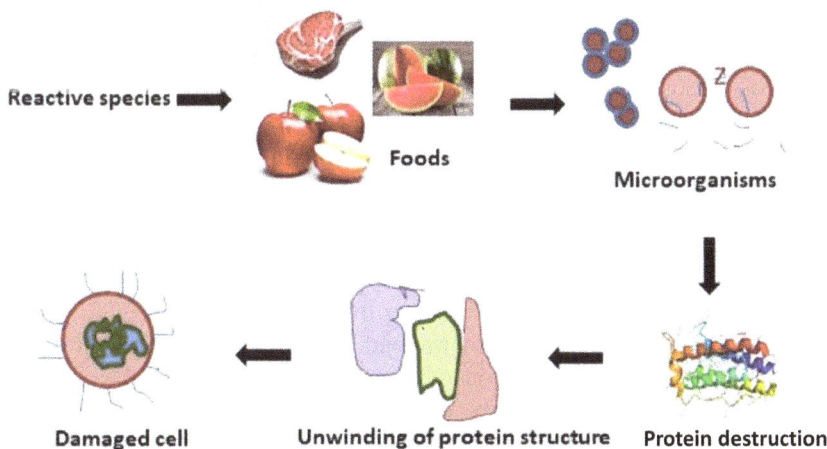

Figure 11.3 Destruction of microorganisms by reactive species of cold plasma.

2012). Being a low temperature process, the plasma treatment aids in the retention of nutritional and sensory parameters, while improving the safety of food. Some studies have shown that UV radiation formed during plasma generation may lead to the formation of phenolic compounds. An increased percentage of protocatechuic acid and luteolin was observed by Grzegorzewski et al. (2011) in the leaves of lamb's lettuce exposed to cold plasma.

11.8.3 Application in Foods

Apart from its application in increasing shelf life by inactivating microorganisms and enzymes in fruits, vegetables, milk products, and meat products, cold plasma technology finds its application in modified starch preparations, alteration of germination in seeds, waste-water treatment, and decontaminating packaging materials. Jiafeng et al. (2014) recorded higher germination in plasma-treated wheat seeds compared with untreated ones. Plasma is widely used in surface treatment of the packaging materials. It can be successfully used in treating heat-sensitive packaging materials like polyethylene and polycarbonate (Pankaj et al., 2014).

11.8.4 Limitations

- In case of foods with high lipid content, cold plasma treatment can lead to increase in lipid oxidation.
- It can also lead to discoloration in some fruits and vegetable.
- The technology is unaccustomed to endogenous enzyme deactivation due to its activity limit and effect on the surface only (surface phenomenon).

11.9 CONCLUSION

As proven by the literature, the non-thermal technologies, when compared with pre-existing conventional ones, showed superior nutritional and organoleptic properties, with same or better degree of safety as that of conventional technologies. Several non-thermal processes like high-pressure processing, pulsed electric field, pulsed white light, ultrasonication, irradiation, cold plasma, and DPCD have been studied and their effect has been proven to render many pathogens inactive, namely *E.coli*, *S. aureus*, and *L. innocua*, as well as decreasing the spoilage. Further, the previously mentioned processes support the inactivation of many endogenous enzymes of foods including peroxidase, polyphenol oxidase, and pectin methyl esterase. Although these technologies have proven to enhance the storage life of foods and their products by reducing possible microflora counts, protecting the bioactive compounds, and enhancing the quality characteristics, their use in food industries is still limited due to high capital involvement in scaling of the processes and huge production volumes in the industries. The advent of non-thermal technologies is a promising field; however, they are in the evolving stages of process development and authentication. In-depth research is needed to check their applicability on various foods and their derived products, following their commercialization. Research literature suggests that the non-thermal technologies, when used in combination with other technologies, produce more effective results than when used alone. The efficacy of non-thermal technologies in producing safe foods when used alone has yet to be established. For selecting the treatment intensities of these processes, target microbes and enzymes should be known, and the effect of the treatment on those target organisms needs to be established. The operating principle of these non-thermal technologies, and their subsequent effect on food quality, storage life, and applicability is summarized in this chapter. The brief description presents an idea of how and on what food types such technologies can be used. Further studies still need to be carried out to make these technologies affordable to be used in the food industries.

REFERENCES

Aguilo-Aguayo, I., Oms-Oliu, G., Martin-Belloso, O., & Soliva-Fortuny, R. (2014). Impact of pulsed light treatments on quality characteristics and oxidative stability of fresh-cut avocado. *LWT – Food Science and Technology, 59*(1), 320–326.

Allende, A., Tomas-Barberan, F. A., & Gil, M. I. (2006). Minimal processing for healthy traditional foods. *Trends in Food Science & Technology, 17*(9), 513–519.

Altuntas, J., Evrendilek, G. A., Sangun, M. K., & Zhang, H. Q. (2010). Effects of pulsed electric field processing on the quality and microbial inactivation of sour cherry juice. *International Journal of Food Science & Technology, 45*(5), 899–905.

Anderson, J. G., Rowan, N. J., MacGregor, S. J., Fouracre, R. A., & Farish, O. (2000). Inactivation of foodborne enteropathogenic bacteria and spoilage fungi using pulsed-light. *IEEE Transactions on Plasma Science, 28*(1), 83–88.

Artíguez, M. L., & de Marañón, I. M. (2015). Improved process for decontamination of whey by a continuous flow-through pulsed light system. *Food Control, 47*, 599–605.

Bank, H. L., John, J., Schmehl, M. K., & Dratch, R. J. (1990). Bactericidal effectiveness of modulated UV light. *Applied and Environmental Microbiology, 56*(12), 3888–3889.

Buckow, R., Semrau, J., Sui, Q., Wan, J., & Knoerzer, K. (2012). Numerical evaluation of lactoperoxidase inactivation during continuous pulsed electric field processing. *Biotechnology Progress, 28*(5), 1363–1375.

Chang, J. C., Ossoff, S. F., Lobe, D. C., Dorfman, M. H., Dumais, C. M., Qualls, R. G., & Johnson, J. D. (1985). UV inactivation of pathogenic and indicator microorganisms. *Applied and Environmental Microbiology, 49*(6), 1361–1365.

Chen, J., Zhang, J., Feng, Z., Song, L., Wu, J., & Hu, X. (2009). Influence of thermal and dense-phase carbon dioxide pasteurization on physicochemical properties and flavor compounds in Hami melon juice. *Journal of Agricultural and Food Chemistry, 57*(13), 5805–5808.

Damar, S., Balaban, M. O., & Sims, C. A. (2009). Continuous dense-phase CO2 processing of a coconut water beverage. *International Journal of Food Science & Technology, 44*(4), 666–673.

Deng, Y., Bi, H., Yin, H., Yu, J., Dong, J., Yang, M., & Ma, Y. (2018). Influence of ultrasound assisted thermal processing on the physicochemical and sensorial properties of beer. *Ultrasonics Sonochemistry, 40*, 166–173.

Dey, A., Rasane, P., Choudhury, A., Singh, J., Maisnam, D., & Rasane, P. (2016). Cold plasma processing: a review. *Journal of Chemical and Pharmaceutical Sciences, 9*(4), 2980–2984.

Dobrynin, D., Fridman, G., Mukhin, Y. V., Wynosky-Dolfi, M. A., Rieger, J., Rest, R. F., … & Fridman, A. (2010). Cold plasma inactivation of Bacillus cereus and Bacillus anthracis (anthrax) spores. *IEEE Transactions on Plasma Science, 38*(8), 1878–1884.

Dunn, J., Ott, T., & Clark, W. (1995). Pulsed-light treatment of food and packaging. *Food Technology (Chicago), 49*(9), 95–98.

Earnshaw, R. (1996). High pressure food processing. *Nutrition & Food Science, 96*(2), 8–11.

Espachs-Barroso, A., Van Loey, A., Hendrickx, M., & Martín-Belloso, O. (2006). Inactivation of plant pectin methylesterase by thermal or high intensity pulsed electric field treatments. *Innovative Food Science & Emerging Technologies, 7*(1–2), 40–48.

Giner, J., Grouberman, P., Gimeno, V., & Martín, O. (2005). Reduction of pectinesterase activity in a commercial enzyme preparation by pulsed electric fields: comparison of inactivation kinetic models. *Journal of the Science of Food and Agriculture, 85*(10), 1613–1621.

Grzegorzewski, F., Ehlbeck, J., Schlüter, O., Kroh, L. W., & Rohn, S. (2011). Treating lamb's lettuce with a cold plasma – influence of atmospheric pressure Ar plasma immanent species on the phenolic profile of *Valerianella locusta*. *LWT – Food Science and Technology, 44*(10), 2285–2289.

Gurol, C., Ekinci, F. Y., Aslan, N., & Korachi, M. (2012). Low temperature plasma for decontamination of *E. coli* in milk. *International Journal of Food Microbiology, 157*(1), 1–5.

Haque, M. A., Hashem, M. A., Mujaffar, M. M., Rima, F. J., & Hossainb, A. (2017). Effect of gamma irradiation on shelf life and quality of beef. *Journal of Meat Science and Technology, 5*(2), 20–28.

Houška, M., Strohalm, J., Kocurová, K., Totušek, J., Lefnerová, D., Tříska, J., … & Paulíčková, I. (2006). High pressure and foods—fruit/vegetable juices. *Journal of Food Engineering, 77*(3), 386–398.

Jan, A., Sood, M., Sofi, S. A., & Norzom, T. (2017). Non-thermal processing in food applications: a review. *International Journal of Food Science and Nutrition, 2*(6), 171–180.

Jiafeng, J., Xin, H., Ling, L. I., Jiangang, L., Hanliang, S., Qilai, X., ... & Yuanhua, D. (2014). Effect of cold plasma treatment on seed germination and growth of wheat. *Plasma Science and Technology*, *16*(1), 54.

Knorr, D. (1999). Process assessment of high-pressure processing of foods: an overview. In *Processing Foods: Quality Optimization and Process Assessment* (pp. 249–267). CRC Press, Boca Raton, FL.

Koh, P. C., Noranizan, M. A., Karim, R., & Hanani, Z. A. N. (2016). Microbiological stability and quality of pulsed light treated cantaloupe (*Cucumis melo* L. *Reticulatus* cv. Glamour) based on cut type and light fluence. *Journal of Food Science and Technology*, *53*(4), 1798–1810.

Krishnamurthy, K., Demirci, A. L. I., & Irudayaraj, J. (2004). Inactivation of *Staphylococcus aureus* by pulsed UV-light sterilization. *Journal of Food Protection*, *67*(5), 1027–1030.

Lebrun, J. P. (2015). Plasma-assisted processes for surface hardening of stainless steel. In *Thermochemical Surface Engineering of Steels* (pp. 615–632). Woodhead Publishing, Waltham, MA.

Manzocco, L., Plazzotta, S., Spilimbergo, S., & Nicoli, M. C. (2017). Impact of high-pressure carbon dioxide on polyphenoloxidase activity and stability of fresh apple juice. *LWT – Food Science and Technology*, *85*, 363–371.

McClements, D. J. (1995). Advances in the application of ultrasound in food analysis and processing. *Trends in Food Science & Technology*, *6*(9), 293–299.

Milani, E. A., & Silva, F. V. (2017). Ultrasound assisted thermal pasteurization of beers with different alcohol levels: inactivation of *Saccharomyces cerevisiae* ascospores. *Journal of Food Engineering*, *198*, 45–53.

Miller, R. V., Jeffrey, W., Mitchell, D., & Elasri, M. (1999). Bacterial responses to ultraviolet light. *ASM News-American Society for Microbiology*, *65*, 535–541.

Min, S., Evrendilek, G. A., & Zhang, H. Q. (2007). Pulsed electric fields: processing system, microbial and enzyme inhibition, and shelf life extension of foods. *IEEE Transactions on Plasma Science*, *35*(1), 59–73.

Min, S., Jin, Z. T., & Zhang, Q. H. (2003). Commercial scale pulsed electric field processing of tomato juice. *Journal of Agricultural and Food Chemistry*, *51*(11), 3338–3344.

Mir, S. A., Shah, M. A., & Mir, M. M. (2016). Understanding the role of plasma technology in food industry. *Food and Bioprocess Technology*, *9*(5), 734–750.

Misra, N. N., Patil, S., Moiseev, T., Bourke, P., Mosnier, J. P., Keener, K. M., & Cullen, P. J. (2014). In-package atmospheric pressure cold plasma treatment of strawberries. *Journal of Food Engineering*, *125*, 131–138.

Misra, N. N., Tiwari, B. K., Raghavarao, K. S. M. S., & Cullen, P. J. (2011). Nonthermal plasma inactivation of food-borne pathogens. *Food Engineering Reviews*, *3*(3–4), 159–170.

Najafabadi, N. S., Sahari, M. A., Barzegar, M., & Esfahani, Z. H. (2017). Effect of gamma irradiation on some physicochemical properties and bioactive compounds of jujube (*Ziziphus jujuba* var vulgaris) fruit. *Radiation Physics and Chemistry*, *130*, 62–68.

Pankaj, S. K., Bueno-Ferrer, C., Misra, N. N., Milosavljević, V., O'donnell, C. P., Bourke, P., ... & Cullen, P. J. (2014). Applications of cold plasma technology in food packaging. *Trends in Food Science & Technology*, *35*(1), 5–17.

Pankaj, S. K., Misra, N. N., & Cullen, P. J. (2013). Kinetics of tomato peroxidase inactivation by atmospheric pressure cold plasma based on dielectric barrier discharge. *Innovative Food Science & Emerging Technologies*, *19*, 153–157.

Patterson, M. F., Linton, M., & Doona, C. J. (2007). Introduction to high pressure processing of foods. In *High Pressure Processing of Foods* (pp. 1–14). Wiley Blackwell Publishing, New York.

Piyasena, P., Mohareb, E., & McKellar, R. C. (2003). Inactivation of microbes using ultrasound: a review. *International Journal of Food Microbiology*, *87*(3), 207–216.

Puppo, M. C., Speroni, F., Chapleau, N., de Lamballerie, M., Añón, M. C., & Anton, M. (2005). Effect of high-pressure treatment on emulsifying properties of soybean proteins. *Food Hydrocolloids*, *19*(2), 289–296.

Raso, J., & Barbosa-Cánovas, G. V. (2003). Nonthermal preservation of foods using combined processing techniques. *Critical Reviews in Food Science and Nutrition*, *43*(3), 265–286.

Sarika, K., & Bindu, J. (2018). *Non-Thermal Technologies for Food Preservation*. ICAR-Central Institute of Fisheries Technology, Cochin, India.

Sentandreu, E., Carbonell, L., Rodrigo, D., & Carbonell, J. V. (2006). Pulsed electric fields versus thermal treatment: equivalent processes to obtain equally acceptable citrus juices. *Journal of Food Protection*, *69*(8), 2016–2018.

Serment-Moreno, V., Barbosa-Cánovas, G., Torres, J. A., & Welti-Chanes, J. (2014). High-pressure processing: kinetic models for microbial and enzyme inactivation. *Food Engineering Reviews*, *6*(3), 56–88.

Surowsky, B., Fischer, A., Schlueter, O., & Knorr, D. (2013). Cold plasma effects on enzyme activity in a model food system. *Innovative Food Science & Emerging Technologies*, *19*, 146–152.

Takeshita, K., Shibato, J., Sameshima, T., Fukunaga, S., Isobe, S., Arihara, K., & Itoh, M. (2003). Damage of yeast cells induced by pulsed light irradiation. *International Journal of Food Microbiology*, *85*(1–2), 151–158.

Tang, Y., Jiang, Y., Jing, P., & Jiao, S. (2021). Dense phase carbon dioxide treatment of mango in syrup: microbial and enzyme inactivation, and associated quality change. *Innovative Food Science & Emerging Technologies*, *70*, 102688

Tekin, Z. H., Başlar, M., Karasu, S., & Kilicli, M. (2017). Dehydration of green beans using ultrasound-assisted vacuum drying as a novel technique: drying kinetics and quality parameters. *Journal of Food Processing and Preservation*, *41*(6), e13227.

Terefe, N. S., Buckow, R., & Versteeg, C. (2014). Quality-related enzymes in fruit and vegetable products: effects of novel food processing technologies, part 1: high-pressure processing. *Critical Reviews in Food Science and Nutrition*, *54*(1), 24–63.

Terefe, N. S., Buckow, R., & Versteeg, C. (2015). Quality-related enzymes in plant-based products: effects of novel food processing technologies part 2: pulsed electric field processing. *Critical Reviews in Food Science and Nutrition*, *55*(1), 1–15.

Thirumdas, R., Sarangapani, C., & Annapure, U. S. (2015). Cold plasma: a novel non-thermal technology for food processing. *Food Biophysics*, *10*(1), 1–11.

Timmermans, R. A. H., Mastwijk, H. C., Knol, J. J., Quataert, M. C. J., Vervoort, L., Van der Plancken, I., … & Matser, A. M. (2011). Comparing equivalent thermal, high pressure and pulsed electric field processes for mild pasteurization of orange juice. Part I: impact on overall quality attributes. *Innovative Food Science & Emerging Technologies*, *12*(3), 235–243.

Torregrosa, F., Esteve, M. J., Frígola, A., & Cortés, C. (2006). Ascorbic acid stability during refrigerated storage of orange–carrot juice treated by high pulsed electric field and comparison with pasteurized juice. *Journal of Food Engineering*, *73*(4), 339–345.

Xu, W., & Wu, C. (2016). The impact of pulsed light on decontamination, quality, and bacterial attachment of fresh raspberries. *Food Microbiology*, *57*, 135–143.

Yeom, H. W., Streaker, C. B., Zhang, Q. H., & Min, D. B. (2000). Effects of pulsed electric fields on the quality of orange juice and comparison with heat pasteurization. *Journal of Agricultural and Food Chemistry*, *48*(10), 4597–4605.

Yordanov, D. G., & Angelova, G. V. (2010). High pressure processing for foods preserving. *Biotechnology & Biotechnological Equipment*, *24*(3), 1940–1945.

Zou, J. J., Liu, C. J., & Eliasson, B. (2004). Modification of starch by glow discharge plasma. *Carbohydrate Polymers*, *55*(1), 23–26.

CHAPTER 12

Chemical Treatment of Foods
Science, Shelf Life, and Quality

Jyoti Nishad, Smruthi Jayarajan, and K. Rama Krishna

CONTENTS

12.1 INTRODUCTION

The relationship that we share with food is tremendous and inexplicable, but the depth can be comprehended by Hippocrates describing food as "let thy food be thy medicine and medicine be thy food." The modern food system evolved over time and the gradual shift of *Homo sapiens* as wanderers to hunters to agriculturists to ultramodern consumers is quite a journey. Healthy and balanced eating is the mantra to healthy life as the food we eat not only affects physical health but also affects mental well-being. The food that we consume, either fresh or processed, undergoes changes with time and often leads to various kinds of deterioration rendering the food unpalatable/unfit for consumption. Hence, it is very important to protect food from further biochemical changes, and to maintain its palatability and safety for consumption. When food is wasted or lost, the resources that are used to produce this food are lost and the demands of waste disposal in landfills further leads to greenhouse gas emissions, which adds to climate change.

The history of preservation of food started with the basic processes like drying, brining, and syruping to the most recent technologies. Food preservation is the process or technique applied to

Email: Jyoti Nishad, bhumi.nishad@gmail.com

DOI: 10.1201/9781003091677-12

stop or gently slow down spoilage and prevent foodborne illness while maintaining the food's nutritional value and organoleptic properties. The process of food preservation can be widely classified as physical (heating, smoking etc.), chemical (butylated hydroxy anisole [BHA], butylated hydroxy toluene [BHT], sorbic acid etc.), and biological (fermentation), among which chemical preservation has found a wider application in food industries because of its efficacy, cost-effectiveness, and ease of availability.

The study of various preservatives that can be utilized to protect food from chemical and biochemical changes dates back to when civilization began. The paradigm shift in food habits of society moved toward the use of preservatives to extend the shelf life of food, and ever since convenience food burst onto the scene, there has been an upsurge in the use of these substances like never seen before. The rise in population also marked the rise in production and consumption of processed and convenience foods worldwide to meet the expectation of market needs and changing lifestyles. The population and its needs are growing and we have few resources for production; therefore, it is imperative to utilize maximum amounts of food products and avoid food losses. The statistics of food loss are scary as one-third of the total produce is wasted or lost, and the cause of the inefficient economy and sustainability issues are out of control. Food losses take place at the production, post-harvest, and processing stages in the food supply chain. To resolve the issue chemical agents could be effectively and efficiently employed at different stages (Parfitt et al., 2010). The global food preservative market is worth $2.4–2.5 billion as per 2019 data and is growing at the rate of 3.7% compound annual growth rate (CAGR). Food preservatives are used in the food system starting from primary processing to secondary and to tertiary processing. The retention of food quality is the main objective behind the use of these preservatives, which in turn prevents/controls the biochemical changes accelerated by intrinsic and extrinsic factors. It also protects food from oxidation, pathogenic microbes, and harsh environmental conditions. The functions of these preservatives can be broadly categorized as antimicrobials, antioxidants, enzyme inhibitors, and sequestrants. In this chapter we will look into various chemical treatments that could be utilized to increase the shelf life of foods to minimize food losses.

12.2 FOOD PRESERVATIVES

Food Safety and Standards Authority of India (FSSAI), under Section 3.1.4 of FSS (Food Product Standards and Food Additives) Regulations (Food Safety and Standards (Food Products Standards and Food Additives) Regulations (2011) defines preservative as "a substance which when added to food, is capable of inhibiting, retarding or arresting the process of fermentation, acidification or other decomposition of food."

The principal mechanism of action of preservatives in extending the shelf life of food involves an increase in acidity of the surroundings, reduction in water availability, and change in redox potential. The International Numbering System (INS), a European-based naming system defined by Codex Alimentarius, is used for designation of different food preservatives. An ideal food preservative is characterized by its non-toxicity; physical and chemical stability; longer shelf life; its ability to maintain physical, chemical, and microbiological food properties; ability to maintain product palatability and wholesomeness; compatibility with all other food ingredients; and it should be non-irritant and potent in action (Dwivedi et al., 2017). Preservatives have become an important component of food with special consideration to processed foods. They increase or maintain the nutritional benefits of food and retain food quality, reduce wastage, increase product availability, and enhance consumer acceptability (Mirza et al., 2017).

The classifications of food preservatives are based on their action mechanisms, chemical characteristics of the compounds, and the sources from which they are obtained. Regarding *mechanism of action*, preservatives are categorized into antimicrobials, antioxidants, anti-enzymatics, and synergists and sequestrants. Food preservatives are classified on the basis of the *chemical characteristics of the compounds* into acids such as lactic acid, benzoic acid, acetic acid, and sorbic acid; alcohols (e.g., ethanol); polyphenols, namely phenols and flavonoids; esters (e.g., esters of *p*-hydroxybenzoic acid); and quaternary ammonium compounds (viz. *n*-dodecyl dimethyl benzyl ammonium chloride and *n*-dodecyl dimethyl ethyl benzyl ammonium chloride) (Mirza et al., 2017). *Sources* of food preservatives are divided into two categories, i.e., Class I and II preservatives, where Class I refers to natural preservatives and Class II depicts artificial preservatives. Natural preservatives could be obtained from plants, animals, insects, microorganisms, or naturally occurring minerals, whereas artificial preservatives are synthesized chemically. In the following sections different categories of food preservatives are discussed in detail focusing on their mechanism of action and source of generation.

12.2.1 Antimicrobial Preservatives

Food, an essential source of energy for humans, is affected by many factors and among them microbial contamination stands as a major factor. Microorganisms such as bacteria, fungi, yeast, molds, etc., can cause serious deterioration in food products. Microorganisms can bring deleterious changes in food by altering its flavor, color, odor, texture, and other sensory attributes, finally making it unfit for consumption, leading to food wastage. Microbial contamination can occur at the field level; during processing, preparation, packaging, storage, distribution, and mishandling by the consumer; and cross-contamination by other foods and water at home. Moreover, food spoilage due to microbial contamination has led to foodborne illness and economic instability among families. It was estimated that in the United States alone every year the outbreak of foodborne illness is approximately 48 million cases. This also impacts the food industry an economic loss of $161.6 billion due to non-consumption of 31% of the available food at the retail and consumer level (Scallan et al., 2011; Buzby et al., 2014). All these concerns have prompted food industries to look for suitable strategies and compounds of both natural and synthetic origin that can prevent microbial food spoilage. Control of microbial food spoilage is generally achieved by the application of antimicrobial compounds and several preservation techniques such as heat treatments, cold treatments, drying, filtration, etc. The use of antimicrobials for food preservation has proven effective in the control of food pathogens and preventing food from microbial toxic contamination. Nevertheless, synthetic chemicals in food have a negative public opinion due to their undesirable aspects, including acute toxicity, carcinogenicity, teratogenicity, and slow degradation (Davidson et al., 2013).

Antimicrobials of natural and synthetic origin, which have a long history of usage in food, proven to have no health adversities, and are Generally Recognized as Safe (GRAS) are extensively used for food preservation. Antimicrobials are compounds that are present in or added to foods (raw or processed) or added to food packaging, food contact surfaces, and the food processing environment to kill or inhibit microbial growth. Food preservation and food safety are considered as two primary functions of antimicrobial. In food preservation antimicrobials are used to control natural spoilage processes, whereas in food safety they are used to prevent or control the growth of microorganisms, including pathogenic microbes (Davidson et al., 2013). Several antimicrobials of synthetic derivatives, plant-based and animal derivatives, and microbial origin are used for food preservation and food safety.

Plant-derived compounds, especially from herbs and spices, have a long history of usage by mankind for their medicinal and antimicrobial characteristics. Research and development activities in the last two decades have shown promising beneficial effects when plant derivatives are used as antimicrobials in food (Cowan, 1999; Gyawali et al., 2015). Among different antimicrobials derived from plants, essential oils extracted from different plant parts such as buds, flowers, barks, roots, leaves, etc., have received an immense amount of attention by the public due to their aromatic and volatile nature apart from antimicrobial activity. The essential oils derived from herbs, spices, and some fruits, such as rosemary, thyme, basil, clove, pepper, cinnamon, ginger, garlic, oregano, coriander, marjoram, lemongrass, and citrus fruits, have shown promising use in food due to their significant antibacterial and antifungal activity (Adelakun et al., 2016). The essential oils and their derivative compounds (terpenes, terpenoids, phenylpropenes, etc.) have been used in food due to their low or negligible toxicity or inability to develop resistance in the microorganisms, thereby they are considered as GRAS compounds. Few commercially available essential oils are marketed as food preservatives despite being permitted to be used as GRAS by the United States and European Union (EU). Nevertheless, the market for essential oils as antimicrobials is increasing because they are perceived as natural by consumers. Very little is known about the interaction of essential oils and microbes. It was found that individual compounds had less impact when compared with the crude application (Hyldgaard et al., 2012). However, it was seen that gram-positive bacteria are more susceptible than gram-negative bacteria (Trombetta et al., 2005). The higher tolerance of gram-negative bacteria is because of hydrophilic lipopolysaccharides in the outer membrane, which create a barrier for macromolecules and hydrophilic compounds similar to that of essential oils (Nikaido, 2003).

Animal-derived compounds such as lactoferrin, lysozyme, chitosan, cathelicidins, etc., have shown antimicrobial activity. These compounds are synthesized in animals in the form of polysaccharides, peptides, terpenes, sterols, and fatty acids as a part of an immunity mechanism to counteract pathogens (Arias-Rios et al., 2017). Some of these compounds have application in the food industry as preservatives. Lysozyme (GRAS) is naturally present in eggs and is commercially extracted from hen eggs. It is used in the preservation of meat and fish products, milk and dairy products (prevent late blowing in cheeses due to *Clostridium* sp.), and fruits and vegetables (Cegielska-Radziejewska et al., 2009). Lactoferrin (U.S. Department of Agriculture [USDA]) is extracted from bovine milk and is commercially used in baby foods (0.66%), energy bars (4.0%), infant formulas (0.1%), dairy products (0.2%), non-alcoholic beverages (0.12%), and sport drinks (0.3%) (García-García & Searle, 2016). Chitosan (GRAS) is a deacetylated derivative of chitin that is extracted from crustacean marine shell animals and some microorganisms such as *Aspergillus niger*. It is extensively used as a film coating on fruits and vegetables due to its film-forming ability and antimicrobial property against a wide range of microbes (Krishna & Rao, 2014; 2017). The combination of lysozyme and lactoferrin showed a bactericidal effect, whereas individual application had a bacteriostatic effect against *Vibrio cholerae*, *Escherichia coli*, and *Salmonella typhimurium* (Ellison & Giehl, 1991). Chitosan has a wide range of antimicrobial action such as on gram-negative (*Pseudomonas fluorescens*, *E. coli*, *Vibrio parahaemolyticus*, *S. typhimurium*) and gram-positive bacteria (*Staphylococcus aureus*, *Listeria monocytogenes*, *Bacillus cereus*, *Bacillus megaterium*) (No et al., 2002). The mode of action of these compounds is by disruption of the cell primarily due to interactions with the cell membrane. Animal-derived antimicrobials have a high concentration of hydrophobic residues containing amino acids with amphipathic and cationic properties, which make them highly effective when interacting with the cell membranes. This ultimately makes the outer membrane more susceptible by disrupting it and leading to leakage of the cell (Ayaad et al., 2012; Brandenburg et al., 2016). Some of the compounds also show non-lytic interactions, which are not well understood. This mechanism may involve the translocation of the compounds to the inner membrane causing intracellular damage, including inhibition of the synthesis of DNA, proteins, cell walls, and inactivation of indispensable enzymes (Scocchi et al., 2016).

The preservation of food using antagonistic microorganisms and their metabolites is called biopreservation (Quinto et al., 2019). Microorganism-derived antimicrobials have gained importance due to their ability to inhibit the growth of pathogenic- and spoilage-causing microbes in food. Among the microorganisms, lactic acid bacteria (LAB) has gained prominence for use in food preservation. LAB has been given the status of GRAS by the USDA and European Food Safety Authority (EFSA). It has application in high pH foods such as vegetables and salads, whereas its activity is limited in low pH food such as fruit juices and condiments. The metabolites derived from

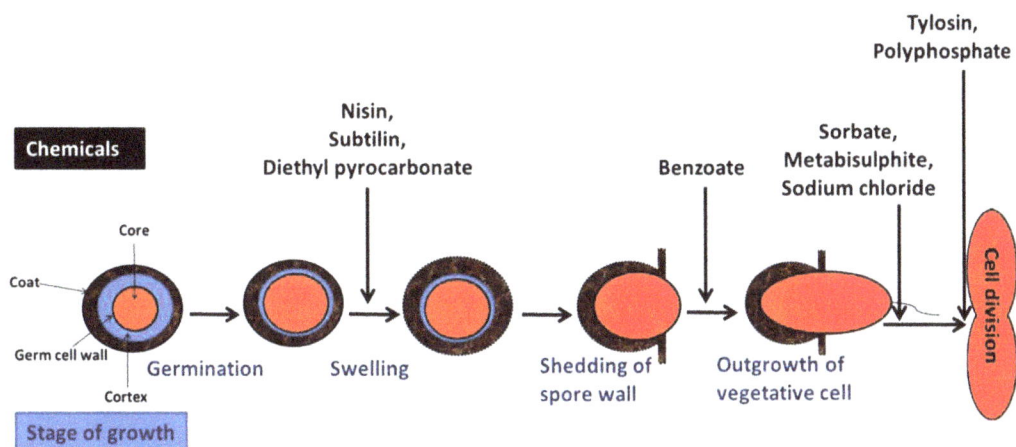

Figure 12.1 Sketch representation of effective inhibition by some antimicrobial compounds at different stages of growth of an endospore into vegetable cells. (Concept adapted from Gould, 1964, with modifications.)

LAB, such as nisin and pediocin, have no such limits for its application as a preservative in acidic food and have proven to be used as bacteriocins. Therefore, the use of non-pathogenic microorganisms and their metabolites for food preservation should be fully harnessed.

The mode of action and stage at which these antimicrobials inhibit the pathogenic microorganisms is different for each group of antimicrobials. The growth stage at which some of the antimicrobials inhibit the microbes is illustrated in Figure 12.1.

12.2.2 Antioxidant Preservatives

Food deterioration causes huge losses to humankind and there are various methods to curb the deteriorating process to extend the shelf life of food. The preservatives that we administer to control the reaction vary ranging from antimicrobial to anti-enzymatic to antioxidative depending on the food matrix and type of degradation process. Antioxidants can be defined as substances/molecules that inhibit or delay the oxidation process either by quenching, chelating, or neutralizing the oxidants generated. Antioxidants are a defense mechanism produced by the living system to neutralize the effects of reactive oxygen species. They can be either enzymatic or nonenzymatic in nature. Nonenzymatic antioxidants include vitamin C, vitamin E, selenium, zinc, beta-carotene, taurine, hypotaurine, and glutathione; whereas enzymatic antioxidants constitute superoxide dismutase, catalase, glutathione reductase, etc.

The antioxidative preservatives are used in the food matrix to control the damage caused by oxidative reactions in food, which leads to the generation of off-flavors, color loss, and textural changes attributed to the degradation of carbohydrate, protein, vitamins, sterols, and lipid peroxidation. Antioxidants play an important role in food preservation by neutralizing the free radicals and protecting the food from further deterioration.

Antioxidant preservatives that are administered in the food matrix can be widely classified as natural and synthetic antioxidants. Most of the natural antioxidants that are commonly used are of plant origin such as polyphenolics; carotenoids; and antioxidant vitamins like A, C, E, etc. On the other hand, synthetic antioxidants are also widely used in the food industry because of higher stability, better performance, cost-effectiveness, and ease of availability. The common synthetic antioxidants in the food industry are BHA, BHT, propyl gallate (PG), butyric acid, and tert-butyl hydroquinone (TBHQ). The possible mechanisms by which antioxidants combat the oxidants include (1) quenching in which antioxidants acts as a quenching agent (Anbazhagan et al., 2008); (2) direct hydrogen transfer, which enables the direct transfer of the hydrogen atom to the radical (Priyadarsini et al., 2003); (3) charge transfer for a doublet radical, which results in anion (closed-shell) and antioxidant cation (Oschman, 2009); and (4) bond-breaking mechanisms, which are prevalent in vitamin E (Roginsky & Lissi, 2005).

12.2.2.1 Quenching

In this case, the antioxidants react with singlet oxygen to form intermediate compounds such as endoperoxides and final products, which are mainly hydroperoxyl dienones. These are responsible for quenching or termination of the propagation process that generates free radicals; examples include vitamin E and carotenoid.

12.2.2.2 Hydrogen Transfer

A complex free radical acceptor is formed between a lipid radical and the antioxidant radical which, in this case, is the free radical acceptor.

12.2.2.3 Charge Transfer

In the charge transfer mechanism, there are two ways in which the charge transfer antioxidation mechanism takes place, both involving the formation of stable radicals. This stops the propagation of reactive species in the biological systems. Examples include flavonoids and phenolic antioxidants.

12.2.2.4 Bond Breaking

α-Tocopherol is a fat-soluble antioxidant that plays an important role in protecting the cytoplasmic membrane against oxidation reactions as induced by lipid radicals, thus, terminating the chain propagation reactions (Herrera & Barbas, 2001).

12.2.3 Anti-Enzymatic Preservatives

The enzyme activity in various foods is imperative in deciding the food's overall quality and consumer acceptability. Various enzymes are involved in the metabolism of plant-based foods and their ripening process. Some enzymes, such as phenolase, continue to metabolize after fruits and vegetables are cut and lead to browning of the cut and exposed surface. These enzymes are also reported as a key factor of the microbial growth. Thus, controlling and inhibiting the enzyme activity during pre- and post-harvest processing will successfully enhance the microbial stability and improved shelf life of foods. For this purpose, the class of preservatives known as anti-enzymatics are utilized for preventing natural metabolic ripening and oxidative deterioration of food by inhibiting fungi, bacteria, and parasites. They can act as metal chelators and remove the metal cofactors required by many enzymes. Erythorbic acid, citric acid, ethylenediaminetetraacetic acid (EDTA), and polyphosphates are few examples of anti-enzymatic preservatives (Abdulmumeen et al., 2012; Amit et al., 2017).

12.2.4 Synergists and Sequestrants

Sequestrants are the organic or inorganic compounds that have preservative action. This group of food preservatives includes lactic acid, lecithin, citric acid, tartaric acid, phosphoric acid, and calcium and sodium salts of EDTA. In addition, phytic acid and its salts; calcium, sodium, and potassium salts of gluconate; calcium polyphosphate; sodium tripolyphosphate; sodium hexam-etaphosphate; and potassium ferrocyanide are also of importance. Sequestrants exhibit preservative effect by forming complexes with metal ions, thus converting them into an inactive form. Inactivation of heavy metal traces such as copper and iron is of special importance as these metal ions catalytically accelerate oxidative changes. Chelating metal ions makes them unavailable for enzyme activity, therefore, causing the inactivation of enzymes and microbial inhibition. This complex formation with metal ions is also useful in preventing the undesirable cloudiness in some beverages, such as those caused by iron and copper in wine. Sequestrants sometimes can be utilized in suppressing the discoloration of food products caused by metal reaction with amines or mercapto compounds. Synergists are the compounds formed with the combination of sequestrants with antioxidants and have the capability of regenerating the spent antioxidants (Lück & von Rymon Lipinski, 2000; Msagati, 2013).

Table 12.1 summarizes the maximum permissible limits along with the food applications of preservatives classified on the basis of mechanism of action.

12.2.5 Natural Preservatives

Among the innumerable preservation techniques, natural preservatives are most preferred by consumers worldwide because of their perceived safety as opposed to artificial preservatives. The living system (plants, animals, insects, crustaceans, microbes) contains various molecules, having immense antimicrobial and antioxidant properties, which have evolved as host defense mechanisms and can be utilized as potential preservatives in the food industry.

Natural preservatives can be defined as any and all substances obtained naturally or directly from a biological system without any change or modification in a laboratory environment and are extracted from various sources including plants, animals, bacteria, algae, and fungi for preserving food items (Gracia et al., 2015). Plant extracts have the advantage of being consumed by humans for thousands of years, and in addition to the antimicrobial, antioxidant, anti-enzymatic properties, several plants are used in different areas of human health, such as traditional medicine, functional foods, food supplements, and production of recombinant protein. Therefore, the natural preservatives obtained from herbs and spices have been the subject of constant research to increase the shelf life and ensure the organoleptic characteristics of food (Table 12.2). Apart from that, herbs and spices have the added benefit of being considered as GRAS foods and can be consumed without risk to health (Viuda-Martos et al., 2010). The wide array of natural preservatives includes phenolics, flavonoids, saponins, lactoperoxidase (milk), lysozyme (egg white), bacteriocins (LAB), microbial-derived compounds (mainly lytic enzymes), chitosan found in shrimp shells, and chitosan-saccharide derivatives. The plant- and animal-based natural preservatives will be discussed in Table 12.2.

12.2.5.1 Plant-Based Natural Preservatives

Natural products obtained from plants, such as herbs, spices, and essential oils, are widely used in foodstuffs as preservatives for their sensory characteristics, ensuring quality and safety throughout the shelf life (Table 12.3) (Negi, 2012). Spices are characterized as products from different parts of certain plants, with the exception of leaves, whereas herbs are those extracted from the leaves of the plant. Herbs and spices can be classified based on the flavor, taxonomy, or part of the plant from which they were extracted. The essential oils are the by-products of secondary metabolism in plants and they are bioactive, volatile, and fragrant compounds formed by a mixture of substances with an oily consistency. The chemical composition of the essential oils is dominated by the presence of a range of compounds including phenolics, terpenoids, aldehydes, ketones, ethers, epoxides, and many others inferring that essential oils must be effective against a wide range of pathogens. Essential oils and their chemical constituents have been reported with well-documented

Table 12.1   Preservatives with Their Maximum Permissible Limits and Food Applications

Category	Preservative	Chemical Formula	E Number	Permissible Limit	Products to Which They Are Added
Antimicrobial	Propionic acid	$C_3H_6O_2$, (m.w. 74.08)	E280	Up to 0.32%	Mold inhibitor in bread, cakes, some cheeses, etc.
	Sodium propionate	$C_3H_5NaO_2$, (m.w. 96.06)	E281		
	Calcium propionate	$C_6H_{10}CaO_4$, (m.w. 186.22)	E282		
	Sorbic acid	$C_6H_8O_2$, (m.w. 112.13)	E200	Up to 0.2%	Mold inhibitor in hard cheeses, syrups, jellies, salad dressing, soft drinks, frozen pizza, cakes, etc.
	Sodium sorbate	$C_6H_7NaO_2$, (m.w. 134.1084)	E201		
	Potassium sorbate	$C_6H_7KO_2$, (m.w. 150.22)	E202		
	Calcium sorbate	$C_{12}H_{14}CaO_4$, (m.w. 262.32)	E203		
	Benzoic acid	$C_7H_6O_2$, (m.w. 122.12)	E210	Up to 0.1%	Yeast and mold inhibitor in pickles, soft drinks, dressings, etc.
	Sodium benzoate	$C_7H_5NaO_2$, (m.w. 144.11)	E211		
	Potassium benzoate	$C_7H_5KO_2$, (m.w. 160.21)	E212		
	Calcium benzoate	$C_{14}H_{10}CaO_4$, (m.w. 282.31)	E213		
	Potassium nitrite	KNO_2, (m.w. 85.11)	E249	Up to 120 ppm	Anti-botulinal agent in cured meats and in cooked meats, bacon, ham and cheese
	Sodium nitrite	$NaNO_2$ (m.w. 69.00)	E250		
	Sodium nitrate	$NaNO_3$ (m.w. 84.9947)	E251	0–5 mg/kg body weight per day	
	Potassium nitrate	KNO_3, (m.w. 101.13)	E252		
	Sodium diacetate	$C_4H_7NaO_4 \cdot xH_2O$, (m.w. of anhydrous form 142.09)	E262	Up to 0.32%	As mold inhibitor in bread and on cereal grains
	Acetic acid	$C_2H_4O_2$, (m.w. 60.052)	E260	No limit	Pickling agent in mayonnaise, pickles, scald water sprays for meat carcasses, sauces, dressings, salads, drinks, fruit juices and concentrates, wines, jams and jellies, candy, cheeses, canned products
	Lactic acid	$C_3H_6O_3$, (m.w. 90.08)	E270	No limit	
	Citric acid	$C_6H_8O_7$ (m.w. 192.12)	E330	No limit	
	Nisin	$C_{143}H_{230}N_{42}O_{37}S_7$, (m.w. 3354.25)	E234	10–100 ppm	Anti-clostridial agent in cheese wraps, pasteurized process cheese spreads; heat adjunct in canned foods
Antioxidant	Ascorbic acid	$C_6H_8O_6$ (m.w. 176.12)	E300	—	All foods
	BHA	$C_{11}H_{16}O_2$ (m.w. 180.24)	E320	0.2–0.75 g/kg	Butter, fat, fish, oil
	BHT	$C_{15}H_{24}O$ (m.w. 220.35)	E321	0.2–0.75 g/kg	Butter, chewing gum, fat, fish, oil, whole meat
	Propyl gallate	$C_{10}H_{12}O_5$ (m.w. 212.2)	E310	0.1–0.2 g/kg	Butter, fats, oils

(Continued)

Table 12.1 Preservatives with Their Maximum Permissible Limits and Food Applications (*Continued*)

Category	Preservative	Chemical Formula	E Number	Permissible Limit	Products to Which They Are Added
	TBHQ	$C_{10}H_{14}O_2$ (m.w. 166.22)	E319	0.02–0.07 g/kg	Crackers, popcorn, butter
	Tocopherols, alpha, gamma, delta	$C_{29}H_{50}O_2$ (m.w. 430.71)	E306, E307, E308, E309	—	All food
Anti-enzymatic	Erythorbic acid (iso-ascorbic acid)	$C_6H_8O_6$ (m.w. 176.13)	E315	250–1500 ppm	Fillings of stuffed pasta (ravioli and similar), heat-treated and non–heat-treated meat products, only cured meat products and preserved meat products, processed fish and fishery products, only preserved and semi-preserved fish products, fish roe
Synergists and sequestrants	Ethylenediaminetetraacetic acid (EDTA)	$C_{10}H_{16}N_2O_8$, (m.w. 292.24)	E385	36–500 ppm expressed as anhydrous calcium disodium EDTA	Mayonnaise, salad dressings, sauces and spreads, canned beans and legumes, canned seafoods, canned carbonated soft drinks, pickled vegetables, distilled alcoholic beverages
	Sodium tripolyphosphate	$Na_5P_3O_{10}$, (m.w. 367.86)	E451	Total added phosphate not to exceed 0.5% calculated as sodium phosphate	Seafood, meats, poultry, and animal feeds
	Tartaric acid	$C_4H_6O_6$, (m.w. 150.087)	E334	No limit with the exception of confectionary, dairy products, fats and oils, and cocoa powder (maximum of 50,000 ppm)	Fruit and vegetable juices, soft drinks, baked goods, confectionary, carbonated beverages, dairy products, edible oils and fats, canned products, fruit jellies, gelatin, cocoa powder, effervescent tablets, cream of tartar, alcoholic drinks
	Lecithin	$C_{35}H_{66}NO_7P$, (m.w. 643.9)	E322	No limit	Whey and whey products, coffee, coffee substitutes, tea, nectars, sugar syrups, fats and oils, candy, chewing gum, pasta and noodles, liquid and frozen egg products, frozen battered fish, fish fillets and fish products including mollusks, crustaceans, and echinoderms

Abbreviations: BHA, butylated hydroxyaniscle; BHT, butylated hydroxytoluene; TBHQ, tert-butylhydroquinone.

Table 12.2 Studies on the Use of Artificial and Natural Preservatives

S. No.	Preservative	Concentration Used	Food Matrix	Shelf Life/ Storage Life	Results	References
1.	Calcium propionate	0.3%	Bread	10–12 days shelf life	Addition of sourdough fermented using a specifically selected antifungal *Lactobacillus amylovorus* DSM 19280 can replace the chemical preservative	Belz et al. (2012)
2.	Sorbic acid and sodium metabisulfate (SM)	0.1% each	Rasogolla	30 days shelf life	Individual application of sorbic acid (0.1%) and sodium metabisulfate (0.1%) extended the shelf life for only 24 and 15 days, respectively	Das et al. (2010)
3.	Raisin paste + citric acid + vinegar	5 g, 0.2 g, and 0.2 mL, respectively	Parottas	9 days shelf life	Sensory evaluation during ambient storage (27 ± 2°C) showed that the parottas were acceptable	Ray and Dasappa, (2017)
4.	Benzoic acid	200 mg/mL	Sobo drink	14 days shelf life	Benzoic acid was most effective against *Aspergillus niger, Aspergillus fumigatus*, and *Bacillus subtilis* and so was organoleptically attractive for 14 days	Doughari et al. (2007)
5.	Potassium sorbate	0.4 %w/v	Fresh catfish (*Clarias gariepinus*)	8 days shelf life	Potassium sorbate followed by smoking showed the lowest microbial load and maximum shelf stability	Efiuvwevwere and Ajiboye (1996)
6.	Sodium acetate (SA)	1%	Catfish fillets	6 days shelf life	Microbial counts of SA-treated fillets remained lower than the control (aerobic plate count [APC] by 0.6–0.7 log units)	Kim et al. (1995)
7.	Sodium lactate + SA	1.74% each	Raw chicken breasts	15 days refrigerated at 2–3°C	SA alone or in combination with sodium lactate had lower APC levels	Tangkham et al. (2012)
8.	SA + vacuum packaging	2%	Rainbow trout (*Oncorhynchus mykiss*)	15 days refrigerated at 2 ± 1°C	SA-treated aerobic packaging showed 10 days of shelf life (2 ± 1°C)	Etemadi et al. (2013)
9.	Biodegradable film containing SM + polycaprolactone	10, 20, and 40 wt%	Apple	24 hours at room temperature	Excellent antimicrobial activities against gram-positive bacteria (*Staphylococcus. aureus*) and gram-negative bacteria (*Escherichia coli*); no fungal growth after 28 days: nearly 45% decrease in browning index	Jeong et al. (2020)
10.	SM	6 g/kg	Fruits and vegetables discarded	7 days at 23.2°C temperature, 64% relative humidity	Decrease in bacteria, yeasts, and molds count	Ahmadi et al. (2019)

(Continued)

Table 12.2 Studies on the Use of Artificial and Natural Preservatives (Continued)

S. No.	Preservative	Concentration Used	Food Matrix	Shelf Life/ Storage Life	Results	References
11.	SO$_2$ fumigation + modified atmosphere packaging	Fumigation for 30 minutes at 20°C (0.66 g/kg of SM)	Blueberry	45 days at 0°C	Nearly 80% decrease in decay, no weight loss, texture remained intact	Rodriguez and Zoffoli (2016)
12.	TBHQ and citrus peel phenolics	100–400 ppm	Mustard oil	3 months at room temperature	Decrease in primary and secondary products in oil compared with control	Nishad et al. (2021)
13.	Tannic acid, erucic acid, ferulic acid, caffeic acid, sinapinic acid, catechin, quercetin, rutin, sesamol, TBHQ, BHT, and BHA	200 ppm	Pecan oil	20 days, 60°C, in the presence of atmospheric air	Caffeic acid depicted on par inhibitory activity with TBHQ, BHT, and BHA	Zhang et al. (2018)
14.	Tocopherol, rosemary essential oil and *Ferulago angulate*, TBHQ	450, 450, 1000, and 150 ppm of tocopherol, rosemary essential oil, *Ferulago angulate*, and TBHQ, respectively	Mayonnaise	6 months at 25°C	Tocopherol and rosemary essential oil were on par with TBHQ in inhibiting primary oxidation	Alizadeh et al. (2019)
15.	Ethylene oxide	—	Black peppercorn	—	5 log reductions of *Salmonella* at 50% RH and 60°C temperature for 20 minutes exposure to ethylene oxide	Wei et al. (2021)
16.	Ethylene oxide	—	Cumin seeds	—	~5 log reductions of *Salmonella enterica* and *Enterococcus faecium* at 20 minutes of ethylene oxide treatment at 50% RH	Chen et al. (2021)
17.	Propylene oxide	—	Cashews and macadamia nuts	—	Reduction in *Salmonella*, *E. faecium*, *Pediococcus acidilactici*, and *Staphylococcus carnosus*	Saunders et al. (2018)
18.	Smoking + antimicrobial agents	25% NaCl and 1% ascorbic acid for 30 minutes or 1 hour, 3% sodium lactate with or without 5% rosemary extract for 30 minutes, and 5% sorbic acid alone for 30 minutes; temperature-10°C; smoking 26–30 hours	Blue catfish steaks	6 weeks at 26–33°C	*Listeria* and *Salmonella* were absent, total plate count was decreased	da Silva et al. (2008)
19.	Rice hull smoke extract	1–0.2%	MICs were determined in nutrient broth	—	MIC of rice hull smoke against *Salmonella typhimurium* was 0.822%	Kim et al. (2012)

(Continued)

Table 12.2 Studies on the Use of Artificial and Natural Preservatives (Continued)

S. No.	Preservative	Concentration Used	Food Matrix	Shelf Life/ Storage Life	Results	References
20.	Smoking	0.9%	Salmon strips	—	Reduction in *Listeria innocua* to <2 log CFU/g after 2 weeks	Montazeri et al. (2013)
21.	Smoking	10, 5, and 2.5%	Frankfurters	—	5% showed bactericidal activity against *Listeria monocytogenes* after 6 weeks and 10% was listericidal after 4 weeks	Morey et al. (2012)
22.	Smoking (liquid smoke)	100%, 60-second dip	Roast beef cuts	—	*L. innocua* was undetectable after 2 and 4 weeks of storage (4°C)	Milly et al. (2008)
23.	Citrus peel and nutmeg extract	0.5% and 1%	Goat meat	−18 ± 1°C for 6 months	1% concentration of mixed extract revealed strong inhibitory activity against lipid and protein oxidation	Nishad et al. (2018)
24.	*Cymbopogon citratus* essential oil	200–300 μL (dip)	Grapes	10 days	Application of 200 and 300 μL of *C. citratus* oil on 1 kg of stored grapes showed enhancement of shelf life up to 10 days	Sonker et al. (2014)
25.	*Artemisia nilagirica* essential oil	200–300 μL (fumigation)	Grapes	9 days	Fumigation of 1 kg of table grapes with 200 and 300 μL dosage of *A. nilagirica* oil enhanced the shelf life for up to 9 days	Sonker et al. (2015)
26.	Clove (*Syzygium aromaticum*), cinnamon (*Cinnamomum verum*), and lavender (*Lavandula stoechas*) essential oil	Combination of the essential oil at 2 mg/mL of cinnamon (59.4%), clove (2.4%), and lavender (38.2%)	UHT milk	—	The constituents such as eugenol, cinnamaldehyde, camphor, and pinene were effective in controlling *E. coli* in UHT milk	Falleh et al. (2019)
27.	*Mentha cardiaca* essential oil	—	Dry fruits	—	Reduced the microbial load of *Aspergillus flavus*	Dwivedy et al. (2017)
28.	Thyme (*Thymus capitatus*) essential oil	1%	Skimmed milk	—	Resulted in higher bacterial growth inhibition, lower minimum inhibitory and bactericidal concentrations, and better time kinetic results	Benjemaa et al. (2018)
29.	*Ocimum basilicum* oil	Alginate coating containing nanoemulsified basil (*O. basilicum.* L) oil	Okra	4 days	Formulations were found effective against spoilage fungi *Penicillium chrysogenum* and *A. flavus* and effectively improved storage life of okra	Gundewadi et al. (2018)
30.	Ajowan (*Trachyspermum ammi*) essential oil	0.8 μL	Wheat and chickpea	—	Effective in controlling the *A. flavus* infection	Kedia et al. (2015)

Abbreviations: BHA, butylated hydroxyanisole; BHT, butylated hydroxytoluene; MIC, minimum inhibitory concentration; RH, relative humidity; SO$_2$, sulfur dioxide; TBHQ, tert-butylhydroquinone; UHT, ultrahigh temperature processing.

Table 12.3 Classification of Natural Preservatives Obtained from Plants

Source	Parts Used	Bioactive Compounds	Plant Source
Herbs	Extracted from leaves	Carnosic acid, carnosol, thymol, carvacrol, geranyl acetate α-pinene, γ-terpinene, rosmarinic acid, sinapinic acid, vanillic acid, ferulic acid	Oregano, rosemary, basil, bay leaf, coriander, marjoram
Spices	Obtained from parts of plants such as fruits, seeds, roots, barks, and shoots	Cinnamaldehyde, cinnamic acid, cinnamate, eugenol, caryophyllene, catechol	Cinnamon, clove, nutmeg, thyme
Essential oils	By-product of secondary metabolism in plants	α-Pinene, 1,8-cineole, camphor, camphene, linalool, caryophyllene, thymol	Coriander seed, clove, rosemary, nutmeg, cinnamon

antimicrobial effects against a range of bacterial, fungal, and viral pathogens. They may be liquid at room temperature, although some of them are solid or resinous, and show different colors ranging from light yellow to emerald green and from blue to dark reddish brown.

Several chemical compounds present in plants have the ability to replace synthetic preservatives, thereby helping to conserve food. Among them are saponin, flavonoids, thiosulfinates, glucosinolates, phenolics, and organic acids. The main active components of plants with antimicrobial action are polyphenolics such as terpenes, aliphatic alcohols, aldehydes, ketones, acids, and isoflavonoids (Campêlo et al., 2019).

12.2.5.2 Animal- and Microbial-Based Natural Preservative

There are numerous arrays of animal-based natural preservatives used in the food industry, including lactoperoxidase (milk), lysozyme (egg white), bacteriocins (LAB), microbial-derived compounds (mainly lytic enzymes), propolis, nisin, chitosan found in shrimp shells, and chitosan-saccharide derivatives. The lysozyme present in eggs and milk can be used as an antimicrobial agent and can be used as a preservative for poultry products, meat, fruits, etc. Mammalian secretions like milk, saliva, etc., contain lactoferrin and is a potential antimicrobial agent. It drastically reduces the amount of iron present in the surrounding environment; this condition will hinder the development of the bacteria cell (Aneja et al., 2014).

12.2.5.2.1 Propolis

Propolis is a natural preservative obtained from bees and it is mainly composed of polyphenols, quinones, coumarins, steroids, amino acids, and inorganic compounds. The percentage of these compounds varies with the botanical nature of the plants from which the vegetable secretions are extracted and the phytogeographical origin (Bankova et al., 2000).

12.2.5.2.2 Chitosan

Chitosan is known for its protective role in the food industry and is obtained from crustacean shells. It is an important biopolymer that has been reported to have a potential use as an additive in the food industry because of its strong antimicrobial properties against a range of foodborne microorganisms (Tharanathan & Kittur, 2003). Apart from that it is reported and proven that chitosan conjugates with other polymeric compounds improves the functionality by forming xylan-chitosan, chitosan-mint (polyphenols), and chitosan-proteins, chitosan-organic acid, which potentially serve as food preservatives (Raafat & Sahl, 2009). It is used with edible coatings and films, which help to reduce transpiration, prevent oxygen transmission, and extend the shelf life of fruits, thereby preventing deterioration of food (Ahmed et al., 2013).

12.2.5.2.3 Nisin

Nisin is an antimicrobial polypeptide produced by some strains of *Lactococcus lactis*. Nisin-like substances are widely produced by LAB. Unlike other bacteriocins, nisin can be used as a broad-spectrum antimicrobial agent (Bonev et al., 2000; Brötz & Sahl, 2000). These bacteriocins form pores in the cytoplasmic membrane of bacteria resulting in depletion of proton motive force and loss of cellular ions, amino acids, and adenosine triphosphate (ATP). Nisin can resist heat and is more effective in diluted acid conditions. Due to its resistance to heat and stability in boiling conditions, nisin is extensively used in canned fruit and vegetable products. Nisin is also used in the preservation of hard cheese and limitedly in meat and meat products. It is very effective against gram-positive bacteria, especially *Clostridium botulinum* and *L. monocytogenes*, which cause spoilage of canned products and hard cheese. The Food Agricultural Organization (FAO)/World Health Organization (WHO) in 1969 permitted utilization of nisin as a food preservative with a daily intake of up to 33,000 units per kg of body weight. The U.S. Food and Drug Administration reported that nisin is "generally recognized as safe" for use as a food preservative (Federal Register: 53 FR 11247, 1988).

Nisin (a 34 amino acid) is a long cation peptide that act by forming pores on the bacterial cell membrane (Class I pore-forming bacteriocin) (Breukink et al., 1999). Nisin comes in contact with bacterial cells and forms a complex with a lipid precursor molecule called "lipid complex II" on the cell wall membrane. This peptide complex lies parallel to the membrane plane initially, but it changes to transmembrane orientation by insertion of the C-terminus, which eventually leads to stable pore formation (Wiedemann et al., 2004). The pore formation is believed to cause rapid degeneracy of transmembrane electrostatic potential resulting in membrane leakage and rapid bacterial cell death. Nisin is also known to interrupt peptidoglycan production leading to the inhibition of cell wall biosynthesis (Wiedemann et al., 2001).

Apart from the natural preservatives discussed earlier, the Class I preservatives like sugar, salt, acids, etc., also belong to the category of natural preservatives as this class does not pose any potential threats to human health and is self-limiting in nature. The classification of this category broadly includes naturally occurring substances, and there is no maximum limit specified under law for their use.

12.2.5.3 Salt

Sodium chloride, commonly known as salt (although sea salt also contains other chemical salts), is an ionic compound with the chemical formula NaCl, representing a 1:1 ratio of sodium and chloride ions. The salt finds its application in food preservation from ancient times and the major addition of sodium to foods was as salt, which acted to prevent spoilage. Prior to ultramodern methods of preservation, salt was one of the best methods for inhibiting the growth and survival of undesirable microorganisms. Although modern-day advances in food preservation have largely replaced other natural methods, salt does remain in widespread use for preventing rapid spoilage by creating an uncongenial environment for pathogens, and promoting the growth of desirable microorganisms in various fermented foods and other products (García-García & Searle, 2016). The hyperosmotic environment created by salt inhibits the growth of microbes such as bacteria, yeast, and molds and thereby extends the shelf life of commodities without posing any harmful side effects. The organisms are not able to survive in brine solution as it causes dehydration of living cells rendering it inhospitable for their survival.

12.2.5.4 Sugar

Similar to salts, sugar also finds its way into food preservation because humans started preserving food for longer periods. The mode of action involves the absorption of excess water, thereby preventing microbial growth. The water activity of the food is reduced to such a level that it does not support any microbial growth.

12.2.5.5 Acids

In the food system, acids like citric acid, malic acid, tartaric acid, lactic acid, sorbic acid, and vinegar are present naturally. These acids have an inherent ability to inhibit the growth of bacterial and fungal cells and inhibit the outgrowth of bacterial spores.

12.2.5.6 Mode of Action of Natural Preservatives

The exact mechanism behind natural preservatives is still hazy. Researchers worldwide rely on the belief that the antimicrobial mechanism of natural preservatives is related to the attack of the cell membrane phospholipid bilayer, hence, the rupture of the enzymatic systems, which leads to compromising the genetic material of the microorganisms. This consequently leads to the formation of hydroperoxidase of fatty acids by the oxygenation of unsaturated fatty acids and coagulation of the cytoplasm, which cause damage to lipids and proteins, and distortion of the proton motive force (FMP), electron flow, and/or active transport (Viuda-Martos et al., 2010). They can also inhibit the activity of protective enzymes and, consequently, block one or more biochemical pathways. Phenolic compounds, when added to foods, act as reducing agents, donating hydrogen and oxygen suppressants, causing an antioxidant effect on the products. Some phenolic compounds also have the ability to chelate metal ions that act as catalysts in oxidation reactions. Flavonoids are natural polyhydroxylated aromatic compounds that are widely distributed in plants (fruits, vegetables, spices, and herbs). Flavonoids have the ability to eliminate free radicals, including hydroxyl, peroxyl and superoxide radicals, and can form complexes with catalytic metal ions, making them inactive. It has also been found that flavonoids can inhibit lipoxygenase and cyclooxygenase enzymes, which are responsible for the development of oxidative rancidity in foods (Karre et al., 2013; Krishnan et al., 2014; Embuscado, 2015). The antioxidant activity might be related to the quantity and quality of phenolic compounds present in the constitution of natural products; therefore, the content of polyphenolic compounds could be used as an indicator of the antioxidant capacity (Miguel, 2010).

Natural preservatives have strong potential to target the various deteriorative reactions and pathogens in the food matrix, hence, they can be inevitably utilized as alternatives to synthetic food preservatives.

12.2.6 Artificial Preservatives

This class of preservative includes all the chemical, semi-synthetic, or synthetic compounds such as benzoates, sorbates, sulfites, lactates, nitrites and nitrates, parabens, etc. Artificial preservatives have a maximum limit of application beyond which they should not be added in food products. Excess use of these preservatives might result in adverse health effects, which is ascribed to their chemical nature. Various artificial preservatives have been extensively studied for their suitability in food applications as detailed in Table 12.2. The understanding of action mechanism of synthetic preservatives, their efficiency in extending shelf life of food, and their effect on human health is imperative in selection of preservatives for specific food application, these factors are discussed in the following sections.

12.2.6.1 Sulfur Dioxide (SO₂) and Sulfites

Sulfur dioxide (SO_2) (E220) and sulfites of sodium, potassium, and calcium (E221 to E224 and E226 to E228) are widely employed as antimicrobial and antioxidant agents, and inhibitors of enzymatic activity and the Maillard reaction. Their activity is pH dependent and these compounds exhibit higher activity at lower pH attributed to the undissociated sulfurous acid (H_2SO_3), predominating at pH < 3. SO_2 is a colorless, nonflammable gas with a strong and suffocating odor, whereas sulfides are available as white colored powder, crystals, or granules with a slight smell of SO_2. These preservatives demonstrate inhibitory activity against yeasts, molds, and bacteria, especially gram-negative bacteria. Food applications of sulfites and SO_2 include fruits; vegetables; pulp; jams; jellies; glucose syrups; pie fillings; dry biscuits; starches; juices and drinks containing fruit juice; some types of beer, wine, and other beverages; meat products; salted and dried fish; and various substitutes for meat, fish, or crustaceans (Voss, 2002). With the high reactivity, the sulfites are very susceptible to easy degradation, which can lead to oxidation into sulfates with no preservative effect.

12.2.6.1.1 Mode of Action

The antimicrobial activity of sulfites is not a result of a single cellular target, but it is reported to be associated with a combined effect on several metabolic systems. The ability of reduction or nucleophilic attack is responsible for the chemical reactivity of sulfites and other associated compounds. The sulfites act as antimicrobial agents and inhibit the enzymes through three mechanisms. First, the preservative modifies the reactive groups of proteins such as carbonyls and inhibits the enzyme activity, although documented evidence does not support the importance of this enzyme inhibition in the antimicrobial potential of sulfites. Moreover, the enzyme inactivation by sulfites could be strongly explained by sulfite-mediated inactivation of oxidized glutathione. This further leads to oxidation of protein thiols, reducing enzyme activity. Second, the mechanism involves the reaction of sulfite with the substrate and/or product of the reaction, having sensitive groups like carbonyl, and thus inhibiting the enzymes. The third mechanism describes the inactivation of enzyme cofactors, viz. thiamine pyrophosphate, pyridoxal phosphate, folate, heme, and flavin, affecting the intermediary metabolism in microorganisms.

When sulfites are used in the approved concentrations they get easily eliminated in the urine as sulfates. Conversely, SO_2 is reported to destroy thiamine leading to loss of vitamin B_1. Allergic reactions (viz. headaches, stomach, or skin irritation; eczema; nausea; diarrhea; and asthma) are also reported in sensitive individuals, although these are not supported by strong evidence (Voss, 2002; Inetianbor et al., 2015). Various researchers have also discussed the mutagenic effect of sulfites in humans, and their role in developing cancer has also discussed (Voss, 2002; Silva & Lidon, 2016).

SO_2, sodium sulfite, sodium bisulfate, sodium metabisulfite, potassium metabisulfite, calcium sulfite, calcium bisulfate, and potassium bisulfate are approved for food applications. Sulfite concentration is depicted in terms of SO_2 equivalents, as SO_2 is released from sulfite salts, and the final concentration in foods depends on the permitted limit for specific products. SO_2 is employed in concentrations ranging from 10 ppm in wines to a maximum of 2000 ppm in dried fruits. The GRAS status has been revoked in the United States for certain food applications of sulfites, mostly in fresh products. Similar regulations are also established in the EU, and to lower the daily intake, the approved limit employed in different foods has been reformed in the past decades. Joint FAO/WHO Expert Committee on Food Additives (JECFA) has allocated an acceptable daily intake (ADI) of 0–0.7 mg/kg body weight to sulfur dioxide and to sulfur dioxide equivalents arising from sodium and potassium metabisulfite, sodium sulfite, and sodium hydrogen sulfite (García-García & Searle, 2016).

12.2.6.2 Organic Acids and Acidulants

Organic acids and their salts have been used extensively in food preservation for the past two decades. Among the organic acids, weak acids such as benzoic acid, propionic acid, and sorbic acids are used extensively in the preservation of various foods such as jams, soups, ketchup, bakery products, etc. These acids are weak due to their inability to completely ionize, and both the undissociated and dissociated forms are present (Msagati, 2013). Weak acids are capable of inhibiting spoilage causing organisms like bacteria and fungi.

12.2.6.2.1 Mode of Action

The change in pH (decrease in pH) in the cytosol of the cell is detrimental to microbial cells. The decrease in pH of the cytosol hinders all the cellular metabolic activity as reported for benzoic acid (Krebs et al., 1983), and the combination of sorbic and benzoic acid (Plumridge, 2005). In the cytoplasm, the pH values are near neutral and when lipophilic weak acids are added they diffuse into the cytoplasm and dissociate, thereby effectively lowering the cytoplasm internal pH. The change in internal pH will cause metabolic alterations, protein misfolding, and enzymatic inhibition. The low internal pH also neutralizes the electrochemical gradient across the plasma membrane, which is responsible for the active transport of many essential nutrients required for metabolic activity (Freese et al., 1973; Thevelein, 1994).

12.2.6.2.2 Benzoic Acid

The sodium salt of benzoic acid (sodium benzoate) is the most common chemical food preservative. Sodium benzoate is extensively used in acidic/acidified foods such as fruit juices, jams, jellies, syrups, fruit cocktails, pickles, preserves, fish sauce, soft drinks, margarine, tomato sauce, etc. Benzoic acid has the most inhibiting action against yeast and to a lesser extent against molds and bacteria. The inhibition action is higher when the pH is lower (most effective at pH 4 at a concentration of 2–6 mM) and becomes ineffective as the pH approaches neutral (Booth & Stratford, 2003). The use of benzoates leads to the development of an aftertaste astringency, which restricts their application. Thus, they often are used in combination with sorbate or parabens to reduce the astringent aftertaste. Benzoate inhibits the uptake of cellular substrate molecules by microorganisms, thereby inhibiting microbial growth. Studies have also suggested its efficiency against bacterial spores where the germination stage of the endospore is found to be the most sensitive. Due to the benzoate detoxifying mechanism, in humans the tolerance to sodium benzoate is high as the benzoate and glycine or glucuronic acid are conjugated and excreted as hippuric acid or benzoyl glucuronide. The maximum permissible level in foods is 0.1%. The mode of action is by inhibiting the growth of microorganisms by inhibiting the action of enzymes that control the acetic acid metabolism and oxidative phosphorylation and various stages in the tricarboxylic acid cycle (Stratford & Eklund, 2003). It also alters the membrane permeability of the microbial cell. The studies have revealed the effect of benzoic acid on health and revealed that it may aggravate asthma and could be a neurotoxin and carcinogen. The preservative may also cause fetal abnormalities and may worsens hyperactivity (Amit et al., 2017)

12.2.6.2.3 Sorbic Acid and Its Salts

Sorbic acid or sorbates are generally used in the form of calcium, sodium, and potassium salts. The advantage of using sorbic acid or its salts is because they are tasteless and odorless when added to food. The effectiveness of sorbates is mostly seen in acid foods with pH below 5.5 and they become ineffective at pH above 6.5 (Booth & Stratford, 2003). Molds and yeasts are effectively inhibited by the use of sorbates in food such as fruit juices, syrups, pickles, jams and jellies, preserves, cheese products, baked foods, fruits, vegetables, soft drinks, wines, salads, margarine, fish products, high moisture dehydrated fruits, etc. The use of sorbate is limited in the bakery industry because it inhibits the growth of yeasts. Sorbate may be added directly to the food or it may be applied by spraying, dusting, dipping, or impregnating wrappers and packing materials. Recent studies showed the effectiveness of sorbate as an anti-botulinal agent in meat products. The maximum permissible levels allowed in food are 0.03–0.2%, and the daily highest acceptable intake through food set by the FAO/WHO is 25 mg/kg body weight, the highest ADI of the common preservatives. In the United States, sorbic acid and sorbates are considered GRAS. The salt of potassium in the form of potassium sorbate is more widely employed than that of sodium and calcium salts because of its white fluffy powder, high water solubility (over 50%), stability, and ease of preparation. The sorbates are more effective on catalase-positive cocci and aerobes than catalase-negative forms and anaerobes. It also prevents the growth of endospores by acting on the vegetative cells germinating from it. The mode of action is by the effect of the proton motive force of microbial cells as the sorbates act on the cytoplasmic membrane of microorganisms and separate H+ ions and hydroxyl ions. Due to this, the intracellular pH lowers, leading to weakening of the transmembrane gradient required for transport of amino acids into the cell, thus adversely affecting membrane transport and causing subsequent cell death (Stratford & Eklund, 2003). Sorbic acid reacts with the low-molecular-weight thiols of food, such as glutathione and cysteine, to form the 5-substituted 3-hexenoic acid. Auto-oxidation of sorbic acid leads to the formation of malonaldehyde, acetaldehyde, and β-carboxyacrolein (Surekha & Reddy, 2014).

12.2.6.2.4 Use of Other Organic Acids

Propionic acid and its salts (Na^{2+}, Ca^{2+}, and K^+ salts) are highly effective against mold, whereas they are less effective against yeasts and most bacteria. In bread, propionic acid inhibits the rope-forming bacteria as a specific target. Propionic acid and its salts have high water solubility and are readily absorbed by the digestive tract. This acid degrades in mammals by linkage with coenzyme-A via methylmalonyl-CoA, succinyl-CoA, and succinate to produce CO_2 and H_2O (Volpe et al., 2017).

Lactic acid is the main product of many food fermentations, which is formed by microbial degradation of sugars in products such as sauerkraut and pickles. Lactic acid is also used as an emulsifier in bakery products and as a flavoring agent in frozen desserts. Lactic acids have only moderate antimicrobial activity and can inhibit the formation of aflatoxin and sterigmatocystin (Surekha & Reddy, 2014). It is also used for faster nitrite depletion and botulinal protection due to lowered pH in meat product processing and inhibiting the growth of putrefactive anaerobes and butyric-acid-producing bacteria.

Formic acid has found its application in livestock feed. It acts as an antibacterial agent when sprayed on fresh hay or other cattle feed by arresting certain decay processes, thereby retaining its nutritive value. Formic acid is also added to feed in the poultry industry to kill *Salmonella* bacteria (Ricke et al., 2020). Application of formic acid directly into food is limited due to its corrosive, irritant, and strong formic flavor development, which makes food unpalatable.

The U.S. Food and Drug Administration (FDA) lists propionic acid and the Na^{2+}, Ca^{2+}, and K^+ salts as preservatives in their summary of GRAS additives, and no upper limits are imposed except on bread, rolls, and cheeses (0.30–0.38%). There is no ADI set by the FAO/WHO for lactic, citric, and acetic acid. These preservatives are permitted for use in many countries.

The antimicrobial action of propionic acid is due to inhibition of growth and nutrient transport by competing substances like alanine and other amino acids required by microorganisms. The mode of action of lactic acid produced in fermentations decreases the pH to levels unfavorable for the growth of spoilage organisms. Yeasts and molds that can grow at such pH levels can be controlled by the inclusion of other preservatives such as sorbate and benzoate. The antimicrobial action of formic acid is similar to any acidulant. Additionally, formic acid inhibits decarboxylase and heme enzymes, especially catalase. The antimicrobial effect of other acids (e.g., tartaric, phosphoric, and succinic) is due to acidification of the microbial cell and inhibiting nutrient transport (Stratford & Eklund, 2003).

12.2.6.3 Nitrates or Nitrites and Their Salts

Sodium nitrite and sodium nitrate for centuries are extensively used in the preservation of meat products such as bacon, hams, hot dogs, bologna, and sausages. They are used not only to inhibit microbial spoilage and food poisoning organisms but also in the curing process of meat because they help in stabilizing the pink color of meat and flavor development. Nitrites prevent the growth of *C. botulinum,* which causes food poisoning and spoilage with microorganisms that grow in anaerobic conditions and in the interior tissues of meat that has been vacuum packed (Lee et al., 2018). The primary objective of adding nitrites is to prevent food poisoning due to *Clostridium* rather than for color and flavor development. The mode of stabilization of pink color by nitrites in meat is by acting as both oxidizing and reducing agent. Nitrites are highly reactive and in acidic conditions they ionize to nitrous acid, which further breaks down into nitric oxide. The nitric oxide formed reacts with the myoglobin (heme protein) under reduced conditions to produce the desired red-pinkish pigment called nitroso-myoglobin and creates a mild antioxidant effect that prevents rancidity and a warmed-over flavor. The effect of nitrite will be synergistic by the addition of acids, acidulant, or glucono delta-lactone. Using nitrite at 50 µg/g helps fix color and flavor development, whereas a concentration of 100 µg/g is required to cause an antimicrobial effect (Govari & Pexara, 2015). So, using lower concentrations of nitrite (40–80 µg/g) in combination with sorbate is more effective for proper color and flavor development as well as antimicrobial action. The mode of antimicrobial action is caused by the inhibition of vegetative cells and spore growth, which results from the heating process or smoking of meat and fish products containing nitrites (Fellows, 2009). The inhibition of iron-sulfur enzymes present in the microorganism leads to the prevention of ATP synthesis from pyruvate. The maximum permitted levels of sodium nitrite (United States and UK) and potassium nitrite (UK) for curing meat is 200 µg/g.

Different salts of nitrates have been widely studied for their toxic health effects. Results reported that potassium nitrate (E249) may lower the oxygen carrying capacity of blood, it may form carcinogenic nitrosamines with other substances, and consumption of high amounts of sodium nitrite and nitrate may cause type 1 diabetes. Further, these effects may lead to irritation of the whole digestive system including the mouth, esophagus, and stomach. Pregnant women and children are more prone to these effects (Amit et al., 2017).

12.2.6.4 Parabens

Parabens are phenol esters used as food preservatives in the form of esterified 4-hydroxybenzoic acid. This group includes methyl paraben (p-hydroxybenzoic acid methyl ester), ethyl paraben (p-hydroxybenzoic acid ethyl ester), propyl paraben (p-hydroxybenzoic acid propyl ester), butyl paraben (p-hydroxybenzoic acid butyl ester), and salicylic acid, among which the methyl and propyl parabens are the most widely used. Naturally occurring parabens in foods include methyl paraben in yellow passion fruit juice, cloudberry, botrytised wine, white wine, and Bourbon vanilla, and propyl paraben in the aerial part of the plant *Stocksia brahuica* (Soni et al., 2005). The pure form of parabens is available as hygroscopic small crystals or crystalline powders and devoid of any odor, taste, and color. Parabens are stable against oxidation and resistant to water and acid hydrolysis. The resistance to hydrolysis is a function of alkyl chain length, where these two are directly correlated; however, substantial hydrolysis occurs at pH above 7. Further, the chain length is associated with the antimicrobial potential, where long-chained compounds possess higher inhibitory activity against microbes. Conversely, lower water solubility is exhibited by these compounds, thus the short-chained methyl and propyl parabens are found suitable for food applications.

Parabens possess multiple bioactivities including antimicrobial potential and can be used in combination with other compounds to achieve an optimal antimicrobial effect (Soni et al., 2005). Parabens are considered to be the ideal preservatives in food because of their unique properties including broad-spectrum antimicrobial activity, stability over a wider pH range (up to a pH of 8.0), sufficient solubility in water for the required effective concentration, heat stability at high temperature of sterilization, biodegradability, and very low toxicity (Maddox, 1982, Msagati, 2013). The activity of parabens is independent of pH, although some studies have reported an increase in antimicrobial activity at low pH levels (Lück & von Rymon Lipinski, 2000).

The methyl and propyl parabens are GRAS as food preservatives as per the FDA regulations with a permissible limit of 0.1%. The use of these compounds in food preservation has steadily increased over the years and are successfully employed in processed vegetables, fruit juices, jams, jellies, preserves, soft drinks, sauces, pickles, fats and oils, seasonings, bakery goods, coffee extracts, sugar substitutes, and frozen dairy products. Generally a combination of methyl and propyl parabens is employed for high antimicrobial efficiency (Soni et al., 2005).

12.2.6.4.1 Mode of Action

Parabens are very efficient in inhibiting the growth of yeasts and molds at lower concentrations when compared with other weak-acid preservatives; however, the activity against bacteria is very low, specifically for those that are gram-negative. The parabens remain undissociated in food under the effect of esterification, and the intact protonated phenol group helps in attaining a pKa value of 8.5.

The broad-spectrum antimicrobial activity manifested by parabens is due to their ability to disrupt cytoplasmic membrane, affecting its permeability, mitochondrial functioning, and enzyme activity. The mechanism includes dissolving parabens in or through the cell membrane and their interaction with membrane proteins and disordering of membrane lipids (García-García & Searle, 2016). Additionally, parabens are reported to inhibit RNA and DNA in bacterial cells and protein synthesis in cell-free systems (Nes & Eklund, 1983). The microbial inhibitory activity of different parabens has been studied widely, reporting that there is no difference in the activity once the compound has reached the target and explained the variability with their ability to reach the target. Also, the difference in lipid solubility of different esters and the presence of outer barrier membranes in case of gram-negative bacteria presents variability in activities (Russell et al., 1985; Lück & von Rymon Lipinski, 2000; García-García & Searle, 2016).

Parabens show rapid metabolization with no accumulation in the body, and they are relatively non-sensitizing and non-irritating with low toxicity. Recent studies have suggested the effect of propyl paraben on estrogenic potential; however, the risk is not evident at current levels of use (Lück & von Rymon Lipinski, 2000). The parabens are often used along with the methyl isothiazoline and methylchloroisothiazolinones, which are potent irritants and allergens, and may cause neurological damage in rats. Intake of these chemicals during pregnancy may adversely affect the brain development in the fetus (Kumari et al., 2019).

The U.S. FDA has given GRAS status to methyl and propyl parabens and allowed their direct addition to food (21 CFR 184.1490) at maximum concentration of 0.1% and for indirect addition via packaging materials (21 CFR 181.23) by prior sanction with no limit or restriction. Methyl, propyl, and butyl parabens are permitted for use as preservatives in synthetic flavoring substances and adjuvants (21 CFR 172.515) in a maximum concentration of 20 ppm. The methyl and propyl parabens are also approved by JECFA and the Flavor and Flavor and Extract Manufacturers Association (FEMA). As per the recommendation of JECFA and EFSA the group ADI for methyl, ethyl, and propyl parabens and their salts is 0–10 mg/kg body weight/day (JECFA, 1974).

Parabens have found major applications in the food products like creams, pastes, olives, and pickles (0.1% of a combination of parabens); syrups (0.07%); cakes, pastries, icings, pie crusts, toppings, and fillings (0.03–0.06% of a 3:1 methyl and propyl parabens); jams, jellies, and preserves (0.07% of a combination of 2:1 of methyl and propyl parabens); soft drinks (0.03–0.05% of 2:1 ratio of methyl and propyl parabens); and fermented malt beverages (heptyl paraben in a maximum concentration of 12 ppm) (Soni et al., 2005; García-García & Searle, 2016).

12.2.6.5 Synthetic Phenolic Compounds

Various synthetically derived polyphenols are known to impart preservative action in different food products. The most commonly used ones are preservative agents BHA, BHT, TBHQ, and caffeine. BHA, BHT, and TBHQ are approved food antioxidants for their application in oils, margarines, and other food products containing high amounts of fat in a concentration of 100–200 ppm. BHA and BHT are generally recommended for fats, and TBHQ has wider application in vegetable oils. Caffeine is a trimethylated xanthine alkaloid derivative, a psychoactive agent, and a flavoring agent in many beverages. It is naturally found in many plant species and can be synthesized chemically. The antimicrobial action of caffeine is not widely investigated and the activity is perhaps associated with the DNA damage of the microbial cells (Ibrahim et al., 2006; García-García, & Searle, 2016).

12.2.6.6 Ethylene and Propylene Oxides

Ethylene and propylene oxides are the gaseous form of preservatives and have antimicrobial potential. Propylene oxide is reported to be less effective compared with the ethylene oxide. Both compounds act as an alkylating agent and are utilized for fumigation of cereals, spices, eggs, dried fruits, dried yeast, etc. It is the hydroxyl ethyl group that blocks the reactive groups of microbial proteins and inhibits their growth. Ethylene oxide is applied as a gaseous sterilant for flexible and semirigid containers in a concentration of 500–700 mg/L. Propylene oxide has found its permitted application in fumigation of packaging for dried fruits, cocoa, gums, spices, and starch. These gaseous preservatives have revealed their inhibitory activities against *C. botulinum*, *Clostridium sporogenes*, *Bacillus coagulans*, *Bacillus stearothermophilus*, and *Deinococcus radiodurans*.

12.2.6.7 Alcohol

Food-grade alcohols such as ethanol are employed for the preservation of fruits due to its antiseptic properties. Fruits are preserved in tightly packed large jars containing the flavorless spirit. Flavoring extracts such as vanilla and lemon are preserved using this technique. Ethanol acts as a coagulant and denatures the microbial cell proteins, thus revealing antimicrobial potential. It has potent germicidal activity at concentrations between 70 and 95%. Preservation in alcohol adversely affects the organoleptic qualities of fruits by dissolving sugar and flavors, therefore, it is recommended to add sugar to the preserve.

12.2.6.8 Packaging Gases

Packaging gases have also been utilized for the preservation of packed foods where these gases protect the foods from oxidative changes by excluding oxygen from the pack. In addition, they restrict the growth of aerobic microorganisms and partially prevent the food from microbial spoilage. Commercially used packaging gases for foods include nitrogen, carbon dioxide, and mixtures of these. At higher concentrations carbon dioxide alone possesses inhibitory activity against some microbes, especially molds. Different combinations of these gases are widely employed for modified atmospheric packaging, controlled atmospheric packaging, and for storage of foods under a controlled gaseous environment (Lück & von Rymon Lipinski, 2000).

12.3 SAFETY CONCERNS OF PRESERVATIVES

To meet the growing needs of food for the ever-growing population it is important to utilize the resources sustainably. The rough quantification of the data elucidates that one-third of edible food is lost or wasted globally. This loss could potentially feed millions of starving people in third-world countries. There are various factors pertaining to the losses; spoilage is one among many important factors that contribute to the loss. Preservatives have become a savior when it comes to preventing food spoilage and they are a real boon when it comes to extending the shelf life of food.

Preservatives are administered in the food matrix with the intent to prevent, suppress, or kill the spoilage-causing pathogenic microbes, enzyme activities, and other physical attributes, thereby enhancing shelf life without compromising the organoleptic properties of food and maintaining high nutritional quality. With the advent of globalization, food industries grew rapidly, making exotic foods accessible to global consumers. The use of preservatives increased with the advancement of industry as in transit it would take more time and the chances of spoilage were higher. As Newton's third law of motion states, every action has an equal and opposite reaction; the savior "preservatives" also caused an adverse effect on health despite their role in increasing the shelf life of food and reducing losses.

As described earlier, the effect of food preservatives on human health may be immediate or occur in the long run with frequent consumption, which leads to accumulation. Quick response to the consumption of these preservatives includes headache, dizziness, energy exhaustion, alteration in mental concentration and behavior, vomiting, and lowered immune responses (Zengin et al., 2011; Mellado-García et al., 2017). Even if most preservatives like sulfites, benzoates, sorbites, and nitrites are considered GRAS, there is concern about their long-term usage and effects (Cao et al., 2020). The study conducted by Hrncirova et al. (2019) provided an insight on the effect of preservatives like sodium benzoate, sodium nitrite, and potassium sorbate on gut microbiota. The continuous and long-term usage of these preservatives led to the reduction of intestinal microbial diversity, a decrease of Clostridiales, and the increase of Proteobacteria.

According to the studies conducted by Pandey and Upadhyay (2012) the risk of cancer and cardiovascular and other degenerative conditions are higher in the long run. Some modern synthetic preservatives have become controversial because they have been shown to cause respiratory or other health problems (Pandey and Upadhyay, 2012).

Boric acid is a common preservative used for preserving meats, caviar, and dairy products (Arslan et al., 2008) and is reported to be toxic to cells (Yiu et al., 2008). Boric acid is harmful to human health if consumed in higher amounts (See et al., 2010), and it was reported by several workers that boric acid suppressed sperm release from the testes as it inhibits DNA synthesis in sperm cells and, hence, impairs fertility (Cox, 2004).

Similarly, nitrites and nitrates, which are commonly used for meat preservation, are reported to have adverse health effects. Higher levels of nitrates or nitrites have been associated with increased incidences of cancer in adults, and possible increased incidences of brain tumors, leukemia, and nasopharyngeal (nose and throat) tumors in children (Volkmer et al., 2005). Reduced oxygenation of hemoglobin (methemoglobinemia) has been reported after exposure to nitrate and nitrite contaminated drinking water. This also called "blue baby syndrome" because of the cyanotic (oxygen-deficient) symptoms that result from reduced oxygenation of the blood. Other health effects following fetal exposure to elevated levels of nitrates in drinking water included intrauterine growth retardation (Bukowski et al., 2001), increased incidence of sudden infant death syndrome (SIDS) (George et al., 2001), cardiac defects (Cedergren et al., 2002), and increased risk of nervous system defects (Croen et al., 2001).

Synthetic antioxidant preservatives such as BHA and BHT, which are utilized commercially in the food sector, are reported to cause hyperactivity in kids. They also have been reported to cause lung, liver, and kidney damage, and, topping it all off, there is a potential role of BHA and BHT in the conversion of benign tumors to malignant ones (or its role as cancer inducer) (Tran, 2013). Williams et al. (1999) found that BHA and BHT can be carcinogenic at high doses, and a concentration greater than 3000 ppm has been known to induce forestomach squamous cell carcinomas in rodents; whereas BHT at 250 mg/kg/day increases spontaneous neoplasms and tumor-promoting activity (Williams et al., 1999). These examples clearly show the ill effects of the use of synthetic food preservatives. The impact of these chemicals vary, as discussed earlier, but if we shift the scenario of chemical preservatives to natural ones, the health benefits associated with food improve drastically along with the action of natural preservatives.

In short, one of the most prominent side effects of sustained use of preservatives is breathing problems, which may arise in forms like asthma, bronchitis, etc. The hyperactivity and weak heart issues in children caused by the sustained use of preservatives in food commodities like jam, jelly, chocolates, and other processed food tops it all. Processed food contains BHA and BHT as food additives, which could cause cancer. BHT is used in cereals and fats, whereas BHA could be present in meats and other baked goods. Preservatives could cause obesity in some as they contain fatty acids, especially in processed foods. Globalization brings the world closer in all aspects including the availability of any kind of food anywhere in the world, so preservation is a must, and avoiding the use of chemical preservatives is not the way out. Thus, for sustainability, advocating the use of natural preservatives, which not only benefits the consumer's health but also increases shelf life and also consumption of fresh fruits and vegetables, could possibly be a good way to decrease the consumption preservative load in processed food.

12.4 ANALYTICAL METHODS FOR PRESERVATIVE RESIDUES

The potential threat to health associated with the application of preservatives in foods requires sophisticated analytical methods for determining the residual effect. The quantitative and qualitative estimation and characterization of different compounds can be successfully done using UV-Visible spectrophotometry, chromatography involving high-performance liquid chromatography (HPLC), gas chromatography (GC), liquid chromatography-mass spectrometry (LCMS), mixed micellar electrokinetic chromatography, and electrophoresis. These analytical methods are validated with respect to limit of detection (LOD), limit of quantitation (LOQ), selectivity, linearity, accuracy, and repeatability (Tfouni and Toledo 2002; García et al., 2003; Juneja et al., 2012; Msagati, 2013; Dwivedi et al., 2017).

12.5 CONCLUSION

As the food industry around the globe is constantly growing, so is the use of food preservatives. Application of natural and artificial (synthetically produced) preservatives has been found to have advantages and disadvantages. This chapter provides an insight into food industries, start-ups, and consumers on the application, microbial stability, and shelf life of products with added preservatives. There is also substantive information on the mode of action of these food preservatives for researchers. Advancements in science and public awareness have emphasized research activity on new, alternative chemicals that have low or no toxicity to humans, but at the same time provide higher antimicrobial stability when used in food. Nanotechnology can be utilized for the development of food preservatives, but careful consideration and research has to follow-up their effect on human health before rolling preservatives out for public usage.

REFERENCES

Abdulmumeen, H. A., Risikat, A. N., & Sururah, A. R. (2012). Food: Its preservatives, additives and applications. *International Journal of Chemical and Biochemical Sciences*, *1*(2012), 36–47.

Adelakun, O. E., Oyelade, O. J., & Olanipekun, B. F. (2016). Use of essential oils in food preservation. In *Essential oils in food preservation, flavor and safety* (pp. 71–84). Academic Press.

Ahmadi, F., Lee, Y. H., Lee, W. H., Oh, Y. K., Park, K., & Kwak, W. S. (2019). Long-term anaerobic conserva-tion of fruit and vegetable discards without or with moisture adjustment after aerobic preservation with sodium metabisulfite. *Waste Management*, *87*, 258–267.

Ahmed, N., Singh, J., Kour, H., & Gupta, P. (2013). Naturally occurring preservatives in food and their role in food preservation. *International Journal of Pharmaceutical and Biological Archives*, *4*(1), 22–30.

Alizadeh, L., Abdolmaleki, K., Nayebzadeh, K., & Shahin, R. (2019). Effects of tocopherol, rosemary essen-tial oil and *Ferulago angulata* extract on oxidative stability of mayonnaise during its shelf life: a com-parative study. *Food Chemistry*, *285*, 46–52.

Amit, S. K., Uddin, M. M., Rahman, R., Islam, S. R., & Khan, M. S. (2017). A review on mechanisms and commercial aspects of food preservation and processing. *Agriculture & Food Security*, *6*(1), 1–22.

Anbazhagan, V., Kalaiselvan, A., Jaccob, M., Venuvanalingam, P., & Renganathan, R. (2008). Investigations on the fluorescence quenching of 2, 3-diazabicyclo [2.2. 2] oct-2-ene by certain flavonoids. *Journal of Photochemistry and Photobiology B: Biology*, *91*(2–3), 143–150.

Aneja, K. R., Dhiman, R., Aggarwal, N. K., & Aneja, A. (2014). Emerging preservation techniques for con-trolling spoilage and pathogenic microorganisms in fruit juices. *International Journal of Microbiology*, *2014*, 758942.

Arias-Rios, E. V., Cabrera-Díaz, E., Márquez-González, M., & Castillo, A. (2017). Natural food antimicrobi-als of animal origin. In *Microbial control and food preservation* (pp. 55–83). Springer.

Arslan, M., Topaktas, M., & Rencuzogullari, E. (2008). The effects of boric acid on sister chromatid exchanges and chromosome aberrations in cultured human lymphocytes. *Cytotechnology*, *56*(2), 91–96.

Ayaad, T. H., Shaker, G. H., & Almuhnaa, A. M. (2012). Isolation of antimicrobial peptides from *Apis florae* and *Apis carnica* in Saudi Arabia and investigation of the antimicrobial properties of natural honey samples. *Journal of King Saud University-Science*, *24*(2), 193–200.

Bankova, V. S., de Castro, S. L., & Marcucci, M. C. (2000). Propolis: recent advances in chemistry and plant origin. *Apidologie*, *31*(1), 3–15.

Belz, M. C., Mairinger, R., Zannini, E., Ryan, L. A., Cashman, K. D., & Arendt, E. K. (2012). The effect of sour-dough and calcium propionate on the microbial shelf-life of salt reduced bread. *Applied Microbiology and Biotechnology*, *96*(2), 493–501.

Benjemaa, M., Neves, M. A., Falleh, H., Isoda, H., Ksouri, R., & Nakajima, M. (2018). Nanoencapsulation of *Thymus capitatus* essential oil: formulation process, physical stability characterization and antibacterial efficiency monitoring. *Industrial Crops and Products*, *113*, 414–421.

Bonev, B. B., Chan, W. C., Bycroft, B. W., Roberts, G. C., & Watts, A. (2000). Interaction of the lantibiotic nisin with mixed lipid bilayers: a 31P and 2H NMR study. *Biochemistry*, *39*(37), 11425–11433.

Booth, I. R., & Stratford, M. (2003). Acidulants and low pH. In *Food preservatives* (pp. 25–47). Springer.

Brandenburg, K., Heinbockel, L., Correa, W., & Lohner, K. (2016). Peptides with dual mode of action: Killing bacteria and preventing endotoxin-induced sepsis. *Biochimica et Biophysica Acta (BBA)-Biomembranes*, *1858*(5), 971–979.

Breukink, E., Wiedemann, I., Van Kraaij, C., Kuipers, O. P., Sahl, H. G., & De Kruijff, B. (1999). Use of the cell wall precursor lipid II by a pore-forming peptide antibiotic. *Science*, *286*(5448), 2361–2364.

Brötz, H., & Sahl, H. G. (2000). New insights into the mechanism of action of lantibiotics—diverse biological effects by binding to the same molecular target. *Journal of Antimicrobial Chemotherapy*, *46*(1), 1–6.

Bukowski J., Somers G., & Bryanton J. (2001). Agricultural contamination of groundwater as a possible risk factor for growth restriction or prematurity. *Journal of Occupational and Environmental Medicine*, *43*, 377–383.

Buzby, J. C., Farah-Wells, H., & Hyman, J. (2014). The estimated amount, value, and calories of postharvest food losses at the retail and consumer levels in the United States. USDA-ERS Economic Information Bulletin No. (EIB-121), USDA-ERS, Washington, DC.

Campêlo, M. C. S., Medeiros, J. M. S., & Silva, J. B. A. (2019). Natural products in food preservation. *International Food Research Journal*, *26*(1), 41–46.

Cao, Y., Liu, H., Qin, N., Ren, X., Zhu, B., & Xia, X. (2020). Impact of food additives on the composition and function of gut microbiota: a review. *Trends in Food Science & Technology*, *99*, 295–310.

Cedergren, M. I., Selbing, A. J., Löfman, O., & Källen, B. A. (2002). Chlorination byproducts and nitrate in drinking water and risk for congenital cardiac defects. *Environmental Research*, *89*(2), 124–130.

Cegielska-Radziejewska, R., Lesnierowski, G., & Kijowski, J. (2009). Antibacterial activity of hen egg white lysozyme modified by thermochemical technique. *European Food Research and Technology*, *228*(5), 841–845.

Chen, L., Wei, X., Chaves, B. D., Jones, D., Ponder, M. A., & Subbiah, J. (2021). Inactivation of Salmonella enterica and Enterococcus faecium NRRL B2354 on cumin seeds using gaseous ethylene oxide. *Food Microbiology*, *94*, 103656.

Cowan, M. M. (1999). Plant products as antimicrobial agents. *Clinical Microbiology Reviews*, *12*(4), 564–582.

Cox, C., 2004. Boric acid and borates. *Journal of Pesticide Reform*, *24*, 10–15.

Croen, L. A., Todoroff, K., & Shaw, G. M. (2001). Maternal exposure to nitrate from drinking water and diet and risk for neural tube defects. *American Journal of Epidemiology*, *153*(4), 325–331.

da Silva, L. V. A., Prinyawiwatkul, W., King, J. M., No, H. K., Bankston Jr, J. D., & Ge, B. (2008). Effect of preservatives on microbial safety and quality of smoked blue catfish (*Ictalurus furcatus*) steaks during room-temperature storage. *Food Microbiology*, *25*(8), 958–963.

Das, D. D., Alauddin, M. A., Rahman, M. A., Nath, K. K., & Rokonuzzaman, M. R. (2010). Effect of preservatives on extending the shelf-life of Rasogolla. *Chemical Engineering Research Bulletin*, *14*(1), 19–24.

Davidson, P. M., Critzer, F. J., & Taylor, T. M. (2013). Naturally occurring antimicrobials for minimally processed foods. *Annual Review of Food Science and Technology*, *4*, 163–190.

Doughari, J. H., Alabi, G., & Elmahmood, A. M. (2007). Effect of some chemical preservatives on the shelf-life of sobo drink. *African Journal Microbiology Research*, *2*, 37–41.

Dwivedi, S., Prajapati, P., Vyas, N., Malviya, S., & Kharia, A. (2017). A review on food preservation: methods, harmful effects and better alternatives. *Asian Journal of Pharmacy and Pharmacology*, *3*(6), 193–199.

Dwivedy, A. K., Prakash, B., Chanotiya, C. S., Bisht, D., & Dubey, N. K. (2017). Chemically characterized *Mentha cardiaca* L. essential oil as plant based preservative in view of efficacy against biodeteriorating fungi of dry fruits, aflatoxin secretion, lipid peroxidation and safety profile assessment. *Food and Chemical Toxicology*, *106*, 175–184.

Efiuvwevwere, B. J. O., & Ajiboye, M. O. (1996). Control of microbiological quality and shelf-life of catfish (*Clarias gariepinus*) by chemical preservatives and smoking. *Journal of Applied Bacteriology*, *80*(5), 465–470.

Ellison, R., & Giehl, T. J. (1991). Killing of gram-negative bacteria by lactoferrin and lysozyme. *Journal of Clinical Investigation*, *88*(4), 1080–1091.

Embuscado, M. E. (2015). Spices and herbs: natural sources of antioxidants–a mini review. *Journal of Functional Foods*, *18*, 811–819.

Etemadi, H., Rezaei, M., Abedian, K. A., & Hosseini, S. F. (2013). Combined effect of vacuum packaging and sodium acetate dip treatment on shelf life extension of rainbow trout (*Oncorhynchus mykiss*) during refrigerated storage. *Journal of Agricultural Science and Technology*, *15*, 929–939.

Falleh, H., Ben Jemaa, M., Djebali, K., Abid, S., Saada, M., & Ksouri, R. (2019). Application of the mixture design for optimum antimicrobial activity: Combined treatment of *Syzygium aromaticum, Cinnamomum zeylanicum, Myrtus communis,* and *Lavandula stoechas* essential oils against *Escherichia coli. Journal of Food Processing and Preservation*, *43*(12), e14257.

Federal Register (1988). Nisin preparation: Affirmation of GRAS status as a direct human food ingredient. *Federal Register*, 53:11247–11251.

Fellows, P. J. (2009). *Food processing technology: principles and practice*. Elsevier.

Food Safety and Standards (Food Products Standards and Food Additives) Regulations (2011). Food Safety and Standards Act, 2006. *Section 3.1.4*. https://www.fssai.gov.in/upload/uploadfiles/files/Food_Additives_Regulations.pdf

Freese, E., Sheu, C. W., & Galliers, E. (1973). Function of lipophilic acids as antimicrobial food additives. *Nature*, *241*(5388), 321–325.

García, I., Ortiz, M. C., Sarabia, L., Vilches, C., & Gredilla, E. (2003). Advances in methodology for the validation of methods according to the International Organization for Standardization: Application to the determination of benzoic and sorbic acids in soft drinks by high-performance liquid chromatography. *Journal of Chromatography A*, *992*(1–2), 11–27.

García-García, R., & Searle, S. S. (2016). Preservatives: food use. In *Encyclopedia of food and health*, 505–509. Elsevier.

George, M., Wiklund, L., Aastrup, M., Pousette, J., Thunholm, B., Saldeen, T., ... & Holmberg, L. (2001). Incidence and geographical distribution of sudden infant death syndrome in relation to content of nitrate in drinking water and groundwater levels. *European Journal of Clinical Investigation*, *31*(12), 1083–1094.

Gould, G. W. (1964) Effect of food preservatives on the growth of bacterial spores. In Molin, N. (ed.), *Microbial inhibitors in food*, 17–24. Almqvist and Wiksell.

Govari, M., & Pexara, A. (2015). Nitrates and nitrites in meat products. *Journal of the Hellenic Veterinary Medical Society*, *66*(3), 127–140.

Gracia, C. M, Gonzalez-Bermudez, C. A., Cabellero-Valcarcel, A. M., Santaella-Pascual, M., & Frontela-Saseta, C. (2015). Use of herbs and spices for food preservation: advantages and limitations. *Food Science 6*, 38–43.

Gundewadi, G., Rudra, S. G., Sarkar, D. J., & Singh, D. (2018). Nanoemulsion based alginate organic coating for shelf-life extension of okra. *Food Packaging and Shelf Life*, *18*, 1–12.

Gyawali, R., Hayek, S. A., & Ibrahim, S. A. (2015). Plant extracts as antimicrobials in food products: Mechanisms of action, extraction methods, and applications. *Handbook of Natural Antimicrobials for Food Safety and Quality*, *49*, 49–62.

Herrera, E., & Barbas, C. (2001). Vitamin E: action, metabolism and perspectives. *Journal of Physiology and Biochemistry*, *57*(1), 43–56.

Hrncirova, L., Machova, V., Trckova, E., Krejsek, J., & Hrncir, T. (2019). Food preservatives induce proteobacteria dysbiosis in human-microbiota associated Nod2-deficient mice. *Microorganisms*, *7*(10), 383.

Hyldgaard, M., Mygind, T., & Meyer, R. L. (2012). Essential oils in food preservation: mode of action, synergies, and interactions with food matrix components. *Frontiers in Microbiology*, *3*, 12.

Ibrahim, S. A., Salameh, M. M., Phetsomphou, S., Yang, H., & Seo, C. W. (2006). Application of caffeine, 1, 3, 7-trimethylxanthine, to control *Escherichia coli* O157: H7. *Food Chemistry*, *99*(4), 645–650.

Inetianbor, J. E., Yakubu, J. M., & Ezeonu, S. C. (2015). Effects of food additives and preservatives on man-a review. *Asian Journal of Science and Technology*, *6*(2), 1118–1135.

JECFA, 1974. 17th Report of the Joint FAO/WHO Expert Committee on Food Additives. World Health Organization Technical Report Series 539.

Jeong, S., Lee, H. G., Cho, C. H., & Yoo, S. (2020). Characterization of multi-functional, biodegradable sodium metabisulfite-incorporated films based on polycarprolactone for active food packaging applications. *Food Packaging and Shelf Life*, *25*, 100512.

Juneja, V. K., Dwivedi, H. P., & Yan, X. (2012). Novel natural food antimicrobials. *Annual Review of Food Science and Technology*, *3*, 381–403.

Karre, L., Lopez, K., & Getty, K. J. (2013). Natural antioxidants in meat and poultry products. *Meat Science*, *94*(2), 220–227.

Kedia, A., Prakash, B., Mishra, P. K., Dwivedy, A. K., & Dubey, N. K. (2015). *Trachyspermum ammi* L. essential oil as plant-based preservative in food system. *Industrial Crops and Products*, *69*, 104–109.

Kim, C. R., Hearnsberger, J. O., Vickery, A. P., White, C. H., & Marshall, D. L. (1995). Extending shelf life of refrigerated catfish fillets using sodium acetate and monopotassium phosphate. *Journal of Food Protection*, *58*(6), 644–647.

Kim, S. P., Kang, M. Y., Park, J. C., Nam, S. H., & Friedman, M. (2012). Rice hull smoke extract inactivates *Salmonella Typhimurium* in laboratory media and protects infected mice against mortality. *Journal of Food Science*, *77*(1), M80–M85.

Krebs, H. A., Wiggins, D., Stubbs, M., Sols, A., & Bedoya, F. (1983). Studies on the mechanism of the antifungal action of benzoate. *Biochemical Journal*, *214*(3), 657–663.

Krishna, K. R., & Rao, D. S. (2014). Effect of chitosan coating on the physiochemical characteristics of guava (*Psidium guajava* L.) fruits during storage at room temperature. *Indian Journal of Science and Technology*, *7*(5), 554.

Krishna, K. R., & Rao, D. S. (2017). Influence of chitosan coating and storage temperatures on postharvest quality of guava. *The Horticultural Society of India (Regd.)*, *74*(3), 466–470.

Krishnan, K. R., Babuskin, S., Babu, P. A. S., Sasikala, M., Sabina, K., Archana, G., Sivarajan, M., & Sukumar, M. 2014. Antimicrobial and antioxidant effects of spice extracts on the shelf life extension of raw chicken meat. *International Journal of Food Microbiology*, *171*, 32–40.

Kumari, P. K., Akhila, S., Rao, Y. S., & Devi, B. R. (2019). Alternative to artificial preservatives. *Systematic Reviews in Pharmacy*, *10*(1), 99–102.

Lee, S., Lee, H., Kim, S., Lee, J., Ha, J., Choi, Y., ... & Yoon, Y. (2018). Microbiological safety of processed meat products formulated with low nitrite concentration—A review. *Asian–Australasian Journal of Animal Sciences*, *31*(8), 1073.

Lück, E., & von Rymon Lipinski, G. W. (2000). Foods, 3. food additives. In *Ullmann's encyclopedia of industrial chemistry*. Wiley.

Maddox, D. N. (1982). The role of p-hydroxybenzoates in modern cosmetics. *Cosmetics and Toiletries, 97*, 85–88.

Mellado-García, P., Maisanaba, S., Puerto, M., Prieto, A. I., Marcos, R., Pichardo, S., & Camean, A. M. (2017). In vitro toxicological assessment of an organosulfur compound from Allium extract: cytotoxicity, mutagenicity and genotoxicity studies. *Food & Chemical Toxicology, 99*, 231–240.

Miguel, M. G. (2010). Antioxidant activity of medicinal and aromatic plants. A review. *Flavour and Fragrance Journal, 25*, 291–312.

Milly, P. J., Toledo, R. T., & Chen, J. (2008). Evaluation of liquid smoke treated ready-to-eat (RTE) meat products for control of Listeria innocua M1. *Journal of Food Science, 73*(4), M179–M183.

Mirza, S. K., Asema, U.K., & Sayyad, S.K. (2017). To study the harmful effects of food preservatives on human health. *Journal of Medicinal Chemistry and Drug Discovery, 2* (2), 610–616.

Montazeri, N., Himelbloom, B. H., Oliveira, A. C., Leigh, M. B., & Crapo, C. A. (2013). Refined liquid smoke: a potential antilisterial additive to cold-smoked sockeye salmon (*Oncorhynchus nerka*). *Journal of Food Protection, 76*(5), 812–819.

Morey, A., Bratcher, C. L., Singh, M., & McKee, S. R. (2012). Effect of liquid smoke as an ingredient in frankfurters on *Listeria monocytogenes* and quality attributes. *Poultry Science, 91*(9), 2341–2350.

Msagati, T. A. M. (2013). *Chemistry of food additives and preservatives: emulsifiers.* Wiley-Blackwell.

Negi, P. S. (2012). Plant extracts for the control of bacterial growth: Efficacy, stability and safety issues for food application. *International Journal of Food Microbiology, 156*(1), 7–17.

Nes, I. F., & Eklund, T. (1983). The effects of paraben on DNA, RNA and protein-synthesis in *Escherichia coli* and *Bacillus subtilis. Journal of Applied Bacteriology, 54*, 237–242.

Nikaido, H. (2003). Molecular basis of bacterial outer membrane permeability revisited. *Microbiology and Molecular Biology Reviews, 67*(4), 593–656.

Nishad, J., Dutta, A., Saha, S., Rudra, S. G., Varghese, E., Sharma, R. R., … & Kaur, C. (2021). Ultrasound-assisted development of stable grapefruit peel polyphenolic nano-emulsion: Optimization and application in improving oxidative stability of mustard oil. *Food Chemistry, 334*, 127561.

Nishad, J., Koley, T. K., Varghese, E., & Kaur, C. (2018). Synergistic effects of nutmeg and citrus peel extracts in imparting oxidative stability in meat balls. *Food Research International, 106*, 1026–1036.

No, H. K., Park, N. Y., Lee, S. H., & Meyers, S. P. (2002). Antibacterial activity of chitosans and chitosan oligomers with different molecular weights. *International Journal of Food Microbiology, 74*(1–2), 65–72.

Oschman, J. L. (2009). Charge transfer in the living matrix. *Journal of Bodywork and Movement Therapy, 13*, (3), 215–228.

Pandey, R. M., & Upadhyay, S. K. (2012). Food additive. In El-Samragy, Y. (ed.), *Food additive.* Intech Open. ISBN: 978–953-51-0067-6.

Parfitt, J., Barthel, M., & Macnaughton, S. (2010). Food waste within food supply chains: quantification and potential for change to 2050. *Philosophical Transactions of the Royal Society B: Biological Sciences, 365*(1554), 3065–3081.

Plumridge, A. (2005) Sorbic acid stress in *Aspergillus niger.* PhD thesis, University of Nottingham, UK.

Priyadarsini, K. I., Maity, D. K., Naik, G. H., Kumar, M. S., Unnikrishnan, M. K., Satav, J. G. & Mohan, H. (2003). Role of phenolic O-H and methylene hydrogen on the free radical reactions and antioxidant activity of curcumin. *Free Radical Biology & Medicine, 35*, (5), 475–484.

Quinto, E. J., Caro, I., Villalobos-Delgado, L. H., Mateo, J., De-Mateo-Silleras, B., & Redondo-Del-Río, M. P. (2019). Food safety through natural antimicrobials. *Antibiotics, 8*(4), 208.

Raafat, D., & Sahl, H. G. (2009). Chitosan and its antimicrobial potential – A critical literature survey. *Microbial Biotechnology, 2*(2), 186–201.

Ray, A., & Dasappa, I. (2017). Effect of natural preservatives on the rheological, physico-sensory, microbiological characteristics and shelf life of south Indian parotta. *Journal of Food Measurement and Characterization, 11*(2), 660–666.

Ricke, S. C., Dittoe, D. K., & Richardson, K. E. (2020). Formic acid as an antimicrobial for poultry production: a review. *Frontiers in Veterinary Science, 7*, 263.

Rodriguez, J., & Zoffoli, J. P. (2016). Effect of sulfur dioxide and modified atmosphere packaging on blueberry postharvest quality. *Postharvest Biology and Technology, 117*, 230–238.

Roginsky, V., & Lissi, E. A. (2005). Review of methods to determine chain-breaking antioxidant activity in food. *Food Chemistry, 92*, (2), 235–254.

Russell, A. D., Furr, J. R., & Pugh, W. J. (1985). Susceptibility of porin-and lipopolysaccharide-deficient mutants of *Escherichia coli* to a homologous series of esters of p-hydroxybenzoic acid. *International Journal of Pharmaceutics*, *27*(2–3), 163–173.

Saunders, T., Wu, J., Williams, R. C., Huang, H., & Ponder, M. A. (2018). Inactivation of Salmonella and surrogate bacteria on cashews and macadamia nuts exposed to commercial propylene oxide processing conditions. *Journal of Food Protection*, *81*(3), 417–423.

Scallan, E., Hoekstra, R. M., Angulo, F. J., Tauxe, R. V., Widdowson, M. A., Roy, S. L., … & Griffin, P. M. (2011). Foodborne illness acquired in the United States—major pathogens. *Emerging Infectious Diseases*, *17*(1), 7.

Scocchi, M., Mardirossian, M., Runti, G., & Benincasa, M. (2016). Non-membrane permeabilizing modes of action of antimicrobial peptides on bacteria. *Current Topics in Medicinal Chemistry*, *16*(1), 76–88.

See, A. S., Salleh, A. B., Bakar, F. A., Yosuf, N. A., Abdulamir, A. S., & Heng, L. Y. (2010). Risk and health effect of boric acid. *American Journal of Applied Sciences*, *7*(5), 620–627.

Silva, M. M., & Lidon, F. (2016). Food preservatives – An overview on applications and side effects. *Emirates Journal of Food and Agriculture*, *28*(6), 366–373.

Soni, M. G., Carabin, I. G., & Burdock, G. A. (2005). Safety assessment of esters of p-hydroxybenzoic acid (parabens). *Food and Chemical Toxicology*, *43*(7), 985–1015.

Sonker, N., Pandey, A. K., & Singh, P. (2015). Efficiency of *Artemisia nilagirica* (Clarke) Pamp. essential oil as a mycotoxicant against postharvest mycobiota of table grapes. *Journal of the Science of Food and Agriculture*, *95*(9), 1932–1939.

Sonker, N., Pandey, A. K., Singh, P., & Tripathi, N. N. (2014). Assessment of *Cymbopogon citratus* (DC.) stapf essential oil as herbal preservatives based on antifungal, antiaflatoxin, and antiochratoxin activities and in vivo efficacy during storage. *Journal of Food Science*, *79*(4), M628–M634.

Stratford, M., & Eklund, T. (2003). Organic acids and esters. In *Food preservatives* (pp. 48–84). Springer.

Surekha, M., & Reddy, S.M. (2014). Preservatives, classification and properties. In Carl, A. B., & Mary, L. T. (eds.), *Encyclopedia of food microbiology* (pp. 69–75). Academic Press.

Tangkham, W., Comeaux, J., Ferguson, C. E., & LeMieux, F. M. (2012). Effect of sodium lactate and sodium acetate on shelf-life of raw chicken breasts. *African Journal of Food Science*, *6*(13), 375–380.

Tfouni, A. S., & Toledo, M. C. (2002). Estimates of the mean per capita daily intake of benzoic and sorbic acids in Brazil. *Food Additives & Contaminants*, *19*(7), 647–654.

Tfouni, S. A. V., & Toledo, M. C. F. (2002). Determination of benzoic and sorbic acids in Brazilian food. *Food Control*, *13*(2), 117–123.

Tharanathan, R. N., & Kittur, F. S. (2003). Chitin—the undisputed biomolecule of great potential. *Critical Reviews in Food Science & Nutrition*, *43*(1):61–87.

Thevelein, J. (1994). Signal-transduction in yeast. *Yeast*, *10*(13), 1753–1790.

Tran, A. V. (2013). Do BHA and BHT Induce Morphological Changes and DNA Double-Strand Breaks in Schizosaccharomyces pombe?

Trombetta, D., Castelli, F., Sarpietro, M. G., Venuti, V., Cristani, M., Daniele, C., … & Bisignano, G. (2005). Mechanisms of antibacterial action of three monoterpenes. *Antimicrobial Agents and Chemotherapy*, *49*(6), 2474–2478.

Viuda-Martos, M., El Gendy, N. G. S., Sendra, E., Fernandez-Lopez, J., El-Razik, K. A. A., El-Sayed, A., & Perez-Alvarez, J. A. (2010). Chemical composition and antioxidant and anti-listeria activities of essential oils obtained from some Egyptian plants. *Journal of Agricultural and Food Chemistry*, *58*, 9063–9070.

Volkmer, B. G., Ernst, B., Simon, J., Kuefer, R., Bartsch Jr, G., Bach, D., & Gschwend, J. E. (2005). Influence of nitrate levels in drinking water on urological malignancies: a community-based cohort study. *BJU International*, *95*(7), 972–976.

Volpe, J. J., Inder, T. E., Darras, B. T., de Vries, L. S., du Plessis, A. J., Neil, J., & Perlman, J. M. (2017). *Volpe's neurology of the newborn e-book*. Elsevier Health Sciences.

Voss, C. (2002). Veneno no seu prato? Utilidades e riscos dos aditivos alimentares. 1ª ed. EDIDECO – Editores Para a defesa do consumidor Lda. Lisboa. *Cad. Saúde Pública, Rio de Janeiro*, *25*(8), 1653–1666.

Wei, X., Chen, L., Chaves, B. D., Ponder, M. A., & Subbiah, J. (2021). Modeling the effect of temperature and relative humidity on the ethylene oxide fumigation of *Salmonella* and *Enterococcus faecium* in whole black peppercorn. *LWT*, *140*, 110742.

Wiedemann, I., Benz, R., & Sahl, H. G. (2004). Lipid II-mediated pore formation by the peptide antibiotic nisin: a black lipid membrane study. *Journal of Bacteriology*, *186*(10), 3259–3261.

Wiedemann, I., Breukink, E., van Kraaij, C., Kuipers, O. P., Bierbaum, G., de Kruijff, B., & Sahl, H. G. (2001). Specific binding of nisin to the peptidoglycan precursor lipid II combines pore formation and inhibition of cell wall biosynthesis for potent antibiotic activity. *Journal of Biological Chemistry*, *276*(3), 1772–1779.

Williams, G. M., Iatropoulos, M. J., & Whysner, J. (1999). Safety assessment of butylated hydroxyanisole and butylated hydroxytoluene as antioxidant food additives. *Food and Chemical Toxicology*, *37*(9–10), 1027–1038.

Yiu, P. H., See, J., Rajan, A., & Bong, C. F. J. (2008). Boric acid levels in fresh noodles and fish ball. *American Journal of Agricultural and Biological Sciences*, *3*(2), 476–481.

Zengin, N., Yüzbasıoglu, D., Ünal, F., Yilmaz, S., & Aksoy, H. (2011). The evaluation of the genotoxicity of two food preservatives: Sodium benzoate and potassium benzoate. *Food & Chemical Toxicology*, *49*(4), 763–769.

Zhang, Y. Y., Zhang, F., Thakur, K., Ci, A. T., Wang, H., Zhang, J. G., & Wei, Z. J. (2018). Effect of natural polyphenol on the oxidative stability of pecan oil. *Food and Chemical Toxicology*, *119*, 489–495.

CHAPTER 13

Natural Preservatives for Improvement of Shelf Life

Sajid Ali, Muhammad Akbar Anjum, Shaghef Ejaz, Mahmood Ul Hasan,
Aamir Nawaz, Sajjad Hussain, and Muhammad Shafique

CONTENTS

13.1 INTRODUCTION

Shelf life is an important factor for the conserved quality and marketing of various commodities including meat, seafood, fruits, vegetables, and processed juices. Fresh vegetables and fruits are an important component of the daily recommended diet of humans. However, fresh horticultural produce including fruits and vegetables contains 80–90% water and is perishable by nature, requiring some sort of protection from spoilage during harvest, handling, packaging, storage and distribution, that eventually affects the potential of shelf life (Dhall, 2013). The said losses of fruits and vegetables are mostly due to a lack of sufficient conservation facilities and inadequate processing. Fruit juices are one of the processed products generally acquired by fruit squeezing with or without specific treatments. The processed juices are alternative ways of consuming vegetables and fruits (Pandey and Negi, 2018). However, juices are prone to deterioration and have a short shelf life. So, some post-harvest strategies are required that can be applied to extend the shelf life of juices.

Seafood includes crustaceans, fishes, echinoderms, and mollusks. Seafood is very popular among people owing to its delicacy and great nutritional value (Viji et al., 2017). Nevertheless, the prompt biochemical changes and infectious developments negatively affect seafood quality and hinder its shelf life potential (Olatunde and Benjakul, 2018). Generally, seafood is susceptible to

Email: Sajid Ali, sajidali@bzu.edu.pk; ch.sajid15@yahoo.com

DOI: 10.1201/9781003091677-13

prompt lipid oxidation. The unpalatable flavor, loss of various nutrients, and color changes are the leading negative influences of lipid oxidations that further negatively influence seafood shelf life (Secci and Parisi, 2016). Likewise, chemical deterioration, microbial spoilage, and browning also affect the shelf life potential of seafood after processing and during the supply chain (Olatunde and Benjakul, 2018).

Meat is an important dietary component of our daily life; it provides protein, lipids, minerals, vitamins, and a low quantity of carbohydrates (Alirezalu et al., 2020). Meat products and meat are prone to protein and lipid oxidations, and the quality deteriorates promptly during processing and subsequent storage. The quality also depends on the processing type and storage conditions (Fernandes et al., 2017). The oxidation reactions mainly affect color pigments, flavor, texture, proteins, and lipids (Devatkal et al., 2014). Different types of preservatives have been reported for quality maintenance and shelf life extension of meat and meat products. The use of chemical-based preservatives is now in question because of the concerns of consumers. Thus, eco-friendly and natural preservatives are in great demand and have been used extensively in recent years (Alirezalu et al., 2020; Baptista et al., 2020).

Different types of preservatives are being used for the quality conservation of meat, meat products, seafood, fruits, vegetables, and juices worldwide. However, among the used preservatives, the majority is chemical based. Some of the chemicals being used are not permitted to be used in the food industry due to allergic reactions and government legislations. So, natural preservatives are required that can conserve the quality and extend the shelf life potential of meat, meat products, seafood, fruits, vegetables, and juices. The natural preservatives are eco-friendly and do not have any potential health hazards to consumers. There are different types of natural preservatives such as plant extracts, edible coatings, essential oils, bacteriocins, and bioactive peptides. This chapter describes the potential applications of various natural preservatives to extend the shelf life and to conserve the quality of different commodities such as fresh fruits, vegetables, juices, seafood, meat, and meat products.

13.2 FACTORS AFFECTING SHELF LIFE OF FRESH PRODUCE

Research for developing technologies to bring fresh food products to the consumer has been conducted for more than a century. Reducing food losses and waste is a critical and dynamic process as the fresh food supply chain is enormously complex mainly due to the perishability and variability of food products originating either from animal or plant sources. Loss of food quality can be defined as an undesirable change in one or more sensory attributes (e.g., texture, appearance, aroma, and flavor) that are generally detectable by sensory organs (Sherratt et al., 2006). Loss of food quality may happen at any step of the food supply chain, especially during its shelf life. The most common reasons for the loss are endogenous enzymes activity; oxidation; damages due to insect, pest, or mechanical injury; too high or low temperature; and ambient gaseous concentration (Abdel-Aziz et al., 2016). Deterioration of food quality is, therefore, a complicated process that involves microbiological, biochemical, physical, and chemical changes.

Consumers assess food quality by analyzing its visual appearance, making it the most important desirable trait. The loss of visual quality due to color change, shrinkage, or shriveling and drying determines the end of the shelf life of a food product. One of the Sustainable Development Goals adopted by 193 member countries of the Food and Agriculture Organization (FAO) of the United Nations is that "By 2030, halve per capita global food waste at the retail and consumer levels and reduce food losses along production and supply chains, including post-harvest losses" (FAO, 2020). Therefore, food research has been focused on increasing the production of food for the rapidly growing global population, on one hand, and developing scientific methods to reduce food loss and waste to meet human nutritional requirements on the other hand. Although the loss of quality of fresh food product is natural, several external factors enhance the rate of spoilage of food. These factors include endogenous enzymes in plants that oxidize phenolic compounds leading to browning or degrade pectin leading to softening; insect and pest damage; parasites in meat or fish; pathogens like bacteria, molds, and yeasts; light degrades color pigments, lipids, and proteins and, thus, produces off-flavors and off-odors; too high or too low temperature cause heat, chilling, or freezing injury; humidity, high or low, cause pathogen growth or tissue dryness, respectively, and gases, particularly oxygen, carbon dioxide, and ethylene (Abdel-Aziz et al., 2016).

Among numerous biotic and abiotic factors that affect food quality, microorganisms are the leading cause of food loss and waste because microorganisms are invisible, grow exponentially, instantly form colonies, disperse through air and water, and secrete toxic compounds responsible for off-flavors and off-odors (Hammond et al., 2015). Airborne transmission of microbes to fresh food products cause contamination and contribute to more food loss. Different microbes have variable tolerance to abiotic factors like temperature, pH, water activity, etc., which determines the type of microbe contaminating a food product. For example, fungi, such as mold and yeast, can tolerate low temperature, pH, and water activity values compared with bacteria (Abdel-Aziz et al., 2016). However, mold and yeast are not heat resistant. Temperature regulates the growth rate of microorganisms as their metabolism and population growth vanishes at sub-zero temperatures (Hammond et al., 2015).

Food moisture content is equally important in determining the extent of microbial spoilage of a fresh food sample. Fresh food with high water content, whether originating from animal or plant, is more prone to bacterial or fungal growth, whereas foods with limited water content are less susceptible to pathogens due to the osmotic physiological stress that reduces microbial growth rates (Pitt and Hocking, 2009). A food may retain high nutritional value and edibility from a few hours to several years depending on its internal water content, ambient temperature, nutrients profile, and presence of antimicrobial compounds (Hammond et al., 2015).

Food spoilage due to fungi has characteristics of visible pigmented growth, formation of slime, off-odors, and off-flavors. Due to the low pH of fruits, spoilage by bacteria is less in fruits than by molds and yeasts, whereas vegetables are prone to spoilage by bacteria and fungi equally. The vegetables most contaminated by fungi include solanaceous vegetables (tomato and potato), legumes, root vegetables (carrot), cucurbits (cucumber), leafy vegetables (lettuce), and asparagus; whereas different species of bacteria, e.g., *Erwinia* and *Pseudomonas*, cause soft rot in fruits and vegetables (Abdel-Aziz et al., 2016). Proteinaceous foods including fish, poultry, meat, and some dairy items permit rapid bacterial growth due to the presence of essential minerals, carbohydrates and lipids, favorable pH (neutral or slightly acidic), and greater water content (Hadawey et al., 2017).

Lipid oxidation by endogenous enzymes is another cause of loss of quality during shelf life of a fresh food item. Lipid oxidation causes a change in the sensory and nutritional profile of fresh food, thus, reducing its shelf life. Lipid peroxidation causes the production of off-odors and off-flavors, loss of nutrients, formation of toxic compounds, and color degradation (Secci and Parisi, 2016). Seafood is more susceptible to lipid oxidation due to the presence of a high amount of long-chain polyunsaturated fatty acids. Lipid-soluble vitamins and other bioactive molecules are also depleted due to lipid oxidation reactions. Similarly, in meat products, oxidative chain reactions deteriorate lipids, proteins, color pigments (myoglobin), flavor, nutritional value, and vitamins (Ribeiro et al., 2019). Overall, indigenous or microbial proteases and lipases, oxygen, microbial load, and moisture cause the spoilage of seafood during storage (Ghaly et al., 2010; Sriket, 2014). Pro-oxidant enzymes also cause oxidation of phenolics leading to surface browning of many fruits and vegetables (Ali et al., 2019b; Ali et al., 2020b).

13.3 TYPES OF PRESERVATIVES FOR SHELF LIFE IMPROVEMENT

There are two types of preservatives, i.e., synthetic/chemical/artificial preservatives and natural preservatives. Preservatives are synthetic and/or natural substances added to food products to suppress deterioration and to enhance shelf life. Preservatives act in several ways; generally, they retard deterioration and microbial contamination ultimately leading to the extended shelf life of a commodity. Some chemical or artificial preservatives are considered hazardous due to their potent health hazards. However, efforts are being made to develop cheap and effective strategies to minimize these negative situations and to provide consumers with better quality and longer shelf life-based products (Pushkala et al., 2013). These include chemical and natural preservatives such as plant extracts, edible coatings, essential oils, bacteriocins, etc.

Some commodities such as fresh and/or fresh-cut fruits and vegetables are highly perishable due to increased biochemical changes with a very short shelf life due to their susceptibility to microbial damage (Qadri et al., 2016). Over the years, artificial preservatives have been used as a tool to increase aesthetic value and shelf life. However, increasing demand for good hygiene along with safe food with the least chemical preservatives and rising consumer's concerns due to health-related issues has coerced scientists into exploring alternatives (Marino et al., 2001). Natural preservatives include plant extracts that contain antioxidative and polyphenol substances that are believed to be safer and may provide additional health benefits. Eugenol, an herbal essential oil from basil, has been reported to exhibit antimicrobial properties against a wide range of microorganisms and is generally recognized as a safe compound (Shah et al., 2013).

Plant extracts are eco-friendly and cost-effective bioactive substances that include essential oils, food coatings, and additives used in the food industry as natural preservatives due to their antimicrobial properties (Dianella, 2012; Zhang and Li, 2012; Chai et al., 2018). Several post-harvest studies have reported the potential role of natural preservatives in extending the shelf life of fresh produce by inhibiting the growth of disease-causing organisms (Lin and Zhao, 2007; Silvestre et al., 2011). Extracts obtained from many plants have recently gained scientific interest in their antifungal activity (Gatto et al., 2011). The antimicrobial properties of plant extracts have been proven to affect *Penicillium brevicompactum* growth in mandarins (Nath et al., 2013). Previously, cistus plant extract reduced the incidence of citrus sour rot by inhibiting the development of *Geotrichum citriaurantii* (Karim et al., 2017); whereas *Orobanche crenata* extract reduced gray mold, brown rot, and green mold incidence on table grapes, apricots, and oranges, respectively (Gatto et al., 2011).

The other natural preservatives are chitosan (Sharif et al., 2015, Adiletta et al., 2018), essential oils (Prakash et al., 2015; Kahramanoğlu, 2019), propolis extract (Kahramanoğlu et al., 2018), and plant extracts (Chen et al., 2019). In addition to plant extracts, some plant oils having high stability and active components of the plants have been used as natural antioxidants for stabilizing polyunsaturated oils. Essential oils of *Prunus persica*, *Ocimum sanctum*, or *Zingiber officinale*; eugenol or thymol; and sprayed with natural thyme (*Thymus vulgaris*) and summer savory (*Satureja hortensis*) oils reduced the number of decayed berries (Romanazzi et al., 2012). These were also found as strong antioxidant sources due to their high content of phenolic compounds; however, these have not been investigated completely through toxicology (Misir et al., 2014). Edible coatings and films as well as the use of bacteriocins and bioactive peptides have also been reported to influence the shelf life potential of various commodities (Bolívar-Monsalve et al., 2019). The use of edible coatings affects CO_2- and O_2-based gas exchange, ultimately leading to the extended shelf life of coated produce (Ali et al., 2019b).

13.4 EDIBLE PRODUCE AND SHELF LIFE IMPROVEMENT

Numerous types of natural preservatives have been reported for the improvement of shelf life potential of different edible produce. However, each natural preservative has its efficacy and potential to conserve the quality and to prolong the shelf life potential of fresh and/or processed produce. As far as edible produce is considered, there are different types of eatable commodities to be considered based on their use for consumers. In this chapter, the impact of natural preservatives on fresh fruits, vegetables, processed juices, meat, meat products, and seafood is discussed.

13.4.1 Shelf Life Improvement of Fruits and Vegetables with Natural Preservatives

13.4.1.1 Plant Extracts

Different extracts including seaweed extract (Banu et al., 2020), garlic extract (Anjum et al., 2020), rhubarb extract (Li et al., 2019), lotus leaf extract (Fan et al., 2019), walnut green husk (Habibie et al., 2019), *Byrsonima crassifolia* extract (González-Saucedo et al., 2019), *Fagonia indica* extract (Khaliq et al., 2019), moringa leaf extract (Tesfay and Magwaza, 2017), neem leaf extract (Jaiswal et al., 2017), *Ficus hirta* fruit extract (Chen et al., 2016), *Bergenia crassifolia* extract (Krásniewska et al., 2015), and grapefruit seed extract (Xu et al., 2007) have been reported to preserve the quality and to extend the shelf life of various fruits and vegetables (Table 13.1). These extracts showed promising results in delaying senescence by inhibiting ethylene biosynthesis, respiration rate, and other metabolic activities. In addition, plant extracts showed higher conservation potential of biochemical and phytochemical quality during shelf life. The plant extracts also suppressed microbial growth, subsequently leading to lower post-harvest rots during storage with maintained quality (Ncama et al., 2018).

13.4.1.2 Essential Oils

Essential oils are characterized as complex volatile compounds with a significant impact on organoleptic attributes of treated produce, and these are obtained from different plant parts such as seeds,

Table 13.1 Impact of Different Plant Extracts on Quality and Shelf Life Potential of Different Fruits and Vegetables

Extract	Concentration (%)	Crop	Inference	References
Ficus hirta fruit	10	Mandarin	Reduced fruit decay, enhanced antioxidants, and defense-related enzyme activities	Chen et al. (2016)
Moringa leaf extract	2	Avocado	Reduced moisture loss, respiration rate, loss in firmness, and delayed ripening	Tesfay and Magwaza (2017)
Neem leaf extract	—	Tomato	Exhibited lower fruit decay and weight loss with maintained fruit quality	Jaiswal et al. (2017)
Rhubarb	—	Peach	Coated fruits had reduced weight loss, firmness, respiration rate, lowered malondialdehyde contents and polyphenol oxidase activity, and inhibited decay	Li et al. (2019)
Lotus leaf extract	0.2	Goji fruits	Significantly maintained ascorbic acid, titratable acidity, soluble solids, and higher antioxidant enzymes activities	Fan et al. (2019)
Byrsonima crassifolia	—	Bell pepper	Maintained higher firmness, delayed ethylene and respiration rate, and reduced microbial activity	González-Saucedo et al. (2019)
Fagonia indica	1	Sapodilla	Had higher ascorbic acid, total flavonoids, total phenolics, and radical scavenging activity with no detrimental effect on the sensory quality	Khaliq et al. (2019)
Seaweed	3	Tomato	Reduced weight loss, suppressed sugar acid ratio, total acidity, and ascorbic acid contents with retained texture quality	Banu et al. (2020)
Garlic and ginger	20	Guava	Extended shelf life and maintained quality with delayed fruit decay	Anjum et al. (2020)

buds, flowers, twigs, stems, bark, wood, roots, and certain fruits as secondary metabolites. The plant-derived essential oils play a key role in functioning as defensive substances and internal messengers, and help to reduce microbial infestations. In addition, these oils are an abundant source of antioxidants and antimicrobial properties, which make them unique from other additives (Brewer, 2011). Essential oils are generally recognized as safer for edible produce use with no risk to human health. These oils were initially utilized in the food industry as additives to increase or supplement the aroma volatiles, but later they were also found to extend the shelf and storage life of fresh or processed produce. With increasing concerns about food safety, the use of organically safe preservatives are preferred by consumers to extend and to preserve eating quality. Essential oils have been used in different products such as meat (Quintavalla and Vicini, 2002), seafood (Kykkidou et al., 2008), cheese (Vazquez et al., 2001), bakery products (Nielsen and Rios, 2000), and minimally processed fruits and vegetables (Patrignani et al., 2015).

It has been reported that various plant-based essential oils have been used to extend shelf and storage life by delaying post-harvest decay of fruits and vegetables either in whole or fresh-cut form. Among the most recently investigated essential oils, *Ruta graveolens* L (Peralta-Ruiz et al., 2020), ginger (Ban et al., 2020), cinnamon (Etemadipoor et al., 2020), *Heracleum persicum* fruit (Taheri et al., 2020), β-cyclodextrin–oregano (Huang et al., 2020a), clove (Hasheminejad and Khodaiyan, 2020), citrus (Das et al., 2020), basil (Khaliq et al., 2020), syringa (Yang et al., 2020), *Satureja khuzistanica* (Nasiri et al., 2019), thyme, peppermint (Chaemsanit et al., 2019), *Foeniculum vulgare* (Rizzo et al., 2019), *Cuminum cyminum* (Karimirad et al., 2019), bergamot (Chi et al., 2019), rosemary (Rizzo et al., 2018), *Mentha spicata* (Shahbazi, 2018), *Eucalyptus staigeriana*, *Zataria multiflora* (Nasiri et al., 2017), lemongrass (Yousuf and Srivastava, 2017), and lavender (Farokhian et al., 2017) have been found to significantly reduce fungal decay with maintained quality during shelf and/or cold storage conditions of fruits and vegetables (Table 13.2).

13.4.1.3 Edible Coatings

Different coating formulations have been developed, but applications of the edible coatings for quality conservation of fresh produce are still limited at the commercial level (Ncama et al., 2018). However, plant-based edible coatings sourced from different plant portions including fruits, seeds, leaves, stem exudates, and plant gums with high antimicrobial contents and potential antioxidant capacities showed a marked difference in coated and uncoated control produce (Florez-López et al., 2016). Plant-based coatings are mostly produced from different parts that can develop the functional layer around the targeted produce. Plant-based coatings are usually categorized into three sections: lipids containing fatty acids and waxes, hydrocolloids consisting of proteins and polysaccharides, and composites (Dhall, 2013; Florez-López et al., 2016). Edible coatings are known as the environmentally friendly and consumer-friendly approaches that work on the principle of developing a barrier between commodity and external environment. This modification of the coated commodity substantially delays the post-harvest senescence during shelf life by inhibiting respiration rate, ethylene biosynthesis, and restricted moisture permeability (Maringgal et al., 2020). It has also been reported that edible coating application significantly conserves phytochemicals, antioxidants, and sensory attributes with the best overall eating quality of the coated produce during storage (Ali et al., 2020b). Edible coatings are generally developed from chemical- and biological-based materials and their thin layer applied to the fresh produce hinders the internal processes and halts the ripening activities, subsequently leading to extended shelf life. The thin layer of edible coatings can be applied through various methods such as spraying, brushing, and dipping depending on the targeted produce (Maringgal et al., 2020).

Different plant-based natural coatings including gum Arabic (Maqbool et al., 2010; Anjum et al., 2020), *Aloe vera* gel (Ali et al., 2019b), guar gum (Abu-Shama et al. 2020), *Cordia myxa* gum

Table 13.2 Essential Oils on Quality and Shelf Life Potential of Different Fruits and Vegetables

Essential Oil	Concentration	Crop	Inference	References
Ruta graveolens L	1.5%	Papaya	Inhibited the growth of *Colletotrichum gloesporioides* and reduced anthracnose disease	Peralta-Ruiz et al. (2020)
Cinnamon	1%	Guava	Delayed browning index, reduced chilling injury weight loss and ion leakage, prevented lipid peroxidation, and maintained fruit firmness	Etemadipoor et al. (2020)
Ginger	99%	Jujube	Extended storage life and maintained quality with delayed fruit decay	Ban et al. (2020)
Heracleum persicum fruit	—	Bell pepper	Showed higher phenolic compounds, flavonoids, and ascorbic acid with maintained quality	Taheri et al. (2020)
Oregano	90% purity	Purple yam	Maintained firmness, total soluble solids, ascorbic acid content and ascorbic acid, and anthocyanins	Huang et al. (2020a)
Clove	100% purity	Pomegranate	Markedly reduced the microbial count, weight loss, firmness loss with maintained sensory quality	Hasheminejad, and Khodaiyan (2020)
Citrus	—	Tomatoes	Coated fruits showed lower *Salmonella* and *Listeria* with maintained higher firmness	Das et al. (2020)
Basil	0.1%	Jamun	Delayed depolymerization of pectin content, inhibited the activity of polygalacturonase and cellulase enzymes	Khaliq et al. (2020)
Syringa	0.67%	Peach	Maintained higher soluble solid contents, firmness, and conserved aroma volatiles and reduced decay	Yang et al. (2020)
Satureja khuzistanica	500 mg/L	Mushroom	Inhibited discoloration, delayed cap opening, reduced respiration rate, and extended shelf life	Nasiri et al. (2019)
Thyme	0.05%	Cantaloupe (fresh cut)	Inhibited growth of *Listeria monocytogenes*, *Staphylococcus aureus*, *Salmonella typhimurium*, and *Escherichia coli*, and increased shelf life	Chen et al. (2019)
Peppermint	700 µL L^{-1}	Brown rice	Reduced mold growth, maintained texture quality during cooking, and extended shelf life for long time	Chaemsanit et al. (2019)
Foeniculum vulgare	0.75%	Globe artichoke	Reduced microbial count, preserved antioxidant capacity, color change, inhibited polyphenol activity, and showed higher sensory scores	Rizzo et al. (2019)
Cuminum cyminum	50 g/500 mL	Mushroom	Maintained color, firmness, overall acceptability, and inhibited mold and yeast growth	Karimirad et al. (2019)
Bergamot	—	Mangoes	Treated fruits had higher firmness and lower weight loss and retained color, acidity, and lower microbe infestation	Chi et al. (2019)

(Haq et al., 2013; El-Mogy et al., 2020), basil seed mucilage (Nourozi and Sayyari, 2020), carnauba wax (Chen et al., 2019), tragacanth gum (Nasiri et al., 2019), pectin (Mannozzi et al., 2017), cassava starch (Aquino et al., 2015), and chitosan (Maqbool et al., 2010) have been found to reduce post-harvest diseases, maintain eating quality, and to extend the shelf life potential of treated fruits and vegetables under various storage conditions (Table 13.3). Overall, natural edible coatings of plant

Table 13.3 Impact of Plant-Based Natural Edible Coatings on Quality and Shelf Life Potential of Different Fruits and Vegetables

Extract	Concentration (%)	Crop	Inference	References
Gum Arabic	10	Banana	Reduced post-harvest anthracnose disease and delayed ripening with retained fruit quality	Maqbool et al. (2010)
Gum Arabic	10	Persimmon	Extended shelf life, delayed softening, reduced oxidative stress, and conserved eating quality	Saleem et al. (2020)
Gum Arabic	10	Apricot	Extended shelf life, reduced fruit softening, alleviated oxidative stress, and preserved eating quality	Ali et al. (2021)
Cassava starch	2	Guava	Reduced microbial count, i.e., molds and yeasts with maintained visual quality and retained color	Aquino et al. (2015)
Tragacanth gum	—	Button mushroom	Reduced microbial growth, maintained higher levels of ascorbic acid and phenolic contents with extended shelf life	Nasiri et al. (2019)
Tragacanth gum	1	Apricot	Extended shelf life, mitigated oxidative stress, decreased softening, and conserved eating quality.	Ali et al. (2020a)
Methylcellulose	1.5	Blueberry	Treated fruits showed higher quality and reduced microbes	Dhital et al. (2017)
Aloe vera	50	Litchi	Reduced pericarp browning with reduced weight loss, retained higher enzymatic and non-enzymatic antioxidants with extended shelf life	Ali et al. (2019b)
Aloe vera	50	Green chilies	Reduced mass loss, conserved green color, reduced oxidative stress, and maintained general eating quality	Hasan et al. (2021)
Lacquer wax	2	Kiwifruit	Decreased respiration and ethylene production resulted in delayed senescence and retained higher antioxidant activities	Hu et al. (2019)
Carnauba wax	15	Indian jujube	Delayed fruit softening, ethylene biosynthesis, respiration rate, and lower activities of polygalacturonase, pectin methylesterase, and cellulase enzymes	Chen et al. (2019)
Guar gum	0.5	Dates	Preserved phytochemical and biochemical quality of fruits, maintained eating quality, and extended storage life	Abu-Shama et al. (2020)
Cordia myxa	0.1	Artichoke bottoms	Coated artichoke bottoms showed reduced weight loss, browning with higher ascorbic acid contents, and inhibited polyphenol oxidase activity	El-Mogy et al. (2020)
Basil seed mucilage	0.1	Apricot	Showed higher antioxidant activity, total phenolic content, and ascorbic acid with no impact on sensory quality during storage	Nourozi and Sayyari (2020)

and/or animal origin could be used to reduce the extent of post-harvest diseases on the product with a lower risk of toxicity compared with synthetic chemicals.

Weight loss, firmness, and phytonutrient quality attributes are the key indicators of maintenance of fruit quality in the post-harvest period either at the shelf or under cold storage conditions. The maintenance of firmness and freshness directly correlates with the textural quality of fruits and vegetables during shelf life conditions (Maringgal et al., 2020). Edible coatings have been reported to conserve the biochemical attributes including soluble solid contents, titratable acidity, and sugar-acid ratio during storage (Ali et al., 2019b; Ali et al., 2020b). Exogenous application of thin layer coatings can also be used to preserve nutritional profile, antioxidants, and enzyme activities during storage. This statement has been supported by the findings of Khaliq et al. (2019) who reported that sapodilla fruits treated with *Aloe vera* gel coating enriched with extract of *Fagonia indica* plant conserved biochemical quality and phytonutrient profile during storage with extended shelf life. Combined applications of coatings and organic acids further resulted in higher total phenolic contents, ascorbic acid content, and flavonoids with no detrimental effect on organoleptic attributes (Ali et al., 2020b).

Post-harvest browning is another important issue that negatively affects the visual appeal of fresh produce during supply chains, and it is considered a major hindrance in extending the shelf life of different fruits and vegetables (Supapvanich and Boonyaritthongchai, 2016a; Ali et al., 2019a, 2020a,c). Some natural plant-based edible coatings such as *Aloe vera* have been applied on litchi (Ali et al., 2019b), lotus roots (Ali et al., 2019a), and fresh-cut apples (Supapvanich et al., 2016b) for browning reduction. Similarly, hydroxypropyl methylcellulose in combination with methanolic extract of propolis (Pastor et al., 2011) and rice starch-ι-carrageenan (Thakur et al., 2019) significantly delayed the oxidative browning of fresh produce by maintaining higher enzymatic and non-enzymatic antioxidant activities during storage. Overall, the edible coatings coupled with cold storage could be a better option to delay post-harvest ripening and fruit decay for a longer period, helping to extend the shelf life potential of edible produce (Ali et al., 2019b).

13.4.1.4 Honey and Propolis Extract

Multiple products including sweet honey, propolis, and royal jelly produced by honeybees have been characterized as functional foods and traditionally been employed as medicines in different cultures around the globe. Honey has been widely used as a sweetener secreted by bees with more than 181 reported compounds with a variety of phytochemicals (Viuda-Martos et al., 2008). Propolis is considered as a mixture of resinous exudates picked by honeybees from different plant portions, i.e., leaves, stem, flower buds, and sweet minor fruits. The composition of collected propolis varied with the season and harvest time of the crops. The propolis extract contains more than 400 active compounds including multiple amino acids, vitamins, esters, aromatic acids, and phenolic compounds due to which the propolis mixture exhibits antimicrobial activities against various pathogens (Pobiega et al., 2020). Due to these reasons both honey and propolis have been used to conserve the quality of fresh and/or fresh-cut fruits and vegetables, and on related products in the processing industry (Kumar et al., 2015; Pobiega et al., 2019).

Honey is considered as a unique edible coating due to its sweet taste, gelling properties, rich source of antioxidants, antifungal properties, and anti-browning compounds that could be considered highly useful in conserving the quality of fresh-cut produce to extend their shelf life potential (Olivas and Barbosa-Canovas, 2005). Honey dip treatment markedly reduced weight loss (Eman et al., 2015), microbial load (Kumar et al., 2015), conserved visual quality (Ergun and Ergu, 2010), color (Wen et al., 2018), and soluble solid contents (Sabir et al., 2011), and maintained higher enzymatic and non-enzymatic antioxidants during storage in the treated commodities (Kumar et al., 2015). The honey dip treatments delayed post-cut browning of persimmon (Son et al., 2001), apple (Jeon and Zhao, 2005), mango (Supapvanich and Boonyaritthongchai, 2016a), nectarine (Wen et al.,

2018), and papaya (Kumar et al., 2015) fruits due to downregulation of pro-oxidant enzyme activities. The delayed post-cut browning ultimately resulted in the extended shelf life of the treated
fresh-cut fruits (Table 13.4). Recently it has been reported that combined application of honey and
isolate of soy protein conserved the quality and extended the shelf life of fresh-cut pineapples for
16 days at 4°C (Yousuf and Srivastava, 2019).

Table 13.4   Impact of Honey and Propolis Extract on Quality and Shelf Life Potential of Different Fruits
and Vegetables

Preservative	Crop	Concentration	Inference	References
Honey	Persimmon	10%	Conserved soluble solid contents, delayed the development of off-aroma, and maintained fruit firmness	Ergun and Ergu (2010)
	Grapes	20%	Extended storage life and maintained quality with less decay	Sabir et al. (2011)
	Guava	15%	Coated guava fruits showed reduced physiological weight loss and fungal decay, and showed extended shelf life	Eman et al. (2015)
	Papaya	10%	Retained higher ascorbic acid contents and total phenolic contents, and reduced microbial growth during storage	Kumar et al. (2015)
	Nectarine	50%	Improved texture, maintained soluble solids and fruit firmness	Wen et al. (2018)
	Apple	25%	Retained quality and delayed post-cut browning	Jeon and Zhao (2005)
	Mango	25%	Significantly inhibited fresh-cut browning and maintained visual quality	Supapvanich and Boonyaritthongchai (2016a)
	Pineapple	100 and 150 mL honey per liter	Conserved soluble solids, acidity, and ascorbic acid, and suppressed weight loss and ripening index	Yousuf and Srivastava (2019)
Propolis	Tomatoes	10%	Reduced fungal growth, increased soluble solids, conserved sensory characters, and retained overall acceptability	Pobiega et al. (2020)
	Raspberries	—	Reduced pathogens, i.e., *Penicillium digitatum, P. expansum, P. italicum*, and *Alternaria alternata* with extended shelf life	Moreno et al. (2020)
	Strawberries	10%	Showed lower weight loss, retained higher firmness and antioxidants, and maintained sensory attributes	Martínez-González et al. (2020)
	Dates	4%	Maintained higher firmness, total phenolics, tannins, and total sugars	Elwahab et al. (2019)
	Celery, leek, and butternut squash	—	Reduced microbial load and suppressed the activity of polyphenol oxidase enzyme with lower browning and extended shelf life	Alvarez et al. (2017)
	Banana	2.5%	Retained higher soluble solids, suppressed sugar acid ratio, delayed acidity loss, and maintained taste	Passos et al. (2015)

Propolis extract has been extensively used as a post-harvest edible coating of whole and/or fresh-cut fruits and vegetables to extend their shelf life by delaying senescence (Zahid et al., 2013). Resinous propolis edible coating significantly reduced moisture loss, maintained firmness, and conserved higher eating quality attributes and during storage (Zahid et al., 2013; Elwahab et al., 2019; Martínez-González et al., 2020). Compared with honey dip treatment, the propolis mixture reduced the microbial spoilage and disease incidence of fresh produce at a higher rate (Moreno et al., 2020). Pobiega et al. (2020) reported that cherry tomatoes treated with propolis ethanolic extract combined with pullulan coating significantly inhibited the proliferation of microorganisms having maintained quality. Similarly, gelatin edible coating enriched with propolis ethanolic extract has been found to reduce the growth of *Penicillium digitatum* and *Botrytis cinerea* in raspberries (Moreno et al., 2020). In addition, propolis extract substantially inhibited microbial growth, delayed enzymatic browning, and maintained the eating quality of fresh-cut mixed vegetables during storage (Alvarez et al., 2017).

13.4.2 Shelf Life Improvement of Juices with Natural Preservatives

Fresh vegetables and fruits are highly perishable and are prone to significantly higher losses during post-harvest supply chains (Shurekha et al., 2010). The losses of fresh vegetables and fruits are estimated as 25–50% worldwide. The said losses are mostly due to a lack of sufficient conservation facilities and insufficient processing. Fruit juices are one of the processed products generally acquired by squeezing with or without particular specific treatments. The processed juices are alternative ways of consuming vegetables and fruits (Pandey and Negi, 2018). However, juices are prone to post-processing deterioration and have a short shelf life. Thus, post-harvest strategies are required that can be applied to extend the shelf life of juices.

13.4.2.1 Natural Preservatives and Shelf Life Extension of Juices

Different types of strategies have been reported for the prevention of juice spoilage. Among the different post-harvest strategies, chemical preservatives and pasteurization are the most common in the juice processing industry (Pandey and Negi, 2018). However, some of the used chemical preservatives are not permitted because they lead to ailments in some consumers. Some natural preservatives are desperately required worldwide. There are different sources of natural preservatives such as plant extracts, essential oils, bacteriocins, and edible coatings. Each natural preservative has its advantages and disadvantages. The following natural preservatives have been reported for the improvement of shelf life in different juices.

13.4.2.2 Plant Extracts

Different spices and herbs, which are generally utilized as food fragrances and flavors, also possess specified antimicrobial activities (Nychas and Skandamis, 2003). The use of tea catechins and extracts of green tea have been reported as natural preservatives in the juice industry (Pandey and Negi, 2018). The shelf life of watermelon juice was markedly extended with the addition of green tea extract as the extract suppressed *Listeria monocytogenes*- and *Staphylococcus aureus*-based infestation, respectively (Lee et al., 2003). The extracts of citrus peel also have strong antimicrobial potential. Numerous plant extracts such as sage, clove, rosemary, garlic, coriander, and onion have also been used as natural preservatives in foods. The microbial load was significantly reduced in apple juice due to the addition of cinnamon extract (Friedman et al., 2004). The highest inhibition was observed in apple juice in response to the addition of clove, lemongrass, and rosemary extracts (Kanako et al., 1998). The application of pepper and squash extracts led to significantly reduced browning and better color retention in apple juice (Iyidogan and Bayindirh, 2004; Roshita et al.,

2004). The activity of the polyphenol oxidase enzyme was significantly suppressed in response to the application of celery, cabbage, and fenugreek leaf extract in apple juice (Eissa and Salama, 2002). The extract of citrus has been used to successfully extend the shelf life of fruit and some vegetable juices (Fisher and Phillips, 2008). It has been reported that the extract of lemon inhibited spoilage-related microorganisms (Conte et al., 2007). The application of mulberry, moringa, and spearmint extracts conserved the quality and extended the shelf life of pineapple juice (Arabshahi-Delouee and Urooj, 2007). However, some more detailed investigation is still required to fully explore the potential of plant extracts as natural preservatives in the juice industry.

13.4.2.3 Essential Oils

Essential oils are oily liquids of an aromatic nature that are generally acquired from seeds, buds, flowers, twig barks, leaves, roots, fruits, and herbs. They are generally removed by extraction, fermentation, or distillation (Massilia et al., 2009). Essential oils have been found effective for the stabilization of juices. Geraniol and lemongrass essential oils were found effective against *Salmonella* spp., *Escherichia coli*, *Listeria* spp. in pear, and apple, and melon juice kept at 35°C (Friedman et al., 2004). The use of cinnamon essential oil has been reported to conserve the quality of apple cider (Ceylan et al., 2004). In a few cases cinnamic aldehyde was significantly more effective compared with potassium sorbate, a common chemical preservative (Muthuswamy et al., 2007). Use of *Mentha spicata* essential oil conserved the quality of mango, guava, cashew, and pineapple juices (da Cruz Almeida et al., 2018). It has also been reported that orange essential oil conserved the quality of orange and apple juices, respectively (Bento et al., 2020).

13.4.2.4 Edible Coatings

Generally, edible coatings have been utilized for the quality preservation of fresh fruits, vegetables, meat, and seafood. However, the use of edible coatings in the quality preservation of juices is limited. Among the different coatings, chitosan has good antimicrobial properties and has been used for the conservation of juices (Rhoades and Roller, 2000). The quality of apple juice was significantly higher and spoilage induced by the yeast population was markedly lower when chitosan glutamate was incorporated (Roller and Covill, 1999). The 0.3 g L^{-1} chitosan addition suppressed yeasts in the pasteurized elderflower-apple juice for 13 days during storage at 7°C (Rhoades and Roller, 2000). The use of edible coatings in the preservation of juices is very limited and needs comprehensive exploration in the near future.

13.4.3 Shelf Life Improvement of Seafood with Natural Preservatives

13.4.3.1 Seafood and Deterioration

Seafood is an important part of the human diet. Seafood includes numerous species of crustaceans, fish, echinoderms, and mollusks. Seafood is very popular among people because it is a delicacy and has high nutritional value as it is an excellent source of fat, protein, minerals, and vitamins (Viji et al., 2017). However, rapid biochemical reactions and microbial infestations negatively affect the quality of seafood and limit its shelf life (Olatunde and Benjakul, 2018). In general, seafood is an abundant source of various polyunsaturated fatty acid contents, which are susceptible to prompt lipid oxidation. The unpalatable flavor and odor as well as loss of certain nutrients, unhealthy molecule production, and color changes are the leading negative effects of lipid oxidation in seafood (Secci and Parisi, 2016). In the same way, microbial spoilage, chemical deterioration, and enzymatic browning also limit the shelf life potential of seafood worldwide (Olatunde and Benjakul, 2018).

13.4.3.2 Natural Preservatives and Shelf Life Potential of Seafood

Various types of natural preservatives have been used to extend the shelf life of seafood. Among the natural preservatives used, essential oils, plant extracts, edible coatings, bacteriocins, and bioactive peptides are the most common for the shelf life extension of seafood (Olatunde and Benjakul, 2018).

13.4.3.3 Essential Oils and Plant Extracts

Due to increased health concerns, the demand for natural antimicrobials, and antioxidants has been greatly increased in recent years (Soto et al., 2015; Olatunde and Benjakul, 2018). The presence of various phenolic compounds and essential oils can disrupt cell membrane integrity due to the interaction with membrane proteins of fungal and bacterial pathogens. Due to the increased cell membrane permeability, the phenolic compounds and essential oils lead to potassium ion leaching and various cytoplasmic structures, eventually resulting in cell death of the spoilage-related microorganism (Bajpai et al., 2008). The leaching of certain ions also leads to the cell death of microorganisms and cell disruption of bacteria (Ajaiyeoba et al., 2003). These effects are very advantageous and can be used as eco-friendly preservation of seafood products without causing any serious health threat to consumers. Different approaches have been reported for the application of essential oils and plant extracts as natural preservatives to extend the shelf life of seafood (Olatunde and Benjakul, 2018). The impact of black seed, rosemary, thyme, bay leaf, grape seed, sage, lemon essential oil, and flaxseed was investigated on chub mackerel during storage at −20°C (Erkan and Bilen, 2010). According to the sensory panelists, grape seed, lemon essential oil, and bay leaf extended the shelf life of chub mackerel for 7 months compared with the control sample as these had a shelf life of 6 months. In the same way, thyme essential oil extended the shelf life of shrimp for 14 days compared with controls whose shelf life was extended for 5 days (Mastromatteo et al., 2010). The essential oil of black cumin, rosemary, bay leaf, and lemon also extended the shelf life of hot smoked rainbow trout (Erkan et al., 2011). The 8% gelatin films with laurel essential oil (1%) extended the shelf life of rainbow trout for 22 days compared with the control, which had a shelf life of only 15 days (Alparslan et al., 2014). The fish protein hydrolysate agar films incorporated with clove essential oil also extended the shelf life of flounder fillets (da Rocha et al., 2018). Coconut husk extract in combination with modified atmospheric packaging (MAP) significantly extended the shelf life of Asian sea bass slices (Olatunde et al., 2019). Similarly, liposomal ethanolic coconut husk extract combined with cold plasma substantially extended the shelf life of Asian sea bass slices during modified atmosphere storage (Olatunde et al., 2020).

13.4.3.4 Edible Coatings

Chito-oligosaccharide and chitosan are the most commonly used edible coatings in the shelf life extension of seafood. These can be applied as a coating agent, additives, and/or wrapping films, and they are considered eco-friendly strategies for the inactivation of microorganisms in seafood (Ganguly, 2013; Olatunde and Benjakul, 2018). Chitosan works as an oxygen barrier and antioxidant, therefore, it lowers the rate of respiration, which can suppress the growth of microorganisms and lead to a longer shelf life for seafood (No et al., 2007; Olatunde and Benjakul, 2018). Both chitooligosaccharides and chitosan have antimicrobial properties against molds, yeasts, and bacteria (Raafat and Sahl, 2009). Microbial infestation was suppressed by chitosan application (in 1.5% lactic acid and 1.5% acetic acid), which led to extended shelf life for 12 days compared with the control shelf life period of only 6 days for brown trout (Alak, 2012). The 2% chitosan film prepared in 1% acetic acid solution reduced microbial infestation of sea bass fillets with a shelf life of 27 days compared with only 3 days for the control (Gunlu and Koyun, 2013). The integrated use of chitosan with other preservatives also has been explored to extend the shelf life of seafood.

Chitosan-coated shrimp along with garlic essential oil had 15 days shelf life compared with only 5 days shelf life for control samples (Asik and Candogan, 2014). The combined use of 0.5–2% orange peel essential oil and chitosan film resulted in a markedly extended shelf life for deep water pink shrimp (Alparslan and Baygar, 2014). The use of Persian lime peel essential oil and chitosan-gelatin coating markedly extended the shelf life of rainbow trout (Sarmast et al., 2019). In the same way, chitosan coating combined with chlorogenic acid suppressed microbial infestation along with conserved sensory quality and showed markedly extended shelf life of snakehead fish during cold storage (Cao et al., 2020).

13.4.3.5 Bacteriocins

Some gram-positive and gram-negative bacteria produce polypeptides or proteins with antimicrobial potential. These antimicrobial substances are known as bacteriocins (Zacharof and Lovitt, 2012), which efficiently combat numerous spoilage and pathogen microorganisms through specific classes and types (Perez et al., 2015; Olatunde and Benjakul, 2018). Bacteriocins have been commonly used to control spoilage and pathogenic bacteria in fish and other types of seafood. The integrated use of bacteriocins and other preservatives has further increased their efficacy to control the spoilage-based deterioration of seafood. Bacteriocins produced by *Bacillus* sp. showed excellent bactericidal potential to suppress the growth of *Vibrio* sp. and *Salmonella* sp. in marine fish (Ashwitha et al., 2017). Application of *Zataria multiflora*, nisin, and *L. monocytogenes* significantly suppressed the growth of *Vibrio parahaemolyticus* on salted fish fillets (Ekhtiarzadeh et al., 2012). The combined application of bacteriocin EFL4 and sodium alginate also significantly extended the shelf life of fresh salmon fillets (Mei et al., 2020).

13.4.3.6 Bioactive Peptides

Certain specific protein fragments are commonly known as bioactive compounds. These are not only potential and resourceful amino acids sources, they also possess antimicrobial properties (Gomez-Guillen et al., 2010). These bioactive peptides can be produced from protein hydrolysis with certain proteases. These peptides consist of 2–20 amino acids, generally encrypted with a specified parent protein sequence (Chalamaiah et al., 2012). The bioactivity of the peptides generally depends on the conformation, size, sequence, and composition of amino acids (Olatunde and Benjakul, 2018). The treatment with grass carp protein hydrolysate suppressed lipid peroxidation, and microbial infestation on fish minced at −10°C (Li et al., 2015). The application of 1.5% fish protein hydrolysate substantially extended the shelf life of minced silver carp meat during storage at 4°C. This treatment of fish protein hydrolysate extended the shelf life for about 12 days compared with only 6 days for the control meat sample (Pezeshk et al., 2017). Overall, the use of bioactive peptides in seafood is very limited but has tremendous potential for shelf life extension, and it should be investigated further in the future.

13.4.4 Shelf Life Improvement of Meat/Meat Products with Natural Preservatives

13.4.4.1 Meat and Shelf Life

Meat is an important dietary component that provides protein, minerals, vitamins, and a low quantity of lipids and carbohydrates. The oxidation of pigments and lipoperoxidation negatively affects the quality of meat and nutrients in the meat products (Alirezalu et al., 2020). On the other side, meat, and meat products also provide saturated fats, salt, and cholesterol. The meat products and meat are prone to protein and lipid oxidation and the quality deteriorates during processing and under storage (Fernandes et al., 2017). The oxidation reactions negatively affect color pigments, texture, flavor, proteins, nutrients, and lipids (Devatkal et al., 2014). Therefore, some eco-friendly preservatives are required that can preserve the nutrients and extend the shelf life of meat and meat products after processing and during supply chains.

13.4.4.2 Plant Extracts and Essential Oils for Meat/Meat Products Shelf Life Extension

Shah et al. (2014, 2015) has revived the use of natural antioxidants in meat and meat products. *Moringa oleifera* leaf extract was used to increase the shelf life of raw beef stored under high-oxygen MAP. Olive leaf extract increased the radical scavenging activity in bovine muscles (Hayes et al., 2009). Similarly, lavender essential oil added in minced beef led to suppressed lipid oxidation and extended the shelf life compared with untreated samples (Djenane et al., 2012). The green tea extract added in frankfurter sausages had significantly higher antioxidants in treated samples compared with untreated control (Alirezalu et al., 2017). Oregano essential oil has strong antioxidant activity and can be used to suppress lipid peroxidation in meat and meat products (Fernandes et al., 2017). Black pepper has also been used to conserve the quality of sausages. Similarly, peppermint contains a significantly higher concentration of menthol, with a strong antioxidant potential to be used in meat and meat products (Ghazaghi et al., 2014). The use of sage essential oil conserved meat quality and suppressed peroxidation (Zhang et al., 2013). Application of garlic essential oil exhibited strong antimicrobial potential, suppressed *L. monocytogenes*, and extended shelf life up to 15 days in ready-to-eat meat (Suet-Yen et al., 2014). One percent olive leaf extract conserved the quality of grilled ground beef patties and resulted in reduced microbial load (Liliana et al., 2013). Similarly, oregano essential oil could also be used as a natural preservative because it has a significantly higher content of thymol and carvacrol. The application of oregano essential oil conserved the quality of beef slices along with extended shelf life (Oussalah et al., 2004). The quality of ground beef was significantly better in response to the use of Mexican oregano essential oil that was used in place of butylated hydroxytoluene and resulted in a significant extension of shelf life during storage at 4°C (Cantú-Valdéz et al., 2020). The use of rosemary has also been reported for the extension of shelf life and maintenance of meat and meat product quality (Fernandez-Lopez et al., 2005). It has also been noted that feeding lamb with grape pomace enhanced the quality and extended the shelf life of lamb meat up to 9 days during post-slaughter packaging conditions (Chikwanha et al., 2019). Detailed use of fruit-based antioxidants of natural origin to conserve the quality of meat products and meat has been reported by Ahmad et al. (2015). The application of litchi pericarp extract conserved the quality of sheep meat nuggets for 12 days compared with control (Das et al., 2016). In the same way, the combined use of grape pomace and citrus dietary pulp extended the beef shelf life with conserved quality for 9 days (Tayengwa et al., 2020). The combined application of grapefruit seed extract along with nisin and cinnamaldehyde suppressed the increase in microbial infestation and conserved the quality of beef with extended shelf life under vacuum skin packaging storage (Yu et al., 2020). The use of *Ephedra alata* conserved the quality along with the extended shelf life of minced beef meat at 4°C conditions for 14 days (Elhadef et al., 2020). The potential of date fruits pits to extend the shelf life and to conserve the quality of beef burgers has also been reported (Sayas-Barberá et al., 2020).

13.4.4.3 Edible Coatings for Shelf Life Extension of Meat/Meat Products

Meat and meat products are prone to rapid microbial-based infestations and oxidation-induced deterioration (Ribeiro et al., 2019). The microbial infections and oxidative deterioration significantly reduce the eating quality and market potential of the meat and meat products. The microorganisms are harmful to humans and may cause some serious ailments in consumers (Torngren et al., 2018). The microorganism-based infestations and oxidation that induces quality deterioration should be managed to extend the shelf life of meat and meat products. Edible coatings and films are being used in various industries for quality maintenance and shelf life prolongation of the commodities. Pectin-fish gelatin-based edible film enriched with 3,4-dihydroxyphenylglycol and hydroxytyrosol preserved the quality of beef meat for 7 days at 4°C (Bermúdez-Oria et al., 2019). Application of alginate-based film developed by using turmeric powder conserved the quality of chicken breast and beef loin with extended shelf life during storage at 4°C (Bojorges et al., 2020). The application of chitosan at a 2% concentration increased the storage stability of Harbin red sausages by decreasing lipid oxidation and conserving water content during ambient conditions (Dong et al., 2020). Chitosan coating enriched with *Zataria multiflora* and *Bunium persicum* conserved the quality of turkey meat due to suppressed microbial infestation that subsequently resulted in the extended shelf life of 18 days (Keykhosravy et al., 2020). Gelatin-chitosan-based non-emulsion coatings enriched with ε-poly-L-lysine and rosemary extract extended the shelf life of carbonado chicken for 16 days at 4°C (Huang et al., 2020b). The use of balangu seed mucilage in combination with cumin essential oil conserved the quality and extended the shelf life of beef slices during 9 days of storage (Behbahani et al., 2020).

13.4.4.4 Bacteriocins

Gram-positive and/or gram-negative bacteria produce polypeptides or proteins with strong antimicrobial potential; the said antimicrobial substances are known as bacteriocins (Zacharof and Lovitt, 2012). Bacteriocins have good antimicrobial action and efficiently combat various spoilage and pathogen-associated microorganisms through specific classes and types of bacteriocins (Perez et al., 2015; Olatunde and Benjakul, 2018). It was noted that bacteriocin application with nisin films decreased *L. monocytogenes* on frankfurters and led to significantly extended shelf life during storage at 4°C for 28 days (Lungu and Johnson, 2005). Similarly, hydrophobic films prepared from soy and whey efficiently worked as excellent delivery carriers of nisin and were found suitable as antimicrobial packages (Benbettaïeb et al., 2018). Collagen and sakacin G casings showed strong bactericidal property against *L. monocytogenes* in meat emulsion at 5°C for 35 days (Rivas et al., 2018).

13.5 CONCLUSION

In conclusion, various natural preservatives such as edible coatings, essential oils, plant extracts, organic acids, bioactive peptides, and bacteriocins possess a good potential for shelf life extension and quality preservation of the treated edible produce. However, the efficacy of natural preservatives is debatable compared with the application of synthetic and/or chemical preservatives. In addition, several potential natural preservatives have been investigated only on limited commodities, and comprehensive research work is still required to further increase the knowledge regarding the use of natural preservatives on a produce-by-produce basis. On the whole, natural preservatives have mostly been used under in vivo conditions, and further research must be varied on a commercial scale.

REFERENCES

Abdel-Aziz S.M., Asker M.M.S., Keera A.A., Mahmoud M.G., 2016. Microbial Food Spoilage: Control Strategies for Shelf Life Extension. In: Garg N., Abdel-Aziz S., Aeron A. (eds) Microbes in Food and Health. Springer, Cham. https://doi.org/10.1007/978-3-319-25277-3_13.

Abu-Shama, H.S., Abou-Zaid, F.O.F., El-Sayed, E.Z., 2020. Effect of using edible coatings on fruit quality of Barhi date cultivar. Sci. Hortic. 265: 109262.

Adiletta, G., Pasquariello, M.S., Zampella, L., Mastrobuoni, F., Scortichini, M., Petriccione, M. 2018. Chitosan coating: a postharvest treatment to delay oxidative stress in loquat fruits during cold storage. Agronomy. 8: 54. https://doi.org/10.3390/agronomy8040054.

Ahmad, S.R., Gokulakrishnan, P., Giriprasad, R., Yatoo, M.A., 2015. Fruit-based natural antioxidants in meat and meat products: A review. Critic. Rev. Food Sci. Nutr. 55(11): 1503–1513.

Ajaiyeoba, E., Onocha, P., Nwozo, S., Sama, W., 2003. Antimicrobial and cytotoxicity evaluation of *Buchholzia coriacea* stem bark. Fitoterapia. 74: 706–709.

Alak, G., 2012. The effect of chitosan prepared in different solvents on the quality parameters of brown trout fillets (*Salmo trutta* fario). Food Nutr. Sci. 3: 1303. https://doi.org/10.4236/fns.2012.39172.

Ali, S., Akbar Anjum, M., Nawaz, A., Naz, S., Ejaz, S., Shahzad Saleem, M., Ul Hasan, M., 2021. Effect of gum Arabic coating on antioxidative enzyme activities and quality of apricot (*Prunus armeniaca* L.) fruit during ambient storage. J. Food Biochem. 45(4): e13656.

Ali, S., Anjum, M.A., Nawaz, A., Naz, S., Ejaz, S., Sardar, H., Saddiq, B., 2020a. Tragacanth gum coating modulates oxidative stress and maintains quality of harvested apricot fruits. Int. J. Biol. Macromol. 163: 2439–2447.

Ali, S., Anjum, M.A., Nawaz, A., Naz, S., Hussain, S., Ejaz, S., Sardar, H., 2020b. Effect of pre-storage ascorbic acid and *Aloe vera* gel coating application on enzymatic browning and quality of lotus root slices. J. Food Biochem. 44(3): e13136. https://doi.org/10.1111/jfbc.13136.

Ali, S., Khan, A.S., Anjum, M. A., Nawaz, A., Naz, S., Ejaz, S., Hussain, S., 2019a. *Aloe vera* gel coating delays post-cut surface browning and maintains quality of cold stored lotus (*Nelumbo nucifera* Gaertn.) root slices. Sci. Hortic. 256: 108612.

Ali, S., Khan, A.S., Anjum, M.A., Nawaz, A., Naz, S., Ejaz, S., Hussain, S., 2020c. Effect of postharvest oxalic acid application on enzymatic browning and quality of lotus (*Nelumbo nucifera* Gaertn.) root slices. Food Chem. 312: 126051.

Ali, S., Khan, A.S., Nawaz, A., Anjum, M.A., Naz, S., Ejaz, S. Hussain, S., 2019b. *Aloe vera* gel coating delays postharvest browning and maintains quality of harvested litchi fruit. Postharvest Biol. Technol. 157: 110960.

Alirezalu, K., Hesari, J., Eskandari, M.H., Valizadeh, H., Sirousazar, M., 2017. Effect of green tea, stinging nettle and olive leaves extracts on the quality and shelf life stability of frankfurter type sausage. J. Food Process. Preserv. 41: 1–11.

Alirezalu, K., Pateiro, M., Yaghoubi, M., Alirezalu, A., Peighambardoust, S.H., Lorenzo, J.M., 2020. Phytochemical constituents, advanced extraction technologies and techno-functional properties of selected Mediterranean plants for use in meat products. A comprehensive review. Trend. Food Sci. Technol. 100: 292–306.

Alparslan, Y., Baygar, T., Baygar, T., Hasanhocaoglu, H., Metin, C., 2014. Effects of gelatin-based edible films enriched with laurel essential oil on the quality of rainbow trout (*Oncorhynchus mykiss*) fillets during refrigerated storage. Food Technol. Biotechnol. 52: 325–333.

Alvarez, M.V., Ponce, A.G., Goyeneche, R., Moreira, M.R., 2017. Physical treatments and propolis extract to enhance quality attributes of fresh-cut mixed vegetables. J. Food Process. Preserv. 41: e13127. https://doi.org/10.1111/jfpp.13127.

Anjum, M.A., Akram, H., Zaidi, M., Ali. S., 2020. Effect of gum Arabic and *Aloe vera* gel based edible coatings in combination with plant extracts on postharvest quality and storability of 'Gola' guava fruits. Sci. Hortic. 271: 109506.

Aquino, A.B.D., Blank, A.F., Santana, L.C.L.D.A., 2015. Impact of edible chitosan–cassava starch coatings enriched with *Lippia gracilis* Schauer genotype mixtures on the shelf life of guavas (*Psidium guajava* L.) during storage at room temperature. Food Chem. 171: 108–116.

Arabshahi-Delouee, S., Urooj, A., 2007. Application of phenolic extracts from selected plants in fruit juice. Int. J. Food Prop. 10(3): 479–488.

Ashwitha, A., Thamizharasan, K., Vithya, V., Karthik, R., 2017. Effectiveness of bacteriocin from *Bacillus subtilis* (KY808492) and its application in biopreservation. J. Fish. Sci. 11: 36–42.

Asik, E., Candogan, K., 2014. Effects of chitosan coatings incorporated with garlic oil on quality characteristics of shrimp. J. Food Qual. 37: 237–246.

Bajpai, V.K., Shukla, S., Kang, S.C., 2008. Chemical composition and antifungal activity of essential oil and various extract of *Silene armeria* L. Bioresour. Technol. 99: 8903–8908.

Ban, Z., Zhang, J., Li, L., Luo, Z., Wang, Y., Yuan, Q., Zhou, B., Liu., H., 2020. Ginger essential oil based microencapsulation as an efficient delivery system for the improvement of Jujube (Ziziphus jujuba Mill.) fruit quality. Food Chem. 306: 125628.

Banu, T.A., Ramani, S.P., School, A.S., 2020. Effect of seaweed coating on quality characteristics and shelf life of tomato (*Lycopersicon esculentum* mill). Food Sci. Human Wellness. 9: 176–183.

Baptista, R.C., Horita, C.N., Sant'Ana, A.S., 2020. Natural products with preservative properties for enhancing the microbiological safety and extending the shelf-life of seafood: a review. Food Res. Int. 127: 108762.

Behbahani, B.A., Noshad, M., Jooyandeh, H., 2020. Improving oxidative and microbial stability of beef using Shahri Balangu seed mucilage loaded with cumin essential oil as a bioactive edible coating. Biocatal. Agric. Biotechnol. 101563. https://doi.org/10.1016/j.bcab.2020.101563.

Benbettaïeb, N., Debeaufort, F., Karbowiak, T., 2018. Bioactive edible films for food applications: Mechanisms of antimicrobial and antioxidant activity. Critic. Rev. Food Sci. Nutr. 1–25. https://doi.org/10.1080/10408398.2018.1494132.

Bento, R., Pagán, E., Berdejo, D., de Carvalho, R.J., García-Embid, S., Maggi, F., Pagán, R., 2020. Chitosan nanoemulsions of cold-pressed orange essential oil to preserve fruit juices. Int. J. Food Microbiol. 331: 108786.

Bermúdez-Oria, A., Rodríguez-Gutiérrez, G., Rubio-Senent, F., Fernández-Prior, Á., Fernández-Bolaños, J., 2019. Effect of edible pectin-fish gelatin films containing the olive antioxidants hydroxytyrosol and 3, 4-dihydroxyphenylglycol on beef meat during refrigerated storage. Meat Sci. 148: 213–218.

Bojorges, H., Ríos-Corripio, M.A., Hernández-Cázares, A.S., Hidalgo-Contreras, J.V., Contreras-Oliva, A., 2020. Effect of the application of an edible film with turmeric (*Curcuma longa* L.) on the oxidative stability of meat. Food Sci. Nutr. https://doi.org/10.1002/fsn3.1728.

Bolívar-Monsalve, J., Ramírez-Toro, C., Bolívar, G., Ceballos-González, C., 2019. Mechanisms of action of novel ingredients used in edible films to preserve microbial quality and oxidative stability in sausages-A review. Trend. Food Sci. Technol. 89: 100–109.

Brewer, M.S., 2011. Natural antioxidants: sources, compounds, mechanisms of action, and potential applications. Comp. Rev. Food Sci. Food Saf. 10: 221–247.

Cantú-Valdéz, J.A., Gutiérrez-Soto, G., Hernández-Martínez, C.A., Sinagawa-García, S.R., Quintero-Ramos, A., Hume, M.E., Méndez-Zamora, G., 2020. Mexican oregano essential oils as alternatives to butylated hydroxytoluene to improve the shelf life of ground beef. Food Sci. Nutr. https://doi.org/10.1002/fsn3.1767.

Cao, X., Islam, M.N., Chitrakar, B., Duan, Z., Xu, W., Zhong, S., 2020. Effect of combined chlorogenic acid and chitosan coating on antioxidant, antimicrobial, and sensory properties of snakehead fish in cold storage. Food Sci. Nutr. 8(2): 973–981.

Ceylan, E., Fung, D.Y.C., Sabah, J.R., 2004. Antimicrobial activity and synergistic effect of cinnamon with sodium benzoate or potassium sorbate in controlling *Escherichia coli* O157:H7 in apple juice. J. Food Sci. 69: 102–106.

Chaemsanit, S., Sukmas S., Matan, N., and Matan, N., 2019. Controlled release of peppermint oil from paraffin-coated activated carbon contained in sachets to inhibit mold growth during long term storage of brown rice. J. Food Sci. 84: 832–840.

Chai, L.Q., Meng, J.H., Gao, J., Xu, Y., Wang, X.W., 2018. Identification of a crustacean -1, 3-glucanase related protein as a pattern recognition protein in antibacterial response. Fish Shellfish Immunol. 80: 155–164.

Chalamaiah, M., Dinesh Kumar, B., Hemalatha, R., Jyothirmayi, T., 2012. Fish protein hydrolysates: Proximate composition, amino acid composition, antioxidant activities and applications: a review. Food Chem. 135: 3020–3038.

Chen, C., Peng, X., Zeng, R., Chen, M., Wan, C., Chen, J., 2016. *Ficus hirta* fruits extract incorporated into an alginate-based edible coating for Nanfeng mandarin preservation. Sci. Hortic. 202: 41–48.

Chen, H., Sun, Z., Yang, H., 2019. Effect of carnauba wax-based coating containing glycerol monolaurate on the quality maintenance and shelf-life of Indian jujube (*Zizyphus mauritiana* Lamk.) fruit during storage. Sci. Hortic. 244: 157–164.

Chi, H., Song S., Luo M., Zhang C., Li, W., Li, L., Qin, Y., 2019. Effect of PLA nanocomposite films containing bergamot essential oil, TiO_2 nanoparticles, and Ag nanoparticles on shelf life of mangoes. Sci. Hortic. 249:192–198.

Chikwanha, O.C., Moelich, E., Gouws, P., Muchenje, V., Nolte, J.V.E., Dugan, M.E., Mapiye, C., 2019. Effects of feeding increasing levels of grape (*Vitis vinifera* cv. Pinotage) pomace on lamb shelf-life and eating quality. Meat Sci. 157: 107887.

Conte, A., Speranza, B., Sinigaglia, M., Nobile, M.A.D., 2007. Effect of lemon extract on foodborne microorganisms. J. Food Prot. 70: 1896–1900.

da Cruz Almeida, E.T., de Medeiros Barbosa, I., Tavares, J.F., Barbosa-Filho, J.M., Magnani, M., de Souza, E.L., 2018. Inactivation of spoilage yeasts by *Mentha spicata* L. and M. × villosa Huds. essential oils in cashew, guava, mango, and pineapple juices. Front. Microbiol. 9: 1111. 10.3389/fmicb.2018.01111.

da Rocha, M., Aleman, A., Romani, V.P., Lopez-Caballero, M.E., Gómez-Guillen, M.C., Montero, P., Prentice, C., 2018. Effects of agar films incorporated with fish protein hydrolysate or clove essential oil on flounder (*Paralichthys orbignyanus*) fillets shelf-life. Food Hydrocoll. 81: 351–363.

Das, A.K., Rajkumar, V., Nanda, P.K., Chauhan, P., Pradhan, S.R., Biswas, S., 2016. Antioxidant efficacy of litchi (*Litchi chinensis* Sonn.) pericarp extract in sheep meat nuggets. Antioxidants. 5(2): 16.

Das, S., Vishakha, K., Banerjee, S., Mondal, S., Ganguli, A., 2020. Sodium alginate-based edible coating containing nanoemulsion of *Citrus sinensis* essential oil eradicates planktonic and sessile cells of food-borne pathogens and increased quality attributes of tomatoes. Int. J. Biol. Macromol. 162: 1770–1777.

Devatkal, S.K., Thorat, P., Manjunatha, M., 2014. Effect of vacuum packaging and pomegranate peel extract on quality aspects of ground goat meat and nuggets. J. Food Sci. Technol. 51: 2685–2691.

Dhall, R.K., 2013. Advances in edible coatings for fresh fruit and vegetables: A review. Critic. Rev. Food Sci. Nutr. 53: 435–450.

Dhital, R., Joshi, P., Becerra-Mora, N., Umagiliyage, A., Chai, T., Kohli, P., Choudhary, R., 2017. Integrity of edible nano-coatings and its effects on quality of strawberries subjected to simulated in-transit vibrations. LWT – Food Sci. Technol. 80: 257–264.

Dianella, S., 2012. Plant-derived antimicrobial compounds: alternatives to antibiotics. Future Microbiol. 7: 979–990.

Djenane, D., Aïder, M., Yangüela, J., Idir, L., Gómez, D., Roncalés, P., 2012. Antioxidant and antibacterial effects of Lavandula and Mentha essential oils in minced beef inoculated with *E. coli* O157:H7 and *S. aureus* during storage at abuse refrigeration temperature. Meat Sci. 92: 667–674.

Dong, C., Wang, B., Li, F., Zhong, Q., Xia, X., Kong, B., 2020. Effects of edible chitosan coating on Harbin red sausage storage stability at room temperature. Meat Sci. 159: 107919.

Eissa, H.A., Salama, M.F., 2002. Inhibition of enzymatic browning by natural leafy vegetables extracts and keeping quality of fresh and dried apple rings. Polish J. Food Nutr. Sci. 52(3): 27–32.

Ekhtiarzadeh, H., Akhondzadeh Basti, A., Misaghi, A., Sari, A., Khanjari, A., Rokni, N., Partovi, R., 2012. Growth response of *Vibrio parahaemolyticus* and *Listeria monocytogenes* in salted fish fillets as affected by *Zataria multiflora* Boiss. essential oil, nisin, and their combination. J. Food Saf. 32: 263–269.

Elhadef, K., Smaoui, S., Hlima, H.B., Ennouri, K., Fourati, M., Mtibaa, A.C., Mellouli, L., 2020. Effects of *Ephedra alata* extract on the quality of minced beef meat during refrigerated storage: a chemometric approach. Meat Sci. 170: 108246.

El-Mogy, M.M., Parmar, A., Alic, M.R., Abdel-Aziz, M.E., Abdeldaym, E.A., 2020. Improving postharvest storage of fresh artichoke bottoms by an edible coating of *Cordia myxa* gum. Postharvest Biol. Technol. 163: 111143.

Elwahab, S.M.A., Allatif, A.M.A., Farid, M.A., Soliman, S.M., 2019. Effect of safe post-harvest alternatives on quality and storage life of "Barhi" date palm. Plant Arch. 19: 3937–3945.

Eman, El-Moneim, A.A. Abd., Kamel, H.M., Zaki, Z.A., Abo Rehab, M.E., 2015. Effect of honey and citric acid treatments on postharvest quality of fruits and fresh cut of guava. World J. Agric. Sci. 11(5): 255–267.

Ergun, M., Ergu, N., 2010. Extending shelf life of fresh-cut persimmon by honey solution dips. J. Food Process. Preserv. 34: 2–14.

Erkan, N., Bilen, G., 2010. Effect of essential oils treatment on the frozen storage stability of chub mackerel fillets. J. Verbr. Lebensm. 5, 101–110. https://doi.org/10.1007/s00003-009-0546-6.

Erkan, N., Ulusoy, S., Tosun, S.Y., 2011. Effect of combined application of plant extract and vacuum packaged treatment on the quality of hot smoked rainbow trout. J. Verbr. Lebensm. 6: 419–426. https://doi.org/10.1007/s00003-011-0665-8.

Etemadipoor, R., Dastjerdi, A.M., Ramezanian, A., Ehteshami, S., 2020. Ameliorative effect of gum Arabic, oleic acid and/or cinnamon essential oil on chilling injury and quality loss of guava fruit. Sci. Hortic. 266: 109255.

Fan, X.J., Zhang, B., Yan, H., Feng, J.T., Ma, Z.Q., Zhang, X., 2019. Effect of lotus leaf extract incorporated composite coating on the postharvest quality of fresh goji (Lycium barbarum L.) fruit. Postharvest Biol. Technol. 148: 132–140.

FAO, 2020. Sustainable development goals. Food and Agricultural Organization of the United Nations. Accessed on August 12, 2020. http://www.fao.org/sustainable-development-goals/indicators/12.3.1/en/

Farokhian, F., Jafarpour, M., Goli, M., Askari-khorasgani, O., 2017. Quality preservation of air-dried sliced button mushroom (*Agaricus bisporus*) by lavender (*Lavendula angustifolia* mill.) essential oil. Food Process. Eng. 40: e12432.

Fernandes, R.P.P., Trindade, M.A., Tonin, F.G., Pugine, S.M.P., Lima, C.G., Lorenzo, J.M., 2017. Evaluation of oxidative stability of lamb burger with *Origanum vulgare* extract. Food Chem. 233: 101–109.

Fernandez-Lopez, J., Zhi, N., Aleson-Carbonell, L., Pérez-Alvarez, J.A., Kuri, V., 2005. Antioxidant and anti-bacterial activities of natural extracts: Application in beef meatballs. Meat Sci. 69: 371–380.

Fisher, K., Phillips, C., 2008. Potential uses of essential oils in food; is citrus the answer. Trend. Food Sci. Technol. 19: 156–164.

Florez-López, M.L., Cerqueira, M.A., de Rodríguez, D.J., Vicente, A.A., 2016. Perspectives on utilization of edible coatings and nano-laminate coatings for extension of postharvest storage of fruit and vegetables. Food Eng. Rev. 8: 292–305.

Friedman, M., Henika, P.R., Levin, C.E., Mandrell, R.E., 2004. Antibacterial activities of plant essential oils and their components against *Escherichia coli* O157:H7 and *Salmonella enterica* in apple juice. J. Agric. Food Chem. 52: 6042–6048.

Ganguly, S., 2013. Antimicrobial properties from naturally derived substances useful for food preservation and shelf-life extension-a review. Int. J. Bioassays. 2: 929–931.

Gatto, M.A., Ippolito, A., Linsalata, V., Cascarano, N.A., Nigro, F., Vanadia, S., Di Venere, D., 2011. Activity of extracts from wild edible herbs against postharvest fungal diseases of fruit and vegetables. Postharvest Biol. Technol. 61: 72–82.

Ghaly, A.E., Dave, D., Budge, S., Brooks, M., 2010. Fish spoilage mechanisms and preservation techniques. Amer. J. Appl. Sci. 7: 859.

Ghazaghi, M., Mehri, M., Bagherzadeh-Kasmani, F., 2014. Effects of dietary *Mentha spicata* on performance, blood metabolites, meat quality and microbial ecosystem of small intestine in growing Japanese quail. Animal Feed Sci. Technol. 194: 89–98.

Gomez-Guillen, M., Lopez-Caballero, M., Aleman, A., Lopez de Lacey, A., Gimenez, B., Montero, P., 2010. Antioxidant and antimicrobial peptide fractions from squid and tuna skin gelatin. In: Bihan, E.L (ed.) Sea By-Products as a Real Material: New Ways of Application (pp. 89–115). Transworld Research Network, Kerala, India.

González-Saucedo, A., Barrera-Necha, L.L., Ventura-Aguilar, R.I., Correa-Pacheco, Z.N., Bautista-Banos, S., Hernandez-Lopez, M., 2019. Extension of the postharvest quality of bell pepper by applying nanostructured coatings of chitosan with *Byrsonima crassifolia* extract (L.) Kunth. Postharvest Biol. Technol. 149:74–82.

Gunlu, A., Koyun, E., 2013. Effects of vacuum packaging and wrapping with chitosan-based edible film on the extension of the shelf life of sea bass (*Dicentrarchus labrax*) fillets in cold storage (4°C). Food Bioprocess Technol. 6: 1713–1719.

Habibie, A., Yazdani, N., Saba, M.K., Vahdati, K., 2019. Ascorbic acid incorporated with walnut green husk extract for preserving the postharvest quality of cold storage fresh walnut kernels. Sci. Hortic. 245: 193–199.

Hadawey, A., Tassou, S.A., Chaer, I., Sundararajan, R., 2017. Unwrapped food product display shelf life assessment. Energy Procedia. 123: 62–69.

Hammond, S.T., Brown, J.H., Burger, J.R., Flanagan, T.P., Fristoe, T.S., Mercado-Silva, N., Okie, J. G., 2015. Food spoilage, storage, and transport: Implications for a sustainable future. BioScience. 65(8): 758–768.

Haq, M.A., Alam, M.J., Hasnain, A., 2013. Gum cordia: A novel edible coating to increase the shelf life of Chilgoza (*Pinus gerardiana*). LWT – Food Sci. Technol. 50: 306–311.

Hasan, M.U., Ullah Malik, A., Anwar, R., Sattar Khan, A., Haider, M. W., Riaz, R., Ali, S., Ziaf, K., 2021. Postharvest *Aloe vera* gel coating application maintains the quality of harvested green chilies during cold storage. J. Food Biochem. 45(4): e13682.

Hasheminejad, N., Khodaiyan, F., 2020. The effect of clove essential oil loaded chitosan nanoparticles on the shelf life and quality of pomegranate arils. Food Chem. 309: 125520.

Hayes, J. E., Stepanyan, V., Allen, P., O'Grady, M. N., O'Brien, N. M., Kerry, J. P., 2009. The effect of lutein, sesamol, ellagic acid and olive leaf extract on lipid oxidation and oxymyoglobin oxidation in bovine and porcine muscle model systems. Meat Sci. 83: 201–208.

Hu, H., Zhou, H., Li, P., 2019. Lacquer wax coating improves the sensory and quality attributes of kiwifruit during ambient storage. Sci. Hortic. 244: 31–41.

Huang, H., Huang, C., Yin, C., Khan, M.R.U., Zhao, H., Xu, Y., Huang, L., Zheng, D., Qi, M., 2020a. Preparation and characterization of β-cyclodextrin–oregano essential oil microcapsule and its effect on storage behavior of purple yam. J. Sci. Food Agric. https://doi.org/10.1002/jsfa.10545.

Huang, M., Wang, H., Xu, X., Lu, X., Song, X., Zhou, G., 2020b. Effects of nanoemulsion-based edible coatings with composite mixture of rosemary extract and ε-poly-l-lysine on the shelf life of ready-to-eat carbonado chicken. Food Hydrocoll. 102: 105576.

Iyidogan, N., Bayindirh, A., 2004. Effect of L-cysteine, kojic acid and 4-hexylresorcinol combination on inhibition of enzymatic browning in Amasya apple juice. J. Food Eng. 62: 299–304.

Jaiswal, A. K., Kumar, S., Bhatnagar, T., 2017. Studies to enhance the shelf life of tomato using *Aloe vera* and neem-based herbal coating. Aust. J. Sci. Technol. 1(2): 67–71.

Jeon, M., Zhao, Y., 2005. Honey in combination with vacuum impregnation to prevent enzymatic browning of fresh-cut apples. Int. J. Food Sci. Nutr. 56: 165–176.

Kahramanoğlu, İ., 2019. Effects of lemongrass oil application and modified atmosphere packaging on the postharvest life and quality of strawberry fruits. Sci. Hortic. 256: 108527.

Kahramanoğlu, İ., Akta, S.M., Gündüz, S., 2018. Effects of fludioxonil, propolis and black seed oil application on the postharvest quality of "Wonderful" pomegranate. PLoS One. 13: e0198411.

Kanako, M., Ogata, M., Sugawara, M., Kiuchi, K., 1998. Anti-fungal activities of essential oils of spice and sucrose ester of lauric acid against saccharophilic fungus *Wallemia sebi*. J Antibact. Antifungal Agent. 26: 3–10.

Karim, H., Boubaker, H., Askarne, L., Cherifi, K., Lakhar, H., Msanda, F., Ait Ben Aoumar, A., 2017. Use of Cistus aqueous extracts as botanical fungicides in the control of Citrus sour rot. Microb. Pathog. 104: 263–267.

Karimirad, R., Behnamian, M., Dezhsetan, S., 2019. Application of chitosan nanoparticles containing *Cuminum cyminum* oil as a delivery system for shelf life extension of *Agaricus bisporus*. LWT – Food Sci. Technol. 106: 218–228.

Keykhosravy, K., Khanzadi, S., Hashemi, M., Azizzadeh, M., 2020. Chitosan-loaded nanoemulsion containing *Zataria multiflora* Boiss and *Bunium persicum* Boiss essential oils as edible coatings: Its impact on microbial quality of turkey meat and fate of inoculated pathogens. Int. J. Biol. Macromol. 150: 904–913.

Khaliq, G., Ramzan, M., Baloch A.H., 2019. Effect of *Aloe vera* gel coating enriched with *Fagonia indica* plant extract on physicochemical and antioxidant activity of sapodilla fruit during postharvest storage. Food Chem. 286: 346–353.

Khaliq, G., Saleh, A., Bugti, G.A., Hakeem, K.R., 2020. Guggul gum incorporated with basil essential oil improves quality and modulates cell wall-degrading enzymes of jamun fruit during storage. Sci. Hortic. 273: 109608.

Krásniewska, K., Gniewosz, M., Synowiec, A., Przybył, J.L., Aczek, K.B., Eglarz, Z.W., 2015. The application of pullulan coating enriched with extracts from *Bergenia crassifolia* to control the growth of food microorganisms and improve the quality of peppers and apples. Food Bioprod. Process. 94: 422–433.

Kumar, U., Sharma, S., Tadapaneni, V.R.R., 2015. *Aloe vera* gel and honey-based edible coatings combined with chemical dip as a safe means for quality maintenance and shelf life extension of fresh-cut papaya. J. Food Qual. 38: 347–358.

Kykkidou, S., Giatrakou, V., Papavergou, A., Kontominas, M.G., Savvaidis, I.N., 2008. Effect of thyme essential oil and packaging treatments on fresh Mediterranean swordfish fillets during storage at 4°C. Food Chem. 115: 169–175.

Lee, D.U., Heinz, V., Knorr, D., 2003. Effects of combination treatments of nisin and high-intensity ultrasound with high pressure on the microbial inactivation in liquid whole egg. Innov. Food Sci. Emerg. Technol. 4: 387–393.

Li, X., Luo, Y., You, J., Shen, H., 2015. Stability of papain-treated grass carp (*Ctenopharyngodon idellus*) protein hydrolysate during food processing and its ability to inhibit lipid oxidation in frozen fish mince. J. Food Sci. Technol. 52: 542–548.

Li, X.Y., Du, X.L., Liu, Y., Tong, L.J., Wang, Q., Li, J.L., 2019. Rhubarb extract incorporated into an alginate-based edible coating for peach preservation. Sci. Hortic. 257:108685.

Liliana, R., Cody, M.H., Yelena, F., Mendel, F., Sadhana, R., 2013. Concentration dependent inhibition of Escherichia coli O157:H7 and heterocyclic amines in heated ground beef patties by apple and olive extracts, onion powder and clove bud oil. Meat Sci. 94: 461–467.

Lin, D., Zhao, Y., 2007. Innovations in the development and application of edible coatings for fresh and minimally processed fruits and vegetables. Comp. Rev. Food Sci. Food Saf. 6: 60–75.

Lungu, B., Johnson, M.G., 2005. Fate of *Listeria monocytogenes* inoculated onto the surface of model Turkey frankfurter pieces treated with zein coatings containing nisin, sodium diacetate, and sodium lactate at 4°C. J. Food Protec. 68(4): 855–859.

Mannozzi, C., Cecchini, J.P., Tylewicz, U., Siroli, L., Patrignani, F., Lanciotti, R., Rocculi, P., Rosa, M.D., Romani, S., 2017. Study on the efficacy of edible coatings on quality of blueberry fruits during shelf-life. LWT – Food Sci. Technol. 85: 440–444.

Maqbool, M., Ali, A., Alderson, P., 2010. A combination of gum Arabic and chitosan can control anthracnose caused by *Colletotrichum musae* and enhance the shelf-life of banana fruit. J. Hortic. Sci. Biotechnol. 85: 432–436.

Maringgal, B., Hashim, N., Tawakkal, I.S.M.A., Mohamed, M.T.M., 2020. Recent advance in edible coating and its effect on fresh/fresh-cut fruits quality. Trend. Food Sci. Technol. 96: 253–267.

Marino, M., Bersani, C., Comi, G., 2001. Impedance measurement to study antimicrobial activity of essential oils from Lamiaceae and Compositae. Int. J. Food Microbiol. 67: 187–195.

Martínez-González, M.D.C., Bautista-Baños, S., Correa-Pacheco, Z.N., Corona-Rangel, M.L., Ventura-Aguilar, R.I., Río-García, J.C.D., Ramos-García, M.D.L., 2020. Effect of nanostructured chitosan/propolis coatings on the quality and antioxidant capacity of strawberries during storage. Coatings. 90: 1–12.

Massilia, R.M.R., Melgar, R.J.M., Fortuny, O.S., Belloso, M., 2009. Control of pathogenic and spoilage micro-organisms in fresh-cut fruits and fruit juices by traditional and alternative natural antimicrobials. Compr. Rev. Food Sci. Food Safety 8: 157–180.

Mastromatteo, M., Danza, A., Conte, A., Muratore, G., Del Nobile, M.A., 2010. Shelf life of ready to use peeled shrimps as affected by thymol essential oil and modified atmosphere packaging. Int. J. Food Microbiol. 144: 250–256.

Mei, J., Shen, Y., Liu, W., Lan, W., Li, N., Xie, J., 2020. Effectiveness of sodium alginate active coatings containing bacteriocin EFL4 for the quality improvement of ready-to-eat fresh salmon Fillets during cold storage. Coatings. 10(6): 506. https://doi.org/10.3390/coatings10060506.

Misir, J., Brishti, F.H., Hoque, M.M., 2014. *Aloe vera* gel as a novel edible coating for fresh fruits: a review. Am. J. Food Sci. Technol. 2: 93–97.

Moreno, M.A., Vallejo, A.M., Ballester, A.R., Zampini, C., InesIsla, M., Lopez-Rubio, M., Fabra, M.J., 2020. Antifungal edible coatings containing Argentinian propolis extract and their application in raspberries. Food Hydrocoll. 107: 105973.

Muthuswamy, S., Rupasinghe, H.P.V., Stratton, G.W., 2007. Antimicrobial effect of cinnamon bark extract on *Escherichia coli* O157:H7, *Listeria innocua* and fresh-cut apple slices. J. Food Saf. 28: 534–549.

Nasiri, M., Barzegar, M., Sahari, M.A., Niakousari, M., 2017. Tragacanth gum containing *Zataria multiflora* Boiss. essential oil as a natural preservative for storage of button mushrooms (*Agaricus bisporus*). Food Hydrocoll. 72: 202–209.

Nasiri, M., Barzegar, M., Sahari, M.A., Niakousari, M., 2019. Efficiency of Tragacanth gum coating enriched with two different essential oils for deceleration of enzymatic browning and senescence of button mushroom (*Agaricus bisporus*). J. Food Sci. Nutr. 7: 1520–1528.

Nath, A., Barman, K., Chandra, S., Baiswar, P., 2013. Effect of plant extracts on quality of Khasi mandarin (*Citrus reticulata Blanco*) fruits during ambient storage. Food Bioprocess Technol. 6: 470–474.

Ncama, K., Magwaza, L.S., Mditshwa, A., Tesfay, S.Z., 2018. Plant-based edible coatings for managing postharvest quality of fresh horticultural produce: a review. Food Packaging Shelf Life. 16: 157–167.

Nielsen, P.V., Rios, R., 2000. Inhibition of fungal growth on bread by volatile components from species and herbs, and the possible application in active packaging, with special emphasis on mustard essential oil. Int. J. Food Microbiol. 60: 219–229.

No, H., Meyers, S., Prinyawiwatkul, W.; Xu, Z., 2007. Applications of chitosan for improvement of quality and shelf life of foods: a review. J. Food Sci. 72: R87–R100. https://doi.org/10.1111/j.1750-3841.2007.00383.x.

Nourozi, F., Sayyari, M., 2020. Enrichment of *Aloe vera* gel with basil seed mucilage preserve bioactive compounds and postharvest quality of apricot fruits. Sci. Hortic. 262: 109041.

Nychas, G.J.E., Skandamis, P.N., 2003. Antimicrobials from herbs and spices. In: Natural Antimicrobials for the Minimal Processing of Foods (pp. 176–200). CRC Press, Boca Raton, FL.

Olatunde, O. O., Benjakul, S., 2018. Natural preservatives for extending the shelf-life of seafood: a revisit. Comp. Rev. Food Sci. Food Saf. 17:(6): 1595–1612.

Olatunde, O. O., Benjakul, S., Vongkamjan, K., 2019. Combined effect of ethanolic coconut husk extract and modified atmospheric packaging (MAP) in extending the shelf life of Asian sea bass slices. J. Aquat. Food Product Technol. 28(6): 689–702.

Olatunde, O. O., Benjakul, S., Vongkamjan, K., 2020. Cold plasma combined with liposomal ethanolic coconut husk extract: A potential hurdle technology for shelf-life extension of Asian sea bass slices packaged under modified atmosphere. Innovat. Food Sci. Emerg. Technol. 65: 102448.

Olivas, G.I., Barbosa-Canovas, G.V., 2005. Edible coatings for fresh-cut fruits. Crit. Rev. Food Sci. Nutr. 45: 657–670.

Oussalah, M., Caillet, S., Salmiéri, S., Saucier, L., Lacroix, M., 2004. Antimicrobial and antioxidant effects of milk protein-based film containing essential oils for the preservation of whole beef muscle. J. Agric. Food Chem. 52: 5598–5605.

Pandey, A., Negi, P.S., 2018. Use of Natural Preservatives for Shelf Life Extension of Fruit Juices. In Fruit Juices (pp. 571–605). Academic Press.

Passos, F.R., Mendes, F.Q., Cunha, M.C.D., Pigozzi, M.T., Carvalho, A.M.X.D., 2015. Propolis extract in postharvest conservation banana 'Prata.' Rev. Bras. Frutic. 38: 1–11.

Pastor, C., Sánchez-González, L., Marcilla, A., Chiralt, A., Cháfer, M., González- Martínez, C., 2011. Quality and safety of table grapes coated with hydroxypropyl methylcellulose edible coatings containing propolis extract. Postharvest Biol. Technol. 6: 64–70.

Patrignani, F., Siroli, L., Serrazanetti, D.I., Gardini, F., Lanciotti, R., 2015. Innovative strategies based on the use of essential oils and their components to improve safety, shelf-life and quality of minimally processed fruits and vegetables. Trend. Food Sci. Technol. 46: 311–319.

Peralta-Ruiz, Y., Tovar, C.G., Mangonez, A.S., Bermont, D., Cordero, A.P., Paparella, A., Chaves-Lopez, C., 2020. *Colletotrichum gloesporioides* inhibition using chitosan-*Ruta graveolens* L essential oil coatings: Studies *in vitro* and *in situ* on *Carica papaya* fruit. Int. J. Food Microbiol. 326: 108649.

Perez, R.H., Perez, M.T.M., Elegado, F.B., 2015. Bacteriocins from lactic acid bacteria: A review of biosynthesis, mode of action, fermentative production, uses, and prospects. Int. J. Philipp. Sci. Technol. 8: 61–67.

Pezeshk, S., Ojagh, S. M., Rezaei, M., Shabanpour, B., 2017. Antioxidant and antibacterial effect of protein hydrolysis of yellow fin tuna waste on flesh quality parameters of minced silver carp. J. Genetic Res. 3: 103–112.

Pitt J.I., Hocking A.D., 2009. Fungi and Food Spoilage (3rd ed.). Springer.

Pobiega, K., Kraśniewska, K., Gniewosz, M., 2019. Application of propolis in antimicrobial and antioxidative protection of food quality – A review. Trend. Food Sci. Technol. 83: 53–62.

Pobiega, K., Przybył, J.L., Żubernik, J., Gniewosz, M., 2020. Prolonging the shelf life of cherry tomatoes by pullulan coating with ethanol extract of propolis during refrigerated storage. Food Biop. Technol. 13: 1447–1461.

Prakash, B., Kedia, A., Mishra, P.K., Dubey, N.K., 2015. Plant essential oils as food preservatives to control moulds, mycotoxin contamination and oxidative deterioration of agri-food commodities—Potentials and challenges. Food Control. 47: 381–391.

Pushkala, R., Raghuram, P.K., Srividya, N., 2013. Chitosan based powder coating technique to enhance phytochemicals and shelf life quality of radish shreds. Postharvest Biol. Technol. 86: 402–408.

Qadri, O.S., Yousuf, B., Srivastava, A.K., 2016. Fresh-cut produce: Advances in preserving quality and ensuring safety. In: Siddiqi, M.W., Ali, A. (eds.) Post-Harvest Management of Horticultural Crops: Practices for Quality Preservation (pp. 265–290). Apple Academic Press, Palm Bay, FL.

Quintavalla, S., Vicini, L., 2002. Antimicrobial food packaging in meat industry. Meat Sci. 62: 373–380.

Raafat, D., Sahl, H.G., 2009. Chitosan and its antimicrobial potential – a critical literature survey. Microb. Biotechnol. 2: 186–201.

Rhoades, J., Roller, S., 2000. Antimicrobial actions of degraded and native chitosan against spoilage organisms in laboratory media and foods. Appl. Environ. Microbiol. 66: 80–86.

Ribeiro, J.S., Santos, M.J.M.C., Silva, L.K.R., Pereira, L.C.L., Santos, I.A., da Silva Lannes, S.C., da Silva, M.V., 2019. Natural antioxidants used in meat products: a brief review. Meat Sci. 148: 181–188.

Rivas, F.P., Cayré, M.E., Campos, C.A., Castro, M.P., 2018. Natural and artificial casings as bacteriocin carriers for the biopreservation of meats products. J. Food Saf. 38(1): e12419. https://doi.org/10.1111/jfs.12419.

Rizzo, V., Amoroso, L., Licciardello, F., Mazzaglia, A., Muratore, G., Restuccia, C., Lombardo, S., Pandino, G., Strano, M.G., Mauromicale, G., 2018. The effect of sous vide packaging with rosemary essential oil on storage quality of fresh-cut potato. LWT – Food Sci. Technol. 94: 111–118.

Rizzo, V., Lombardo, S., Pandino, G., Barbagallo, R.N., Mazzaglia, A., Restuccia, C., Mauromicale, G., Muratore, G., 2019. Shelf-life study of ready-to-cook slices of globe artichoke 'Spinoso sardo': effects of anti-browning solutions and edible coating enriched with *Foeniculum vulgare* essential oil. J. Sci. Food Agric. 99: 5219–5228.

Roller, S., Covill, N., 1999. The antifungal properties of chitosan in laboratory media and apple juice. Int. J. Food Microbiol. 47: 67–77.

Romanazzi, G., Lichter, A., Gabler, F.M., Smilanick, J.L., 2012. Recent advances on the use of natural and safe alternatives to conventional methods to control postharvest gray mold of table grapes. Postharvest Biol. Technol. 63: 141–147.

Roshita, I., Azizah, O., Nazamid, S., Russly, A.R., 2004. Effects of anti-browning treatments on the storage quality of minimally processed shredded cabbage. Food Agric. Environ. 2: 54–58.

Sabir, A., Sabir, F.K., Kara, Z., 2011. Effects of modified atmosphere packing and honey dip treatments on quality maintenance of minimally processed grape cv. Razaki (*V. vinifera* L.) during cold storage. J. Food Sci. Technol. 48(3): 312–318.

Saleem, M.S., Ejaz, S., Anjum, M.A., Nawaz, A., Naz, S., Hussain, S., Ali, S., Canan, İ., 2020. Postharvest application of gum Arabic edible coating delays ripening and maintains quality of persimmon fruits during storage. J. Food Process. Preserv. 44(8): e14583.

Sarmast, E., Fallah, A. A., Dehkordi, S. H., Rafieian-Kopaei, M., 2019. Impact of glazing based on chitosan-gelatin incorporated with Persian lime (*Citrus latifolia*) peel essential oil on quality of rainbow trout fillets stored at superchilled condition. Int. J. Biol. Macromol. 136: 316–323.

Sayas-Barberá, E., Martín-Sánchez, A.M., Cherif, S., Ben-Abda, J., Pérez-Álvarez, J.Á., 2020. Effect of date (*Phoenix dactylifera* L.) pits on the shelf life of beef burgers. Foods. 9(1): 102.

Secci, G., Parisi, G., 2016. From farm to fork: lipid oxidation in fish products. A review. Italian J. Animal Sci. 15: 124–136.

Shahbazi, Y., 2018. Application of carboxymethyl cellulose and chitosan coatings containing *Mentha spicata* essential oil in fresh strawberries. Int. J. Biol. Macromol. 112: 264–272.

Shah, M.A., Bosco, S.J.D., Mir, S.A., 2014. Plant extracts as natural antioxidants in meat and meat products. Meat Sci. 98: 21–33.

Shah, M.A., Bosco, S.J.D., Mir, S.A., 2015. Effect of *Moringa oleifera* leaf extract on the physicochemical properties of modified atmosphere packaged raw beef. Food Packaging Shelf Life, 3: 31–38.

Shah, B., Davidson, P.M., Zhong, Q., 2013. Nanodispersed eugenol has improved antimicrobial activity against *Escherichia coli* O157: H7 and Listeria monocytogenes in bovine milk. Int. J. Food Microbiol. 161: 53–59.

Sharif, R., Mujtaba, M., Ur Rahman, M., Shalmani, A., Ahmad, H., Anwar, T., Tianchan, D., Wang, X., 2015. The multifunctional role of chitosan in horticultural crops: a Review. Molecules. 23: 872.

Sherratt, T.N., Wilkinson, D.M., Bain, R.S., 2006. Why fruits rot, seeds mold and meat spoils: a reappraisal. Ecol. Model. 192: 618–626

Shurekha, C., Shethankumar, K.V., Lakshmipathy, R., 2010. Enhancement of shelf life of tomatoes using herbal extracts. J. Phytopathol. 2: 13–17.

Silvestre, C., Duraccio, D., Cimmino, S., 2011. Food packaging based on polymer nanomaterials. Prog. Polym. Sci. 36: 1766–1782.

Son, S.M., Moon, K.D., Lee, C.Y., 2001. Inhibitory effects of various antibrowning agents on apple slices. Food Chem. 73: 23–30.

Soto, M.L., Falque, E., Domınguez, H., 2015. Relevance of natural phenolics from grape and derivative products in the formulation of cosmetics. Cosmetics. 2: 259–276.

Sriket, C., 2014. Proteases in fish and shellfish: Role on muscle softening and prevention. Int. Food Res. J. 21: 433–445.

Suet-Yen, S., Lee, T.S., Tiam-Ting, T., Soo-Tueen, B., Rahmat, A.R., Rahman, W.A.W.A., 2014. Control of bacteria growth on ready-to-eat beef loaves by antimicrobial plastic packaging incorporated with garlic oil. Food Control. 39: 214–221.

Supapvanich, S., Boonyaritthongchai, P., 2016a. Visual appearance maintenance of fresh-cut 'Nam Dok Mai' mango fruit by honey dip. Int. Food Res. J. 23: 389–394.

Supapvanich, S., Mitrsang, P., Srinorkham, P., Boonyaritthongchai, P., Wongs-Aree, C., 2016b. Effects of fresh *Aloe vera* gel coating on browning alleviation of fresh cut wax apple (*Syzygium samarangenese*) fruit cv. Taaptimjaan. J. Food Sci. Technol. 53(6): 2844–2850.

Taheri, A., Behnamian, M., Dezhsetan, S., Karimirad, R., 2020. Shelf life extension of bell pepperby application of chitosan nanoparticles containing *Heracleum persicum* fruit essential oil. Postharvest Biol. Technol. 170: 111313.

Tayengwa, T., Chikwanha, O.C., Gouws, P., Dugan, M.E., Mutsvangwa, T., Mapiye, C., 2020. Dietary citrus pulp and grape pomace as potential natural preservatives for extending beef shelf life. Meat Sci. 162: 108029.

Tesfay, S.Z., Magwaza, L.S., 2017. Evaluating the efficacy of moringa leaf extract, chitosan andcarboxymethyl cellulose as edible coatings for enhancing quality and extending postharvest life of avocado (*Persea americana* Mill.) fruit. Food Packaging Shelf Life. 11: 40–48.

Thakur, R., Pristijono, P., Bowyer, M., Singh, S.P., Scarlett, C.J., Stathopoulos, C.E., 2019. A starch edible surface coating delays banana fruit ripening. LWT – Food Sci. Technol. 100: 341–347.

Torngren M.A., Darre M., Gunvig A., Bardenshtein A., 2018. Case studies of packaging and processing solutions to improve meat quality and safety. Meat Sci. 144: 149–158.

Vazquez, B.I., Fente, C., Franco, C.M., Vazquez, M.J., Cepeda, A., 2001. Inhibitory effects of eugenol and thymol on *Penicillium citrinum* strains in culture media and cheese. Int. J. Food Microbiol. 67: 157–163.

Viji, P., Venkateshwarlu, G., Ravishankar, C., Gopal, T.S., 2017. Role of plant extracts as natural additives in fish and fish products-A review. Fish. Technol. 54: 145–154.

Viuda-Martos, M., Ruiz-Navajas, Y., Fernández-López, J., Pérez-Alvarez, J.A., 2008. Functional properties of honey, propolis, and royal jelly. J. Food Sci. 73: 117–124.

Wen, B., Wu, X., Boon-Ek, Y., Xu, L., Pan, H., Xu, P. Wu, Y., Supapvanich, S., 2018. Effect of honey and calcium dips on quality of fresh-cut nectarine (*Prunus persica* L. Batsch). Agric. Nat. Res. 52: 140–145.

Xu, W., Qu, W., Huang, K., Guo, F., Yang, J., Zhao, H., 2007. Antibacterial effect of grapefruit seed extract on food-borne pathogens and its application in the preservation of minimally processed vegetables. Postharvest Biol. Technol. 45: 126–133.

Yang, W., Wang, L., Ban, Z., Yan, J., Lu, H., Zhang, X., Wu, Q., Aghdam, M.S., Luo, Z., Li, L., 2020. Efficient microencapsulation of Syringa essential oil; the valuable potential on quality maintenance and storage behavior of peach. Food Hydrocoll. 95: 177–185.

Yousuf, B., Srivastava, A.K., 2017. Flaxseed gum in combination with lemongrass essential oil as an effective edible coating for ready-to-eat pomegranate arils. Int. J. Biolog. Macromol. 104: 1030–1038.

Yousuf, B., Srivastava, A.K., 2019. Impact of honey treatments and soy protein isolate-based coating on fresh-cut pineapple during storage at 4°C. Food Packaging Shelf Life. 21: 100361.

Yu, H.H., Kim, Y.J., Park, Y.J., Shin, D.M., Choi, Y.S., Lee, N.K., Paik, H.D., 2020. Application of mixed natural preservatives to improve the quality of vacuum skin packaged beef during refrigerated storage. Meat Sci. 169: 108219.

Zacharof, M.P., Lovitt, R.W., 2012. Bacteriocins produced by lactic acid bacteria a review article. APCBEE Procedia. 2: 50–56. https://doi.org/10.1016/j.apcbee.2012.06.010.

Zahid, N., Ali, A., Siddiqui, Y., Maqbool, M., 2013. Efficacy of ethanolic extract of propolis in maintaining postharvest quality of dragon fruit during storage. Postharvest Biol. Technol. 79: 69–72.

Zhang, H., Li, M., 2012. Transcriptional profiling of ESTs from the biocontrol fungus *Chaetomium cupreum*. Sci. World J. https://doi:10.1100/2012/340565.

Zhang, L., Lin, Y.H., Leng, X.J., Huang, M., Zhou, G.H., 2013. Effect of sage (*Salvia officinalis*) on the oxidative stability of Chinese-style sausage during refrigerated storage. Meat Sci. 95: 145–150.

CHAPTER 14

Active Packaging of Foods

Mahmoud Said Rashed, Mabrouk Sobhy, and Shivani Pathania

CONTENTS

14.1 INTRODUCTION

Packaging plays a critical role in the food supply chain. The primary function of packaging is to serve as a container for the food enabling efficient transport within the whole supply chain, preventing any physical damage, and protecting against manipulation and theft. Packaging meets the fundamental need to maintain food quality and safety from production to final consumption by preventing any unwanted chemical and biological changes. The protective role of the packaging is primarily passive, acting as a barrier between the food, the atmosphere surrounding the food, and the external environment such as oxygen, moisture, light, dust, pests, volatiles, and both chemical and microbiological contamination (Coles et al. 2008; Yildirim 2011; Arvanitoyannis and Oikonomou 2012; Pereira de Abreu et al. 2012).

Additionally, consumer concern about the safety and additive content of food has also received a great deal of attention over the last few decades. There is an increasing trend to natural, unprocessed

E-mail: Shivani Pathania, Shivani.pathania@teagasc.ie

DOI: 10.1201/9781003091677-14

and minimally processed high-quality foods with minimal to no preservatives, while exhibiting an acceptable shelf life (Singh et al. 2011; Gerez et al. 2013). In response to this, the protective function of packaging has been refined and improved, leading to the development of new packaging technologies, such as modified atmosphere packaging (MAP) (Ohlsson and Bengtsson 2002; Rodriguez-Aguilera and Oliveira 2009; Sandhya 2010; Cooksey 2014; Zhuang et al. 2014), active packaging (AP) (Singh et al. 2011; Yildirim 2011; Arvanitoyannis and Oikonomou 2012; Pereira de Abreu et al. 2012; Dobrucka and Cierpiszewski 2014; Realini and Marcos 2014; Kuorwel et al. 2015; Brockgreitens and Abbas 2016), smart and intelligent packaging (SP/IP) (Kerry and Butler 2008; Lee and Rahman 2014; Realini and Marcos 2014; Biji et al. 2015; Brockgreitens and Abbas 2016), and the application of nanomaterials (Imran et al. 2010; Llorens et al. 2012; Rhim et al. 2013; Reig et al. 2014; Rhim and Kim 2014). An example of AP is oxygen absorbers, antimicrobial packaging, and intelligent packaging with a maturity indicator and pathogenic bacteria indicator. Thanks to their active role in the conservation of food products, these packaging solutions make it possible to anticipate very positive benefits, linked to the reduction of food loss and waste. Significant investments in research have been devoted to the technical development of this new packaging. Much more recently, research has turned to the quantitative evaluation of the positive repercussions that are expected on the reduction of loss and waste with new strategic and methodological challenges to be met.

AP refers to the incorporation of certain additives into packaging film or within packaging containers to maintain and extend product shelf life. AP can, therefore, be used to voluntarily modify the internal atmosphere of the packaging to improve product conservation (Guillaume et al. 2008). For example, iron-based oxygen absorbers are thus commonly used to reduce, without additives or treatment, the oxidation reactions of vitamins or essential fatty acids. They also delay microbial development and make it possible to significantly increase the shelf life without the addition of preservatives or the use of treatments likely to alter the taste or nutritional qualities of food. Many other types of AP are now on the market, such as moisture absorbers for the preservation of moisture-sensitive products or ethylene for the preservation of climacteric fruits. AP is subject to specific European regulations (450/2009/EC; European Commission 2009) aimed at supporting innovations in the European market. This regulation establishes specific requirements concerning the use and authorization of intelligent and active materials intended to come into contact with food. It also establishes a European Union (EU)-wide list of substances that can be used in the manufacture of these materials.

14.2 TYPES OF ACTIVE PACKAGING

AP is an innovative approach to maintaining or prolonging the shelf life of food products while ensuring their quality, safety, and integrity. AP systems can be divided into active-releasing systems (emitters) and active scavenging systems (absorbers). Classification of the primary AP technologies used in food applications are represented in Figure 14.1.

14.2.1 Active Releasing Systems

There has been increased activity in the development of releasing packaging systems for food applications during recent years. Because of the diversity of food product characteristics, the conditions needed for maintaining food quality are varied. Generally, active-releasing packaging systems are categorized into two main categories, antioxidant releasers and carbon dioxide emitters. The concept behind the incorporation of active-releasing systems is that the embedded active molecules will subsequently volatilize from the package matrix diffusing their way into the food, promoting the stability and retaining the organoleptic properties of food products (Van Aardt et al. 2007).

Different Types of Active Packaging

Active Releasing Systems

Antioxidants releasers

Carbon Dioxide Emitters

Active Scavenging Systems

Oxygen Scavengers

Moisture Scavengers

Ethylene Absorbers

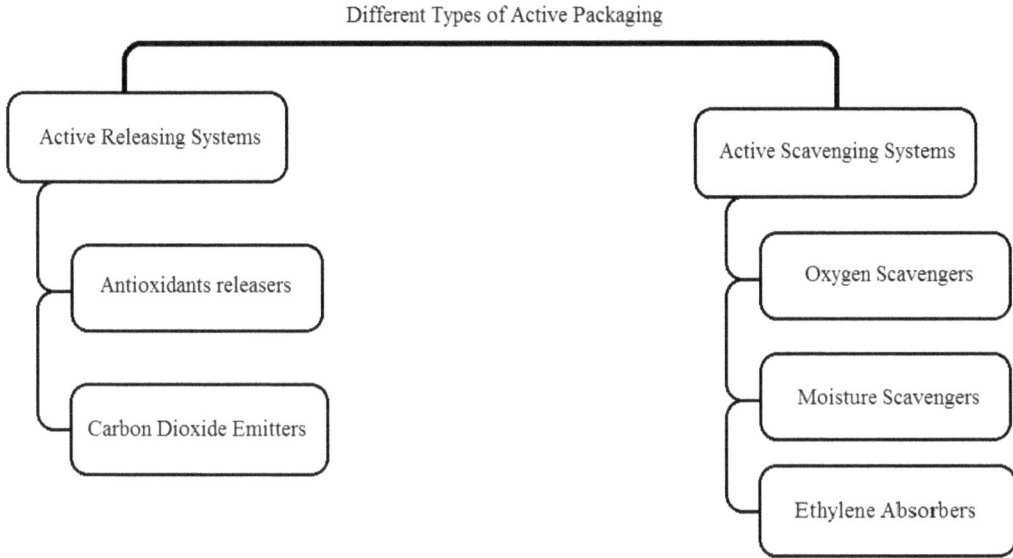

Figure 14.1 Classification of the primary active packaging technologies used in food applications.

14.2.1.1 Antioxidant Releasers

The role of antioxidants in food and human health, as pro-oxidants in preventing lipid oxidation and as anti-inflammatory agents, has been well documented in scientific literature (Dornic et al. 2016; Eghbaliferiz and Iranshahi 2016; Rivaroli et al. 2016). It has been recognized that the introduction of antioxidants through food packaging undergoes a steady decrease as opposed to the rapid degradation observed with an initial single addition (Wessling et al. 1998). The most commonly used antioxidants in the food industry are usually synthetically derived dodecyl gallate, propyl gallate, nordihydroguaiaretic acid, thiodipropionic acid, butylated hydroxy anisole (BHA), and butylated hydroxy toluene (BHT) (Torres-Arreola et al. 2007; Van Aardt et al. 2007).

Lower values of lipid oxidation expressed as peroxide index values (PV), thiobarbituric acid (TBA) values, and free fatty acid (FFA) contents were found in fresh sierra fish fillet samples packaged in BHT incorporated into low-density polyethylene (LDPE) films when compared with normal LDPE (Torres-Arreola et al. 2007). Furthermore, texture characteristics such as tissue damage and retained firmness were better in case the samples baked in BHT-LDPE packaging films compared with those packed in LDPE films.

Growing concerns related to the presence of synthetic antioxidants in food products drive substitutional approaches and as a new alternative, the use of natural antioxidants like α-tocopherol was assessed. Many authors have incorporated natural antioxidants in plant extracts like polyphenols, tocopherols, and essential oils (EOs) and developed AP solutions (Nerín et al. 2006; Park et al. 2012; Barbosa-Pereira et al. 2014; Marcos et al. 2014). Wessing et al (2000) compared the stability of two antioxidants (BHT and α-tocopherol) impregnated in LDPE and polypropylene (PP)-based films. They observed that the retention of antioxidant was dependent on the contact medium. α-Tocopherol-impregnated LDPE film enhanced the stability of a linoleic acid emulsion stored at 6°C, whereas BHT-impregnated LDPE film reduced the odor changes in an oatmeal product stored at 20°C. Overall, BHT was rapidly lost from the LDPE film compared with α-tocopherol impregnated in PP and LDPE films. Van Aardt et al. (2007) developed poly(lactide-co-glycolide) (PLGA) (50:50)-based active film using 1% BHA with 1% BHT and a second type with 2% α-tocopherol to assess their functionality in maintaining the stability of dry whole milk and dry buttermilk. The results revealed that the volatility of films impregnated with BHA and BHT is higher than the volatility of α-tocopherol. Hence, it was suggested that α-tocopherol-impregnated packaging films should be utilized in dry food applications.

Graciano-Verdugo et al. (2010) developed two α-tocopherol loaded 2 and 4% in LDPE films to test the migration of the antioxidant to corn oil as well as its effect on the oil's oxidative stability. The study showed that α-tocopherol concentration from 1.9 to 3% can maintain the oxidative stability of oil during 16 weeks of storage compared with 12 weeks observed for the control at 30°C. Granda-Restrepo et al. (2009b) also compared multilayer films with different antioxidant loadings and their role in whole milk powder's oxidative stability and found that high-density polyethylene (HDPE) + TiO$_2$/ethylene vinyl alcohol (EVOH)/LDPE + 4% α-tocopherol film showed gradual release compared with BHT- and BHA-impregnated films. They also observed that (HDPE + TiO$_2$/ EVOH/LDPE + 1.5% BHA + 4% α-tocopherol) BHA migrated from the BHA–α-tocopherol film, contributing to the protection of both whole milk powder and α-tocopherol.

Polylactic acid (PLA) film impregnated with 2.58% α-tocopherol can delay the induction of the oxidation measured as peroxide value compared with control at the same temperature (Manzanarez-López et al. 2011). Torrieri et al. (2011) revealed that the combination of the application of using MAP and LDPE-embedded α-tocopherol packaging can reduce fat oxidation in fresh tuna fillets. Barbosa-Pereira et al. (2013) studied the effect of naturally derived tocopherols incorporated into LDPE films on the oxidative shelf life of salmon, and the results showed a 40% reduction of lipid oxidation in salmon muscles. Therefore, results concluded that natural antioxidant products have a great potential to be a member of antioxidants incorporated into packaging films.

Incorporating natural flavonoid quercetin into the EVOH matrix has been carried out by López-de-Dicastillo et al. (2012b) as 4.6% loading showed an enhancement in lipid oxidative stability expressed in lowering the peroxide index along with the reduction in TBA values by 25% along the storage time. Green tea is an important source of antioxidants; therefore, catechin is considered as one of the best candidates to be incorporated into active food packaging systems as it has similar properties to quercetin. López-de-Dicastillo et al. (2012a) demonstrated the potential usage of catechin in retarding the oxidation of sunflower oil and fried peanuts. Results also revealed that not only the antioxidant activity but also the solubility of the antioxidants have key roles in the efficacy of the antioxidant. Quercetin-containing films showed a higher efficiency compared with those catechin-containing films. However, these results were obtained under the usage of accelerated shelf life testing, therefore, the real assessment of the antioxidant performance under real storage conditions needs to be evaluated before commercial implementation.

The studies carried out by Siripatrawan and Noipha (2012) and Yang et al. (2016) showed that there was great improvement in oxidative stability of pork meat products when green tea extract was impregnated into packaging films. Studies performed by Yang et al. (2016) confirmed these findings, as the antioxidants produced from green tea extract showed a stronger antioxidant capacity compared with those compounds obtained from oolong tea and black tea extracts. Specifically, in the context of reducing the rancidity of pork meat after 10 days of storage, the TBA values of the samples wrapped using films containing green tea extract were lower than the values obtained from samples wrapped using film with oolong tea extract and film with black tea extract. Another investigation performed by Carrizo et al. (2016) used peanuts and cereals covered with chocolate for evaluating the radical scavenging capacity of green tea extract incorporated into a multilayer film used with no direct contact with packaged food. The results showed that this packaging system can be utilized to protect food against oxidation for 16 months of storage. These findings could be utilized for facilitating the industrial implementations of this technology as it has great potential as an industrial relevance.

In recent years, there has been a growing interest in using food industry by-products as a sustainable source of antioxidants for food packaging and add value to these residuals. An example of using food by-products in packaging materials is the study conducted by Barbosa-Pereira et al. (2014) using LDPE film containing a brewery residual waste extract for packaging of beef. It was observed that lipids were reduced by 80% during cold storage. In the same context, using barley husk obtained from the brewery industry has proved to be effective not only in slowing down lipid hydrolysis but also improving the oxidative stability in blue shark muscle (Pereira de Abreu et al. 2011). Meanwhile, many different residuals such as anthocyanins from wine grape pomace, mango and acerola pulp, and beet root residue powder (Souza et al. 2011; Oliveira et al. 2016; Stoll et al. 2016) were incorporated into sealable biodegradable films, which showed a protective effect on palm oil and sunflower oxidation. Films produced by coupling cassava starch and encapsulated anthocyanins used for storing sunflower oil represented a decrease in the peroxide index of the oil exposed to the air and light compared with the control (Stoll et al. 2016). Souza et al. (2011) packed palm oil in cassava starch films with high concentrations of mango and acerola pulp additives and, interestingly, the results showed a significantly lower peroxide index compared with the control. It was also observed that acerola pulp contains vitamin C, which could act as a pro-oxidant agent.

14.2.1.2 *Carbon Dioxide Emitters*

The antimicrobial effect of CO_2 has been widely discussed (Kolbe 1882; Valley 1928; Haas et al. 1989; Debs-Louka et al. 1999). Solubility is the key factor governing the efficacy of CO_2 as an antimicrobial agent, and the solubility of CO_2 is almost the same in both aqueous and fatty phases of food products. However, this property is affected by a number of factors (Devlieghere et al. 1998; Devlieghere and Debevere 2000): solubility increases with decreasing temperature and food properties such as food composition (namely, protein, water, fat), surface area, and pH of the food (Chaix et al. 2015) and environmental conditions, i.e., the antimicrobial effect of CO_2, is proportional to the partial pressure of the gas (Blickstad et al. 1981).

From the point of view of food packaging, the total amount of CO_2 that exists in the headspace plays an important role to exert its effect. The concepts explaining the theories of the CO_2-releasing device in MAP packages for maintaining a specific level of CO_2 in the headspace along the storage for prolonging the shelf life was originally illustrated in the 1990s. CO_2 emitters in MAP packaging systems improve the efficiency of transportation, increasing the filling degree; minimize environmental impact; and reduce packaging size. Furthermore, the release of CO_2 could be used to prevent packaging deformation, especially when the CO_2 tune emitter is used as the level of CO_2 decreases as a result of absorption into the food product in the initial stage of storage. One advantage of this property, as described by Holck et al. (2014), is its ability to counteract the effect of negative pressure in MAP packages, which increases the drip loss of the product that is responsible for giving the package an unattractive appearance. Along with the inhibition of the growth of spoilage bacteria responsible for prolonging the shelf life of fresh food products, this system offers a sustained high CO_2 level in the package that will have great potential to reduce food waste, which is gaining great interest and increasing attention nowadays.

In most cases, CO_2 emitters are produced as a pad or sachet with liquid absorbers. The CO_2 is released as the pad absorbs the liquid seeping from the product. The development of CO_2 emitters over the last decade represents an increase in research and a boom in sales of commercial CO_2 emitters. According to Rooney (1995), Sivertsvik (2003), and Restuccia et al. (2010), one of the most widely documented and discussed methodologies for the development of CO_2 emitters is the utilization of ferrous carbonate. Also, another highly documented technology using the CO_2 releasing system is based on the combination of sodium bicarbonate and citric acid. This system is quite old as Bjerkeng et al. (1995) first presented it in 1995. This study used a noncommercial CO_2 emitter in a vacuum package to evaluate its effect on the microbial and sensory shelf life of cod fillets packed in MAP. CO_2 was used in combination with N_2 at a percentage of 70 and 30%, respectively.

Hansen et al. (2009a,b,c) assessed various strategies to prepare different combinations of using CO_2 with a liquid absorber on a laboratory scale. Hansen et al. (2009c) studied their impact and evaluated the effectiveness of using the laboratory-scale type CO_2 emitter for reducing the gas volume to product volume ratio (g/p ratio) of MAPs (60% CO_2, 40% N_2) of salmon fillets without compromising the shelf life (based on microbial, sensory, and textural analysis). For the MAP packages (g/p 3/1) without emitters, the CO_2 level dropped to 40% 4 days after packaging and then stabilized. For the MAPs (g/p 1/1) with an emitter, the CO_2 level displayed an initial drop to about 45% (day 1), but, subsequently, the level increased and reached 65–70% during the storage time. The measured total viable count (TVC) levels for the salmon packaged in MAPs with and without emitters were comparable, and the obtained shelf life for the two packaging methods was the same. The TVC of the MA-packaged samples (with and without emitter) reached a level of log 5 to 6 CFU/g after 15 days of storage, whereas for the vacuum-packaged salmon, the same bacterial counts were measured 7–10 days into storage. The results illustrated that a CO_2 emitter can allow for more sustainable packaging of fresh fish products with a significant reduction in package sizes and, hence, the amount of packaging material because a comparable shelf life can be obtained at a significantly reduced g/p ratio.

A study performed by Pettersen et al. (2014) aimed to investigate the effect of using a laboratory-type CO_2 emitter with different packaging systems on the freshness of fresh reindeer meat. The results revealed that sensory shelf life, in terms of odor evaluation, was prolonged for the meat packaged in MAP (60% CO_2, 40% N_2) in both cases with or without CO_2 emitters for 21 days compared with findings from 17 days in vacuum packaging. Interesting results were found in those samples that used a CO_2 emitter as they showed a lower level of TVCs (log 3 to 4 CFU/g) after 17 days compared with those samples packed in MAP without CO_2 emitters at the same time with a TVC(s) of log 4 to 5 CFU/g. Along these results, it was noted that there was a significant reduction in drip loss for reindeer meat packed in MAPs with an emitter. The drip loss was one-third the amount of drip loss for the samples packed in MAPs without an emitter. This article stated that the capacity of the emitter used in this study was not sufficient, in terms of capacity, for the product, but the performance of the emitter is much higher if adequate emitters are used.

Similar to the previously mentioned study, Holck et al. (2014) conducted a study using an emitter accurately tuned for CO_2 release based on the food product. This emitter was designed for well recompensing the CO_2 absorbed by chicken fillets in MAP (100% CO_2). Interestingly, a dramatic decrease in drip loss was observed and emitters could maintain a 100% level of CO_2 in the package. On the other hand, CO_2 levels decreased in other packages with different percentages, according to the g/p ratios. Furthermore, packages containing 100% CO_2 showed significant inhibition of bacterial growth compared with the commonly applied gas composition of 60% CO_2 and 40% N_2 in other packages. It was observed that packaging in 100% CO_2 is not possible without emitters.

An interesting study was conducted by Chen and Brody (2013) in which a CO_2 emitter was used for packaging cooked ham and stored for 4 weeks. The results of the study showed that controlled proliferation of artificially inoculated *Listeria monocytogenes* in addition to Enterobacteriaceae and TVC. The comparison of different CO_2 emitters concepts is a very challenging field because of the many different aspects associated with the pre-handling processes, storage condition, temperature, the type of packages and package size, g/p ratio, and gas composition. All of these variables greatly affect the contribution of CO_2 emitters in extending the shelf life of food products. Furthermore, other important factors that could contribute to the evaluation process of CO_2 emitters and rarely evaluated in the literature is the material and density of the emitter substrate. The type and structure of the packaging films in which the active ingredients are incorporated play an important role in controlling the amount of CO_2 release and liquid absorption. There are many examples of materials that are used in different layers of CO_2 emitters/liquid absorbers such as fiber-based materials, superabsorbent polymers (SPAs), hydrogels, perforated plastic films, and in this sector cellulosic fiber is the most famous type.

The results of these studies summarized the optimum conditions for the optimal effect of CO_2. This can only be achieved when the emitter capacity is optimized, which is related to the physiological properties and the weight of the food product. An emitter with an optimal capacity will ensure an adequate CO_2 level, counteract the formation of negative pressure within the package, ensure sufficient liquid absorption, and extend shelf life. The study performed by Hansen et al. (2009a) aimed to optimize the emitter capacity used in salmon fillets MAP in different sizes and different gram/package ratios. The results of the study were able to calculate the required amounts of citric acid and sodium bicarbonate based on the surface area and weight (gram/package ratio) of the salmon fillets. However, there are still niche research points related to the effect of the product type and widening the utilization of this technology to include a broader spectrum of products.

14.2.2 Active Scavenging Systems

One of the keys in the development of AP for fresh products is to identify the food degradation mechanisms and act on them to reduce or delay these processes, thus increasing the shelf life of the product. Generally, there are three main systems commonly used in the sector of food AP that are associated with scavenging techniques: oxygen scavenging systems, moisture scavengers, and ethylene absorbers. The coming section will summarize those three different systems.

14.2.2.1 Oxygen Scavengers

Oxygen scavengers (OS) are considered one of the main AP technologies. The aim of using OS is to remove the residual of oxygen present in the food package (Solovyov 2010; Arvanitoyannis and Oikonomou 2012; Realini and Marcos 2014) or acting as an active barrier by improving barrier properties (Sängerlaub et al. 2013b). Because of the sensitivity of several food products to the existence of oxygen, the demand for different technologies concerning oxygen removal is quite high. Different technologies used for applying different OS technologies are represented in Figure 14.2. However, MAP is considered an OS technology because the existence of concentration varies between 0.5% and 5% of oxygen residual with a possibility to further increase it in the package and storage decreases its efficacy (Solovyov 2010; Gibis and Rieblinger 2011; Pereira de Abreu et al. 2012). This increase in oxygen concentration (Pereira de Abreu et al. 2012) could be a result of insufficient evacuation during the packaging process, poor sealing properties, the permeation of oxygen through packaging materials, or other reasons described by Pénicaud et al. (2012) such as

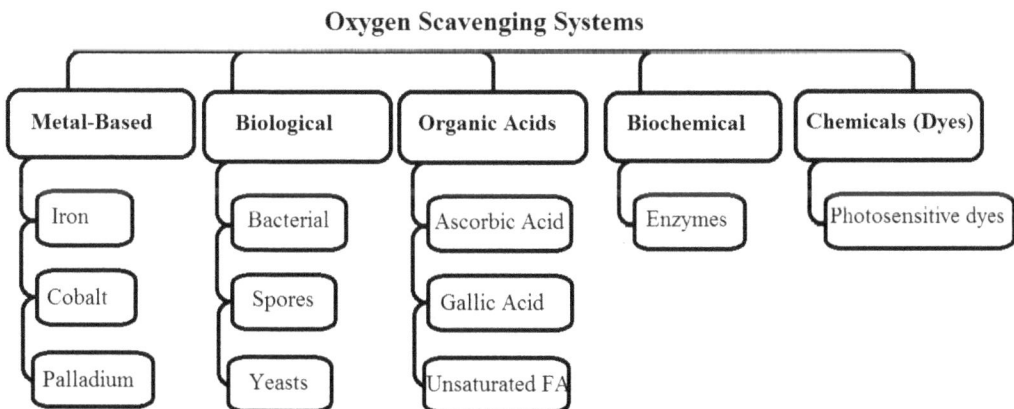

Figure 14.2   Different technologies used for applying different OS technologies.

dissolved oxygen in the food itself being liberated into the headspace of the package reaching equilibrium with the gas phase. The existence of oxygen negatively affects both the quality and shelf life of several food products as it leads to oxidation of the product (Choe and Min 2006), color modifications (Møller et al. 2000; Nannerup et al. 2004; Larsen et al. 2006; Gibis and Rieblinger 2011; Hutter et al. 2016), promoting the growth of aerobic microorganisms (Lee 2010; Solovyov 2010), changing sensory quality attributes (Jacobsen 1999; Granda-Restrepo et al. 2009a; Li et al. 2013), or causing nutritional losses (Chung et al. 2004; Lopez-Gomez and Ros-Chumillas 2010; Van Bree et al. 2012). The application of OS could reduce the residual oxygen level of a food package down to <0.01 volume percentage.

Different studies concerning oxygen scavenging technologies suggest its potential in various food applications. However, the literature indicates that most of the research has been performed using OS sachets containing iron powder (Charles et al. 2003; Solovyov 2010; Antunez et al. 2012; Cruz et al. 2012; Kartal et al. 2012; Chounou et al. 2013; Cichello 2015). Contrary to their commercial use in sachet-based applications in the countries like the United States or Asia, these technologies are not accepted in European countries (Restuccia et al. 2010) as these components are considered as foreign bodies in food containers. One of the major drawbacks of using such a sachet is the possibility of accidental breakage, which could lead to accidental consumption of the content. Furthermore, these technologies need additional requirements for an additional packaging operation and their unsuitability in moist foods due to their moisture sensitivity is a limiting factor (Suppakul et al. 2003; Day 2008; Pereira de Abreu et al. 2012). Alternatively, there are several new developments associated with the improvement of OS technologies, including incorporating active substances directly into the containers or within packaging films. However, few of these technologies are implemented successfully in real food systems. In the same context, the benefits of other technologies, which could be considered as an alternative to oxygen scavenging, systems are rather rare.

14.2.2.1.1 Iron

Shin et al. (2009) used an iron-based OS approach for its ability to extend the shelf life of meat-based products. Meatballs were packed into a tray made of a multilayer of PP and the middle layer of the trays contained different levels of iron-based OS substances (40%, 80%, 100% w/w). The results indicated that the oxidative-induced color and flavor changes of the product packed using 100% w/w had lower levels of oxygen compared with the samples packed into other passive packs for 9 months of shelf life studies. These findings were further confirmed by measuring the TBA value, which indicated a lower level of lipid oxidation in the samples packed with the OS system.

Gomes et al. (2009) studied the effect of iron-based OS materials in hot-filled ready-to-eat (MRE) cheese spread in the context of producing longer shelf-stable food products for military purposes. The oxygen concertation in the container reduced by 67.44 vol.% of its original level during the first day and reached below 1% of the original volume during storage for 11 days, and it was maintained at that level for 1 year. The clear effect of the O_2-absorbing system was represented in decreasing the rancidity and increasing the sensory acceptance of MRE cheese spread compared with those stored without the OS system. In addition, the recovery of vitamin C content in the samples was around 1.5 times higher in the samples packed using OS absorbers. Hence, OS-absorption systems illuminate the effect of oxygen in the packs before its availability for further reactions with food components.

Further work on iron-based OS systems was carried out by Mu et al. (2013). An OS system was developed using iron in the form of nanoparticles (110 nm), carbon, sodium chloride, and calcium chloride. The study aimed to evaluate the O_2 adsorption ability and evaluate the effect of using this system on lipid antioxidant properties. The results indicated that these particles showed a 1.4 times higher oxygen scavenging rate compared with OS systems containing micro iron powder. The importance of these results was the ability to utilize this approach to mitigate lipid peroxidation in products containing a high level of lipids.

From another point of view, maintaining food quality under MAP conditions could be challenging, especially if the sealing materials of the packs contain defects. According to Sasaki and Kamimura (1997) and Sängerlaub et al. (2013b), the effect must be considered especially when the size of these defects is lower than the detection limits of a standard leak tester (10 μm). The study conducted by Sängerlaub et al. (2013b) aimed to simulate the effect of the existence of an iron-based multilayer OS system in food packs with pinhole defect sizes of 10 μm stored for 300 days with two different food products (snack food product, salami in a baked bread roll). The results indicated that the total stoppage of oxygen reaction is not possible; on the other hand, it was reduced significantly in terms of the samples used iron-based OS systems. These differences appeared in the color of salami samples (expressed as total color difference ▲E). Furthermore, it was noted that a reduction in lipid peroxidation in the products was reduced four times more than control samples without iron-based OS systems. Thus, these results indicated the possibility of using OS systems to exceed the seal defects along with providing an extended defense.

14.2.2.1.2 Palladium

OS systems suffer from the slowness of their effect. These systems are used for boiled meat products that are sensitive to oxygen, and its elimination is generally not achievable by conventional methods. Additionally, in such products, discoloration induced by light, occurs within hours even at levels of oxygen between 0.5 and 0.1% volume according to the g/p ratio (Andersen and Rasmussen 1992; Møller et al. 2000; Nannerup et al. 2004; Larsen et al. 2006; Gibis and Rieblinger 2011; Böhner et al. 2014; Hutter et al. 2016). Most OS systems also need several days to remove initial headspace oxygen (Matche et al. 2011).

Many developments have been carried out on a catalytic system with palladium (CSP), where magnetron sputtering technology was used for coating palladium on polyethylene terephthalic silicon oxide (PET/SiOx) film technology (Lohwasser and Wanner 2005; Yildirim et al. 2010, 2015). This product of OS films was able to remove oxygen residual in food packages reaching 2.5 vol.%, especially when hydrogen was introduced in MAP (Yildirim et al. 2015). Because of the high efficiency of the film, it is suitable for use in food products very susceptible to oxygen and the oxidation reactions are very fast. One of the applications of OS film was carried out by Hutter et al. (2016) for packaging cooked cured ham to prevent the discoloration. The results illustrated that OS films removed around 2 vol.% of initial oxygen in the headspace (160 cm³) within 35 minutes after packaging. Discoloration was, therefore, prevented for 21 days of storage, and the color of the ham was preserved even when the packages were exposed to light 24 hours a day. On the other hand, samples packed without OS film lost their redness within 2 hours of storage.

In bakery products, mold is the key factor responsible for limiting shelf life; therefore, the potential of these films was assessed in bread packaging. A study conducted by Rüegg et al. (2016) used MAP coupled with palladium-based OS film for packing gluten-free bread slices, partially baked buns, and toasted bread slices. For the control group, after 2 days in parallel with a decrease in headspace oxygen concentration, visible mold growth was detected for all samples. In contrast, shelf life for the samples packed in MAP packages with OS film was extended 3- to 4-fold for all the tested samples and mold growth was retarded up to 8–10 days.

One of the greatest challenges faced by this technology is the need for hydrogen; thus, this technology is limited to products packed using MAP. Also, Röcker et al. (2017) suggested that volatile sulfur compounds present in the headspace of packaged food products can inhibit or even have the potential to inactivate the catalytic activity of the palladium. In terms of safety, a scientific opinion was published by EFSA concerning the assessment of the safety of palladium metal and hydrogen gas for use in active food contact materials. The EFSA Panel on Food Contact Materials, Enzymes,

Flavorings and Processing Aids (CEF) concluded that "the active substances palladium and hydrogen do not raise a safety concern for the consumer when used as an OS in packages for foods and beverages at room temperatures or below. Palladium should not be in direct contact with food and should be incorporated into a passive structure impermeable to liquids which prevent the migration at detectable levels" (EFSA CEF Panel 2014).

The characteristics of OS technologies, their sensitivity to environmental conditions such as temperature and pH (Galdi et al. 2008; Solovyov 2010; Damaj et al. 2014), need for humidity, need for ultraviolet (UV) light to trigger the oxidative reaction (Rooney 1999; Miller et al. 2003; Zerdin et al. 2003), or its ability to interact with food volatiles (Röcker et al. 2017) pose challenges to their implementation into food packaging processes. One of the major points to be taken into consideration is how the integration of OS materials into polymers could affect the film properties, such as oxygen transmission rates (OTRs) or water vapor transmission rates (WVTR) (Matche et al. 2011). The changes in such properties may affect the reduction in the quality of the food. The process of incorporation of any substances, not only active substances, into the packaging should be safe. Conclusively, any OS system must have no negative effect on the sensory properties of the food.

14.2.2.1.3 Ascorbic Acid

Ascorbic acid (AA) acts as a reducing agent when it is mixed with iron or with zinc to produce (Fe/AA) or (Zn/AA); these mixtures are used for modification of linear low-density polyethylene (LLDPE) films as the transition metals were used to catalyze the oxidation reaction (Graf 1994; Matche et al. 2011). These active films can be formed into a bag and assessed for their OS performance as well as shelf life extension properties on bakery products. Matche et al. (2011) indicated that microbial growth retarded from 2 to 5 days in the samples packed with OS film. Furthermore, analysis of moisture content and instrumental texture measurement indicated that samples stored in OS films showed a significant difference in terms of firmness and dryness values, respectively, after 4 days. The reason behind this was related to the lower WVTR of both Fe/AA (17.2) and Zn/AA films (17.4) compared with the pure LLDPE film (20 g/m² 100 gauge/day), which was used as a control. These modifications also led to lower OTRs. From the point of view of sensory testing, the bread slices packed in OS film were acceptable (softness and taste) up to 5 days, whereas the control samples were not acceptable on the second day. The authors, however, did not document the package size and the development of oxygen concentration in the headspace of the packed bread. Therefore, more research work is required to understand the capacity of the scavengers and the connection between the OS activity and the extension of the mold-free shelf life of the product.

14.2.2.1.4 Photosensitive Dyes

Using photosensitive OS film to improve the oxidative stability of sunflower oil was carried out by Maloba et al. (1996). Moreover, many different organic dyes such as eosin and curcumin and the synthesized polyether polyfuryloxirane (PFO) have been applied to ethyl cellulose polymer films. The mechanism behind the use of these dyes is explained by Rooney (1995). The polymers, when exposed to light, uses light energy to convert the triplet oxygen (3O₂) to highly reactive singlet oxygen (1/2 O₂), which is absorbed irreversibly by PFO. The functional effect of these films on the oxidative stability, peroxide values, and gas-chromatographic (GC) measurement of headspace hexanal in sunflower oil was determined.

Another successful application of OS films is to prevent degradation and oxidation processes during the storage of orange juices. Zerdin et al. (2003) carried out a storage experiment for 1 year in which orange juice was filled in vacuum-sealed OS packages. It was discovered that UV illumination before packaging was the key factor for the activation of OS films. These findings were confirmed by the results obtained from AA retention, which was significantly higher in juices packed in

OS films compared with the control after 1 year of storage. Additionally, Johnson et al. (1995) discovered that the loss in AA is associated with the increase in non-enzymatic browning of the juice. According to Kennedy et al. (1990) the rate of formation of 5-hydroxymethylfurfural (5-HMF), the major compound responsible for fruit juice browning, is dependent on the loss of L-AA and is likewise directly related to storage temperature. The results indicated that the browning index of juice samples packed in the OS pouches and stored at 4°C was below 0.15 during the entire 1-year storing period; the values are closer to freshly pressed orange juice. Also, the study revealed that storage at 25°C led to an increase in juice browning, although values of the browning index of samples filled in OS pouches were significantly lower compared with control samples. Thus, rapid oxygen removal could be considered as a key factor for minimizing the higher levels of AA and color retention in orange juice over long storage periods. Application of OS films in juice packaging can be one of the strategies to reduce or omit the use of the antioxidant substances in juice processing; however, the addition of another step in the production, UV illumination, must be taken into consideration.

Other potential applications of OS could be its use in specific products such as probiotic yogurt. Shah et al. (1995) explained that OS application can be beneficial in yogurt products as the viability of probiotics like *Lactobacillus acidophilus* and *Bifidobacterium* species, used in yogurt production, is negatively impacted by dissolved oxygen. Rooney (1999) developed an OS film containing a reducible organic compound such as substituted anthraquinone, and its application in controlling the amounts of dissolved oxygen was studied by Miller et al. (2003).

The film was completely different from the one introduced by Maloba et al. (1996) as it required only UV light exposure to trigger the scavenging process instead of using a constant source of light. The study carried out by Maloba et al. (1996) used PFO/ethyl cellulose polymer film and stored it in plastic pouches made of high barrier film (nylon/EVOH/PE) for storing sunflower oil samples, but the samples were wrapped by aluminum foil to avoid the effect of the light. In this study, Miller et al. (2003) tested various manufacturing methods as well as different packaging systems for probiotic yogurt and observed their effect on the dissolved oxygen content during yogurt shelf life studies of 42 days. The results revealed that in the normal container the level of dissolved oxygen concentration after the first day was reduced from 16 to 3 parts per million (ppm), whereas it reached 1.7 ppm in the OS container. Furthermore, it decreased to 0.2 and 0 ppm, respectively, after 42 days. The importance of the initial rapid oxygen concentration reduction is its role in providing low oxygen concentration during post-fermentation, leading to a product with elevated health benefit. The applications of these OS films could be used successfully with high-oxygen barrier containers. Baiano et al. (2004) proposed the incorporation of unsaturated hydrocarbon dienes as an oxygen scavenging co-polymer into PET bottles. Cobalt is used as a transition metal to catalyze the reaction between oxygen and unsaturated hydrocarbon dienes, which are linked to the polyester polymer (Cahill and Chen 2000). The influence of OS-PET bottles on AA degradation was evaluated using a model system simulating one of citrus juice over 16 weeks of storage at 5 and 35°C. The results indicated that the use of OS-PET bottles showed a significant slowdown in the degradation kinetics of AA and browning reactions. Comparing these results with the results obtained from juice filled in glass jars and conventional PET bottles, the loss in vitamin C content in these OS-PET bottles was half for the samples stored at 35°C and was between three and four times lower for the samples stored refrigerated at 5°C. The reason behind these findings, as described by the authors, is the oxygen permeability of PET and the existence of pro-oxidant substances in glass containers. These results greatly support the thoughts concerning the replacement of glass bottles by polymeric bottles containing an OS, especially when packing beverages containing AA. In contrast, the use of real fruit juice instead of a model system would be preferable. In the context of using the OS polymer with the same approach in the form of a cast-extruded monolayer PET film, Galdi and Incarnato (2011) revealed that it helped prevent the browning of bananas. The results showed that there was a significantly less color difference (▲E) (approximately 50%) after 3 days of wrapping in the OS films compared with the conventional PET films. The optimization of OS-PET films carried out by

Di Maio et al. (2015) represented in a form of co-extruded multilayer films as the internal active layer was protected from fast oxidation by two external layers of pure PET (PET/OS-PET/PET) to increase the reaction time.

14.2.2.2 Moisture Scavengers

Water activity is one of the most critical factors affecting the quality and safety of different food products (Labuza and Hyman 1998). The importance of this factor appears in dried products as these products are very sensitive to humidity during storage, and product quality could deteriorate even under low levels of relative humidity (RH) inside the packages. According to Labuza and Hyman (1998) and Day (2008), an increase in moisture content not only leads to microbial spoilage but also could cause an alteration in the product texture, consequently, reducing the shelf life of a food product. However, in the case of fresh fish, meat, and fruit/vegetables, maintaining controlled high RH inside the package is important for preventing product deterioration.

On the other hand, Droval et al. (2012) stated that the presence of some excess liquid in some fresh products, such as meat and fish as a result of drip loss, decreases consumer appeal. The process of controlling the moisture in the package could be carried out through two different strategies. The first strategy is moisture reduction, which is an approach that can be applied using MAP where the atmospheric air in the headspace is replaced by MA gas or vacuum packaging. Second is the moisture prevention (by barrier packaging) approach, which aims at moisture elimination through the use of a desiccant/absorber. The second approach could be considered active. Furthermore, levels of humidity could be controlled by the use of the appropriate packaging materials with a specific property with a high barrier against water vapor.

Active moisture scavengers can be categorized into two different classes (Brody et al. 2001): introduced RH controllers that scavenge humidity in the headspace, such as desiccants, and the moisture removers that absorb liquids. The moisture removers could be used in different forms such as pads, sheets, or blankets, and are usually placed underneath fresh products in different packaging strategies (e.g., MAP, vacuum, skin pack). McMillin (2008) stated that the increase in the surface area and longitudinal cutting of the muscle fibers are responsible for increasing the drip loss during storage time. Along with fish, meat, and poultry products, product packing for fruits and vegetables (especially cut products) also use moisture-remover technologies (Vermeiren et al. 1999; Day 2008).

Ščetar et al. (2010) described these pads as mostly composed of porous materials, polymers (PP or PE), foamed and perforated PS sheets, or cellulose, combined with SAPs/minerals/salts (polyacrylate salts, carboxymethyl cellulose, starch copolymers, silica/silicates). According to the EU Guidance to the Commission Regulation (EC) No 450/2009: "Materials and articles functioning on the basis of the natural constituents only, such as pads composed of 100% cellulose, do not fall under the definition of active materials because they are not designed to deliberately incorporate components that would release or absorb the substance." Thereby, moisture-absorbing pads are not often considered to be AP. However, moisture-absorbing pads containing components that "are intentionally designed to absorb moisture from the food" can be considered as AP (European Commission 2009). Absorbing pads can also be used in combination with antimicrobials, pH control agents, and/or carbon dioxide generators/OS to avoid certain shortcomings, such as odor generation or leakage.

14.2.2.2.1 Desiccants

Desiccants are commonly placed into packages in the form of sachets and microporous bags or are integrated into pads. The absorption capacity of desiccants depends on their water vapor sorption isotherm (Sängerlaub et al. 2013a). These moisture absorbers include silica gel, clays, molecular sieves (synthetic crystalline version, such as from zeolite, sodium, potassium, calcium alumina silicate), humectant salts (such as sodium chloride, magnesium chloride, calcium sulfate), and other

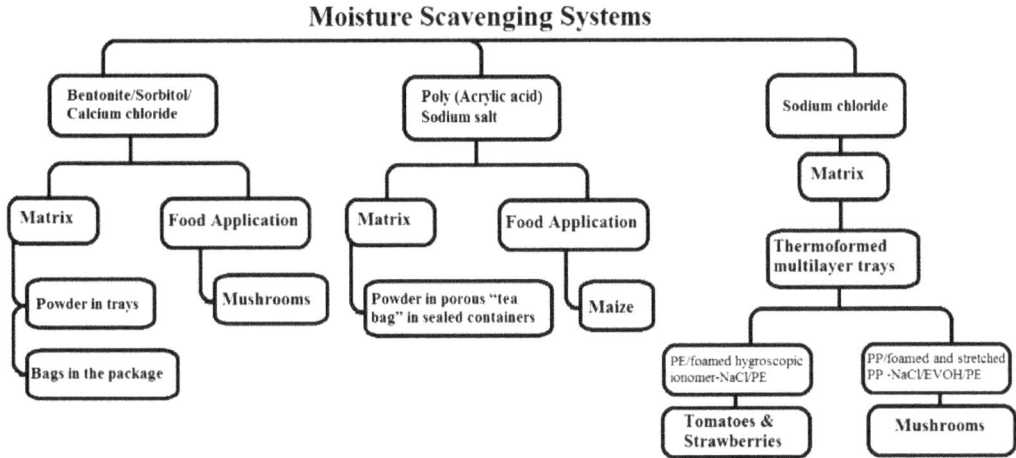

Figure 14.3 Moisture scavenging systems applied to food products.

humectant compounds (such as sorbitol), as well as calcium oxide (Day 2008). Some examples applied to food products are given in Figure 14.3.

14.2.2.2.2 Zeolites

Julkapli and Bagheri (2016) stated that zeolites can absorb the release of the absorbed water without any modulation in their characteristics such as crystalline structure or moisture absorption. Thanks to these characteristics, it could be added to pulp/paper and plastic films for regulating the amount of moisture absorbed. Wu et al. (2010) stated that its application is still limited to non-food packaging.

14.2.2.2.3 Bentonite/Sorbitol/Calcium Chloride

Some food products are sensitive to the RH levels in the package and could affect the product quality; mushroom is one of these products. An RH value below 86% could increase moisture loss and, consequently, weight loss and textural changes. On the other hand, 100% RH promotes the growth of psychrophilic bacteria, which can cause discoloration of mushrooms. Mahajan et al. (2008) assessed that 96% is the optimal level of RH required to be in the headspace of the package used for packing mushrooms. Mahajan et al. (2008) aimed to test new packaging concepts for fresh mushrooms (*Agaricus bisporus*). The study used a mixture of fast-absorbing and slow-absorbing moisture absorbers in different combinations. It revealed that the moisture-holding capacity of the scavenger is dependent on the RH, and the change in storage temperature (from 4 to 16°C) did not affect the moisture-holding capacity of the absorbers. Also, it was observed that the moisture condensation into the package was decreased, which improved packaging film transparency. The study also stated that the importance of adjusting the optimal levels of desiccants, as at a higher level desiccants caused an increase in browning indices as a result of browning and excessive moisture loss. The results revealed that the moisture absorbers could be precisely adjusted to fulfill the optimal effect.

14.2.2.2.4 Poly (Acrylic Acid) Sodium Salt

A study performed by Mbuge et al. (2016) aimed to reduce the moisture content in maize to inhibit mold growth, therefore reducing contamination by aflatoxin. The experiment used a food-grade

SAP as a desiccant. This method was able to reduce the level of moisture content reaching to 13% at 40°C drying temperature. The experiment revealed that the 1:5 SAP to maize ratio is the optimum ratio for reducing aflatoxin content. Aflatoxin contamination could be reduced to <4 ng/g in this experiment, complying with current European legislation that limits aflatoxin content to 4 ng/g. Hence, poly (acrylic acid) sodium salt showed possible potential to be used in drying grains for reducing aflatoxin contamination. Furthermore, it could be considered as an economically efficient method as it is a cheap and reusable solution.

14.2.2.2.5 Sodium Chloride and Hygroscopic Ionomer

A method aimed to prepare a salt-embedded, humidity-regulating tray has been carried out by Langowski et al. (2006) and Sängerlaub et al. (2013a). The method consisted of a three-layer structure, including a barrier layer, active layer with NaCl, and a sealing layer, to control the humidity in food packages. Thereby, the active layer was foamed and stretched to form cavities around the salt particles. The same humidity-regulating trays were used by Rux et al. (2016) for packing tomatoes and strawberries; these trays were produced in the form of a thermoforming multilayer structure containing a foaming hygroscopic ionomer as an active layer. The results stated that no condensation was observed in the package as the humidity produced by the products was effectively absorbed by the trays. However, Rux et al. (2015) reported water loss in packaged mushrooms using similar humidity-regulating trays. The RH value remained stable at 93% during storage for 6 days at 7°C. The packed mushroom samples showed better color appearance and less incidence of decay compared with control samples.

The study performed by Singh et al. (2010a,b) further confirmed the findings of water loss in packed mushrooms with humidity-regulating trays. The study revealed that the increase in the amount of NaCl in the trays from 6 to 18 wt% led to increased water loss in packed mushrooms stored at 5°C. Although, it was stated that the weight loss could also be a result of the difference in the physiological state of the product, storing temperature, and the type of packaging film.

14.2.2.2.6 Tamarind Seed Galactoxyloglucan

Moisture-absorbing aerogels can be prepared from the extracts of polysaccharides such as tamarind seeds. According to Gracanac (2015), this aerogel proved its capacities to absorb water and saline solution 40 times its original weight when it was coupled with enzyme-based (galactose oxidase) OS systems. However, in terms of absorbing the drip, this capacity dramatically decreased 20-folds its original weight. Nonetheless, it showed its potential to be used for meat packaging applications.

There are different moisture scavenging systems that could be utilized for maintaining the quality and extending the shelf life of food products. It is, therefore, important to maintain product-adjusted moisture scavenging systems for different food applications. Also, there are many different commercial technologies that use moisture scavengers in different forms such as sachets, pouches, and pads, all devices that are not part of the structure of external packaging materials. On the other hand, the research trends aiming to incorporate moisture absorbers into packaging films for commercial applications of different food products (e.g. fresh-cut fruits, vegetables, fish, and meat) are still under development (Restuccia et al. 2010).

14.2.2.3 Ethylene Absorbers

Ethylene (C_2H_4) is a growth-stimulating hormone (plant growth regulator) responsible for quickening the ripening and senescence of fruits. It increases the respiration rate of fresh fruit and shortens shelf life during postharvest storage. According to the literature (Saltveit 1999; Ozdemir and Floros

2004), ethylene is responsible for accelerating chlorophyll degradation rates, particularly in leafy products as well as excessive softening of fruits, thereby, the removal of ethylene from the environment of food products is very important. For this purpose, many researchers aimed to produce and develop different commercial products called ethylene scavengers for enhancing the quality and prolonging the shelf life of food products

14.2.2.3.1 Potassium Permanganate

Different ethylene scavenging systems involve the inclusion of a small sachet containing specific scavengers into the package, which are highly permeable to ethylene, allowing its diffusion or integration of ethylene scavengers into the film structure. Potassium permanganate ($KMnO_4$) is one of the most widely utilized active compounds in a sachet for oxidizing ethylene (Floros et al. 1997; Ayhan 2011; Llorens et al. 2012). According to Martínez-Romero et al. (2007), $KMnO_4$ is never used in direct contact with food because of its toxicity.

14.2.2.3.2 Minerals

Another category of ethylene scavenging system uses minerals in the form of fine powders, such as zeolite, active carbon, or pumice. According to De Kruijf et al. (2002), these compounds could be integrated with plastic films to be used in the production of fresh produce packaging. These minerals not only scavenge ethylene but are also able to modify the gas permeability of the film, enabling an increase in CO_2 diffusivity where oxygen can enter more easily than through pure PE to secure an equilibrium atmosphere (De Kruijf et al. 2002; Esturk et al. 2014). Metals and metal oxides are also good candidates for ethylene removal. TiO_2 is one of the well-known metal oxides commonly used to oxidize ethylene into water and carbon dioxide. It must be taken into account that metal oxides need to be activated through visible light or UV light, and the effect of the exposure of the food to the UV light is debatable and must be considered.

The use of nanosilver as an ethylene blocker was tested (Hu and Fu 2003; Fernández et al. 2010) on fresh-cut melons in the form of absorbent pads. Also, palladium-based scavengers were widely studied in the form of sachets in packages (Abe and Watada 1991; Bailén et al. 2006, 2007; Cao et al. 2015) and in storage rooms (Martínez-Romero et al. 2009) and showed good ethylene adsorption capacity. According to Martínez-Romero et al. (2007), the limitation of the use of palladium in industrial application is associated with the high cost. Abe and Watada (1991) studied the effect of coupling palladium chloride and charcoal integrated into a paper sachet. The results revealed that this mixture is effective not only in the prevention of ethylene accumulation and subsequent reduction of fresh-cut softening but showed its effectiveness in preventing chlorophyll loss in spinach leaves. Furthermore, this mixture showed its effectiveness in absorbing most of the ethylene during 3 days of storage at 20°C.

Also, the experiment conducted by Sothornvit and Sampoompuang (2012) tested the possibility of coupling activated carbon with polysaccharide glucomannan into a paper made from rice straw. The scavenging capacity of ethylene was 77%. Thus, suggestions related to spreading carbon into the packages that contain products such as bananas, mangoes, tomatoes, and apples might have the potential of extending their shelf life as those products are categorized as ethylene-sensitive products. The incorporation of carbon into packages has many different forms such as wrapping the product or laminating the inside surface of the packages. Ozdemir and Floros (2004) stated that the integration of scavengers into packaging films could be a great alternative to exceed the hurdle associated with sachet-related problems. Furthermore, another advantage is associated with the ability of ethylene scavengers to be integrated into different layers of packaging. Nonetheless, the research studies associated with the applications of ethylene absorbers in packaging film structure is quite limited.

14.3 ANTIMICROBIAL ACTIVE PACKAGING SYSTEM

Antimicrobial food packaging exhibits a system designed to inhibit the growth of pathogenic micro-organisms and spoilage. Figure 14.4 shows the most studied antimicrobial food packaging systems according to their active substance/material.

14.3.1 Essential Oils

Recent interest in reducing the use of petroleum-based additives as active materials for food preservation has led to the application of natural additives both for the benefit of the individual as well as for the environment (Alves-Silva et al. 2013). EOs play an important role in plant defense, consequently, some of them possess strong antimicrobial properties. In addition to this, most of them are classified as GRAS (Ruiz-Navajas et al. 2013).

Cinnamon EO is among the most studied EOs in active materials. Gherardi et al. (2016) mentioned that a multilayer material containing 18% cinnamaldehyde as the major compound of cinnamon EO provides high activity against *Saccharomyces cerevisiae* and *Escherichia coli* O157:H7, as the material reduced both microorganisms by 3 log CFU/mL. Wen et al. (2016), developed an antimicrobial packaging material by incorporating a cinnamon EO/β-cyclodextrin inclusion complex into PLA nanofibers via an electrospinning technique to control loss of volatile substances and maintain greater cinnamon EO in the film. Application of cinnamon EO in pork preservation at 25°C shows that in the third day the control packed with fresh-keeping film had decayed, whereas the sample packed with the nanofilm decayed on the eighth day (Higueras et al. 2015).

Another EO is carvacrol, which is used as a bio-based bioactive compound. Campos-Requena et al. (2015) studied the synergistic antimicrobial effect of different EOs on food based on carvacrol and thymol, with both included in HDPE/modified montmorillonite nanocomposite films. The films were applied through indirect contact with strawberries to study the synergistic antimicrobial effect against *Botrytis cinereal*. The IC_{50} of the EOs in the film was 40.4 mg/g in the case of carvacrol only, whereas it was 13.2 mg/g in the case of both EOs 50:50. Rodriguez-Garcia et al. (2016) studied

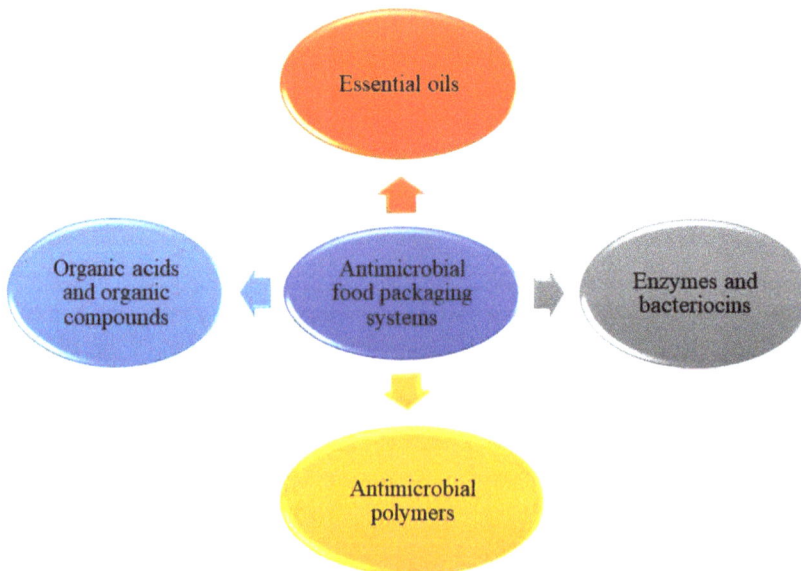

Figure 14.4 Antimicrobial packaging systems.

the effect of oregano EO applied within pectin coatings on the inhibition of *Alternaria alternata* on tomatoes and found that 25.9 g/L was effective in inhibiting microbial growth.

Safety and shelf life enhancement studies of cooked ham revealed that 1–2% of EOs extracted from *Thymus piperella* and *Thymus moroderi* in the packaging films is effective in lowering lactic acid bacteria count and aerobic mesophilic count; the higher effect was exhibited by *T. piperella* (Ruiz-Navajas et al. 2015). A new AP system has been designed with thyme EOs by Quesada et al. (2016) to extend the shelf life of sliced ready-to-eat cooked pork during refrigerated storage. Packages containing thyme EOs (0%, 0.5%, 1%, and 2%) were coated with chitosan film on the inner surface to ensure indirect contact and preserve the modification of organoleptic properties of meat. Thyme EOs had a negative effect on the yeast populations, and the yeast counts decreased with an increased dose of EOs in the film, especially during the first 21 days of storage.

Basil leaf EO was incorporated into fish protein/fish gelatin composite films and the efficacy of the film was tested by wrapping sea bass slices and studying any microbiological and sensory changes for 12 days at 4°C storage temperature (Arfat et al. 2015). Control samples exhibited 6 days shelf life, whereas the sea bass wrapped in active film demonstrated a shelf life of 10–12 days. Vanillin and *Allium* spp. extract have also been proposed as bioactive EOs. Llana-Ruiz-Cabello et al. (2015) reported that *Allium* spp. had an efficiency against molds in lettuce during storage for 7 days. On the other hand, Lee et al. (2016) mentioned that vanillin showed antimicrobial activity against *L. monocytogenes* in crab sticks packed with starfish gelatin films containing 0.05% of vanillin.

14.3.2 Enzymes and Bacteriocins

Applications of bacteriocins, enzymes, and proteins in food packaging to prevent food spoilage have been an area of research for several decades (Figure 14.5).

Physically entrapped or chemically bonded to packaging films, enzymes can serve as effective antimicrobials in food packaging. Lysozyme is a popular antimicrobial enzyme that can destroy the glycosidic bonds of the gram-positive bacterial peptidoglycans. Min et al. (2005) reported that the shelf life of smoked salmon could be extended, and the growth of *L. monocytogenes* was inhibited to 4.4 log CFU/cm^2 when lysozyme (204 mg/g of film) was incorporated into whey protein films.

The incorporation of enzymes into a matrix can modify its properties. Lactoferrin and/or lysozyme alter the carrier paper matrix (containing carboxymethyl cellulose), allowing it to bind positively charged proteins (Barbiroli et al. 2012) These paper sheets containing lysozyme, lactoferrin,

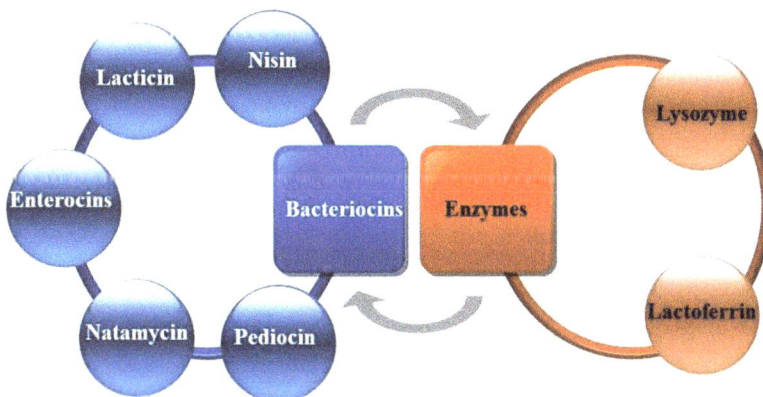

Figure 14.5   Food packaging enzymes and bacteriocins.

or both had prevented the growth of aerobic bacteria in the meat sample, and lysozyme was the most effective.

The growth of food spoilage bacteria, fundamentally gram-positive bacteria, could be inhibited by bacteriocins, which are peptides or small proteins produced by some species of lactic acid bacteria. The bacteriocin, nisin, has been successfully integrated into methylcellulose/hydroxypropyl methylcellulose coatings (Franklin et al. 2004) and coated on LDPE films (Mauriello et al. 2005; Neetoo et al. 2008) or paperboard (Lee et al. 2004). Franklin et al. (2004) mentioned that the packaging of hot dogs in packaging films coated with a cellulose derivative-based solution containing 10,000 IU/mL nisin during refrigerated storage significantly decreased *L. monocytogenes* by greater than 2 log CFU per package after 60 days.

The integration of nisin into PE/polyethylene oxide films or polyamide coatings effectively inhibited coliform bacteria, *Brochothrix thermosphacta* growth, and extended the shelf life of beef (Cutter et al. 2001; Kim et al. 2002). Khan et al. (2016) reported that the immobilization of nisin and ethylenediaminetetraacetic acid (EDTA) on the surface of the cellulose nanocrystal/chitosan films reduced the growth of mesophiles, psychrotrophs, and *Lactobacillus* spp. in fresh pork loin meats. Other bacteriocins such as lactocins (Massani et al. 2014), pediocin (Santiago-Silva et al. 2009), enterocins (Marcos et al. 2007), and natamycin (De Oliveira et al. 2007) have been used as coatings on various substrates or integrated into biopolymer-based films. They decrease the growth of *Lactobacillus plantarum*; *Salmonella* spp.; *L. monocytogenes*; and *Listeria innocua* on wieners, sliced ham, cooked ham, and Gorgonzola cheese, respectively.

14.4 NANOPARTICLES

Until now, at the microscale or nanoscale level, research studies related to encapsulation concept applications in food are mainly focused on liposomal composition, methodology, and prolonged stability, whereas the release mechanism of diverse active agents from liposomes seem less of a focus. Figure 14.6 summarizes the critical features of food active agents (antioxidant, antimicrobials, enzymes, fortifiers) by encapsulation (liposome), with release mechanisms. The actual phenomenon of the release from liposomes had been proved by a few studies. Colas et al. (2007) reported that in the case of nisin release, a transmission electron microscope reveals the fusion of liposome with the bacterial cell envelope to discharge the encapsulated nisin in a liposome.

14.4.1 Release Efficiency of Bioactive Compounds at Nanoscale

Targeting the bioactive agents and their protection from degradation and inactivation using encapsulation techniques has been investigated extensively in the scientific literature (Mozafari et al. 2006, 2008). According to Champagne and Fustier (2007), the term release contains two major phenomena: diffusion of active agents from the liposome core through unilamellar/multilamellar bilayers and desorption from liposome into the medium. Figure 14.7 shows the release process of lipophilic agents in aqueous liposomal dispersions. According to Fahr et al. (2005) these steps may differ at high phospholipid concentrations, where it is believed that collision between the lipid vesicles is the main transferring mechanism.

Because of the novelty of this field and/or the lack of advanced tools to study nanostructured materials, the literature shows no significant work done to elaborate on the release process of food active agents from liposomes. Moreover, the research in this field did not progress as AP materials were not allowed for food packaging applications in the European region and regulations were stringent and unrevised for a long time for food contact materials. However, the recently launched EU Green Deal 2050 pushes research on AP materials and proposed that regulations will be revisited and revised for food packaging applications (640/2020/EC; European Commission 2020).

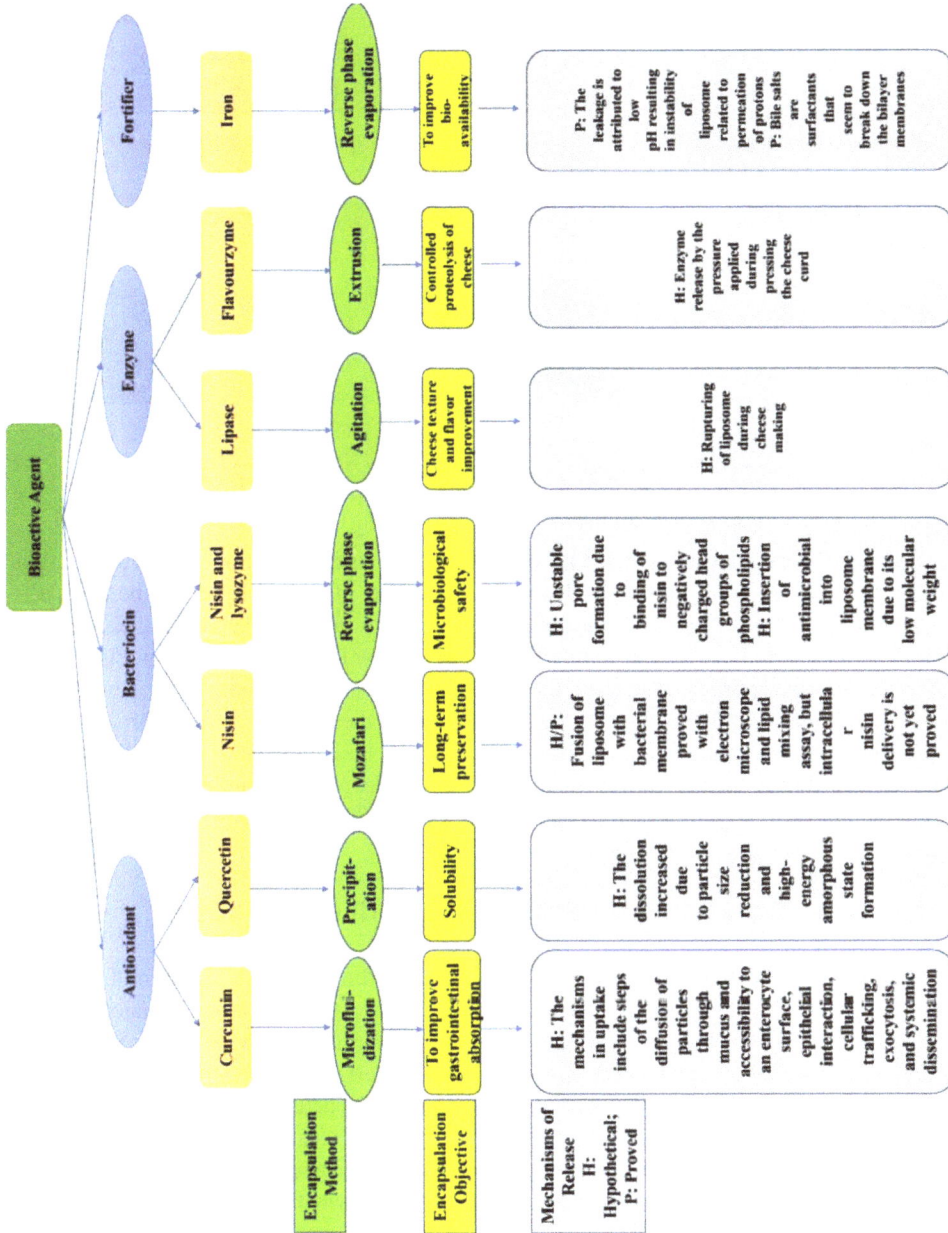

Figure 14.6 Nanoencapsulation of various bioactive agents in food systems. (From Takahashi et al. 2009; Wu et al. 2008; Colas et al. 2007; Were et al. 2003; Kheadr et al. 2003; Anjani et al. 2007; and Ding et al. 2009.)

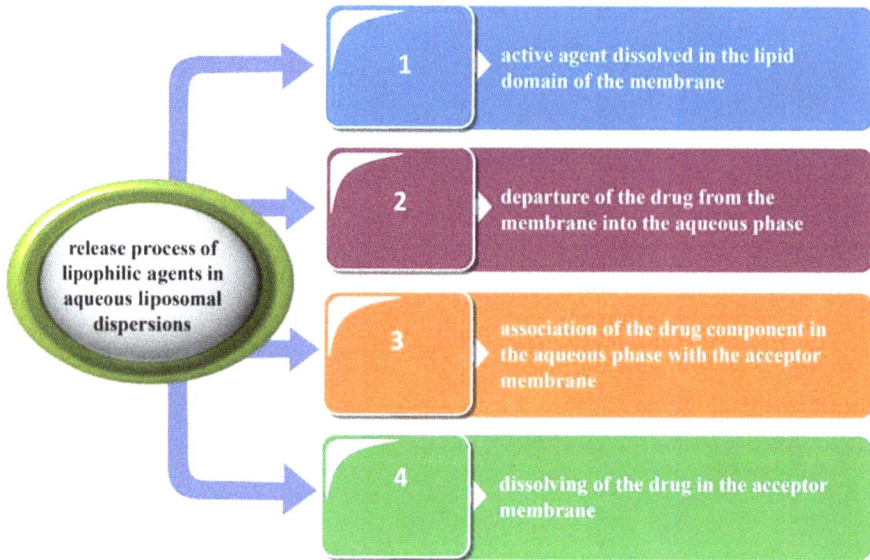

Figure 14.7   Release process of lipophilic agents. (From Fahr et al. 2005.)

Were et al. (2004) had analyzed the release phenomenon by using a fluorescent dye (calcein) in their work on nisin release from liposomes. However, the results of nisin release will be quite different from fluorescent dye because of its capability to form pores in liposome membrane, as a high concentration of nisin would disrupt the membrane instead of slow inside-out migration. Figure 14.8 summarizes the different approaches utilized for the quantification of the release efficiency of some encapsulated active agents employed in the food sector. Most of the research data comprise an indirect evaluation of bioactive agents through antimicrobial activity measurement or residual enzymatic measurement. With improvements in manufacturing technologies, encapsulation carriers will play an important role in increasing the efficacy of functional foods (Chen et al. 2006).

14.5 SHELF LIFE EXTENSION USING ACTIVE PACKAGING

Although the interest concerning the utilization of AP for extending the shelf life of food products is growing, the practice of this approach still needs some considerations. One of the major barriers is the regulatory issues associated with the type of packaging. Because of the differences between the regulatory systems between the United States and EU, this barrier could be the major hurdle for the international implementation of these systems. The complex structure of these different strategies, which focused on the protection and the safety of the customers, causes conflict of understanding and classification. For example, in the United States AP categorized under the Food Additive Petition or Food Contact Notification is processed according to whether it may intentionally or unintentionally become part of the food, and if so, at what level. Other factors associated with the functional barriers and non-migration may apply. The EU commands are to some extent similar but have some important variances concerning food simulators and relocation levels.

Further problems include the cost-to-benefit ratio, manufacturing capability, customer acceptance, and most importantly the sensory effects on the food. Although the existence of some AP systems is commercially feasible, it suffers in terms of cost-effective production issues. Adding to that, numerous approaches established in laboratories have not been on a commercial-scale apparatus

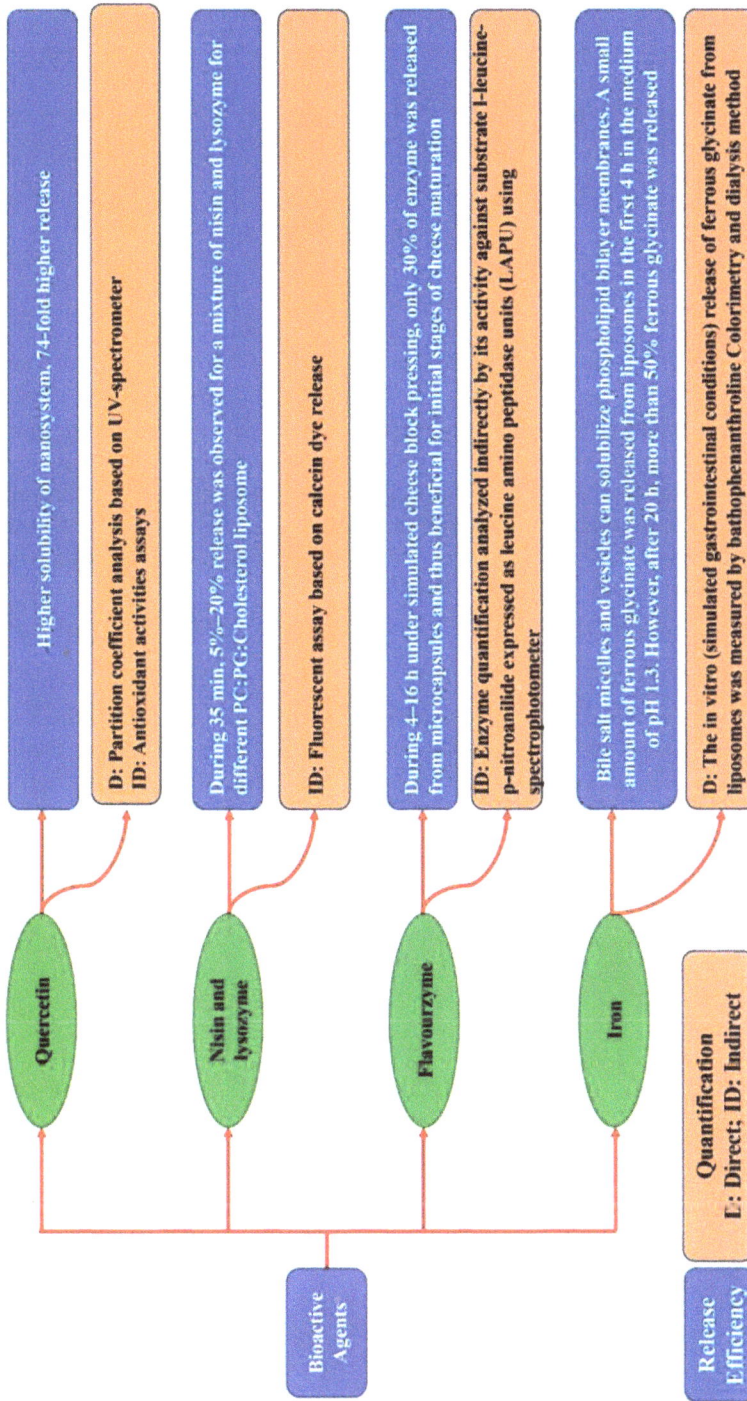

Figure 14.8 Different approaches of quantification of the release efficiency of encapsulated agents. (From Wu et al. 2008; Were et al. 2003; Anjani et al. 2007; and Ding et al. 2009.)

and a technique such as drying overnight at room temperature is not applicable. Technologies such as oxygen scavenging films are not very common, therefore, they do not cause any consumer anxiety, but other approaches such as antimicrobial-releasing materials may be noticeable to customers and cause misperception or doubt, particularly if the food is changed in any way. Additionally, if the active device does not work as planned, obligation regarding the cause of disaster may be an issue. Generally, it is easy to say that the implementation of AP will contribute very effectively to extending the shelf life of the food products. Mixtures of systems along with new knowledge to be further established will continue to improve the quality and safety of foods around the world.

14.6 FUTURE CONSIDERATIONS

14.6.1 Global Trends

The annual market growth was approximately 14% in the period between 2001 and 2005 and around 10% between 2005 and 2010 reaching the market at $3 billion, as illustrated in Figure 14.9. Notably, extending the shelf life and improving food convenience are the major drivers for this growth. This appears in the growth of using OS and moisture scavengers in many different products, which are closely related to extending the products shelf life. The development of the AP technology is also associated with the market characteristics as well; for example, the Japanese market accepts more than 50% of the global market of AP because of consumer sympathy toward minimally processed food and for the latest technology implementations.

14.6.2 Europe

One of the major players in spreading and influencing the technologies throughout the marketplace could be concluded in legislation differences around the world. The regularity of the legislation around the EU could help in this sector. However, it could be a source of a delay drawing up the legislation that has hindered this packaging technology, especially in western European markets. There is a delay because active packages were only approved after the formulation of EU Framework Regulation 1935/200425, which permitted the use of active releasing materials. Before 2004, only absorbing systems were allowed along with susceptor laminates (microwave heating), subsequently, they did not hinder the legislation at that period.

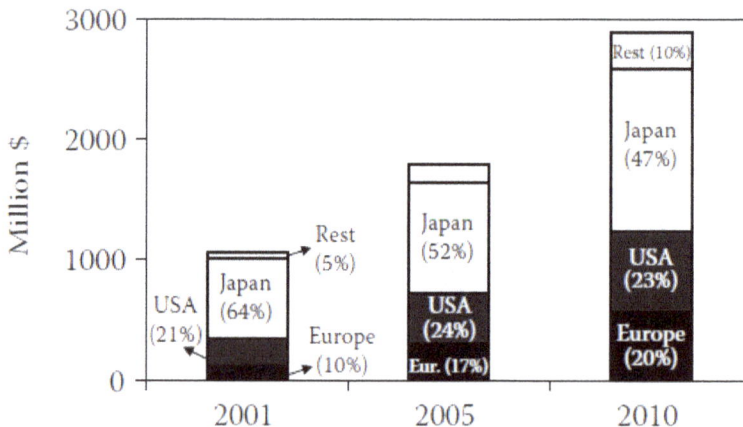

Figure 14.9   Global markets for active packaging in 2001, 2005, and 2010. (From Climpson, J., The Future of Active Packaging in Food and Drink Markets, Pira International Leatherhead, Surrey, UK, 2005.)

The conventional customer attitude in European countries toward AP technologies could be another challenging point. This barrier is facing AP penetration into the European market, and unlike the Japanese consumer, the European customer has less affinity to pay for this technology and toward the technology overall. Furthermore, consumer insights into the role of AP, even though its role is to extend food shelf life, contribute to decreasing the quality and freshness. Therefore, it is expected that the tendency, particularly in Europe, is to get the approval for current AP technologies and to do further research focusing on different strategies for cost reduction and in finding further benefits that will rationalize the extra packaging costs. Because of the growing demands for convenience, it could be easily expected that there is a growing interest in susceptor laminates for microwave packaging.

Different surveys stated that, although the importance of releasing chemicals is very important for food product acceptance, consumer acceptance is relatively high toward the biological origin compounds such as vitamin C as a releasing antioxidant compared with other nonbiological origin compounds. In contrast, the industry's need to pack foods efficiently and securely for transportation and storing while preserving the quality, along with increasing demands from customers for fresh, convenient, slightly processed, and safe products with an extended shelf life, offers a bright future for AP in the mid and long term. Interest in AP and the number of food-related applications are expected to increase significantly in the years to come.

14.7 CONCLUSION

The need for AP is driven by consumers' inclination toward the "natural" and "additive-free" alternatives to extend the shelf life of foods. However, their inclusion can impact the organoleptic properties of the food; therefore, natural antimicrobials are incorporated into food packaging materials to exploit their functional properties without imparting any undesirable characteristics to the food product. Research studies have documented and proved successful utilization of packaging systems for preventing oxidation reactions in food, but the gap concerning the validation of the efficiency of the antioxidant packages in commercial applications and different considerations concerning target market and legal status still exists. Hence, for the industrial implementation of this technology it is important to study real food packaging systems produced through scalable film processing techniques (such as coating compared with solvent-casting or extrusion), to examine the validity of the films under storage conditions, to study the effect of food sensory properties, and to understand the suitability for different industrial packaging techniques (MAP and vacuum compared with wrapping) and formats of the studied food product. Nonetheless, it is undisputable that the practice of AP is becoming gradually popular and many new opportunities in the food and non-food industries will open for applying this technology in the future.

REFERENCES

Abe K, Watada AE. 1991. Ethylene absorbent to maintain quality of lightly processed fruits and vegetables. J Food Sci 56(6):1589–92. https://doi.org/10.1111/j.1365-2621.1991.tb08647.x.

Al-Nabulsi AA, Holley RA. 2006. Enhancing the antimicrobial effects of bovine lactoferrin against *Escherichia coli* O157:H7 by cation chelation, NaCl and temperature. J Appl Microbiol 100:244–55.

Alves-Silva JM, dos Santos SMD, Pintado ME, Pérez-Álvarez JA, Fernáandez-Lóopez J, Viuda-Martos M. 2013. Chemical composition and in vitro antimicrobial, antifungal and antioxidant properties of essential oils obtained from some herbs widely used in Portugal. Food Control 32(2):371–78. https://doi.org/10.1016/j.foodcont.2012.12.022.

Anal AK, Singh, H. 2007. Recent advances in microencapsulation of probiotics for industrial applications and targeted delivery. Trends Food Sci Technol 18:240–51.

Andersen HJ, Rasmussen MA. 1992. Interactive packaging as protection against photodegradation of the colour of pasteurized, sliced ham. Int J Food Sci Technol 27(1):1–8. https://doi.org/10.1111/j.1365-2621.

Anjani K, Kailasapathy K, Phillips M. 2007. Microencapsulation of enzymes for potential application in acceleration of cheese ripening. Int Dairy J 17:79–86.

Antunez PD, Botero, Omary M, Rosentrater KA, Pascall M, Winstone L. 2012. Effect of an oxygen scavenger on the stability of preservative-free flour tortillas. J Food Sci 77(1):1–9. https://doi.org/10.1111/j.1750-3841.2011.02470.x.

Arfat YA, Benjakul S, Vongkamjan K, Sumpavapol P, Yarnpakdee S. 2015. Shelf-life extension of refrigerated sea bass slices wrapped with fish protein isolate/fish skin gelatin-ZnO nanocomposite film incorporated with basil leaf essential oil. J Food Sci Technol 52(10):6182–93. https://doi.org/10.1007/s13197-014-1706-y.

Arvanitoyannis IS, Oikonomou G. 2012. Active and intelligent packaging. In: Arvanitoyannis IS, editor. Modified atmosphere and active packaging technologies. Boca Raton, FL: CRC Press. pp. 628–54.

Ayhan Z. 2011. Effect of packaging on the quality and shelf-life of minimally processed/ready to eat foods. Acad Food J 9(4):36–41.

Baiano A, Marchitelli V, Tamagnone P, Nobile MAD. 2004. Use of active packaging for increasing ascorbic acid retention in food beverages. J Food Sci 69(9):E502–8. https://doi.org/10.1111/j.1365-2621.2004.tb09936.x.

Bailén G, Guillén F, Castillo S, Serrano M, Valero D, Martínez-Romero D. 2006. Use of activated carbon inside modified atmosphere packages to maintain tomato fruit quality during cold storage. J Agric Food Chem 54(6):2229–35. https://doi.org/10.1021/jf0528761.

Bailén G, Guillén F, Castillo S, Zapata PJ, Serrano M, Valero D, Martíínez-Romero D. 2007. Use of a palladium catalyst to improve the capacity of activated carbon to absorb ethylene, and its effect on tomato ripening. Span J Agric Res 5(4):579–86. https://doi.org/10.5424/sjar/2007054-5359.

Barbiroli A, Bonomi F, Capretti G, Iametti S, Manzoni M, Piergiovanni L, Rollini M. 2012. Antimicrobial activity of lysozyme and lactoferrin incorporated in cellulose-based food packaging. Food Control 26(2):387–92. https://doi.org/10.1016/j.foodcont.2012.01.046.

Barbosa-Pereira L, Aurrekoetxea GP, Angulo I, Paseiro-Losada P, Cruz JM. 2014. Development of new active packaging films coated with natural phenolic compounds to improve the oxidative stability of beef. Meat Sci 97(2):249–54. https://doi.org/10.1016/j.meatsci.2014.02.006.

Barbosa-Pereira L, Cruz JM, Sendón R, Rodríguez Bernaldo de Quirós A, Ares A, Castro-López M, Abad MJ, Maroto J, Paseiro-Losada P. 2013. Development of antioxidant active films containing tocopherols to extend the shelf life of fish. Food Control 31(1):236–43. https://doi.org/10.1016/j.foodcont.2012.09.036.

Biji KB, Ravishankar CN, Mohan CO, Srinivasa Gopal TK. 2015. Smart packaging systems for food applications: a review. J Food Sci Technol 52(10):6125–35. https://doi.org/10.1007/s13197-015-1766-7.

Bjerkeng B, Sivertsvik M, Rosnes JT, Bergslien H. 1995. Reducing package deformation and increasing filling degree in packages of cod fillets in CO_2-enriched atmospheres by adding sodium carbonate and citric acid to an exudate absorber. In: Ackermann P, Jagerstad M, Ohlsson T, editors. Foods and packaging materials – Chemical interactions. Cambridge, England: Royal Society of Chemistry. pp. 222–7.

Blickstad E, Enfors SO, Molin G. 1981. Effect of hyperbaric carbon dioxide pressure on the microbial flora of pork stored at 4 or 14°C. J Appl Bacteriol 50(3):493–504. https://doi.org/10.1111/j.1365-2672.1981.tb04252.x.

Böhner N, Hösl F, Rieblinger K, Danzl W. 2014. Effect of retail display illumination and headspace oxygen concentration on cured boiled sausages. Food Pack Shelf Life 1(2):131–9. https://doi.org/10.1016/j.fpsl.2014.04.003.

Brockgreitens J, Abbas A. 2016. Responsive food packaging: recent progress and technological prospects. Comp Rev Food Sci Food Safety 15(1):3–15. https://doi.org/10.1111/1541-4337.12174.

Brody AL, Strupinsky E, Kline LR. 2001. Active packaging for food applications. Boca Raton, FL: CRC Press. pp. 101–8.

Bumbudsanpharoke N, Choi J, Ko S. 2015. Applications of nanomaterials in food packaging. J Nanosci Nanotechnol 15(9):6357–72. https://doi.org/10.1166/jnn.2015.10847.

Cahill PJ, Chen SY, inventors. 04 July 2000. Oxygen scavenging condensation copolymers for bottles and packaging articles. U.S. Patent 6083585.

Campos-Requena VH, Rivas BL, Pérez MA, Figueroa CR, Sanfuentes EA. 2015. The synergistic antimicrobial effect of carvacrol and thymol in clay/polymer nanocomposite films over strawberry gray mold. LWT – Food Sci Technol 64(1):390–6. https://doi.org/10.1016/j.lwt.2015.06.006.

Cao J, Li X, Wu K, Jiang W, Qu G. 2015. Preparation of a novel PdCl2–CuSO4–based ethylene scavenger supported by acidified activated carbon powder and its effects on quality and ethylene metabolism of broccoli during shelf-life. Postharvest Biol Technol 99:50–7. https://doi.org/10.1016/j.postharvbio.2014.07.017.

Carrizo D, Taborda G, Neríin C, Bosetti O. 2016. Extension of shelf life of two fatty foods using a new antioxidant multilayer packaging containing green tea extract. Innov Food Sci Emerg 33:534–41. https://doi.org/10.1016/j.ifset.2015.10.018.

Chaix E, Guillaime C, Gontard N, Guillard V. 2015. Diffusivity and solubility of CO_2 in dense solid food products. J Food Eng 166:1–9. https://doi.org/10.1016/j.jfoodeng.2015.05.023.

Champagne CP, Fustier P. 2007. Microencapsulation for the improved delivery of bioactive compounds into foods. Curr Opin Biotechnol 18:184–190.

Charles F, Sanchez J, Gontard N. 2003. Active modified atmosphere packaging of fresh fruits and vegetables: modeling with tomatoes and oxygen absorber. J Food Sci 68(5):1736–42. https://doi.org/10.1111/j.1365-2621.2003.tb12321.x.

Chen J, Brody AL. 2013. Use of active packaging structures to control the microbial quality of a ready-to-eat meat product. Food Control 30(1):306–10. https://doi.org/10.1016/j.foodcont.2012.07.002.

Chen L, Remondetto GE, Subirade M. 2006. Food protein-based materials as nutraceutical delivery systems. Trends Food Sci Technol 17:272–83.

Chounou N, Chouliara E, Mexis SF, Stavros K, Georgantelis D, Kontominas MG. 2013. Shelf life extension of ground meat stored at 4°C using chitosan and an oxygen absorber. Int J Food Sci Technol 48(1):89–95. https://doi.org/10.1111/j.1365-2621.2012.03162.x.

Chung HJ, Colakoglu AS, Min DB. 2004. Relationships among headspace oxygen, peroxide value, and conjugated diene content of soybean oil oxidation. J Food Sci 69(2):83–8. https://doi.org/10.1111/j.1365-2621.2004.tb15507.x.

Cichello SA. 2015. Oxygen absorbers in food preservation: a review. J Food Sci Technol 52(4):1889–95. https://doi.org/10.1007/s13197-014-1265-2.

Colas JC, Shi W, Rao VSNM, Omri A, Mozafari MR, Singh H. 2007. Microscopical investigations of nisin-loaded nanoliposomes prepared by Mozafari method and their bacterial targeting, Micron 38:841–47.

Coles R, McDowell D, Kirwan MJ, editors. 2008. Food packaging technology. Boca Raton, FL: CRC Press.

Coma V. 2008. Bioactive packaging technologies for extended shelf life of meat-based products. Meat Sci 78(1–2):90–103. https://doi.org/10.1016/j.meatsci.2007.07.035.

Cooksey K. 2014. Modified atmosphere packaging of meat, poultry, fish. In: Han JH, editor. Innovations in food packaging (2nd ed.). San Diego, CA: Academic Press. pp. 475–93.

Cruz RS, Camilloto GP, dos Santos Pires AC. 2012. Oxygen scavengers: an approach on food preservation. In: Eissa AA, editor. Structure and function of food engineering. Rijeka, Croatia: Intech.

Cutter CN, Willett JL, Siragusa GR. 2001. Improved antimicrobial activity of nisin-incorporated polymer films by formulation change and addition of food grade chelator. Lett Appl Microbiol 33(4):325–8. https://doi.org/10.1046/j.1472-765X.2001.01005.x.

Damaj Z, Joly C, Guillon E. 2014. Toward new polymeric oxygen scavenging systems: formation of poly (vinyl alcohol) oxygen scavenger film. Packaging Technol Sci 28(4):293–302. https://doi.org/10.1002/pts.2112.

Day B. 2008. Active packaging of food. In: Kerry J, Butler P, editors. Smart packaging technologies for fast moving consumer goods. Chichester, England: John Wiley & Sons Ltd. pp. 1–18.

De Kruijf N, Van Beest M, Rijk R, Sipiläinen-Malm T, Losada PP, Meulenaer BD. 2002. Active and intelligent packaging: applications and regulatory aspects. Food Addit Contam 19(suppl):144–62. https://doi.org/10.1080/02652030110072722.

De Oliveira TM, de Fátima Ferreira Soares N, Pereira RM, de Freitas Fraga K. 2007. Development and evaluation of antimicrobial natamycin-incorporated film in gorgonzola cheese conservation. Packaging Technol Sci 20(2):147–53. https://doi.org/10.1002/pts.756.

Debs-Louka E, Louka N, Abraham G, Chabot V, Allaf K. 1999. Effect of compressed carbon dioxide on microbial cell viability. Appl Environ Microbiol 65(2):626–31.

Devlieghere F, Debevere J. 2000. Influence of dissolved carbon dioxide on the growth of spoilage bacteria. LWT – Food Sci Technol 33(8):531–7. https://doi.org/10.1006/fstl.2000.0705.

Devlieghere F, Debevere J, Van Impe J. 1998. Concentration of carbon dioxide in the water-phase as a parameter to model the effect of a modified atmosphere on microorganisms. Int J Food Microbiol 43(1–2):105–13. https://doi.org/10.1016/S0168-1605(98)00101-9.

Di Maio L, Scarfato P, Galdi MR, Incarnato L. 2015. Development and oxygen scavenging performance of three-layer active PET films for food packaging. J Appl Polym Sci 132(7). https://doi.org/10.1002/app.41465.

Ding, B., Xia, S., Hayat, K, Zhang, X. 2009. Preparation and pH stability of ferrous glycinate liposomes. J Agric Food Chem 57:2938–44.

Dobrucka R, Cierpiszewski R. 2014. Active and intelligent packaging food – research and development – a review. Pol J Food Nutr Sci 64(1):7–15. https://doi.org/10.2478/v10222-012-0091-3.

Dornic N, Ficheux AS, Roudot AC. 2016. Qualitative and quantitative composition of essential oils: a literature-based database on contact allergens used for safety assessment. Regul Toxicol Pharmacol 80:226–32. https://doi.org/10.1016/j.yrtph.2016.06.016.

Droval AA, Benassi VT, Rossa A, Prudencio SH, Paiao FG, Shimokomak IM. 2012. Consumer attitudes and preferences regarding pale, soft, and exudative broiler breast meat. J Appl Poult Res 21(3):502–7. https://doi.org/10.3382/japr.2011-00392.

EFSA CEF Panel. 2014. Scientific opinion on the safety assessment of the active substances, palladium metal and hydrogen gas, for use in active food contact materials. EFSA J 12(2):3558. https://doi.org/10.2903/j.efsa.2014.3558.

Eghbaliferiz S, Iranshahi M. 2016. Prooxidant activity of polyphenols, flavonoids, anthocyanins and carotenoids: updated review of mechanisms and catalyzing metals. Phytother Res 30(9):1379–91. https://doi.org/10.1002/ptr.5643.

Esturk O, Ayhan Z, Gokkurt T. 2014. Production and application of active packaging film with ethylene adsorber to increase the shelf life of broccoli (Brassica oleracea L. var. Italica). Packaging Technol Sci 27(3):179–91. https://doi.org/10.1002/pts.2023.

European Commission. 2009. EU Guidance to the Commission Regulation (EC) No 450/2009 of 29 May 2009 on active and intelligent materials and articles intended to come into the contact with food (version 1.0). Available from: https://ec.europa.eu/food/safety/chemical_safety/food_contact_materials. Accessed 2020 July 20.

European Commission. 2020. EU Guidance to the Commission Regulation (EC) No 640/2020 of March 2020 A Farm to Fork Strategy For a fair, healthy and environmentally friendly food system. Available from: politico.eu/vp-content/uploads/2020/03/FINAL-FINAL-F2F-MARCH-CLEAN.pdf. Accessed 2020 July 20.

Fahr A, Hoogevest PV, May S, Bergstrand N, Leigh MLS. 2005. Transfer of lipophilic drugs between liposomal membranes and biological interfaces: consequences for drug delivery. Eur J Pharm Sci 26:251–65.

Fernández A, Picouet P, Lloret E. 2010. Cellulose-silver nanoparticle hybrid materials to control spoilage-related microflora in absorbent pads located in trays of fresh-cut melon. Int J Food Microbiol 142(1–2):222–8. https://doi.org/10.1016/j.ijfoodmicro.2010.07.001.

Floros J, Dock L, Han J. 1997. Active packaging technologies and applications. Food Cos Drug Package 20(1):10–17.

Franklin NB, Cooksey KD, Getty KJK. 2004. Inhibition of Listeria monocytogenes on the surface of individually packaged hot dogs with a packaging film coating containing nisin. J Food Prot 67(3):480–85. https://doi.org/10.4315/0362-028X-67.3.480.

Galdi MR, Incarnato L. 2011. Influence of composition on structure and barrier properties of active PET films for food packaging applications. Packaging Technol Sci 24(2):89–102. https://doi.org/10.1002/pts.917.

Galdi MR, Nicolais V, Di Maio L, Incarnato L. 2008. Production of active PET films: evaluation of scavenging activity. Packaging Technol Sci 21(5):257–68. https://doi.org/10.1002/pts.794.

Gerez CL, Torres MJ, Font de Valdez G, Rollán G. 2013. Control of spoilage fungi by lactic acid bacteria. Biol Control 64(3):231–37. https://doi.org/10.1016/j.biocontrol.2012.10.009.

Gherardi R, Becerril R, Nerin C, Bosetti O. 2016. Development of a multilayer antimicrobial packaging material for tomato puree using an innovative technology. LWT – Food Sci Technol 72:361–7. https://doi.org/10.1016/j.lwt.2016.04.063.

Gibis D, Rieblinger K. 2011. Oxygen scavenging films for food application. Proc Food Sci 1:229–34. https://doi.org/10.1016/j.profoo.2011.09.036.

Gomes C, Castell-Perez ME, Chimbombi E, Barros F, Sun D, Liu J, Sue H-J, Sherman P, Dunne P, Wright AO. 2009. Effect of oxygen-absorbing packaging on the shelf life of a liquid-based component of military operational rations. J Food Sci 74(4):E167–76. https://doi.org/10.1111/j.1750-3841.2009.01120.x.

Gracanac B. 2015. Polysaccharide-based aerogels as water absorbent and oxygen scavenger in meat packaging, Master's thesis, University of Helsinki, Finland.

Graciano-Verdugo AZ, Soto-Valdez H, Peralta E, Cruz-Zárate P, Islas-Rubio AR, Sánchez-Valdes S, Sánchez-Escalante A, González-Méndez N, González-Ríos H. 2010. Migration of α-tocopherol from LDPE films to corn oil and its effect on the oxidative stability. Food Res Int 43 (4):1073–78. https://doi.org/10.1016/j.foodres.2010.01.019.

Graf E, inventor. 08 February 1994. Oxygen removal. U.S. Patent 5284871.

Granda-Restrepo DM, Peralta E, Troncoso-Rojas R, Soto-Valdez H. 2009a. Release of antioxidants from co-extruded active packaging developed for whole milk powder. Int Dairy J 19(8):481–8. https://doi.org/10.1016/j.idairyj.2009.01.002.

Granda-Restrepo DM, Soto-Valdez H, Peralta E, Troncoso-Rojas R, Vallejo-Córdoba B, Gómez-Meza N, Graciano-Verdugo AZ. 2009b. Migration of α-tocopherol from an active multilayer film into whole milk powder. Food Res Int 42(10):1396–402. https://doi.org/10.1016/j.foodres.2009.07.007.

Haas GJ, Prescott HE, Dudley E, Dik R, Hintlian C, Keane L. 1989. Inactivation of microorganisms by carbon dioxide under pressure. J Food Saf 9(4):253–65. https://doi.org/10.1111/j.1745-4565.1989.tb00525.x.

Hansen AA, Høy M, Pettersen MK. 2009a. Prediction of optimal CO2 emitter capacity developed for modified atmosphere packaging of fresh salmon fillets (*Salmo salar* L.). Packaging Technol Sci 22(4):199–208. https://doi.org/10.1002/pts.843.

Hansen AA, Mørkøre T, Rudi K, Langsrud Ø, Eie T. 2009b. The combined effect of super chilling and modified atmosphere packaging using CO_2 emitter on quality during chilled storage of pre-rigor salmon fillets (*Salmo salar*). J Sci Food Agric 89(10):1625–33. https://doi.org/10.1002/jsfa.3599.

Hansen AA, Mørkøre T, Rudi K, Rødbotten M, Bjerke F, Eie T. 2009c. Quality changes of prerigor filleted Atlantic salmon (*Salmo salar* L.) packaged in modified atmosphere using CO_2 emitter, traditional MAP, and vacuum. J Food Sci 74(6):M242–49. https://doi.org/10.1111/j.1750-3841.2009.01233.x.

Higueras L, López-Carballo G, Gavara R, Hernández-Muñoz P. 2015. Reversible covalent immobilization of cinnamaldehyde on chitosan films via Schiff base formation and their application in active food packaging. Food Bioprocess Technol 8(3):526–38. https://doi.org/10.1007/s11947-014-1421-8.

Holck AL, Pettersen MK, Moen MH, Sørheim O. 2014. Prolonged shelf life and reduced drip loss of chicken filets by the use of carbon dioxide emitters and modified atmosphere packaging. J Food Prot 77(7): 1133–41. https://doi.org/10.4315/0362-028X.JFP-13-428.

Hu AW, Fu ZH. 2003. Nanotechnology and its application in packaging and packaging machinery. Packaging Eng 24:22–4.

Hutter S, Rüegg N, Yildirim S. 2016. Use of palladium-based oxygen scavenger to prevent discoloration of ham. Food Packaging Shelf Life 8:56–62. https://doi.org/10.1016/j.fpsl.2016.02.004.

Imran M, Revol-Junelles A-M, Martyn A, Tehrany EA, Jacquot M, Linder M, Desobry S. 2010. Active food packaging evolution: transformation from micro- to nanotechnology. Crit Rev Food Sci Nutr 50(9): 799–821. https://doi.org/10.1080/10408398.2010.503694.

Jacobsen C. 1999. Sensory impact of lipid oxidation in complex food systems. Eur J Lipid Sci 101(12):484–92. https://doi.org/10.1002/(SICI)1521-4133(199912) 101:12<484::AID-LIPI484>3.0.CO;2-H.

Jiang H, Manolache S, Wong ACL, Denes FS. 2004. Plasma-enhanced deposition of silver nanoparticles onto polymer and metal surfaces for the generation of antimicrobial characteristics. J Appl Polym Sci 93:1411–22.

Johnson RL, Htoon AK, Shaw KJ. 1995. Detection of orange peel extract in orange juice. Food Aust 47(9):426–32.

Julkapli NM, Bagheri S. 2016. Developments in nano-additives for paper industry. J Wood Sci 62(2):117–30. https://doi.org/10.1007/s10086-015-1532-5.

Kartal S, Aday MS, Caner C. 2012. Use of microperforated films and oxygen scavengers to maintain storage stability of fresh strawberries. Postharvest Biol Technol 71:32–40. https://doi.org/10.1016/j.postharvbio.2012.04.009.

Kennedy JF, Rivera ZS, Lloyd LL, Warner FP, Jumel K. 1990. Studies on non-enzymic browning in orange juice using a model system based on freshly squeezed orange juice. J Sci Food Agric 52:85–95.

Kerry J, Butler P. 2008. Smart packaging technologies for fast moving consumer goods. Chichester, England: John Wiley & Sons, Ltd.

Khan A, Gallah H, Riedl B, Bouchard J, Safrany A, Lacroix M. 2016. Genipin cross-linked antimicrobial nanocomposite films and gamma irradiation to prevent the surface growth of bacteria in fresh meats. Innov Food Sci Emerg 35:96–102. https://doi.org/10.1016/j.ifset.2016.03.011.

Kheadr EE, Vuillemard JC, El-Deeb SA. 2003. Impact of liposome-encapsulated enzyme cocktails on cheddar cheese ripening. Food Res Int 36:241–52.

Kim H-Y, Gornsawun G, Shin I-S. 2015. Antibacterial activities of isothiocyanates (ITCs) extracted from horseradish (*Armoracia rusticana*) root in liquid and vapor phases against 5 dominant bacteria isolated from low-salt Jeotgal, a Korean salted and fermented seafood. Food Sci Biotechnol 24(4):1405–12. https://doi.org/10.1007/s10068-015-0180-2.

Kim Y-M, Paik H-D, Lee D-S. 2002. Shelf-life characteristics of fresh oysters and ground beef as affected by bacteriocin-coated plastic packaging film. J Sci Food Agric 82(9):998–1002. https://doi.org/10.1002/jsfa.1125.

Kolbe H. 1882. Antiseptische Eigenschaften der Kohlensäure. J Praktische Chem 26(1):249–55. https://doi.org/10.1002/prac.18820260116.

Kuorwel KK, Cran MJ, Orbell JD, Buddhadasa S, Bigger SW. 2015. Review of mechanical properties, migration, and potential applications in active food packaging systems containing nanoclays and nanosilver. Comp Rev Food Sci Food Saf 14(4):411–30. https://doi.org/10.1111/1541-4337.12139.

Labuza TP, Hyman CR. 1998. Moisture migration and control in multi-domain foods. Trends Food Sci Technol 9(2):47–55. https://doi.org/10.1016/S0924-2244(98)00005-3.

Langowski HC, Sängerlaub S, Wanner T, inventors. 24 April 2006. Humidity regulating packaging material for packing e.g. liquid food, has material made of polymer matrix which is provided with cavities, and packaging material has polymer, copolymer and their blends or alloys. World Patent 2007121909.

Laridi R, Kheadr EE, Benech RO, Vuillemard JC, Lacroix C, Fliss I. 2003. Liposome encapsulated nisin Z: optimization, stability and release during milk fermentation. Int Dairy J 13:325–36.

Larsen H, Westad F, Sorheim O, Nilsen LH. 2006. Determination of critical oxygen level in packages for cooked sliced ham to prevent color fading during illuminated retail display. J Food Sci 71(5):S407–13. https://doi.org/10.1111/j.1750-3841.2006.00048.x.

Lee CH, An DS, Lee SC, Park HJ, Lee DS. 2004. A coating for use as an antimicrobial and antioxidative packaging material incorporating nisin and α-tocopherol. J Food Eng 62(4):323–9. https://doi.org/10.1016/S0260-8774(03)00246-2.

Lee K-Y, Lee J-H, Yang H-J, Song KB. 2016. Characterization of a starfish gelatin film containing vanillin and its application in the packaging of crab stick. Food Sci Biotechnol 25(4):1023–8. https://doi.org/10.1007/s10068-016-0165-9.

Lee SJ, Rahman ATMM. 2014. Intelligent packaging for food products. In: Han JH, editor. Innovations in food packaging (2nd ed.). San Diego, CA: Academic Press. pp. 171–209.

Llana-Ruiz-Cabello M, Pichardo S, Baños A, Núñez C, Bermúdez JM, Guillamón E, Aucejo S, Cameán AM. 2015. Characterisation and evaluation of PLA films containing an extract of *Allium* spp. to be used in the packaging of ready-to-eat salads under controlled atmospheres. LWT – Food Sci Technol 64(2):1354–61. https://doi.org/10.1016/j.lwt.2015.07.057.

Llorens A, Lloret E, Picouet PA, Trbojevich R, Fernandez A. 2012. Metallic-based micro and nanocomposites in food contact materials and active food packaging. Trends Food Sci Technol 24(1):19–29. https://doi.org/10.1016/j.tifs.2011.10.001.

Lohwasser W, Wanner T, inventors. 22 November 2005. Composite system, used as packaging foil, bag foil or partially applied single foil, comprises substrate foil thin film and catalyst for reduction of oxygen. European Patent 1917139.

López-de-Dicastillo C, Gómez-Estaca J, Catalá R, Gavara R, Hernández-Muñoz P. 2012a. Active antioxidant packaging films: development and effect on lipid stability of brined sardines. Food Chem 131(4):1376–84. https://doi.org/10.1016/j.foodchem.2011.10.002.

López-de-Dicastillo C, Pezo D, Nerín C, López-Carballo G, Catalá R, Gavara R, Hernández-Muñoz P. 2012b. Reducing oxidation of foods through antioxidant active packaging based on ethyl vinyl alcohol and natural flavonoids. Packaging Technol Sci 25(8):457–66. https://doi.org/10.1002/pts.992.

Lopez-Gomez A, Ros-Chumillas M. 2010. Packaging and shelf life of orange juice. In: Robertson GL, editor. Food packaging and shelf life. Boca Raton, FL: CRC Press. pp. 179–98.

Mahajan PV, Rodrigues FA, Motel A, Leonhard A. 2008. Development of a moisture absorber for packaging of fresh mushrooms (*Agaricus bisporous*). Postharvest Biol Technol 48(3):408–14. https://doi.org/10.1016/j.postharvbio.2007.11.007.

Maloba FW, Rooney ML, Wormell P, Nguyen M. 1996. Improved oxidative stability of sunflower oil in the presence of an oxygen-scavenging film. J Am Oil Chem Soc 73(2):181–5. https://doi.org/10.1007/bf02523892.

Manzanarez-López F, Soto-Valdez H, Auras R, Peralta E. 2011. Release of α-tocopherol from poly (lactic acid) films, and its effect on the oxidative stability of soybean oil. J Food Eng 104(4):508–17. https://doi.org/10.1016/j.jfoodeng.2010.12.029.

Marcos B, Aymerich T, Monfort JM, Garriga M. 2007. Use of antimicrobial biodegradable packaging to control *Listeria monocytogenes* during storage of cooked ham. Int J Food Microbiol 120(1–2):152–8. https://doi.org/10.1016/j.ijfoodmicro.2007.06.003.

Marcos B, Sáarraga C, Castellari M, Kappen F, Schennink G, Arnau J. 2014. Development of biodegradable films with antioxidant properties based on polyesters containing α-tocopherol and olive leaf extract for food packaging applications. Food Packaging Shelf Life 1(2):140–50. https://doi.org/10.1016/j.fpsl.2014.04.002.

Martínez-Romero D, Bailén G, Serrano M, Guillén F, Valverde JM, Zapata P, Castillo S, Valero D. 2007. Tools to maintain postharvest fruit and vegetable quality through the inhibition of ethylene action: a review. Crit Rev Food Sci Nutr 47(6):543–60. https://doi.org/10.1080/10408390600846390.

Martínez-Romero D, Guillén F, Castillo S, Zapata PJ, Valero D, Serrano M. 2009. Effect of ethylene concentration on quality parameters of fresh tomatoes stored using a carbon-heat hybrid ethylene scrubber. Postharvest Biol Technol 51(2):206–11. https://doi.org/10.1016/j.postharvbio.2008.07.011.

Massani MB, Molina V, Sanchez M, Renaud V, Eisenberg P, Vignolo G. 2014. Active polymers containing *Lactobacillus curvatus* CRL705 bacteriocins: effectiveness assessment in Wieners. Int J Food Microbiol 178:7–12. https://doi.org/10.1016/j.ijfoodmicro.2014.02.013.

Matche RS, Sreekumar RK, Raj B. 2011. Modification of linear low-density polyethylene film using oxygen scavengers for its application in storage of bun and bread. J Appl Polym Sci 122(1):55–63. https://doi.org/10.1002/app.33718.

Mauriello G, De Luca E, La Storia A, Villani F, Ercolini D. 2005. Antimicrobial activity of a nisin-activated plastic film for food packaging. Lett Appl Microbiol 41(6):464–9. https://doi.org/10.1111/j.1472-765X.2005.01796.x.

Mbuge DO, Negrini R, Nyakundi LO, Kuate SP, Bandyopadhyay R, Muiru WM, Torto B, Mezzenga R. 2016. Application of superabsorbent polymers (SAP) as desiccants to dry maize and reduce aflatoxin contamination. J Food Sci Technol 53(8):3157–65. https://doi.org/10.1007/s13197-016-2289-6.

McMillin KW. 2008. Where is MAP going? A review and future potential of modified atmosphere packaging for meat. Meat Sci 80(1):43–65. https://doi.org/10.1016/j.meatsci.2008.05.028.

Miller CW, Nguyen MH, Rooney M, Kailasapathy K. 2003. The control of dissolved oxygen content in probiotic yogurts by alternative packaging materials. Packaging Technol Sci 16(2):61–7. https://doi.org/10.1002/pts.612.

Min S, Harris LJ, Krochta JM. 2005. Antimicrobial effects of lactoferrin, lysozyme, and the lactoperoxidase system and edible whey protein films incorporating the lactoperoxidase system against *Salmonella enterica* and *Escherichia coli* O157:H7. J Food Sci 70(7):m332–8. https://doi.org/10.1111/j.1365-2621.2005.tb11476.x.

Møller JKS, Jensen JS, Olsen MB, Skibsted LH, Bertelsen G. 2000. Effect of residual oxygen on colour stability during chill storage of sliced, pasteurized ham packaged in modified atmosphere. Meat Sci 54(4):399–405. https://doi.org/10.1016/S0309-1740(99)00116-3.

Mozafari MR, Flanagan J, Matia-Merino L, Awati A, Omri A, Suntres ZE, Singh H. 2006. Recent trends in the lipid-based nanoencapsulation of antioxidants and their role in foods. J Sci Food Agric 86:2038–45.

Mozafari MR, Johnson C, Hatziantoniou S, Demetzos C. 2008. Nanoliposomes and their applications in food nanotechnology. J Liposome Res 18:309–27.

Mu H, Gao H, Chen H, Tao F, Fang X, Ge L. 2013. A nanosised oxygen scavenger: preparation and antioxidant application to roasted sunflower seeds and walnuts. Food Chem 136(1):245–50. https://doi.org/10.1016/j.foodchem.2012.07.121.

Nannerup LD, Jakobsen M, van den Berg F, Jensen JS, Møller JKS, Bertelsen G. 2004. Optimizing colour quality of modified atmosphere packed sliced meat products by control of critical packaging parameters. Meat Sci 68(4):577–85. https://doi.org/10.1016/j.meatsci.2004.05.009.

Neetoo H, Ye M, Chen H, Joerger RD, Hicks DT, Hoover DG. 2008. Use of nisin-coated plastic films to control *Listeria monocytogenes* on vacuum-packaged cold-smoked salmon. Int J Food Microbiol 122(1–2):8–15. https://doi.org/10.1016/j.ijfoodmicro.2007.11.043.

Nerín C, Tovar L, Djenane D, Camo J, Salafranca J, Beltrán JA, Roncalés P. 2006. Stabilization of beef meat by a new active packaging containing natural antioxidants. J Agric Food Chem 54(20):7840–46. https://doi.org/10.1021/jf060775c.

Ohlsson T, Bengtsson N. 2002. Minimal processing technologies in the food industry. Cambridge, England: Woodhead Publishing Ltd.

Oliveira A, Iahnke S, Costa TMH, de Oliveira Rios A, Flôres SH. 2016. Antioxidant films based on gelatin capsules and minimally processed beet root (*Beta vulgaris* L. var. *Conditiva*) residues. J Appl Polym Sci 133(10):43094. https://doi.org/10.1002/app.43094.

Ozdemir M, Floros JD. 2004. Active food packaging technologies. Crit Rev Food Sci Nutr 44(3):185–93. https://doi.org/10.1080/10408690490441578.

Park H-Y, Kim S-J, Kim KM, You Y-S, Kim SY, Han J. 2012. Development of antioxidant packaging material by applying corn-zein to LLDPE Film in combination with phenolic compounds. J Food Sci 77(10):E273–9. https://doi.org/10.1111/j.1750-3841.2012.02906.x.

Pénicaud C, Peyron S, Gontard N, Guillard V. 2012. Oxygen quantification methods and application to the determination of oxygen diffusion and solubility coefficients in food. Food Rev Int 28(2):113–45. https://doi.org/10.1080/87559129.2011.595021.

Pereira de Abreu DA, Cruz JM, Losada PP. 2012. Active and intelligent packaging for the food industry. Food Rev Int 28(2):146–87. https://doi.org/10.1080/87559129.2011.595022.

Pereira de Abreu DA, Paseiro Losada P, Maroto J, Cruz JM. 2011. Natural antioxidant active packaging film and its effect on lipid damage in frozen blue shark (*Prionace glauca*). Innov Food Sci Emerg 12(1): 50–55. https://doi.org/10.1016/j.ifset.2010.12.006.

Pettersen MK, Hansen AA, Mielnik M. 2014. Effect of different packaging methods on quality and shelf life of fresh Reindeer meat. Packaging Technol Sci 27(12):987–97. https://doi.org/10.1002/pts.2075.

Quesada J, Sendra E, Navarro C, Sayas-Barberá E. 2016. Antimicrobial active packaging including chitosan films with *Thymus vulgaris L.* essential oil for ready-to-eat meat. Foods 5(3):57. https://doi.org/10.3390/foods5030057.

Realini CE, Marcos B. 2014. Active and intelligent packaging systems for a modern society. Meat Sci 98(3):404–19. https://doi.org/10.1016/j.meatsci.2014.06.031.

Reig CS, Lopez AD, Ramos MH, Ballester VAC. 2014. Nanomaterials: a map for their selection in food packaging applications. Packaging Technol Sci 27(11):839–66. https://doi.org/10.1002/pts.2076.

Restuccia D, Spizzirri UG, Parisi OI, Cirillo G, Curcio M, Iemma F, Puoci F, Vinci G, Picci N. 2010. New EU regulation aspects and global market of active and intelligent packaging for food industry applications. Food Control 21(11):1425–35. https://doi.org/10.1016/j.foodcont.2010.04.028.

Rhim J-W, Kim Y-T. 2014. Biopolymer-based composite packaging materials with nanoparticles. In: Han JH, editor. Innovations in food packaging (2 ed.). San Diego, CA: Academic Press. pp. 413–42.

Rhim J-W, Park H-M, Ha C-S. 2013. Bio-nanocomposites for food packaging applications. Prog Polym Sci 38(10–11):1629–52. https://doi.org/10.1016/j.progpolymsci.2013.05.008.

Rivaroli DC, Guerrero A, Velandia Valero M, Zawadzki F, Eiras CE, Campo MdM, Sañudo C, Mendes Jorge A, Nunes do Prado I. 2016. Effect of essential oils on meat and fat qualities of crossbred young bulls finished in feedlots. Meat Sci 121:278–84. https://doi.org/10.1016/j.meatsci.2016.06.017.

Röcker B, Rüegg N, Glöss AN, Yeretzian C, Yildirim S. 2017. Inactivation of palladium-based oxygen scavenger system by volatile sulfur compounds present in the headspace of packaged food. Packaging Technol Sci 30(8):427–42. https://doi.org/10.1002/pts.2220.

Rodriguez-Aguilera R, Oliveira JC. 2009. Review of design engineering methods and applications of active and modified atmosphere packaging systems. Food Eng Rev 1(1):66–83. https://doi.org/10.1007/s12393-009-9001-9.

Rodriguez-Garcia I, Cruz-Valenzuela MR, Silva-Espinoza BA, Gonzalez-Aguilar GA, Moctezuma E, Gutierrez-Pacheco MM, Tapia-Rodriguez MR, Ortega-Ramirez LA, Ayala-Zavala JF. 2016. Oregano (*Lippia graveolens*) essential oil added within pectin edible coatings prevents fungal decay and increases the antioxidant capacity of treated tomatoes. J Sci Food Agric 96(11):3772–8. https://doi.org/10.1002/jsfa.7568.

Rooney ML. 1995. Active packaging in polymer films. In: Rooney ML, editor. Active packaging. London, England: Blackie Academic and Professional. pp. 74–110.

Rooney ML, inventor. 28 September 1999. Oxygen scavengers independent of transition metal catalysts. U.S. Patent 5958254.

Rüegg N, Blum T, Röcker B, Kleinert M, Yildirim S. 2016. Application of palladium-based oxygen scavenger to extend the shelf life of bakery products. Book of abstracts of 6th international symposium on food packaging; Barcelona, Spain. Brussels, Belgium: ILSI Europe.

Ruiz-Navajas Y, Viuda-Martos M, Barber X, Sendra E, Perez-Alvarez JA, Fernández-López J. 2015. Effect of chitosan edible films added with *Thymus moroderi* and *Thymus piperella* essential oil on shelf-life of cooked cured ham. J Food Sci Technol 52(10):6493–501. https://doi.org/10.1007/s13197-015-1733-3.

Ruiz-Navajas Y, Viuda-Martos M, Sendra E, Perez-Alvarez JA, Fernández-López J. 2013. In vitro antioxidant and antifungal properties of essential oils obtained from aromatic herbs endemic to the southeast of Spain. J Food Prot 76(7):1218–25. https://doi.org/10.4315/0362-028X.JFP-12-554.

Rux G, Mahajan PV, Geyer M, Linke M, Pant A, Saengerlaub S, Caleb OJ. 2015. Application of humidity-regulating tray for packaging of mushrooms. Postharvest Biol Technol 108:102–10. https://doi.org/10.1016/j.postharvbio.2015.06.010.

Rux G, Mahajan PV, Linke M, Pant A, Sängerlaub S, Caleb OJ, Geyer M. 2016. Humidity-regulating trays: moisture absorption kinetics and applications for fresh produce packaging. Food Bioprocess Technol 9(4):709–16. https://doi.org/10.1007/s11947-015-1671-0.

Saltveit ME. 1999. Effect of ethylene on quality of fresh fruits and vegetables. Postharvest Biol Technol 15 (3):279–92. https://doi.org/10.1016/S0925-5214(98)00091-X.

Sandhya. 2010. Modified atmosphere packaging of fresh produce: current status and future needs. LWT – Food Sci Technol 43(3):381–92. https://doi.org/10.1016/j.lwt.2009.05.018.

Sängerlaub S, Böhmer M, Stramm C. 2013a. Influence of stretching ratio and salt concentration on the porosity of polypropylene films containing sodium chloride particles. J Appl Polym Sci 129(3):1238–45. https://doi.org/10.1002/app.38793.

Sängerlaub S, Gibis D, Kirchhoff E, Tittjung M, Schmid M, Müller K. 2013b. Compensation of pinhole defects in food packages by application of iron-based oxygen scavenging multilayer films. Packaging Technol Sci 26(1):17–30. https://doi.org/10.1002/pts.1962.

Santiago-Silva P, Soares NFF, Nóbrega JE, Júnior MAW, Barbosa KBF, Volp ACP, Zerdas ERMA, Würlitzer NJ. 2009. Antimicrobial efficiency of film incorporated with pediocin (ALTAR _ 2351) on preservation of sliced ham. Food Control 20(1):85–9. https://doi.org/10.1016/j.foodcont.2008.02.006.

Sasaki H, Kamimura K. 1997. Pinhole inspection machine for sealed packages: for detection of pinholes of 0.5μm or below. Packaging Technol Sci 10(2):109–18. https://doi.org/10.1002/(SICI)1099-1522(199703/04)10:2<109:AID-PTS394>3.0.CO;2-S.

Ščetar M, Kurek M, Galić K. 2010. Trends in meat and meat products packaging – a review. Croat J Food Sci Technol 2(1):32–48.

Shah NP, Lankaputhra WE, Britz ML, Kyle WS. 1995. Survival of *Lactobacillus acidophilus* and *Bifidobacterium bifidum* in commercial yogurt during refrigerated storage. Int Dairy J 5(5):515–21. https://doi.org/10.1016/0958-6946(95)00028-2.

Shin Y, Shin J, Lee Y. 2009. Effects of oxygen scavenging package on the quality changes of processed meatball product. Food Sci Biotechnol 18(1):73–8.

Singh P, Abas Wani A, Saengerlaub S. 2011. Active packaging of food products: recent trends. J Nutr Food Sci 41(4):249–60. https://doi.org/10.1108/00346651111151384.

Singh P, Langowski H-C, Wani AA, Saengerlaub S. 2010a. Recent advances in extending the shelf life of fresh *Agaricus* mushrooms: a review. J Sci Food Agric 90(9):1393–402. https://doi.org/10.1002/jsfa.3971.

Singh P, Saengerlaub S, Stramm C, Langowski H. 2010b. Humidity regulating packages containing sodium chloride as active substance for packing of fresh raw *Agaricus* mushrooms. Proceeding of the 4th international workshop cold chain management; Bonn, Germany, Bonn, Germany: University of Bonn.

Siripatrawan U, Noipha S. 2012. Active film from chitosan incorporating green tea extract for shelf life extension of pork sausages. Food Hydrocoll 27(1):102–8. https://doi.org/10.1016/j.foodhyd.2011.08.011.

Sivertsvik M. 2003. Active packaging in practice: fish. In: Ahvenainen R, editor. Novel food packaging techniques. Cambridge, England: Woodhead Publishing Ltd. pp. 384–400.

Solovyov SE. 2010. Oxygen scavengers. In: Yam KL, editor. The Wiley encyclopedia of packaging technology (3 ed.). Hoboken, NJ: John Wiley & Sons Ltd. pp. 841–50.

Sothornvit R, Sampoompuang C. 2012. Rice straw paper incorporated with activated carbon as an ethylene scavenger in a paper-making process. Int J Food Sci Technol 47(3):511–7. https://doi.org/10.1111/j.1365-2621.2011.02871.x.

Souza CO, Silva LT, Silva JR, López JA, Veiga-Santos P, Druzian JI. 2011. Mango and acerola pulps as antioxidant additives in cassava starch bio-based film. J Agric Food Chem 59(6):2248–54. https://doi.org/10.1021/jf1040405.

Stoll L, Haas Costa TM, Jablonski A, Hickmann Flôres S, de Oliveira Rios A. 2016. Microencapsulation of anthocyanins with different wall materials and its application in active biodegradable films. Food Bioprocess Technol 9(1):172–81. https://doi.org/10.1007/s11947-015-1610-0.

Suppakul P, Miltz J, Sonneveld K, Bigger SW. 2003. Active packaging technologies with an emphasis on antimicrobial packaging and its applications. J Food Sci 68(2):408–20. https://doi.org/10.1111/j.1365-2621.2003.tb05687.x.

Takahashi M, Uechi S, Takara K, Asikin Y, Wada K. 2009. Evaluation of an oral carrier system in rats: bioavailability and antioxidant properties of liposome-encapsulated curcumin, J Agric Food Chem, 57:9141–46.

Taylor TM, Davidson PM, Bruce BD, Weiss J. 2005. Liposomal nanocapsules in food science and agriculture. Crit Rev Food Sci Nutr 45:587–605.

Torres-Arreola W, Soto-Valdez H, Peralta E, Cárdenas-López JL, Ezquerra-Brauer JM. 2007. Effect of a low-density polyethylene film containing butylated hydroxytoluene on lipid oxidation and protein quality of Sierra fish (*Scomberomorus sierra*) muscle during frozen storage. J Agric Food Chem 55(15):6140–6. https://doi.org/10.1021/jf070418h.

Torrieri E, Carlino PA, Cavella S, Fogliano V, Attianese I, Buonocore GG, Masi P. 2011. Effect of modified atmosphere and active packaging on the shelf-life of fresh bluefin tuna fillets. J Food Eng 105(3):429–35. https://doi.org/10.1016/j.jfoodeng.2011.02.038.

Valley G. 1928. The effect of carbon dioxide on bacteria. Q Rev Biol 3(2):209–24.

Van Aardt M, Duncan SE, Marcy JE, Long TE, O'Keefe SF, Sims SR. 2007. Release of antioxidants from poly(lactide-co-glycolide) films into dry milk products and food simulating liquids. Int J Food Sci Technol 42(11):1327–37. https://doi.org/10.1111/j.1365-2621.2006.01329.x.

Vermeiren L, Devlieghere F, Debevere J. 2002. Effectiveness of some recent antimicrobial packaging concepts. Food Addit Contam, 19:163–71.

Vermeiren L, Devlieghere F, van Beest M, de Kruijf N, Debevere J. 1999. Developments in the active packaging of foods. Trends Food Sci Technol 10(3):77–86. https://doi.org/10.1016/S0924-2244(99)00032-1.

Wagener M. 2006. Antimicrobial coatings. Polym Paint Colour J 196:34–7.

Wen P, Zhu D-H, Feng K, Liu F-J, Lou W-Y, Li N, Zong M-H, Wu H. 2016. Fabrication of electrospun poly-lactic acid nanofilm incorporating cinnamon essential oil/β-cyclodextrin inclusion complex for antimicrobial packaging. Food Chem 196:996–1004. https://doi.org/10.1016/j.foodchem.2015.10.043.

Were LM, Bruce BD, Davidson PM, Weiss J. 2003. Size, stability, and entrapment efficiency of phospholipid nano capsules containing polypeptide antimicrobials. J Agric Food Chem 51:8073–79.

Were LM, Bruce B, Davidson PM, Weiss J. 2004. Encapsulation of nisin and lysozyme in liposomes enhances efficacy against Listeria monocytogenes. J Food Prot 67:922–7.

Wu C-S, Liao J-Y, Fang S-Y, Chiang AST. 2010. Flexible and transparent moisture getter film containing zeolite. Adsorption 16(1):69–74. https://doi.org/10.1007/s10450-009-9196-3.

Wu TH, Yen FL, Lin LT, Tsai TR, Lin CC, Cham TM. 2008. Preparation, physicochemical characterization, and antioxidant effects of quercetin nanoparticles. Int J Pharm 346:160–8.

Yang H-J, Lee J-H, Won M, Song KB. 2016. Antioxidant activities of distiller dried grains with solubles as protein films containing tea extracts and their application in the packaging of pork meat. Food Chem 196:174–9. https://doi.org/10.1016/j.foodchem.2015.09.020.

Yildirim S. 2011. Active packaging for food bio preservation. In: Lacroix C, editor. Protective cultures, antimicrobial metabolites and bacteriophages for food and beverage bio preservation. Cambridge, England: Woodhead Publishing Ltd. pp. 460–89.

Yildirim S, Jammet JC, Lohwasser W, inventors. 06 October 2010. Multi-layer film. European Patent 2236284.

Yildirim S, Röcker B, Rüegg N, Lohwasser W. 2015. Development of palladium-based oxygen scavenger: optimization of substrate and palladium layer thickness. Packaging Technol Sci 28(8):710–18. https://doi.org/10.1002/pts.2134.

Zerdin K, Rooney ML, Vermuë J. 2003. The vitamin C content of orange juice packed in an oxygen scavenger material. Food Chem 82(3):387–95. https://doi.org/10.1016/S0308-8146(02)00559-9.

Zhuang H, Barth MM, Cisneros-Zevallos L. 2014. Modified atmosphere packaging for fresh fruits and vegetables. In: Han JH, editor. Innovations in food packaging (2nd ed.). San Diego, CA: Academic Press. pp. 45–73.

CHAPTER 15

Smart Packaging for Managing and Monitoring Shelf Life and Food Safety

Mamta Thakur, Ishrat Majid, and Vikas Nanda

CONTENTS

15.1 INTRODUCTION

The customer today demands quality foods that should be minimally processed, convenient, safe, ready-to-eat, and easily prepared following the modern trend of clean labeling, thus requiring innovative progress both in food processing and food packaging (Silberbauer and Schmid, 2017; Majid et al., 2018). Packaging refers to enclosing the food materials to protect them from the harsh conditions of the external environment. To ensure the safety and quality in the food supply chain, the core feature of food packaging is specifically related to its application. The protection, connectivity, ease of use, and necessary containment of packaged foods are offered by traditional packaging technologies (Mariusz, 2019; Müller and Schmid, 2019). They offer food composition information, cooking instructions, warning labels, storage guidelines, manufacturing address, best use dates, monitoring, etc. However, the temperature of transportation and storage, relative humidity, and gas concentration influence perishable foods directly. Thus, modern the "smart packaging" approach regulates the conditions of food throughout the supply chain (storage and manufacturing) (Ahmed et al., 2018; Schaefer and Cheung, 2018). Smart packaging is an emerging solution that leads to the expansion of shelf life and offers useful knowledge about food product quality and safety for food waste management and improved customer security. The "smartness" of packaging implies that it

E-mail: Ishrat Majid, ishratmajid89@gmail.com

DOI: 10.1201/9781003091677-15

can communicate information about the quality of product, integrity of the product (leak indicators), and track the product's time temperature, which simply means it gives product quality details (Alfiya et al., 2017; Biegańska, 2017).

Smart packaging is simply the combination of active and intelligent packaging systems. Active packaging concentrated directly on the product shelf life by eliminating or retaining the particular chemical compounds either in or out of packaged food or its surroundings (Wyrwa and Barska, 2017). Intelligent packaging systems, on the other hand, widen the communication feature of the package to facilitate the effective management of their supply chain and increase the shelf life of food products via real-time monitoring (Müller and Schmid, 2019). Smart packaging is an extended feature of the basic protection and communication function of conventional packaging. It alerts customers about their ability to identify, detect, and monitor any changes or spoilage within the product package and warns against any possible transport and storage issues (Realini and Marcos, 2014; Ghaani et al., 2016; Fang et al., 2017). Smart packaging solutions put together benefits from both technologies and simplify the automatic data collection or other interaction in the food supply chain by using technological devices like radiofrequency identification (RFID) technology (Rai et al., 2019). An example of smart packaging is the combination of ethanol emitters (active form) and oxygen sensors (intelligent form) during the modified atmospheric packaging (MAP) of bread. In this instance, the oxygen sensor detected the integrity of packaging and ethanol acted as an antimicrobial preservative (Hempel et al., 2013). These primary or indirect quality measures are focused on an acknowledgment and detailed review of the degrading phenomenon that distinguishes food and beverage spoilage processes during their expected shelf life. Economically, specific active and intelligent packaging solutions are not feasible as active packaging increases the product's shelf life, but the compromised product becomes difficult to identify manually in a lot (Mohan and Ravishankar, 2019). In contrast, it is possible to know the quality or safety of the product without any intervention by humans through intelligent packaging; however, if adverse circumstances occur during product storage, it cannot take all the required steps (Yam et al., 2005; Sohail et al., 2018). To know the quality of the products in real time and take the appropriate steps to prolong the product life without any human interference, smart packaging is the method to use.

15.2 SMART PACKAGING: FUNDAMENTALS AND PRINCIPLES

The basis of smart packaging is effective interaction between the packaging environment and food (Biji et al., 2015). According to Vanderroost et al. (2014), "smart packaging features a comprehensive packaging solution that tracks the alterations in a product or the environment (intelligent) and on the other hand, acts on these changes (active)." The fundamentals of smart packaging are explained in depth in Figure 15.1. The cornerstone of smart packaging is the use of chemical sensor, electric sensor, electronic sensor, mechanical sensor, or a combination thereof. The bio sensors or chemical sensors in smart packaging enable the tracking of safety and quality of food (Nicoletti and Del Serrone, 2017). The smart packaging technology via use of suitable sensors is used to control food safety and food quality (freshness, microbial contamination, carbon dioxide, oxygen, pH, temperature, or time). It helps customers, authorities, and food producers ensure the desired digital quality and safety control by using smart tags, mobile apps, and sensors together to properly track and monitor food products (Maksimović et al., 2015; Lydekaityte and Tambo, 2020). The shelf life of a product is extended by sensors and innovative safe and stable radiant energy systems that destroy pathogens. It further facilitates industry-wide transparency, resulting in more economies of scale, which reduces product recall risks (Wilder, 2015). Smart packaging has a huge potential for development of innovative food packaging sensor systems that work beyond traditional technology like regulating the volume, weight, color, and presentation of the package (Kuswandi et al., 2011).

Figure 15.1 Fundamentals of smart packaging.

Smart packaging bridges the physical world with the digital realm, creating the ideal "digital bridge" between retailers, producers, consumers, and social media channels (Lydekaityte and Tambo, 2020; Young et al., 2020). It focuses primarily on reporting the product's safety (showing that food is unsafe or safe) and quality indices (freshness, maturity, or firmness). Smart packaging is therefore capable of tracing the product, sensing the environment outside or inside the package, and informing the producer, customer, and retailer about the product status (Yezza, 2009). "Smartness" of packages must include the properties shown in Figure 15.2. However, the smart packaging of a single food commodity does not display all of these features.

15.3 ELEMENTS OF SMART PACKAGING

The two basic elements of a smart packaging system are (1) active and (2) intelligent systems. The integration of packaging material, product, and the environment inside the package enhances the product shelf life and produces desirable features (Biji et al., 2015). The active compound integration into packaging materials is done by either incorporating a surface coating or immobilizing or changing the chemical composition of the package. Active packaging relies primarily on the properties of the polymer, adding the substance during polymer manufacturing, or incorporating materials over a polymer's surface as a layer or containing pads, containers, sticks, etc. (Gontard,

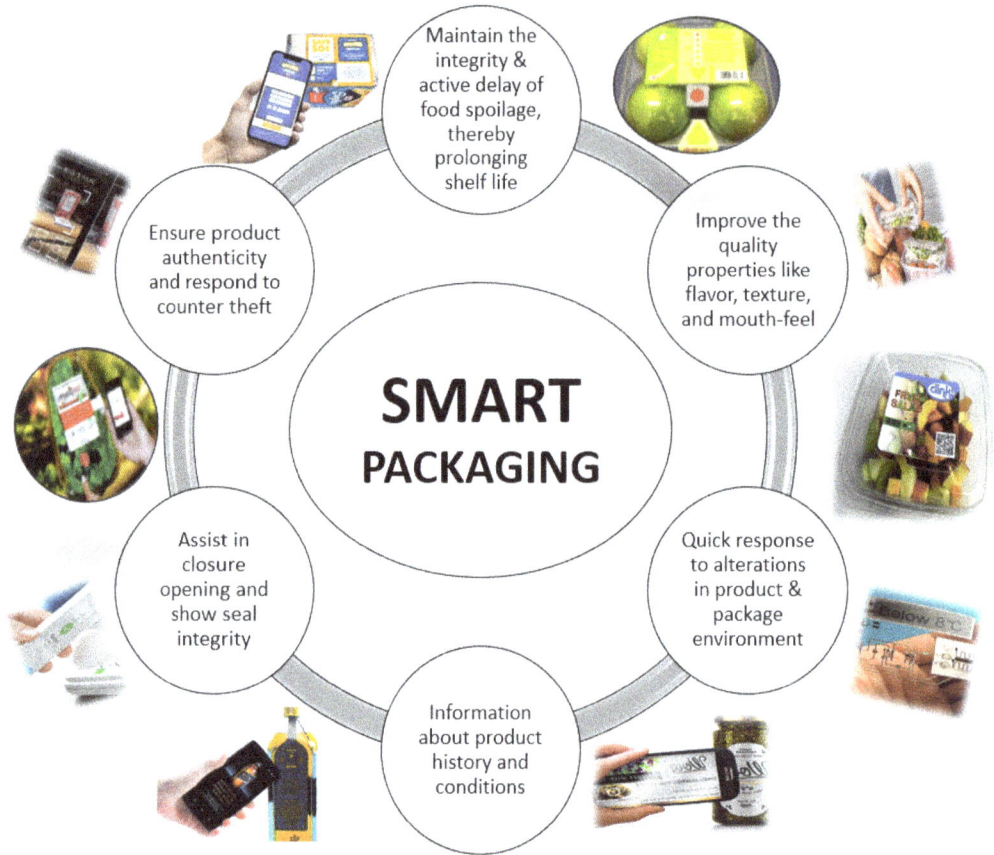

Figure 15.2 Summary of attributes of a smart packaging system.

2006; Mlalila et al., 2016). For existing and newly developed food products, the applications of active packaging-based technologies are innovative and promise that food is provided to customers with their initial or improved sensory properties, with extended shelf life and safety, which can help reduce food wastage (Dainelli et al., 2008). This also decreases additional processing (for example, MAP) of foods to prolong their shelf life. Active packaging may be particularly useful for perishable foods (Yildirim et al., 2018). Oxygen scavengers, moisture scavengers, carbon dioxide (CO_2) emitters, ethylene scavengers, flavor and odor absorbers or releasers, and antimicrobial agents are the most common active systems used for food packaging (Robertson, 2006; Restuccia et al., 2010). The detailed information about active packaging systems is presented in Table 15.1. Such advancements in active packaging have led to advances in many fields such as delayed muscle food oxidation, regulated respiration in horticultural products, moisture growth, and microbial growth in dry products. However, active packaging also modifies the specificity by micro-perforation, coating, laminating, polymer blending, or co-extrusion to alter the environment of gaseous compounds within the product (Brody et al., 2008). Although there has been extensive work on novel active packaging, several commercial active packaging systems exist in the market ranging from basic moisture adsorbent pads to sophisticated systems for absorbing or removing different chemicals (Realini and Marcos, 2014; Fang et al., 2017). Oxygen scavengers such as Ageless® (Mitsubishi Gas Chemical, Japan) or FreshMax® (Multisorb Technologies, USA), FreshPax type M (Multisorb Technologies, USA), and carbon dioxide emitter Cryovac® and DryLoc® (Sealed Air Corporation, USA) are widely known

Table 15.1 Summary of Active Packaging Systems

Type	Properties	Commercial Active Systems (Trade Name)	Manufacturing Company	Food Applications	References
Oxygen scavengers	1. Decrease the oxygen permeation through the package 2. Remove oxygen below 0.01% 3. Works on oxidation of iron powder or glucose oxidase or antioxidant compounds 4. Offer benefits like inhibiting formation of microbial growth, maintaining the quality of lipid-containing foods, inhibiting rancidity, avoiding discoloration, and avoiding oxidation 5. Examples: catechol, ascorbic acid, quinones and polyunsaturated fatty acids, and microorganisms like *Pichia subpelliculosa* and *Kocuria varians*	Ageless® ATCO® Bioka® Cryovac® OS2000 FreshPax® FreshMax, and Fresh Pack Freshilizer Keplon™ OMAC® Oxyeater™ Oxy-Guard™ Oxysorb® Secule™-OxySorb™ Zero₂	Mitsubishi Gas Chemical, Japan STANDA Industries, France Bioka, Finland Sealed Air Corporation, USA Multisorb Technologies, Inc., USA Toppan Printing Co. Ltd., Japan Keplon Mitsubishi Gas Chemical, Japan Ueno Seiyaku, Japan Clariant Ltd., Switzerland Pillsbury Company, USA Nippon Soda, Japan Food Science Australia, Australia	Milk powder, packaged pasta, biscuits and other baked foods, cheese, muscle foods, frozen foods, fruit juices, dry products like coffee, nuts, etc.	Clariant (2017); EFSA Panel on Food Contact Materials, Enzymes, Flavourings and Processing Aids (CEF) (2014); Laboratoires STANDA (2017a); Mitsubishi Gas Chemical (2017a,b); Sealed Air (2017); Biji et al. (2015)
Moisture scavengers	1. Regulate the moisture formation to prevent microbial growth and improve the product appearance 2. Available as pads, sachets, sheets, or blankets 3. Control the water activity of product to suppress the microbial growth 4. Absorbers react with moisture via physical adsorption 5. Examples: activated clays, silica gel, calcium oxide (CaO), and other minerals	Activ-Film™ Dri-Fresh DryLoc® Fresh-R-Pax® Luquasorb® MeatGuard® Nor®Absorbit SOCO® TenderPac®	CSP Technologies, USA Sirane Ltd, UK Sealed Air Corporation, USA Aptar Food and Beverage, USA BASF Chemical Company, Germany McAirlaid, Inc., USA Nordenia International AG, Germany SOCO Chemical, China SEALPAC, Germany	Fruits, vegetables, meats, poultry, fish, and dry and crispy products	Sealed Air (2017); McAirlaid (2017); Nordenia (2011); Aptar Food and Beverage (2017); SEALPAC (2014); Gaikwad et al. (2020); Biji et al. (2015)

(*Continued*)

Table 15.1 Summary of Active Packaging Systems (*Continued*)

Type	Properties	Commercial Active Systems (Trade Name)	Manufacturing Company	Food Applications	References
Ethylene scavengers	1. Absorb emitted ethylene and preserve the ethylene-susceptible vegetables and fruits 2. Excellent alternative use of chemicals and disinfectants 3. Change in color of potassium permanganate from purple to brown on oxidation to show ethylene scavenging capacity by color change 4. Available as sachets, films, and corrugated boxes 5. Examples: potassium permanganate, activated charcoal, zeolite, crystalline aluminosilicates, hexylene glycol, squalene, silica gel, aluminum oxide	Biofresh BION Ethyl Stopper BIOPAC Evert-Fresh Green Bags® Hatofresh Lipmen® Neupalon PEAKfresh® Purafil PURETHYL Retarder® Ryan® Sendo-Mate	Grofit Plastics, Israel EMCO Packaging, UK Biopac Pty Ltd., Australia Evert-Fresh Corporation, USA Evert-Fresh Corporation, USA Honshu paper Ltd, Japan Lipmen Company Ltd., Korea Sekisui Jushi Ltd., Japan PEAKfresh Products, Australia Purafil, Inc., USA Isolcell Spa, Italy Retarder S.r.l., Italy Sensitech, Inc., USA Mitsubishi Gas Chemical, Japan	Mangoes, apples, onions, tomatoes, bananas, guava, peach, kiwifruits, carrots, broccoli florets, lettuce, iceberg, fresh-cut apple, and other horticultural products	Abreu et al. (2011); Gaikwad et al. (2020); EMCO Packaging (2020a); Lipmen (2020); Biji et al. (2015)
Carbon dioxide (CO_2) scavengers	1. Remove CO_2 from package headspace to ensure food preservation 2. High CO_2 into foods is sometimes detrimental, causing package collapse and undesirable product quality (flavor and texture) 3. Removal of CO_2 can be achieved by chemical reaction with an alkaline solution, physical adsorption, membrane separation, and cryogenic condensation 4. Examples: calcium/sodium/potassium hydroxide, silica gel and calcium oxide	Freshock Ageless E Evert-Fresh green bags Evert-Fresh type G Oxyfresh	Mitsubishi Gas Chemical, Japan Evert-fresh Corporation, USA Evert-Fresh Corporation, USA EMCO Packaging, UK	Coffee, fruits and vegetables like strawberries, grapes, spinach, brinjal lettuce, green onions, etc., fermented foods like kimchi, and carbonated beverages	Biji et al. (2015); Lee (2016)
CO_2 emitters	1. Protect foods from oxidation 2. Keeps produce fresh by reducing physiological activities such as respiration and ethylene production 3. Extension of microbiological shelf life 4. Examples: Na_2CO_3, $FeCO_3$	Active CO_2 pad Ageless G CO_2-Pad FreshPax M Fresh Pads™ SUPERFRESH system Verifrais™ UltraZap® XtendaPak	CellComb AB, Sweden Mitsubishi Gas Chemical, Japan McAirlaid's GmbH, Germany Multisorb Technologies, Inc., USA CO_2 Technologies, USA Vartdal Plastindustri AS, Norway SARL Codimer, France Paper Pak Industries, USA	Fresh meat, poultry, seafoods, and cheese	Kerry (2014); Yildirim et al. (2018)

(Continued)

Table 15.1 Summary of Active Packaging Systems (*Continued*)

Type	Properties	Commercial Active Systems (Trade Name)	Manufacturing Company	Food Applications	References
Flavor/odor absorbers or releasers	1. Desirable flavors can be slowly distributed/emitted inside the package, which may balance the inherent loss of smell or taste during storage 2. Off-odors like sulfurous compounds are obtained from breakdown of protein/amino acid and can be removed 3. Some plastics may interact with food flavors resulting in loss of flavors or flavors are degraded after processing foods at high temperatures; 4. Such flavors must be removed by flavor absorbers 5. Examples: cellulose triacetate, acetylated paper, citric acid, ferrous salt/ascorbate, activated carbon/clays/zeolites, baking soda, etc	—	—	Cereal products, fresh poultry, fruit juices, fried snack foods, fish, dairy products, and fruit	Majid et al. (2018)
Use of antimicrobial agents (antimicrobial packaging)	1. Addition of antimicrobial agent into a polymer film to reduce the microbial activities 2. Decrease the surface contamination of processed food and reduce the microbial growth rate by extending the lag phase of microbes or inactivating them 3. Examples: nisin, essential oils, ethanol, chlorine dioxide, silver ions, antibiotics, organic acids, peptides, spices, plant extracts, etc.	Agion® Bactiblock® Biomaster® d2p® antimicrobial Danisco, BioVia®, Natamax®, Nisaplin® Delvo®Nis Food-touch® IonPure® Irgaguard® Sanic films SANICO®	Life Materials Technology, USA Laboratorios Argenol, Spain Addmaster Limited, USA Symphony Environmental, UK IFF, Denmark DSM Food Specialties, Netherlands MicrobeGuard Corporation, USA Ishizuka Glass Co., Ltd., Japan BASF, Germany Atlas Filtri, Italy Laboratoires STANDA, France	Meat, fish, poultry, and horticultural produce; bread, cheese, dairy desserts, bakery products, and snack foods	Biji et al. (2015); Laboratoires STANDA (2017b); Malhotra et al. (2015); Symphony Environmental (2020); IFF (2020); DSM (2020); MicrobeGuard (2020); Ishizuka Glass (2020); Atlas Filtri (2020)

active packaging systems (Mohan and Ravishankar, 2019). The usage of biodegradable substances, bio-preservatives, and natural antimicrobial agents in food products (Wojciechowska et al., 2018) is a promising strategy for the growth of active packaging. This is aligned with the global movement toward evolving green innovations and is well received by food consumers.

Intelligent packaging monitors the package's condition and the package's viability (Ghaani et al., 2016). Intelligent packaging systems are designed to monitor the state of the packaged product or the environment in real time (Poyatos-Racionero et al., 2018). This is achieved by sensors (chemical, biosensor, and gas sensor), indicators (freshness, time-temperature indicators [TTIs], and integrity indicators), barcodes, and RFIDs (Kerry, 2014). The intelligent packaging systems can be categorized into four groups: sensors, indicators, RFID, and barcode systems (Figure 15.3).

Sensors address the adverse changes in food products and indicate the impairment of the commodity being packed. They are comprised of a receptor and transducer with continuous signal output. The function of receptors is to transform the chemical or physical data into energy form and the transducer transmits this energy in electrical signals (Fuertes et al., 2016). Food packaging sensors help in the detection of toxins, pathogens, chemicals, and allergens from different food sources, like peanuts, tree nuts, and fish, that are potentially detrimental to human health (Kuswandi et al., 2011; Morsy et al., 2016). There are three types of indicators: (a) internal indicators, such as indicators of atmospheric leakage indicators, which are put inside the packaging or correlated with the closure; (b) external indicators placed onto the exterior of the package (TTIs); and (c) indicators that enhance data processing and interact efficiently with the product and customers, for example, special barcodes hold information about the food commodity (i.e. expiry dates) (Ahvenainen, 2003). These are designed using an indicator or sensor that responds explicitly to identified phenomena including the existence of identifiable (target) chemicals (e.g., oxygen indicators) or microbiological contaminants (e.g., microbial metabolites indicator) (Schaefer and Cheung, 2018). Indicators involve tamper-proof, product traceability and anti-counterfeiting devices (Coles et al., 2003). The common colorimetric indicators provide details about, for example, exceeding the specified temperature limit or identifying target chemical compounds in a simple way. The safety and freshness of food materials are determined by TTIs, which communicate and monitor if the food items are safe for consumption. When food is stored under conditions that are not ideal, such as excessive heat or cold, this becomes essential (Wang et al., 2015). With non-frozen foodstuffs, a TTI may show that the food was exposed to cold temperatures inappropriately. In comparison, a TTI may assess the exposure period and when heat-sensitive foods were exposed to unnaturally high temperatures. TTI indicators using temperature-dependent chemical reactions, enzyme activity, or physical processes are the most applicable and simple intelligent packaging systems (Sohail et al., 2018). The most common TTI indicators for food product labeling include 3M™ MonitorMark™ (3M Company, USA), OnVu™ (Freshpoint, USA), and Fresh-Check® (Temptime Co., USA) (Müller and Schmid, 2019). Several kinds of TTI indicators are shown in Figure 15.4. More insights about the present condition of packaged food and its surrounding are demonstrated by integrity or freshness indicators, which quickly respond to the presence of particular gaseous compounds within the package (Fuertes et al., 2016). The different kind of metabolites released in the packaging environment make freshness indicators susceptible to compounds like ethyl alcohol, carbon dioxide, ester, organic acids, volatile sulfur, or nitrogen (Nopwinyuwong et al., 2010; Kuswandi et al., 2011). The vast groups of these food freshness indicators are colorimetric pH-sensitive labels that change their color with a target volatile substance linked to extreme microbial growth, thereby damaging the organoleptic properties and reducing shelf life. A number of these intelligent systems are presently on the market, e.g., FreshTag® (COX Technologies) (Kuswandi et al., 2011). Integrity indicators sense the gas in packaged products or leaky packaging, where the packaging is particularly necessary for food safety. Ageless Eye® (Mitsubishi Gas Chemical), an integrity indicator, can recognize the oxygen level rise in the packaging environment caused by leakage (the lack of a barrier to atmospheric oxygen) and must be accompanied with an oxygen

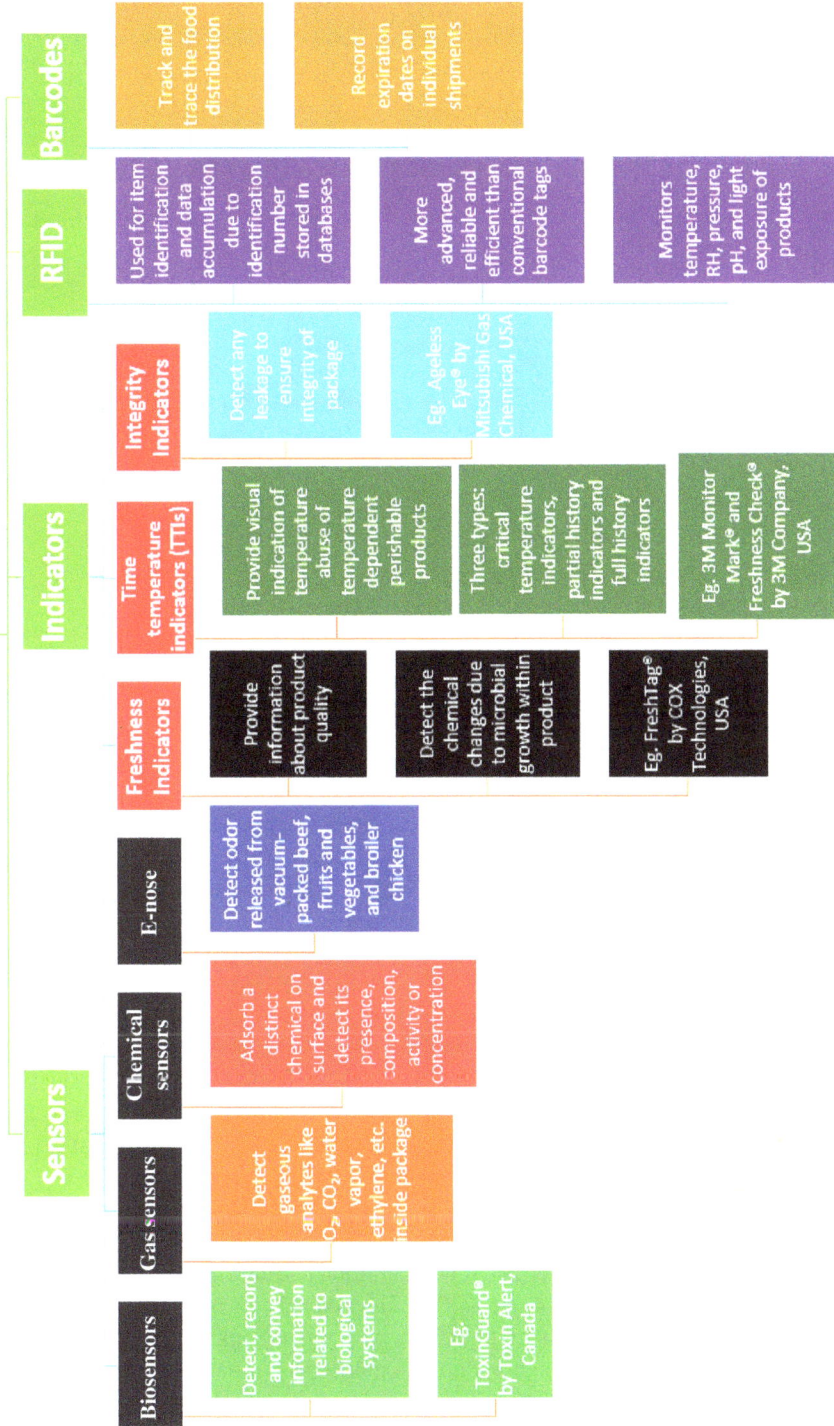

INTELLIGENT PACKAGING SYSTEMS

Sensors

Biosensors
- Detect, record and convey information related to biological systems
- Eg. ToxinGuard® by Toxin Alert, Canada

Gas sensors
- Detect gaseous analytes like O_2, CO_2, water vapor, ethylene, etc. inside package

Chemical sensors
- Adsorb a distinct chemical on surface and detect its presence, composition, activity or concentration

E-nose
- Detect odor released from vacuum-packed beef, fruits and vegetables, and broiler chicken

Indicators

Freshness Indicators
- Provide information about product quality
- Detect the chemical changes due to microbial growth within product
- Eg. FreshTag® by COX Technologies, USA

Time temperature indicators (TTIs)
- Provide visual indication of temperature abuse of temperature dependent perishable products
- Three types: critical temperature indicators, partial history indicators and full history indicators
- Eg. 3M Monitor Mark® and Freshness Check® by 3M Company, USA

Integrity Indicators
- Detect any leakage to ensure integrity of package
- Eg. Ageless Eye® by Mitsubishi Gas Chemical, USA

RFID
- Used for item identification and data accumulation due to identification number stored in databases
- More advanced, reliable and efficient than conventional barcode tags
- Monitors temperature, RH, pressure, pH, and light exposure of products

Barcodes
- Track and trace the food distribution
- Record expiration dates on individual shipments

Figure 15.3 Overview of intelligent packaging systems.

Fresh-Check®   3M™ MonitorMark™   OnVu™

CheckPoint®   Keep-it®

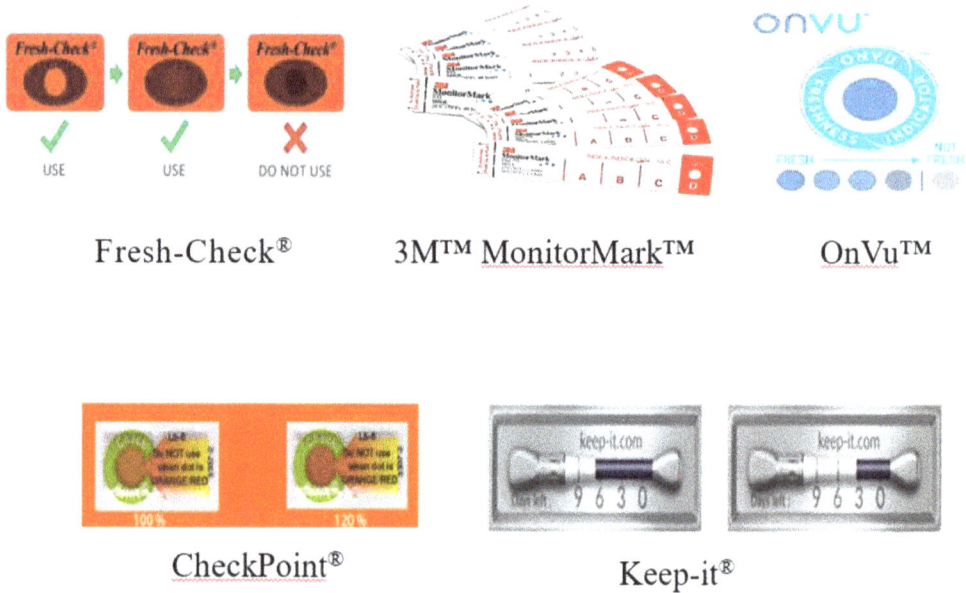

Figure 15.4   Commercially available TTI indicators. (From Vitsab, 2014; Biegańska, 2017; Temptime, 2018; 3M, 2020.)

absorber in a package to prevent interruption with residual oxygen released from the packaged product (Mitsubishi Gas Chemical, 2020). The RFID system provides wireless surveillance of food products through readers, tags, and computer systems. There are numerous applications in the food industry, ranging from simple food traceability to increased supply chain performance. Yet then, the real benefit of RFID in food packaging is that it speeds stock rotation and increases tracking. In certain food and supply chains, RFID solutions are helpful but must be integrated into the food science processes (Yam et al., 2005). Building the reliability and security benefits, retailers are now exploring how to implement RFID technologies in the management of their supply chains. Supermarket stores like Walmart and Home Depot are employing RFID systems. These are commonly used in the food industry to promote traceability and to provide instant monitoring and stock rotation information to enhance supply chain performance (Regattieri et al., 2014; Bibi et al., 2017).

One type of intelligent system is a barcode, which is based on an optical machine attached to an object for automatic product detection. The first ever commercial barcode was adopted in the 1970s by Universal Product Code (UPC) and now it is used in almost every plant to promote inventory control and checkout (Manthou and Vlachopoulou, 2001). It comprises bar and space patterns that represent 12 digits of numerical data with limited details, including product identification numbers and item numbers, etc. (Yam et al., 2005; Ahmed et al., 2018). Barcodes further provide information about the date of the lot/batch number, food packaging, nutrition profile, weight, guidance on cooking, and supplier information. The barcode permits precise and fast reading and prevents unnecessary expenditure in equipment and workforce (Šenk et al., 2013). Various categories of symbology have been established, including composite symbology, GS1 Data Bar Band, two-dimensional (2D) and reduced space symbology, to meet the increasing demand for encoding more data in a limited area (Katuk et al., 2018). The Internet-detectable smart packaging device is a 2D barcode that consist of graphics, lines, and characters and designed to decode the product details (Wang et al., 2015).

15.4 FOOD APPLICATIONS OF SMART PACKAGING

The idea of smart packaging is quickly gaining fame during this era of synthetic chemical-based preservation. It is the next level in developing the previously mentioned intelligent and active technologies. Smart packaging materials with established microstructures or nanostructures, possessing specific properties like self-healing or self-cleaning or modern data transfer technologies, are also used in smart packaging (Biji et al., 2015; Mlalila et al., 2016). This packaging provides constant monitoring of changes to the packed product or its environment and, concurrently, interacts with these changes. Therefore, smart packaging indicators, which show good reliability and stability during distribution and inventory, are interactive (Kuswandi, 2011; Pavelková, 2012). However, increased commercial use has been achieved by ripeness indicators, TTIs, chemical sensors, biosensors, and RFIDs. The comprehensive lists of applications of smart packaging systems in the food processing sector are noted in the following sections.

15.4.1 Animal-Based Food Products

Animal foods such as beef, milk, and eggs play a major role in maintaining safe human health. Meat is spoiled as a result of oxidation, microbial growth, and enzyme autolysis (Olaoye and Ntuen, 2011). A number of intrinsic factors such as processing, oxygen supply, chemical characteristics, microflora, temperature, type of storage, and the environment effect the quality characteristics of food during the storage of poultry and meat products (Rahman, 1999). On the other hand, extrinsic factors can be regulated during storage and transport only to preserve freshness for a longer period.

After slaughtering, the quality changes can be efficiently monitored using a CO_2 indicator, which protects the MAP fillet gilthead (*Sparus aurata*) with the help of a CO_2 bicarbonate buffers system (Giannoglou et al., 2012). Such smart labels based on a CO_2-permeable mini-pouch buffer with a pH indicator achieve a visible color signal. The results have demonstrated that the MAP shelf life extension can easily be maintained during chilling using such CO_2 indicators (Smolander et al., 2002). The combination of an indicator with a CO_2 detection function and a TTI would provide more data about the possibility of quality degradation of a packaged product so that the cold chain from manufacturing up to consumption can be better managed and optimized. A tablet version of an oxygen indicator, called the Tell-Tab, has been developed by the IMPAK Corporation (2020). The indicator is preserved until used in a cold dark location without oxygen, which is typically achieved using oxygen-absorbing bags. The oxygen exposure to the packaged product is checked when the tablet color is changed to blue or purple from its original pink. As the amount of oxygen is decreased, color reversal is seen (Lee and Rahman, 2014). Ageless Eye® from Mitsubishi Gas Chemical offers in-package monitoring of O_2 levels and is available in the form of tablets that change color from pink ($\leq 0.1\% \ O_2$) to blue ($\geq 0.5\% \ O_2$). It is also temperature dependent and its applications range from controlling the packaging and sealing to the breathability of the package (Mitsubishi Gas Chemical, 2020). These tablets work together with AGELESS oxygen absorbers. Similarly, OxySense, a luminescence-based O_2 indicator, shows leakage in a MAP. A non-reversible O_2 indicator was also presented by EMCO Packaging (2020b). Another similar device called the Freshness Timer has been developed by Insignia Technologies. It is a film plate that is triggered on inspection of the package and color changes when the food has lost its freshness (Insignia Technologies, 2020). The microbial safety of meat and other products can be ensured with ToxinGuard® (Toxin Alert, Canada), which is a visual diagnosis program focused on the antibodies displayed on the plastic material. This guard is used for the identification of target pathogens, like *Salmonella* sp., *Campylobacter* sp., *Escherichia coli*, *Listeria* sp., etc. (Chowdhury and Morey, 2019).

Several concepts of the freshness indicator based on volatile amines (Smolander 2008), such as trimethylamine, ammonia, and dimethylamine, have been identified as useful seafood spoilage indicators as per Commission Reg. No. 2074/2005. One color indicator is FreshTag® (COX

Technologies, USA), which senses the generation of volatile amines creating the "fishy odor" in seafood (Williams et al., 2006). The odor-generating compounds react with the safe food grade dye indicator, which shows a color response by gradually changing from yellow to dark blue indicating that seafoods have crossed the point of usable freshness. FreshQ™, introduced by U.S. Food Quality Sensor International, was identified as a freshness sensor that is applied outside of freshly wrapped meat and poultry packages (Williams et al., 2006). The sensitivity of FreshQ™ is adjusted in accordance with the original label pH. The indicator can be employed to monitor the production of acidic compounds by pH tuning during microbial spoilage. DSM NV, in cooperation with Food Quality Sensor International, Inc., announced the development of SensorQ™, which is an anthocyanin-based pH-sensing technique capable of recognizing synthesis of biogenic amines in packed meats and poultry from the microbiological origin (Realini and Marcos, 2014). The sensory quality of meat products is greatly compromised by sulfuric compounds with its common odor and low-odor threshold. A freshness indicator based on myoglobin's color change by H_2S has been developed by Smolander et al. (2002). This indicator was evaluated in the quality assurance of fresh and unmarinated broiler cuts under MAP, and it was found that the alteration in color of myoglobin indicators was in line with product quality degradation. It may also be inferred that indicators based on myoglobin tend to be satisfactory for quality assurance of packaged poultry products (Smolander, 2008). UPM Raflatac, developed by the UPM Corporation, Finland, is an opaque light brown tag/label based on a nanolayer of silver to signify poultry meat freshness. It reacts with hydrogen sulfide to create silver sulfide and provide transparent colors (UPM Corporation, 2020). Work by Dalkiran et al. (2014) led to the development of innovative biosensors (amperometric xanthine) based on nanoparticles of cobalt oxide, chitosan, and multiwall carbon electrode by xanthine oxidase/horseradish peroxidase co-immobilization and xanthine oxidase immobilization. The designed biosensors have been successfully used in fish samples to detect xanthine. An amperometric xanthine biosensor through glutaraldehyde from covalently immobilized crude xanthine oxidase on a hybrid nanocomposite layer was developed by Sadeghi et al. (2014). A film based on Fe_3O_4/polyaniline solution together with H_2PtCl_6 and chitosan was dispersed and thereafter electronically deposited on the surface of the prepared carbon paste electrode. The designed hybrid biosensor was employed to detect amperometric xanthine based on the measurement of decreased hydrogen peroxide. The biosensor has a rapid reaction time (8 seconds) to xanthine and has been used to test the freshness of chicken and fish. Many scientists have also been working to develop and utilize nanocomposite sensors as a very reliable approach to regulate freshness in meat (Dervisevic et al., 2016). Pospiskova et al. (2013) have created a sensor to detect the biogenic amine synthesized from amino acid decarboxylation or amination and transamination of aldehydes and ketones by microbial action. Arvanitoyannis and Stratakos (2012) developed biosensors to detect xanthine (adenine nucleotide product in animal tissue) by immobilizing xanthine oxide onto electrodes made from substances like platinum, silver, and pencil graphites (Dolmacı et al. 2012; Devi et al. 2013).

CheckPoint® TTI (Vitsab International, Sweden) is a color change-based enzymatic system that responds to pH decline by microbial lipase (*Rhizopus oryzae*) of a lipid substrate (methyl myristate) (Vitsab, 2014). Once the substrate and the enzyme are activated, a separation wall inside the TTI is mechanically broken. The initially green TTI slowly turns yellow/orange and finally red. A number of variations of enzyme substrate and concentrations can be used to provide different response lives and temperature dependence (Vitsab, 2014). A sensory color change scale allows the visual detection and assessment of the degree of color change and its importance. The consistent alteration in color can also be assessed with instruments and the data can be used in a shelf life management system. Fresh-Check TTI (Temptime Corporation, USA) works based on a solid-state polymerization reaction (Temptime, 2018). The TTI function is based on the property of di-substituted diacetylene crystals to be polymerized, which results in a highly colored polymer through a lattice-monitored solid-state reaction. The TTI reaction is a change in color as measured by reduced reflection. The color of the TTI "active" core is compared with a surrounding ring's reference color.

This can be linked with the packaging of perishable foodstuffs like meat and poultry (Kuswandi et al., 2011; Temptime, 2018). The newly launched solid-state-based TTI called OnVu (Chemische Industrie Basel–CIBA, Switzerland) works because of the reproducibility of crystal phase reactions. The exposure to low-wavelength light excites and colors photosensitive compounds such as benzylpyridines (Eichen et al., 2008). This colored state reverses at temperature-dependent rates back to its initial colorless condition. The temperature sensitivity and length of the TTI can be fixed by regulating the photochromic compound and light-exposure time during activation. This TTI can be photosensitive ink and can be used very flexibly (Eichen et al., 2008). The (eO)® TTI developed by CRYOLOG, France, reacts to the time- and temperature-dependent pH change, which is expressed in terms of color change in the appropriate pH indicators when the change in pH is caused by microbial growth in the TTI gel (Louvet et al., 2005; Ellouze et al., 2008). For certain microorganisms, the TTI parameters are modified by subsequent variations in gel composition. The response of a TTI is believed to imitate the microbiological spoilage of food items being tracked, provided their reaction is focused on growth from related microorganisms including *Carnobacterium piscicola*, *Lactobacillus fuchuensis*, and *Leuconostoc mesenteroides*. The pH drop is exhibited by a color change from green to red in the pH indicator. The Sensor™ TTI (CCL Design, USA) is based on the concept of diffusion reaction, wherein the diffusion of a polar compound occurs between two polymeric films and alteration in its amount changes the color of a fluorescent indicator from yellow to bright pink (Subramaniam and Wareing, 2016). The 3M MonitorMark® (3M Company, USA) is based on proprietary polymer diffusion in which a viscoelastic substance shifts to a transparent porous matrix at a temperature-dependent rate (3M, 2020). It is prepared from fatty acid esters with particular melting points that are added into a blotting paper. This TTI consists of two forms: one is dedicated to controlling the temperature, distribution, and time via revealing the threshold level, and the other offers the relevant data to customers. On exceeding the threshold level by storage temperature, the polymeric substance starts to melt creating a blue color on the passive indicator. The type and amount of fatty acid esters determine the temperature range (–15 to 26°C) and life of the MonitorMark (3M, 2020). A wholesome breach indicator (Timestrip, developed by Timestrip Ltd., UK) displays the value for excursions at more than 10°C within 2 weeks (Timestrip, 2017). This label is small, light, clear, and easy to understand, and it is cost-efficient to monitor many batches for temperature abuse identification. It evaluates the time interval of opening or utilization of a product. This is a consumer-enabled smart product that is capable of measuring and monitoring the time of the perishable products in freezing, cold, and elevated temperatures from minutes to more than a year (Timestrip, 2017). The label is intended for automatic activation after opening the package. Further, it can be manually activated as an external label to meet consumer requirements (Kuswandi et al., 2011). This is designed to stick easily to any substance without harmful contaminants and is eco-friendly. Thermax (Thermographic Measurements, UK) offers varieties of color-changing strips and labels indicating the maximum attained temperature. It is also available in several temperatures and size ranges (Thermographic Measurements, 2015). The product temperature can also be tracked using WarmMark (designed by DeltaTrak, USA) when the range of response temperatures varies from –18 to 37°C. It was developed for precise monitoring of time and temperature and offers permanent temperature exposure at the lowest possible cost (DeltaTrak, 2020).

The introduction of TT sensors into RFID products has been described as a substantial advancement in the food packaging sector, which has permitted product temperature control across the entire food supply chain. The incorporation of sensor and RFID technology into the cold chain substantially enhanced the effectiveness of the food supply chain and savings resulted from reduced waste (Buisman et al., 2019). Some examples of reusable TTI tags include Easy2log (CAEN RFID S.r.l., Italy), TempTRIP sensor tags (TempTRIP LLC, USA), and sensor tag CS8304 (Convergence Systems Ltd., Hong Kong, China), which are particularly employed for recording the temperature history of food products in the cold chain (CAEN RFID S.r.l., 2018, Convergence Systems, 2020; TempTRIP, 2020). As a result of new European Union (EU) regulations that specify fish should be labeled and completely traceable, the Craemer Group GmbH launched a smart fish case. This fish case is incorporated with an RFID transponder, which monitors and traces the details about fish and fishing areas and even about the quality and the quantity of the catch (Craemer, 2020). Atlantic Beef Products (ABP), a Canadian beef manufacturer, used the RFID chips for tracking beef along the production chain. This system includes connections from the integrator, barcode scanner, RFID tags, and Merit-Trax software that are integrated into the ABP database. This provides details about the real-time location of any carcass in the factory to the Canadian Food Inspection Agency and also maintains an online record of the carcasses slaughtered in any package that exits the factory (Swedberg, 2006).

15.4.2 Plant-Based Food Products

With the exception of fruits and vegetables that demand significant post-harvest attention, plant-based foods are generally shelf stable. Major changes in quality during storage of horticultural products include sensory qualities (texture, taste, color), nutritional value (proteins, carbohydrates, vitamins, flavonoids, etc.), and microbial growth. Any such details predict the packaged product quality. The shelf life of fruit also depends on the rate of respiration and is expressed as the concentration of oxygen used or the amount of CO_2 released. The color of a product also changes due to respiration (Mangaraj and Goswami, 2009). With increased respiration rate, the production of ethylene increases. These characteristics are used as indicators of product quality. Certain indicators based on CO_2, ethylene, and color changes have been developed (Kim and Shiratori, 2006; Lang and Hübert, 2012; Meng et al., 2015). The quality of products marketed can also be revealed by the flavor of fruits and vegetables.

The volatile compounds generated during tomato growth were studied by Gómez et al. (2006) using PEN 2 e-nose, which distinguishes the various ripeness stages of a tomato. The freshness indicator, RipeSense® (RipeSense Limited, New Zealand), is a label that responds to the aromatic compounds produced during storage/transport (RipeSense, 2004). This label changes color as per the concentration of aroma components produced by fruit during ripening; the label is first red and finally turns to orange and then yellow (Vanderroost et al., 2014). The fruit can be kept in the fridge to significantly decrease the ripening, after achieving the required ripeness color by a sensor. This sensor was utilized to check the ripeness of kiwifruit, mangoes, melons, pears, avocado, etc. (Kuswandi et al., 2011; Lee and Rahman, 2014). For this purpose, ethanol spraying on foodstuffs or ethanol sachets may be used. Ethanol-emitting sachets like Ethicap® and Antimold-Mild® work by moisture absorption and emit ethanol vapors (Mexis and Kontominas, 2014; Freund, 2020). Ethanol vapor generators are used for baking items with high moisture content. When heated before use, the product should be free of ethanol. Chlorine dioxide, which is an effective antimicrobial agent against fungi, bacteria, and viruses, is used. It can be produced from acid precursors and sodium chlorite, which are combined into a copolymer of hydrophilic and hydrophobic phases. The contact between food-derived moisture and the hydrophobic phase releases the acid, further interacting with sodium chlorite and emitting chlorine dioxide (Young et al., 2020).

The use of ethylene scavengers or emitters is another way to improve shelf life. Such techniques are effective in delivering the packages but have restricted use in primary or retail packaging films and overwraps (Gaikwad et al., 2020). Apart from the preservation of product quality, mold growth prevention is also a goal. Contamination from food storage has been relevant to bacteria-related food-borne outbreaks. An analysis by Bailén et al. (2006) has shown an increase in sensory qualities of fresh tomatoes by joining the MAP with activated carbon, which showed decreased spoilage and improved quality compared with using just MAP. The reported result was due to the activated carbon's ability to act as an ethylene absorber. The effects of the use of activated carbon together with palladium were more prominent. Oxygen enhances the respiratory metabolism and inhibits the safe storage of biological materials; therefore, the smart food packages must be integrated with oxygen indicators (Mills, 2005). This indicator offers details about leakage because oxygen can access the package when tempered or damaged during shipping or storage. A redox dye (for example, methylene blue), alkaline (for example, sodium hydroxide), and reducing compound (for example, reducing sugar) are the typical examples of an oxygen indicator (Eaton, 2002; Jang and Won, 2014; Huang et al., 2018). Furthermore, the indicator can be accompanied by key components, like a solvent (water or alcohol) and a bulking agent (silica gel, polymers, cellulose substrate, zeolite) (Otles and Sahyar, 2016). A label, a printed sheet, a tablet, or a laminated polymer film can be utilized to construct the indicator. As freshness detectors for fermented foods, O_2 indicators may also be used. Commercially packed kimchi products are an example of products whose maturity/over-ripeness cannot be measured without damaging the materials in packaging through general testing methods (Hong and Park, 2000). Therefore, a sensor can be employed that can track the CO_2 gas constantly and display levels accurately by displaying various colors. Changes in CO_2 levels in the kimchi pack indicated a sigmoid increase during fermentation. The heterocyclic aromatic compound, methylene blue ($C_{16}H_{18}N_3SC$), is generally employed as an indicator of oxygen, which oxidizes to a blue color in the presence of oxygen (Hong, 2002; Hong and Park, 2000). The oxygen indicators are usually water soluble, non-toxic, and irreversible. Desiccants such as molecular sieves, silica gel, and calcium oxide are used in dry foodstuffs like cookies, nuts, candy, etc., whereas pads of inorganic salt or microporous bags are used to balance humidity within cartons with covered layers of solid polymeric humectants (Brody et al., 2008).

15.4.3 Other Applications

Active microwave packaging is designed to enhance food heating behavior by shielding, changing the field, and using susceptors (Regier, 2014). Microwave susceptors are made of stainless steel or aluminum that is mounted on substrates such as paper boards or on polyester film that result in equal warming, crisping, and browning of surfaces (Ahvenainen, 2003; Perry and Lentz 2009; Kerry et al., 2014). SmartPouch® (VacPac, Inc., USA) and Sira Crisp™ (Sirane Ltd., UK) are commercially accessible susceptors based on microwaves (Sirane, 2011; VacPac, 2014). Steam valves that allow fast steam release during microwave cooking are attached along with active microwave packs. For the purpose of heating or steaming convenience food in a microwave oven, the Flexis™ Steam Valve (Avery Dennison Corporation, USA), an industrial pressure-sensitive steam valve, can be used for the majority of flexible food packaging applications. This gives the product a hermetic seal that initially protects the product material and becomes self-venting during the cooking process. To ensure food quality, it monitors the incremental temperature balance during the cooking process (Avery Dennison Corporation, 2010).

15.5 FUTURE TRENDS

Smart packaging is focused on sensing and detecting performance mechanisms including the detecting limit, sensitivity, and process range. To identify microorganisms and pathogens at the initial stage of decomposition and degradation, smart packaging systems must be adequately effective, thereby enabling rapid preventive or curative reactions. Currently, smart packaging systems are normally costly when used to identify and sensitize several risks at a time. In addition, the industrial-scale responsive packaging system would be capable of handling stresses from packaging production, i.e., temperature, pressure, mechanical vibration, and other aids in manufacturing. The labels must be incorporated economically and be printed or added directly to the packaging materials. TTI-based systems can substitute the present first-in, first-out (FIFO) method by improving delivery logistics and food cold chain management, leading to risk prevention and quality improvement. This facilitates the rotation of stock in specified cool chain locations, which ensures that the food is consumed before hitting an unreasonable danger in terms of temperature. However, TTI-based systems can manage the chain in case of incidental temperature abuse by diverting the abused products to prevent the risk and the final rejection. Using RFID tags for responsive substrates provides useful information about the content of the food package and food quality within the package. RFID tags have substituted traditional barcodes, enabling them to track multiple items instantly and to store a wide variety of information. These tools may contribute to the safety and quality monitoring as well as product control and traceability. In tandem with other technologies, including nanomaterial packaging, edible films, and active and biodegradable packaging, smart packaging systems can perform even better. Along with that, significant advancement and progress in smart packaging has been done over the past two decades, while further research and improvement is still possible. In recent years, demand for smart packaging increased rapidly worldwide because of the growing need for specialty packaging used to inform consumers about the freshness and quality of foodstuffs.

15.6 CONCLUSION

Through real-time tracking and effectively impacting the quality and safety, smart packaging systems can efficiently improve the shelf stability of food items. The advantages of intelligent and active packaging technologies are combined in smart packaging to provide efficient communication about the state of a commodity across the entire supply chain. Sensors are the most sophisticated devices that provide simple, accurate, and detailed information on product safety and quality and are capable of attempting to resolve the adverse changes in food items. On the other hand, barcodes and RFID tags offer real-time information on identification and product traceability, while increasing efficiency and distribution speeds. Smart packaging created a massive transformation in the packaged sector in terms of shelf life, safety, quality, traceability, authenticity, fraud prevention, sustainability, and convenience. Numerous research works are, however, already ongoing to boost quality and reliability through the convergence of other recent digital innovations, and new approaches and innovations in food packaging will certainly be seen in the future.

REFERENCES

3M, 2020. 3M™ MonitorMark™ Time Temperature Indicators. 3M Science, USA. https://www.3mindia.in/3M/en_IN/p/d/v000177167/. Accessed 30 July 2020.

Abreu, D.P., Losada, P.P., Maroto, J. and Cruz, J.M., 2011. Natural antioxidant active packaging film and its effect on lipid damage in frozen blue shark (*Prionace glauca*). *Innovative Food Science & Emerging Technologies*, *12*(1), pp. 50–55.

Ahmed, I., Lin, H., Zou, L., Li, Z., Brody, A.L., Qazi, I.M., Lv, L., Pavase, T.R., Khan, M.U., Khan, S. and Sun, L., 2018. An overview of smart packaging technologies for monitoring safety and quality of meat and meat products. *Packaging Technology and Science*, *31*(7), pp. 449–471.

Ahvenainen, R., 2003. Active and intelligent packaging: an introduction. In *Novel Food Packaging Techniques* (pp. 5–21). Woodhead Publishing.

Alfiya, P.V., Murali, S., Delfiya, A. and Samuel, M.P., 2017. Smart packaging: revolutionizing the era of food packaging. http://210.212.228.207/bitstream/handle/123456789/4019/ Smart%20Packaging.pdf?sequence=1&isAllowed=y. Accessed 13 July 2020.

Aptar Food and Beverage, 2017. Fresh-R-Pax® trays. https://www.aptarfoodprotection.com/product/absorbent-packaging-systems/. Accessed 29 July 2020.

Arvanitoyannis, I.S. and Stratakos, A.C., 2012. Application of modified atmosphere packaging and active/smart technologies to red meat and poultry: a review. *Food and Bioprocess Technology*, *5*(5), pp. 1423–1446.

Atlas Filtri, 2020. Antimicrobial Systems SANIC. Atlas Filtri, Italy https://www.atlasfiltri.com/en/products/antimicrobial-systems-sanic. Accessed 26 July 2020.

Avery Dennison Corporation, 2010. Avery Dennison Corporation: Heat-Activated Steam Release Valve Provides Precise Control During Microwave Cooking. Packaging World, USA. https://www.packworld.com/design/flexible-packaging/product/13347469/avery-dennison-corporation-heatactivated-steam-release-valve-provides-precise-control-during-microwave-cooking. Accessed 11 August 2020.

Bailén, G., Guillén, F., Castillo, S., Serrano, M., Valero, D. and Martínez-Romero, D., 2006. Use of acti-vated carbon inside modified atmosphere packages to maintain tomato fruit quality during cold storage. *Journal of Agricultural and Food Chemistry*, 54(6), pp. 2229–2235.

Bibi, F., Guillaume, C., Gontard, N. and Sorli, B., 2017. A review: RFID technology having sensing aptitudes for food industry and their contribution to tracking and monitoring of food products. *Trends in Food Science & Technology*, 62, pp. 91–103.

Biegańska, M., 2017. Shelf-life monitoring of food using time-temperature indicators (TTI) for application in intelligent packaging. *Towaroznawcze Problemy Jakości*, 2, pp. 75–85.

Biji, K.B., Ravishankar, C.N., Mohan, C.O. and Gopal, T.S., 2015. Smart packaging systems for food applications: a review. *Journal of Food Science and Technology*, 52(10), pp. 6125–6135.

Brody, A.L., Bugusu, B., Han, J.H., Sand, C.K. and McHugh, T.H., 2008. Innovative food packaging solutions. *Journal of Food Science*, 73(8), pp. 107–116.

Buisman, M.E., Haijema, R. and Bloemhof-Ruwaard, J.M., 2019. Discounting and dynamic shelf life to reduce fresh food waste at retailers. *International Journal of Production Economics*, 209, pp. 274–284.

CAEN RFID S.r.l., 2018. RT0005 Easy2Log Low Cost, Semi-Passive UHF Logger Tag. CAEN RFID S.r.l., Italy. https://www.caenrfid.com/en/products/rt0005/. Accessed 8 August 2020.

Chowdhury, E.U. and Morey, A., 2019. Intelligent packaging for poultry industry. *Journal of Applied Poultry Research*, 28(4), pp. 791–800.

Clariant, 2017. OXY-GURARD Oxygen Absorbers. Preserve Freshness, Minimize the Need for Preservatives, and Help Prolong Shelf Life in Packaged Foods. http://www.clariant.com/oxy-guard-oxygen-scavenger. Accessed 16 July 2020.

Coles, R., McDowell, D. and Kirwan, M.J. (Eds.), 2003. *Food Packaging Technology* (Vol. 5). CRC Press.

Convergence Systems, 2020. CS8304 Cold Chain Temperature Logging Tag. Convergence Systems Ltd., Hong Kong, China. http://www.veryfields.net/rfid-tags-Convergence+Systems+Limited+CS8304+Cold+Chain+Temperature+Logging+Tag.html. Accessed 8 August 2020.

Craemer, 2020. Storage and Transport Containers – Fish Boxes. Craemer Group GmbH. https://en.e-catalog.craemer.com/storage-and-transport-containers/fish-boxes. Accessed 8 August 2020.

Dainelli, D., Gontard, N., Spyropoulos, D., Zondervan-van den Beuken, E. and Tobback, P., 2008. Active and intelligent food packaging: legal aspects and safety concerns. *Trends in Food Science & Technology*, 19, pp. S103–S112.

Dalkiran, B., Kacar, C., Erden, P.E. and Kilic, E., 2014. Amperometric xanthine biosensors based on chitosan-Co3O4-multiwall carbon nanotube modified glassy carbon electrode. *Sensors and Actuators B: Chemical*, 200, pp. 83–91.

DeltaTrak, 2020. WarmMark® Time-Temperature Indicator. DeltaTrak, USA. https://www.deltatrak.com/warmmark-labels. Accessed 8 August 2020.

Dervisevic, M., Dervisevic, E., Azak, H., Cevik, E., Şenel, M. and Yildiz, H.B., 2016. Novel amperomet-ric xanthine biosensor based on xanthine oxidase immobilized on electrochemically polymerized 10-[4H-dithieno (3, 2-b: 2′, 3′-d) pyrrole-4-yl] decane-1-amine film. *Sensors and Actuators B: Chemical*, 225, pp. 181–187.

Devi, R., Yadav, S., Nehra, R., Yadav, S. and Pundir, C.S., 2013. Electrochemical biosensor based on gold coated iron nanoparticles/chitosan composite bound xanthine oxidase for detection of xanthine in fish meat. *Journal of Food Engineering*, 115(2), pp. 207–214.

Dolmacı, N., Çete, S., Arslan, F. and Yaşar, A., 2012. An amperometric biosensor for fish freshness detection from xanthine oxidase immobilized in polypyrrole-polyvinylsulphonate film. *Artificial Cells, Blood Substitutes, and Biotechnology*, 40(4), pp. 275–279.

DSM, 2020. Delvo®Nis Preserving Processed Foods with Delvo®Nis. DSM Food Specialties, Netherlands. https://www.dsm.com/food-specialties/en_US/products/dairy/delvonis.html. Accessed 28 July 2020.

Eaton, K., 2002. A novel colorimetric oxygen sensor: dye redox chemistry in a thin polymer film. *Sensors and Actuators B: Chemical*, 85(1–2), pp. 42–51.

EFSA Panel on Food Contact Materials, Enzymes, Flavourings and Processing Aids (CEF), 2014. Scientific opinion on the safety assessment of the active substances, palladium metal and hydrogen gas, for use in active food contact materials. *EFSA Journal*, 12(2), p. 3558.

Eichen, Y., Haarer, D., Feuerstack, M. and Jannasch, U., 2008, June. What it takes to make it work: the On-Vu TTI. Cold Chain-Management, 3rd International Workshop, Bonn, Germany, pp. 106–113.

Ellouze, M., Pichaud, M., Bonaiti, C., Coroller, L., Couvert, O., Thuault, D. and Vaillant, R., 2008. Modelling pH evolution and lactic acid production in the growth medium of a lactic acid bacterium: application to set a biological TTI. *International Journal of Food Microbiology*, *128*(1), pp. 101–107.

EMCO Packaging, 2020a. Ethylene Absorbers Sachets. EMCO Packaging, UK. https://www.emcotechnologies. co.uk/products/ethylene-absorbers-sachets. Accessed 24 July 2020.

EMCO Packaging, 2020b. Oxygen Indicator Labels. Oxygen Indicating Color Change Chemistry. https:// www.emcotechnologies.co.uk/category/oxygen-indicator-labels. Accessed 12 July 2020.

Fang, Z., Zhao, Y., Warner, R.D. and Johnson, S.K., 2017. Active and intelligent packaging in meat industry. *Trends in Food Science & Technology*, *61*, pp. 60–71.

Freund (2020). Antimold-Mild®. Freund Corp, Japan. https://www.freund.co.jp/english/product/preservation/ preservation_ethanol/antimoldmild.html. Accessed 10 August 2020.

Fuertes, G., Soto, I., Carrasco, R., Vargas, M., Sabattin, J. and Lagos, C., 2016. Intelligent packaging systems: sensors and nanosensors to monitor food quality and safety. *Journal of Sensors*, *2016*, 4046061.

Gaikwad, K.K., Singh, S. and Negi, Y.S., 2020. Ethylene scavengers for active packaging of fresh food produce. *Environmental Chemistry Letters*, pp. 1–16.

Ghaani, M., Cozzolino, C.A., Castelli, G. and Farris, S., 2016. An overview of the intelligent packaging technologies in the food sector. *Trends in Food Science & Technology*, *51*, pp. 1–11.

Giannoglou, M., Ronnow, P., Tsironi, T., Platakou, E., Metaxa, I. and Taoukis, P., 2012, June. Development and application of CO2 indicating smart labels for monitoring shelf life of MAP fish fillets. In IFT Annual Meeting, Las Vegas, NV, p. 255.

Gómez, A.H., Hu, G., Wang, J. and Pereira, A.G., 2006. Evaluation of tomato maturity by electronic nose. *Computers and Electronics in Agriculture*, *54*(1), pp. 44–52.

Gontard, N., 2006, September. Tailor made food packaging concept. In IUFoST, 13th World Congress of Food Science and Technology, Food is Life, Nantes, France, pp. 17–21.

Hempel, A.W., O'Sullivan, M.G., Papkovsky, D.B. and Kerry, J.P., 2013. Use of smart packaging technologies for monitoring and extending the shelf-life quality of modified atmosphere packaged (MAP) bread: application of intelligent oxygen sensors and active ethanol emitters. *European Food Research and Technology*, *237*(2), pp. 117–124.

Hong, S.I., 2002. Gravure-printed colour indicators for monitoring Kimchi fermentation as a novel intelligent packaging. *Packaging Technology and Science: An International Journal*, *15*(3), pp. 155–160.

Hong, S.I. and Park, W.S., 2000. Use of color indicators as an active packaging system for evaluating kimchi fermentation. *Journal of Food Engineering*, *46*(1), pp. 67–72.

Huang, S., Li, H., Wang, Y., Liu, X., Li, H., Zhan, Z., Jia, L. and Chen, L., 2018. Monitoring of oxygen using colorimetric indicator based on graphene/TiO_2 composite with first-order kinetics of methylene blue for modified atmosphere packaging. *Packaging Technology and Science*, *31*(9), pp. 575–584.

IFF, 2020. Antimicrobials and Fermentates. IFF Nutrition and Biosciences, Denmark. https://www. dupontnutritionandbiosciences.com/product-range/antimicrobials.html. Accessed 25 July 2020.

IMPAK Corporation, 2020. Oxygen Indicating Tablets. IMPAK Corporation, USA. https://www. impakcorporation.com/oxygen_absorbers/oxygen_indicating_tablets. Accessed 03 June 2020.

Insignia Technologies, 2020. Insignia after Opening Freshness Timer. Insignia Technologies Ltd., UK. https:// www.pac.ca/wp-content/uploads/2020/06/fwcasestudy-insignia.pdf. Accessed 12 July 2020.

Ishizuka Glass, 2020. IonPure® – Inorganic Silver Antimicrobial. Ishizuka Glass Co., Ltd., Japan. https:// www.ishizuka.co.jp/english/material/antimicrobial/. Accessed 19 July 2020.

Jang, N.Y. and Won, K., 2014. New pressure-activated compartmented oxygen indicator for intelligent food packaging. *International Journal of Food Science & Technology*, *49*(2), pp. 650–654.

Katuk, N., Mahamud, K. R. K., Zakaria, N. H., 2018. A review of the current trends and future directions of camera barcode reading. *Journal of Theoretical and Applied Information Technology*, *97*(8), pp. 2268–2288.

Kerry, J. P., 2014. New packaging technologies, materials and formats for fast-moving consumer products. In *Innovations in food packaging* (2nd ed.; pp. 549–584). Academic Press, San Diego, CA.

Kim, J.H. and Shiratori, S., 2006. Fabrication of color changeable film to detect ethylene gas. *Japanese Journal of Applied Physics*, *45*(5R), p. 4274.

Kuswandi, B., Wicaksono, Y., Abdullah, A., Heng, L.Y. and Ahmad, M., 2011. Smart packaging: sensors for monitoring of food quality and safety. *Sensing and Instrumentation for Food Quality and Safety*, *5*(3–4), pp. 137–146.

Laboratoires STANDA, 2017a. Active Packaging ATCO®. http://www.standa-fr.com/eng/laboratoiresstanda/acto/. Accessed 4 August 2020.

Laboratoires STANDA, 2017b. SANICO® Is Our Range of Antifungal Coatings for the Agro-Food Industry. http://www.standa-fr.com/eng/laboratoires-standa/sanico/. Accessed 4 August 2020.

Lang, C. and Hübert, T., 2012. A colour ripeness indicator for apples. *Food and Bioprocess Technology*, 5(8), pp. 3244–3249.

Lee, D.S., 2016. Carbon dioxide absorbers for food packaging applications. *Trends in Food Science & Technology*, 57, pp. 146–155.

Lee, S.J. and Rahman, A.M., 2014. Intelligent packaging for food products. In *Innovations in Food Packaging* (pp. 171–209). Academic Press.

Lipmen, 2020. Ethylene Gas Absorber (ETS-TYPE). Lipmen Company Ltd., Korea. http://www.lipmen.co.kr/eng/product/ets.asp. Accessed 17 July 2020.

Louvet, O., Thuault, D. and Vaillant, R., 2005. Method and device for determining whether or not a product is in condition to be used or consumed. Patent. International Publication Number WO, 26383, p.A1.

Lydekaityte, J. and Tambo, T., 2020. Smart packaging: definitions, models and packaging as an intermediator between digital and physical product management. *The International Review of Retail, Distribution and Consumer Research*, 30(4), pp. 377–410.

Majid, I., Thakur, M. and Nanda, V., 2018. Innovative and safe packaging technologies for food and beverages: updated Review. In *Innovations in Technologies for Fermented Food and Beverage Industries* (pp. 257–287). Springer.

Maksimović, M., Vujović, V. and Omanović-Miklićanin, E., 2015. Application of internet of things in food packaging and transportation. *International Journal of Sustainable Agricultural Management and Informatics*, 1(4), pp. 333–350.

Malhotra, B., Keshwani, A., and Kharkwal, H., 2015. Antimicrobial food packaging: potential and pitfalls. *Frontiers in Microbiology*, 6, 611.

Mangaraj, S. and Goswami, T.K., 2009. Modified atmosphere packaging of fruits and vegetables for extending shelf-life-A review. *Fresh Produce*, 3(1), pp. 1–31.

Manthou, V. and Vlachopoulou, M., 2001. Bar-code technology for inventory and marketing management systems: A model for its development and implementation. *International Journal of Production Economics*, 71(1–3), pp. 157–164.

Mariusz, T., 2019. Smart packaging improving shelf life of food product. *Agro Food Industry Hi Tech*, 30(5), pp. 22–24.

McAirlaid, 2017. MeatPad. https://www.mcairlaids.net/de/produkte/lebensmittel/meatpad.html. Accessed 25 July 2020.

Meng, X., Lee, K., Kang, T.Y. and Ko, S., 2015. An irreversible ripeness indicator to monitor the CO_2 concentration in the headspace of packaged kimchi during storage. *Food Science and Biotechnology*, 24(1), pp. 91–97.

Mexis, S.F. and Kontominas, M.G., 2014. Packaging| Active Food Packaging. *Reference Module in Food Science—Encyclopedia of Food Microbiology* (2nd ed., pp. 999–1005). Elsevier.

MicrobeGuard, 2020. FoodTouch® Products. MicrobeGuard Corporation, USA. https://www.microbeguard.com/index.php/products/foodtouch/products. Accessed 24 July 2020.

Mills, A., 2005. Oxygen indicators and intelligent inks for packaging food. *Chemical Society Reviews*, 34(12), pp. 1003–1011.

Mitsubishi Gas Chemical, 2017a. AGELESS OMAC® Oxygen Absorbing Film. http://ageless.mgc-a.com/product/nutrasave/#:~:text=AGELESS%20OMAC%20is%20the%20next,the%20package%20itself%20the%20absorber. Accessed 15 July 2020.

Mitsubishi Gas Chemical, 2017b. AGELESS® Oxygen Absorber. http://www.mgc.co.jp/eng/products/abc/ageless/index.html. Accessed 15 July 2020.

Mitsubishi Gas Chemical, 2020. AGELESS EYE Oxygen Indicator. Mitsubishi Gas Chemical, Japan. https://www.mgc.co.jp/eng/products/sc/ageless-eye.html. Accessed 26 July 2020.

Mlalila, N., Kadam, D.M., Swai, H. and Hilonga, A., 2016. Transformation of food packaging from passive to innovative via nanotechnology: concepts and critiques. *Journal of Food Science and Technology*, 53(9), pp. 3395–3407.

Mohan, C.O. and Ravishankar, C.N., 2019. Active and intelligent packaging systems-application in seafood. *World Journal of Aquaculture Research and Development*, 1, pp. 10–16.

Morsy, M.K., Zor, K., Kostesha, N., Alstrøm, T.S., Heiskanen, A., El-Tanahi, H., Sharoba, A., Papkovsky, D., Larsen, J., Khalaf, H. and Jakobsen, M.H., 2016. Development and validation of a colorimetric sensor array for fish spoilage monitoring. *Food Control*, *60*, pp. 346–352.

Müller, P. and Schmid, M., 2019. Intelligent packaging in the food sector: a brief overview. *Foods*, *8*(1), p. 16.

Nicoletti, M. and Del Serrone, P., 2017. Intelligent and smart packaging. In *Future Foods*. IntechOpen.

Nopwinyuwong, A., Trevanich, S. and Suppakul, P., 2010. Development of a novel colorimetric indicator label for monitoring freshness of intermediate-moisture dessert spoilage. *Talanta*, *81*(3), pp. 1126–1132.

Nordenia, 2011. Nor®Absorbit Makes Your Food Nice and Crispy. https://news.thomasnet.com/companystory/nor-absorbit-makes-your-food-nice-and-crispy-850018. Accessed 12 July 2020.

Olaoye, O.A. and Ntuen, I.G., 2011. Spoilage and preservation of meat: a general appraisal and potential of lactic acid bacteria as biological preservatives. *International Research Journal of Biotechnology*, *2*(1), pp. 33–46.

Otles, S. and Sahyar, B.Y., 2016. Intelligent food packaging. In *Comprehensive Analytical Chemistry* (Vol. 74, pp. 377–387). Elsevier.

Pavelková, A., 2012. Intelligent packaging as device for monitoring of risk factors in food. *Journal of Microbiology, Biotechnology and Food Sciences*, *2*(1), p. 282.

Perry, M.R. and Lentz, R.R., 2009. Susceptors in microwave packaging. In *Development of Packaging and Products for Use in Microwave Ovens* (pp. 207–236). Woodhead Publishing.

Pospiskova, K., Safarik, I., Sebela, M. and Kuncova, G., 2013. Magnetic particles–based biosensor for biogenic amines using an optical oxygen sensor as a transducer. *Microchimica Acta*, *180*(3–4), pp. 311–318.

Poyatos-Racionero, E., Ros-Lis, J.V., Vivancos, J.L. and Martínez-Máñez, R., 2018. Recent advances on intelligent packaging as tools to reduce food waste. *Journal of Cleaner Production*, *172*, pp. 3398–3409.

Rahman, M.S., 1999. Postharvest handling of foods of animal origin. In *Handbook of Food Preservation* (pp. 47–74). Marcel Dekker.

Rai, M., Ingle, A.P., Gupta, I., Pandit, R., Paralikar, P., Gade, A., Chaud, M.V. and dos Santos, C.A., 2019. Smart nanopackaging for the enhancement of food shelf life. *Environmental Chemistry Letters*, *17*(1), pp. 277–290.

Realini, C.E. and Marcos, B., 2014. Active and intelligent packaging systems for a modern society. *Meat Science*, *98*(3), pp. 404–419.

Regattieri, A., Santarelli, G., Gamberi, M. and Gamberini, R., 2014. The use of radio frequency identification technology in packaging systems: experimental research on traceability. *Packaging Technology and Science*, *27*(8), pp. 591–608.

Regier, M., 2014. Microwavable food packaging. In *Innovations in Food Packaging* (pp. 495–514). Academic Press.

Restuccia, D., Spizzirri, U.G., Parisi, O.I., Cirillo, G., Curcio, M., Iemma, F., Puoci, F., Vinci, G. and Picci, N., 2010. New EU regulation aspects and global market of active and intelligent packaging for food industry applications. *Food Control*, *21*(11), pp. 1425–1435.

RipeSense, 2004. RipeSense®. The Next Revolution in Fresh Produce Marketing. Ripesense Limited, New Zealand. http://www.ripesense.co.nz/. Accessed 10 August 2020.

Robertson, G.L., 2006. Food packaging principles and practice. In *Food Science and Technology* (2nd ed.). Marcel Dekker.

Sadeghi, S., Fooladi, E. and Malekaneh, M., 2014. A nanocomposite/crude extract enzyme-based xanthine biosensor. *Analytical Biochemistry*, *464*, pp. 51–59.

Schaefer, D. and Cheung, W.M., 2018. Smart packaging: opportunities and challenges. *Procedia CIRP*, *72*, pp.1022–1027.

Sealed Air, 2017. Cryovac® OS Films-Rapid Headspace. http://www.cryovac.com/NA/EN/pdf/osfilms.pdf. Accessed 10 August 2020.

SEALPAC, 2014. TenderPac-Best Meat Quality, Appetizing Appearance. http://www.sealpac.de/fileadmin/user_upload/media/innovations/verpackungsloesungen/TenderPac_2014_online-EN.pdf. Accessed 13 July 2020.

Šenk, I., Ostojić, G., Tarjan, L., Stankovski, S. and Lazarević, M., 2013, April. Food product traceability by using automated identification technologies. In *Doctoral Conference on Computing, Electrical and Industrial Systems* (pp. 155–163). Springer.

Silberbauer, A. and Schmid, M., 2017. Packaging concepts for ready-to-eat food: recent progress. *Journal of Packaging Technology and Research*, *1*(3), pp. 113–126.

Sirane Group. 2011. Sira-Crisp™ Boxes, Boards, Sleeves and Liners for Crisping in a Microwave. Sirane Group, England. https://www.sirane.com/food-packaging-products/microwave-susceptors-crisp-it-range/sira-cook-crisp-it-boards.html. Accessed 10 August 2020.

Smolander, M., 2008. Freshness indicators for food packaging. In *Smart Packaging Technologies for Fast Moving Consumer Goods* (pp. 111–127). Wiley.

Smolander, M., Hurme, E., Latva-Kala, K., Luoma, T., Alakomi, H.L. and Ahvenainen, R., 2002. Myoglobin-based indicators for the evaluation of freshness of unmarinated broiler cuts. *Innovative Food Science & Emerging Technologies*, 3(3), pp. 279–288.

Sohail, M., Sun, D.W. and Zhu, Z., 2018. Recent developments in intelligent packaging for enhancing food quality and safety. *Critical Reviews in Food Science and Nutrition*, 58(15), pp. 2650–2662.

Subramaniam, P. and Wareing, P. (Eds.), 2016. *The Stability and Shelf Life of Food*. Woodhead Publishing.

Swedberg, C., 2006. Beef Tracking, the RFID Way. *RFID Journal*. https://www.rfidjournal.com/beef-tracking-the-rfid-way. Accessed 29 Jan 2022.

Symphony Environmental, 2020. d2p® Antimicrobial – Antimicrobial Plastic and Rubber Technology. Symphony Environmental, UK. https://www.symphonyenvironmental.com/additives/antimicrobial. Accessed 25 July 2020.

Temptime, 2018. Fresh-Check® Indicator Temperature Intelligence™. Temptime, USA. http://fresh-check.com/. Accessed 3 August 2020.

TempTRIP, 2020. Temperature loggers. TempTRIP LLC, USA. http://temptrip.com/products#temp_loggers. Accessed 8 August 2020.

Thermographic Measurements, 2015. THERMAX. Thermographic Measurements, UK. https://tmchallcrest.com/thermax/. Accessed 8 August 2020.

Timestrip, 2017. eTimestrip. Timestrip Ltd., UK. https://timestrip.com/electronic-temperature-indicator/. Accessed 7 August 2020.

UPM corporation, 2020. Food labels. UPM Corporation, Finland. https://www.upmraflatac.com/labels-by-industry/food/. Accessed 19 July 2020.

VacPac, 2014. VacPac-Products. VacPac, Inc., USA. https://vacpacinc.com/products/. Accessed 11 August 2020.

Vanderroost, M., Ragaert, P., Devlieghere, F. and De Meulenaer, B., 2014. Intelligent food packaging: the next generation *Trends In Food Science & Technology*, 39(1), pp. 47–62.

Vitsab, 2014. TTI Label. Vitsab International Sweden. http://www.vitsab.com/?page_id=1983#. Accessed 10 July 2020.

Wang, S., Liu, X., Yang, M., Zhang, Y., Xiang, K. and Tang, R., 2015. Review of time temperature indicators as quality monitors in food packaging. *Packaging Technology and Science*, 28(10), pp. 839–867.

Wilder, C., 2015. What does Food Packaging have to do With Big Data and the Internet of Things?. https://www.forbes.com/sites/moorinsights/2015/10/01/what-does-food-packaging-have-to-do-with-big-data-and-the-internet-of-things/. Accessed 21 July 2020.

Williams, J., Myers, K., Owens, M. and Bonne, M., Food Quality Sensor International, Inc., 2006. Food quality indicator. U.S. Patent Application 11/225,410.

Wojciechowska, P., Tichoniuk, M., Gwiazdowska, D., Maciejewski, H. and Nowicki, M., 2018. Antimicrobial activity of organic–inorganic hybrid films based on gelatin and organomodified silicones. *Advances in Polymer Technology*, 37(8), pp. 2958–2970.

Wyrwa, J. and Barska, A., 2017. Innovations in the food packaging market: active packaging. *European Food Research and Technology*, 243(10), pp. 1681–1692.

Yam, K.L., Takhistov, P.T. and Miltz, J., 2005. Intelligent packaging: concepts and applications. *Journal of Food Science*, 70(1), pp. R1–R10.

Yezza, I.A., 2009. Printed intelligence in packaging: current and potential applications of nanotechnology. In *2009 Symposium of Nanomaterials for Flexible Packaging*, Columbus, Code (Vol. 80863).

Yildirim, S., Röcker, B., Pettersen, M.K., Nilsen-Nygaard, J., Ayhan, Z., Rutkaite, R., Radusin, T., Suminska, P., Marcos, B. and Coma, V., 2018. Active packaging applications for food. *Comprehensive Reviews in Food Science and Food Safety*, 17(1), pp. 165–199.

Young, E., Mirosa, M. and Bremer, P., 2020. A systematic review of consumer perceptions of smart packaging technologies for food. *Frontiers in Sustainable Food Systems*, 4, 63.

CHAPTER 16

Aseptic Packaged Food
Science, Shelf Life, and Quality

Omar Bashir, Sumira Rashid, Nusrat Jan, Abira Umam, Tawheed Amin,
Abida Jabeen, Sajad Mohd Wani, Syed Afiya, and Haamiyah Sidiq

CONTENTS

Email: Tawheed Amin, Email: tawheed.amin@gmail.com

DOI: 10.1201/9781003091677-16

16.1 INTRODUCTION

Aseptic packaging techniques have been used for years to develop particulate foods and low-acid food products. The word aseptic is derived from the Greek word *septicos*, which means the non-appearance of microorganisms that cause spoilage (Gotz et al., 2014). Aseptic packaging involves the filling of a commercially pre-sterile product, sterilization by suitable heat treatment into a pre-sterilized container under sterile conditions, and hermetically sealing the product to avoid any reinfection (Ansari and Datta, 2003). The difference between conventional and aseptic packaging technology is that aseptic packaging permits a comparable shelf life with canning or retort processing; whereas, in conventional packaging, the food product is filled into pre-cleaned packaging material, followed by immense heat treatment so that the desired commercial sterility is obtained. This technology is restricted to packaging materials like tin plates, glass and aluminum laminates that can with stand high temperature ($\geq 127^\circ$C), high pressure, and acidic environments. The packaging materials used are metallic, making them costly and susceptible to corrosion. Also, the extent of the heat treatment destroys important nutritional and organoleptic properties.

Aseptic packaging systems comprise a sterilizer, product filling assembly, sterile air and gas pipes, aseptic packaging material, sterilization areas to sterilize laminates of packaging materials, and a sterile zone where product filling and sealing are done. High-pressure inert gas is kept in the aseptic zone to assist in the removal of the product and to reject any contamination from the surroundings. After sterilization, food products are carried to the product filling and sealing zone via aseptic pipes where the product is put into pre-sterilized containers under sterile conditions. This extends the shelf life for 3–6 months, even if stored at room temperature. Food products like milk and milk products, tea, concentrates, drinks, fruit and vegetable juices, nutritional beverages, mineral water, tomato products, sauces, wines, particulate foods, and ready-to-eat foods are processed by ultrahigh temperature (UHT) treatment followed by filling under aseptic conditions. In aseptic packaging systems, the Tetra Pak is the most commonly used package; they are brick-shaped cartons made up of aluminum foil/polyethylene (PE)/paperboard. The food and pharmaceutical industry alone spends approximately $3.5 billion on aseptic packaging. Aseptic packaging of foods in cans, plastic cups, bottles, paperboards, and pouches has gained a place on every retail shelf across the world (Pillai and Shayanfar, 2015).

Although the aseptic packaging technology is more than 70 years old, the main breakthrough in marketing this technology was the U.S. Food and Drug Administration (FDA) approval in 1981. Initially, approval was given for the use of hydrogen peroxide (H_2O_2) as a sterilizer for packaging material that has direct interaction with food surfaces. A great deal of progress has been achieved in the aseptic packaging field in the areas of sterilization, filling, the packaging itself, and locking or sealing over the last 30 years (Pillai and Shayanfar, 2015). A major advantage of this food packaging technique is that the whole process involves food-friendly materials and operations. The food processors and their retailers are more concerned about the food items that have prolonged shelf life and that can resist ambient storage environments. Various research and development projects are under way to enhance the sustainability of aseptic packaging. Research involves the development of innovative and microbial-resistant materials for a simple and economical packaging process. Consequently, a number of exclusive plastic polymers, including polyethylene terephthalate (PET) and PE, were used early on for foods with high-acid value. Now they are used to pack shelf-stable beverages with low-acid value, such as sports drinks, flavored coffee, and so on. Currently, in the United States there are more than 500 distinct aseptic systems available for the production of retail packs as well as bulk packs and two dozen manufacturing units for the aseptic filling equipment globally (Nelson, 2010; David et al., 2013; Pillai and Shayanfar, 2015).

Numerous methods are accessible by which food can be packed, and the most imperative among them is aseptic packaging. The objective of the aseptic packaging is to preserve the food items to be stored without refrigeration for a few months keeping their physical or chemical properties intact.

16.2 ASEPTIC PACKAGING TECHNOLOGY

Aseptic packaging technology initiates the utilization of packaging materials that are otherwise treated as unsuitable for sterilization (in-packing). The high-temperature short-time (HTST) sterilization procedures in aseptic packaging are thermally operative and usually increase the manufacturing of supreme quality products, unlike those manufactured at lower temperatures for longer times (LTLT). The product shelf life is extended even at ambient temperatures when food products are aseptically processed and packaged. Aseptic packaging ensures that the container does not allow the entry of microorganisms before or after filling. Filling of the product is done in a bounded area that has a sterile environment (sterilization of air is done by heating or filtration). The capability of sealing hermetically in place to prohibit the entry of contaminating microorganisms aids in maintaining the zone of sterilization during handling and distribution. Aseptic packaging ensures the efficient filling of the product processed in an aseptic way (by the UHT or HTST system) followed by sealing the container hermetically so that sterility of the product is maintained throughout the processing (Gotz et al., 2014).

16.2.1 Basic Principles of Aseptic Packaging

The major purpose of food packaging is to hold, safeguard, and preserve the food. This is similar for both conventional and aseptic packaging. In a nutshell, food packaging is anticipated to save and uphold the safety as well as the quality of the food product. In addition to this, various other desired characteristics of packaging are customer expediency, aesthetic appeal, and low environmental effects. Designing of an aseptic packaging system involves a complete and systematic knowledge of food chemistry, food engineering, and food microbiology. The goal of the sterilization technique used for attaining food sterility in aseptic and other UHT processes is to achieve a stable balance between reducing microbes and influencing the sensory attributes of the food. As soon as the thermal or non-thermal sterilization procedure is augmented, this process has to be certified in a commercial facility and executed. This final step needs a complete understanding of food process engineering principles to confirm and augment the procedure commercially. The difference between the aseptic packaging and traditional packaging is represented schematically by Pillai and Shayanfar (2015) in Figure 16.1. In case of aseptic packaging, the packaging materials and food are regularly sterilized independently (Floros, 1993). Various advantages of aseptic packaging are listed as follows:

- Maintaining the sensory and nutritional qualities of the foods at excellent levels because the sterilization environments are designed precisely for the food.
- Foods possess enhanced shelf lives as a result of sterility, high-ranking packaging materials, and product quality.
- Food packages possess superior aesthetic appeal due to their customer-friendly sizes and shapes.
- Expenses are diminished due to less energy demands for sterilization, processing, packaging, transportation, and ambient-temperature storage.

16.2.2 Sterilization and Ultrahigh Temperature (UHT) Aseptic Process

Sterilization effectively eradicates any kind of microorganism existing in the food product. Therefore, a sterilized food product is free of any microbe, and microbial damage may be caused by recontamination from the surrounding atmosphere. Autoclaving done in hermetically sealed containers at a temperature of 121.1°C and pressure of 2.013×10^5 Pa for a period of more than 15 minute is the only procedure that ensures a sterile condition. This effectively leads to the broad elimination of all colony forming units (CFUs) in the initial product, but it destroys various important sensory and nutritional characteristics of the food. HTST processing is used for aseptic packaging to retain

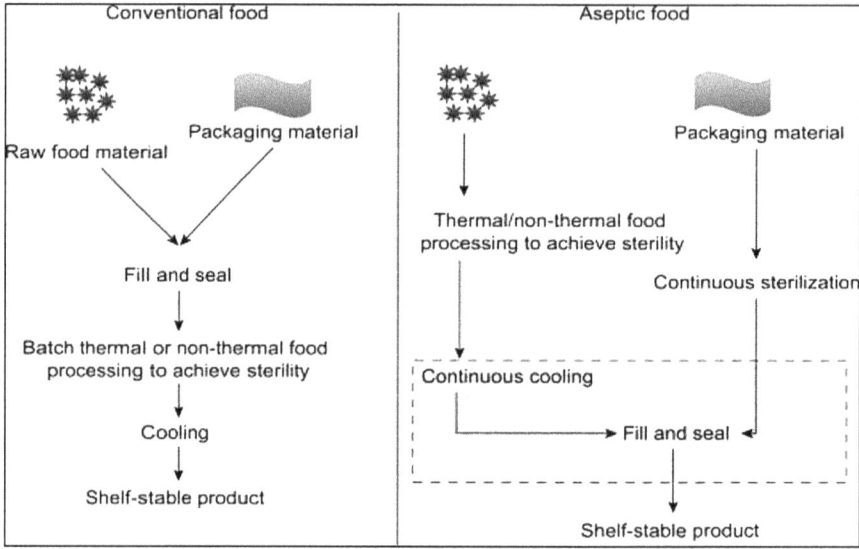

Figure 16.1   Conventional and aseptic packaging technology. (Adapted from Pillai and Shayanfar, 2015.)

the sensory and nutritional characteristics in addition to attaining a commercially sterile product. In this case, the reduction of CFUs follows algorithmic function, as shown in Equation (16.1).

$$\log S(t) = \log N_t / N_0 \tag{16.1}$$

Equation (16.1) depicts that time-dependent survival rate $S(t)$ is a function of cell count after a defined treatment time t (N_t) compared with the cell count at initial time $t = 0$ (N_0). This shows that S can never reach zero. A low number of microorganism CFUs (bacteria, molds, and yeast) are able to survive and multiply in the product is achievable, which ensures the microbial stability of the closed product over a defined period of storage time (in general, approximately 3–6 months at ambient conditions). Thus, the number of live cells is reduced to 1 CFU/1000 L or even lower. This means that the risk per package to include one CFU equals 1:1000. Typical z-values for microbes have been approximately 10°C until the proposal of typical values for quality factors C_0 to compare the heating effects on appropriate compounds. It is similar to the F_0 value, except the reference temperature is 100°C:

$$C_0 = 10^{(T-100)/z} T \tag{16.2}$$

Equation (16.2) shows that high temperatures have a minimal effect on the destruction of quality parameters because of their high z-values. UHT processing has a very low thermal effect on food at temperatures of 130–150°C for 2–6seconds of the heating period but enough for the destruction of microbes. For this reason, such methods are frequently defined as "mild" methods. One drawback of UHT processing is that several enzymes, which are heat resistant (such as lipases, proteases, and peroxidases) are still live even after high heat treatment, which leads to deterioration of the food quality, resulting in the development of off-flavors and age gelation during storage.

UHT processing is performed in four important steps for the sterilization of the food product:

1. Preheating of the product to a well-defined temperature
2. Final heating up to sterilization temperature

3. Temperature holding over a defined period
4. Instant adequate cooling at ambient or chilled conditions to avoid product damage

The processing steps should be as short as possible to minimize the loss of food quality. Different types of manufacturing plants have been developed for this process that fall into two classes: class one is the indirect heating method and class two is direct heating via infusion of steam into the food product. Both of these systems have benefits and drawbacks. Direct methods offer better food quality by using HTST. Indirect methods involve a longer heating time, which may disturb the quality of the food and there is a buildup of material on the heating surface. Cost-efficiency and higher throughput than direct methods add to the advantages of indirect methods. Today, companies are focusing on the use of indirect heating procedures (direct heating systems are not used as often).

16.3 ELEMENTS OF THE ASEPTIC PROCESS

Aseptic packaging is divided into two discrete but unified procedures, i.e., processing and packaging systems in which the packaging stresses the processor to produce a product with superior quality while maintaining its sterility. The pre-sterilization of food products includes heating the food to the required UHT and preserving this temperature for the desired period to gain sterility. Food is then left to cool at ambient temperature or an increased temperature to attain the desired thickness for filling. Methods for indirect heating of liquids containing particulates consist of ultrahigh pressure (UHP) sterilization, tube-type heat exchangers, Rota holders, scrape-type heat exchangers, and microwave sterilization (Buchner, 1993). Recent advanced methods include the use of saturated steam, superheated steam, pulsed light, dry heat, ozone, ultraviolet (UV) light, ethylene, and hydrogen peroxide (Walsh and Kerry, 2012). The influence of different processing techniques on the packaging materials has been studied (Ozen and Floros, 2001). For instance, the form-fill-seal system used in the aseptic packaging technique depends on the temperatures raised by thermoplastic resin producing the multilayer packaging material during the co-extrusion process to form a sterile surface for the food that can stay fresh for a longer time (Han and Scanlon, 2014).

16.3.1 Containment Material in Aseptic Packaging

The main aseptic packaging involves the use of a carton that is specifically made up of 70% paper, 24% low-density PE (LDPE), and 6%aluminum with an interior layer of tight PE. Other forms of packaging including cups, pouches, plastic cans, and trays can also be packaged aseptically. Strength, toughness, and a brick-like stiff and tough shape are delivered by the paper material of the package. PE is used on both the exterior printable surface and the interior layer forming a stiff seal. The use of extremely thin aluminum foil offers a barricade against gases and light and eradicates the requirement of refrigeration. It also protects the food from deterioration, omitting the usage of food additives and restoring food safety (Voicu et al., 2019). In 2005, the very first microwaveable aseptic packaging was created by Tetra Pak under the name, Tetra Wedge Aseptic (TWA) microwaveable 200-S. This inventive package uses a polyethylene terephthalic silicon oxide (PET SiOx) barrier against oxygen and is offers integrity and safety to the product as well as retaining the original color, texture, and flavor for about 6 months without requiring preservatives or refrigeration. The ease of handling, uniform heating, and accurate filling provided by the unique shape are distinct features of the microwaveable TWA 200-S when compared with conventional plastic pouches. Aseptically processed and packaged products like ready-to-serve meals are the latest aseptic technologies available on the market. For instance, Vetetee Rice Ltd. (Japan) started a Dine-In range of cooked rice packed in microwaveable plastic trays (shelf stable) with a transparent plastic lid.

Table 16.1 Aseptic Packaging Systems Available on the Market

Packaging Material	Materials	Subcategories	Properties
Cartons	Laminates comprised of PE, paperboard, and AL foil	Form-fill-seal Prefabricated	Commonly sterilized by hydrogen peroxide (H_2O_2) in deep bath, wetting systems as H_2O_2 aerosol, H_2O_2 steam, or in combination with UV Uninterrupted filling of product until sealing Filling machine must consist of several separate areas Commonly sterilized by H_2O_2 in deep bath, wetting systems, as H_2O_2 aerosol, H_2O_2 steam, or in combination with UV Residual peroxide might be a problem
Cans	Tinplate	Metal cans Composite cans	Existing in capacity from 125 mL to 22 L Superheated steam is used to ensure aseptic conditions during filling Filling with headspace Heat resistant Similar to metal cans Sterilization by hot air at 143°C for 3 minutes
Bottles	Glass, HDPE and PP	Glass Plastic	Sterilized with all methods Threat of breaking Not used commercially now Little or no protection against light, however, can be mixed with pigments for light protection Available as non-sterile and single-station systems (blowing, filling, and sealing in one step) H_2O_2 or peracetic acid used for sterilization Only headspace filling possible
Sachets and pouches	Laminates of LDPE, a center layer of EVOH, and carbon black	Form-fill-seal Lay-flat tubing	Production is similar to laminated cartons Material is pre-sterilized by bathing in H_2O_2 Blown polymer film tubes Only transversal seal required
Cups	High-impact PS, PP, PVC and EVOH	Pre-formed Form-fill-seal	Sterilization by H_2O_2 One step process Sterilization by H_2O_2, saturated steam

Abbreviations: AL, aluminum laminates; EVOH, ethylene vinyl alcohol; HDPE, high-density polyethylene; LDPE, low-density polyethylene; PE, polyethylene; PP, polypropylene; PS, polystyrene; PVC, polyvinylidene chloride; UV, ultraviolet.
Source: Adapted from Gotz, A., Wani, A. A., Langowski, H. C., & Wunderlich, J. (2014). *Food technologies: aseptic packaging encyclopedia of food safety*, Vol. 3 (pp. 124–134). Elsevier, Amsterdam.

16.3.2 Types of Aseptic Packaging Material

Aseptic packages used in commercial manufacturing include the Dole aseptic canning system, Scholle, and Tetra Pak. The different kinds of aseptic packaging materials used for containment of food products (Smith et al., 2004; Rahman, 2007; Sanjana et al., 2019)are mentioned in Table 16.1.

16.3.3 Aseptic Packaging Line

Generally, there are five types of aseptic packaging lines (Smith et al., 2004; Rahman, 2007):

1. *Film and seal:* Pre-made packages that are formed of glass, thermoformed plastic, or metal are sterilized and filled in sterile conditions and sealed.
2. *Form, fill, and seal:* Includes sterilization of a roll of packaging, produced in sterile conditions, filled, and sealed, such as Tetra Pak.
3. *Erect, fill, and seal:* Uses knocked, erected, and down blanks that are sterilized, filled, and sealed such as cambri-block and gable-top cartons.

4. *Thermoform, fill:* Sterilization of sealed roll stock, that is thermoformed, filled, and sealed aseptically like plastic soup cans and creamers.
5. *Blow, mold, fill, and seal:* Pack forms utilized in aseptic UHT processing are paperboards, plastic laminates, cans, thermoformed plastic containers, plastic, foil, flexible pouches, and bag inboxes.

16.4 BRIEF METHODOLOGY OF ASEPTIC PACKAGING

Aseptic processing includes the following broad steps (Fellows, 2000; Sanjana et al., 2019):

- Sterilization of the food commodity before filling
- Sterilization of containment or packaging material and seals before filling
- Sterilization of aseptic installations and machinery prior to procedure (UHT unit, sterile gases and air, product lines, filler, and appropriate machine areas)
- Maintaining sterility in the entire system throughout the operation
- Hermetic package production
- Sterilization of aseptic packaging materials and equipment

16.5 STERILIZATION AGENTS IN ASEPTIC PACKAGING

Several agents have been utilized for the sterilization of aseptic packaging materials and equipment (Gotz et al., 2014; Chavan et al., 2016) including:

Heat: Initially, for the sterilization of the aseptic systems, heat was utilized as a natural way to thermal process food products. Sterilization using heat is implemented two ways, as "dry heat" and "moist heat." Dry heat can be implemented to sterilize the machinery in the form of hot air or superheated steam. Moist heat in the form of saturated steam under pressure or hot water is used to sterilize the product supply line and fillers. Heat resilience of bacterial endospores to the dry heat demands a higher time temperature for sterilization compared with the sterilization temperatures for moist heat. It is important to ensure that sterilization has properly occurred, which is achieved by higher temperatures and longer holding time. The sterilization temperature for systems ranges from 121 to 129°C, whereas for dry heat sterilization the temperature ranges from 176 to 232°C. The temperature range of 260–315°C is utilized for air sterilization by means of incineration.

Chemicals: Hydrogen peroxide is a suitable choice as a chemical agent for sterilization purposes. Various other chemical agents like ethylene oxide, acids, and peracetic acid have been implemented in acidic food items to attain sterilization. The sporicidal activity of hydrogen peroxide is achieved as well as enhanced by surging the temperatures. In several aseptic packaging systems, hydrogen peroxide is utilized as a sterilizing agent with concentrations ranging from 30 to 35% for packaging materials followed by treatment with temperatures ranging from 60 to 125°C to dissolve hydrogen peroxide residues.

Irradiation: The radiation process has been utilized over the years for sterilization of packaging materials to implement them in aseptic systems for acidified foods. The greater penetration power of gamma radiation makes it possible to treat the packages in bulk with the help of commercial irradiators. Approximately, a 1.5-Mrad dose is usually applied to decontaminate acidic foods. The procedures for the aseptic filling and packaging systems for low-acidic food has been applied recently. Low-acid foods require higher doses of radiation to sterilize their packages compared with the doses needed to sterilize high-acid and acidified foods. Several types of radiations are not widely acknowledged in the aseptic systems. Decontamination of food contact surfaces by means of UV-C light is in trend. Limitations of the using UV-C in the aseptic packaging of low acid food is due to less penetration and issues allied with shadowing. Restrictions for the use of electron beam (e-Beam) irradiators is a result of equipment size, speed, and costs.

e-Beam technology: The e-Beam technology is a non-thermal method utilized in aseptic packaging to achieve extremely high levels of microbial death without adversely affecting the food product.

Relying on the bio-burden, e-Beam doses can be used to attain the essential kill. For sterilization of space foods, the FDA has certified the utilization of ionizing radiation using e-beam technology up to 44 KGy, which is extreme commercial sterilization of foods. The shelf life, safety, and acceptability of space foods for the NASA missions are of utmost significance, but will lead to more implications for long-duration missions, like the Mars mission. Because remarkable nutrient loss and foodborne illness could be a hurdle for the success of such a mission, their exclusion is a high priority for NASA. Food and suppertimes play an intense role in relieving the boredom and stress of protracted missions, and the suitability of food concerning texture, appearance, aroma, and flavor can also have noteworthy psychological influences on the routine of the astronaut. Because NASA emphasizes the prolonged Mars and lunar missions, there is a requirement for innovative food processing techniques that can confirm the nutrition, acceptability, and safety of food systems with shelf life extending up to 5 years. NASA is already implementing techniques, including irradiation and thermostabilization, and is exploring progressive techniques such as microwave sterilization and high-pressure processing.

16.5.1 Control Points and Control Parameters for Effective Sterilization of Packaging Material

Control points and procedures are important to guarantee the sterility of packaging materials. The safety of the chemical to be utilized as a sterilization medium can be checked by measuring the concentration by the staff of the laboratory or by the machine operator, typically at the start of the packaging process. The packaging area must be sterilized before the start of production where the packages are designed, filled, and sealed. Several control parameters are involved in the entire procedure, all of which are important for the safety and suitability of the operation.

16.6 NEW TECHNOLOGIES FOR PRODUCT STERILIZATION

During the past years, various new prospective techniques have emerged in which no heat is used for the destruction of microbes. Some of these technologies include:

- Radio heating frequency
- Pulsed electric field (PEF)
- Ohmic heating
- UV radiation
- UHP
- Pulsed light

PEF and UHP are by now being used and show favorable outcomes in terms of the quality of food products. These technologies are restricted in use because of high operational cost. Furthermore, with the use of such technologies, the issues of sterilization are still of concern.

16.7 TETRA PAK ASEPTIC CARTONS

The most utilized aseptic packaging material is the Tetra Pak aseptic carton, which is the buildup of three basic substances that in combination result in a very competent, lightweight, and safe package. Each substance has a particular function. The Tetra Pak carton is composed of six layers, including the following (Figure 16.2):

1. PE: Moisture barrier
2. Paper: Firmness and strength

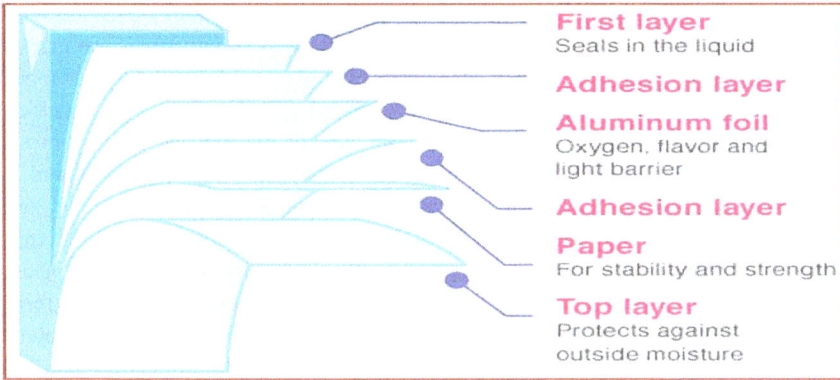

Figure 16.2 Composition of Tetra Pak. (Adapted from Voicu et al., 2019.)

3. PE: Adhesion
4. Aluminum: Oxygen and light barrier
5. PE: Adhesion
6. PE: Seals in the liquid

PE (25%) is used to make packages liquid tight and to offer a barrier to microbes. Paper (85%) provides strength and stiffness. Aluminum foil (5%) creates a barricade against light, air, and off-flavors. It, therefore, maintains the nutritional quality of the food and prevents deterioration (Voicu et al., 2019) (see Figure 16.2).

A combination of PE, paper, and aluminum has made Tetra Pak an outstanding packaging material with optimal performance. The package provides a greater degree of hygiene, safety, and nutritional superiority in food, thereby preserving the original freshness, and prolongs the shelf life without any need for preservatives or refrigeration.

Four types of Tetra Paks are available that are employed in the packaging systems: Tetra Classic Aseptic (TCA), TWA, Tetra Brik Aseptic (TBA), and Tetra FinoAseptic (TFA). These packages are accessible in size and shape variations with appropriate opening and closing in accordance with the product and customer requirements. Bulk aseptic bags are also employed as aseptic packaging material for the filling of high- to low-acid foods with a capacity of 25–1140 L. A bulk aseptic bag comprises an exterior barrier laminate and an interior bag in contact with the food product. Gamma radiation is employed for the pre-sterilization of bags and has three different features, including a highly secure spout and outer and inner bag barriers. The bags are categorized as supreme and medium barriers depending on the choice of the barrier. For the medium barrier bags, metalized polyester is used as a conventional barrier, whereas in supreme bags, aluminum foil is used as a barrier material if the food products are susceptible to oxidation and aroma loss (Willhoft, 1993; Ranganna, 2000; Paine, 2012).

16.8 SIGNIFICANCE OF ASEPTIC PACKAGING

Regulatory and concerned advisory bodies have devised strict regulations in different segments of food processing that have forced the food industries to attain and sustain food quality standards. The outbreak of foodborne diseases due to lack of hygienic conditions during food processing and food packaging was the main reason for framing such rules and regulations. Aseptic packaging has been in use for over a century. In 1913, Denmark was the first to develop aseptically packaged food (milk packed in metal cans), and the development of laminate paperboard cartons (Tetra Paks) by Sweden dating back to 1961 led to the first commercially successful product with a massive market to date. Emphasis was given to developing a system that permits the reasonable production of food products with a long shelf life at ambient conditions. Aseptic processing has become a commercially viable and effective technique (i.e., UHT). Aseptic packaging extends the shelf life of food products even at room temperature. For example, commonly pasteurized milk stored under refrigerated conditions isstableforapproximately5–8days, whereas milk processed and packed under aseptic conditions is stable up to 6 months at ambient conditions. All of these characteristics reveal the importance of aseptic packaging for foods. Aseptically processed foods are highly accepted by consumers with respect to shelf life extension without the involvement of chemical preservatives and refrigeration. A huge variety of aseptically filled commodities are packaged in distinct packaging materials offering the following distinct characteristics to the product (Smith et al., 2004; Rahman, 2007; Gotz et al., 2014; Sanjana et al., 2019):

Containment of the product: The requirement for enclosing the product means that liquids or powders do not seep out of the packaging material.

Physical protection: This is mandatory while dealing with brittle foods like snack foods or eggs and minor effects on fresh fruits will liberate enzymes causing softening and browning. Even though the product is in good condition, the damaged package has an effect on sales.

Safety/security of the food: It is essential that the package should remain sterile by preventing accidental contamination from microorganisms. Tamper resistance is also a necessary prerequisite in terms of hostile contamination circumstances. Another characteristic of food safety is the preventing long-term chronic influences from the food packaging materials.

Shelf life of the food: A key factor in evaluating the shelf life of dried foods is moisture. For the aseptically packed foods including milk, fruit juices, soups, and cream, light catalyzed atmospheric oxidation is more perilous. Therefore, a good barricade against light and oxygen furnished by aluminum foil or a tinplate is required for the shelf stability of such products.

Interconnection of data regarding product and manufacturer: The package is required to convey to the customer what kind of food is in the interior and whose product it is. In addition to this, various other data should be conveyed to the customer, like a list of ingredients, net weight, expiry date, batch number, nutritional data, etc.

Aesthetic appeal: The package should be appealing to customers and should be opening and dispensing the commodity should be trouble free.

Cost-effective: In packaging, the value is more essential than focusing on the lowest price. A cost-efficient but dimensionally flexible container could result in longer downtime during the production or a surge of "leakers" in the market place, thus, refuting the apparent cost saving. In contrast, higher expenditure in making a package appears more attractive and could be defensible in the surged sales.

16.9 PREREQUISITES FOR ASEPTIC PACKAGING

The prerequisites required for aseptic packaging include containment of the product/commodity, prohibition of physical impairment to the packaged product, smooth processing on filling lines, tolerance for packaging procedures, controllable during the distribution process, prevention of dirt and other types of contamination, safeguarding the commodity from taints and odors, resilient to rodent attacks, prevention from insect invasion, and biologically safe (i.e., non-toxic). The packaging must be extremely compatible with food stuffs and must monitor the sterility of the product, thus preventing transmission of microorganisms. It should also display signs of tampering, should monitor gain or loss of moisture, must offer a barricade to oxygen and light, and must keep a check on gaseous atmospheres (i.e., CO_2/N_2). It should also interconnect all the information related to the product and manufacturer, must be alluring, should be convenient to open and handle, and should be economical (Rahman, 2007; Fellows, 2016).

16.10 QUALITY AND SAFETY OF ASEPTICALLY PACKAGED FOODS

16.10.1 Aseptic Foods Products over In-Container Sterilized Products

Undoubtedly, the quality of aseptically processed food is improved because of the UHT sterilization process. This is effective with respect to larger packs as heat dissemination is time-taking in the retorting process and, subsequently, quality suffers from overheating. For thin layers of food product in the heat exchanger, UHT treatment is mostly achieved by procedures that permit fast heating and cooling. During high UHT temperatures, microorganisms are more sensitive to high temperatures compared with food ingredients. Although the Q_{10}-value (factor for catalyzing reactions when increasing the temperature by 10°C) for chemical reactions of food products is generally between 2 and 4, the assessed value for the destruction of bacteria is 10. If the temperature from the typical retorting value of 121°C is increased by 10°C, the time for the comparable death rate for bacteria will be 10%, and only 1% with a 20°C increase. If for chemical reactions, a Q_{10} of 3 is presumed, they will be accelerated by only a factor of $3 \times 3 = 9$ when increasing the temperature by 20°C. Consequently, a temperature surge from 121 to 141°C for sterilization of foods will hypothetically be only about 9% of the impairment to quality, in contrast with retorting at 121°C (Buchner, 1993; Willhoft, 1993).

UHT sterilization requires less energy, which may be due to the heat gained with the help of heat exchangers, with the recovery rates greater than 90%. Additionally, the aseptic process is a continuous flow method, requiring fewer operators and space than retorting. At the discharge of the aseptic filler, the packages are storage ready. Packs do not need to be transferred into retorts where, depending on the attribute of the retorts, certain expenses are important for handling the packages. Milder conditions are utilized for packaging material and packs sterilized in aseptic packaging than those prevailing in retorts; thus, a broad choice of packaging materials and packs are possible. Packs may be cost-effective, lighter, less space-consuming, and offer more convenience.

Compared with retorted trays and cups of polypropylene/ethylene vinyl alcohol/polypropylene (PP/EVOH/PP) mainly seen in the use of soups and other meals, aseptically filled packs from these materials are not damaged during storage from water vapor absorption by EVOH during retorting and increased oxygen transmission from the atmosphere. Even after drying, the penetrating power of retorted EVOH may continue to be greater due to "retortshock" (Buchner, 1993; Willhoft, 1993).

16.10.2 Aseptic Products over Pasteurized Chilled Products

The superiority of aseptically filled products compared with pasteurized chilled foods can be visualized from the final properties and the nature of various operations involved in the process. The aseptically filled products have various benefits if filled and sealed effectively as they are germ-free and require no post-packing refrigeration. Compared with the pasteurized products, the aseptic products do not need any temperature regulation against possible fluctuations and, therefore, are less susceptible to quality damage.

As aseptic products do not need refrigeration, special storage, and distribution, the process becomes cost-effective and selling tracks may be extensive and permit more distinction. The large production or the so-called "economies of scale" makes the process and the products inexpensive. In the case of chilled commodities, a modified atmosphere with an adequate amount of CO_2 is used to maintain the controlled medium.

16.10.3 Aseptic Filling over "Clean-Fill"

Clean-fill is a method analogous to aseptic packing. The presence of a small number of microorganisms, even after total commercial sterilization and sealing, and the need to chill the product to increase its shelf life prevent the commercialization of the process. Although there is a slightly lower cost and quality damage due to chilling during clean-fill, the process is limited to high-acid foods only due to the presence of fewer microorganisms.

16.11 LOGISTIC ADVANTAGES

Aseptic packaging helps lower the transport weight, for instance, a 1-Lbeverage carton weighs only 28 g versus a glass container, which weighs up to 380 g. In addition to warehousing efficacy, minimum storage costs, better stack capability, and enhanced pallet application configuration, unfilled packs are stored flat until ready to be employed (unlike glass or cans). Comparatively, the transportation of 1.5 million unoccupied drink boxes by a standard semi-trailer truck versus only 150,000 aseptically packed glass bottles depicts the huge difference in transportation cost. The transportation of aseptic packaging is space effective from the source to the market. The brick shape permits spare products capable of being stacked in the truck. Steam sterilization ensures safety and sterility can be well monitored, measured, and recorded. The aseptic packaging process does not involve the use of chemical sprays for chamber sterilization. The spout employed is tamper-evident and the process of sterilization is safe and easy to monitor. The process does not use any chemical spray on food items and the use of laminate material eliminates the chance of any future contamination. Aseptic packaging employs a simple filling machine without the need fora sterile chamber. The weight of the outgoing packed products regulates the process of filling. Therefore, no modifications for specific gravity are required to achieve high accuracy (Smith et al., 2004; Pillai and Shayanfar, 2015; Smolin et al., 2015).

16.12 IMPACT ON THE ENVIRONMENT

Aseptic cartons rely on an inexhaustible raw material (i.e., paperboard obtained from wood). The transportation efficacy, lightweight, storage without cooling, and energy-effective production are the main reasons for its proficiency. Aseptic cartons are made from a green material that is easy to recycle and are being recycled at the corresponding collection systems. The aseptic cartons reduce the consumption of fossil resources, thereby reducing the production of greenhouse gases, which influences global warming (Willhoft, 1993).

16.13 CONCLUSION

Aseptic packaging has successfully gained its place in global food applications as a safer and superior quality packaging choice. Aseptic processing involves the sterilization of food commodities by eliminating microorganisms, employing a tightly precise thermal procedure, and linking the germ-free commodity with the disinfected packaging material in a hygienic atmosphere. The output is a shelf-stable product that does not require refrigeration or the addition of preservatives. Aseptic packaging leads to an extended shelf life, greater seal integrity, and a good aesthetic appeal. The enhancement in the non-refrigerated shelf life can be attributed to the use of plastics in packaging. Today, this is principally employed in the deepest contact surfaces of the package. Aseptic packaging is used for baby foods, soup, wine, and water products, among others.

REFERENCES

Ansari, I. A., & Datta, A. K. (2003). An overview of sterilization methods for packaging materials used in Aseptic packaging systems. Food and Bioproducts Processing, 81(1), 57–65.
Buchner, N. (1993). Aseptic processing and packaging of food particulates. In Aseptic processing and packaging of particulate foods (pp. 1–22). Springer, Boston, MA.

Chavan, R. S., Ansari, M. I., & Bhatt, S. (2016). Packaging: aseptic filling. In Encyclopedia of food and health (pp. 191–198). Apple Academic Press, Palm Bay, FL.

David, J. R., Graves, R. H., & Szemplenski, T. (2013). Handbook of aseptic processing and packaging. Taylor & Francis, Boca Raton, FL.

Fellows, P. (2016). Food processing technology: principles and practice (4th ed.). Woodhead Publishing/ Elsevier Science, Kent, UK.

Fellows, P. (2000). Food processing technology: Principles and practice. Woodhead Publishing Limited, Cambridge, UK.

Floros, J. D. (1993). Aseptic packaging technology. In Nelson, P. E. (Ed.). Principles of aseptic processing and packaging (3rd ed.). Purdue University Press, West Lafayette, IN.

Gotz A., Wani, A. A., Langowski, H. C., &Wunderlich, J. (2014). Food technologies: aseptic packaging encyclopedia of food safety, Vol. 3 (pp. 124–134). Elsevier, Amsterdam.

Han, J. H., & Scanlon, M. G. (2014). Mass transfer of gas and solute through packaging materials. In Innovations in food packaging (pp. 37–49). Academic Press, London.

Nelson, P. E. (Ed.). (2010). Principles of aseptic processing and packaging (2nd ed.) (pp. 12–19). Purdue University Press, West Lafayette, IN.

Ozen, B. F., &Floros, J. D. (2001). Effects of emerging food processing techniques on the packaging materials. Trends in Food Science &Technology, 12(2), 60–67.

Paine, F. A. (Ed.). (2012). Modern processing, packaging and distribution systems for food. Springer Science & Business Media, New York.

Pillai, S. D., & Shayanfar, S. (2015). Aseptic packaging of foods and its combination with electron beam processing. In Electron beam pasteurization and complementary food processing technologies (pp. 83–93). Woodhead Publishing, Cambridge, UK.

Rahman, M. S. (Ed.). (2007). Handbook of food preservation. CRC Press, Boca Raton, FL.

Ranganna, S. (2000). Handbook of canning and aseptic packaging. Tata McGraw-Hill Publishing, New Delhi.

Robertson, G. L. (2019). History of food packaging. Elsevier

Sanjana, M. C., Hemegowda, R., & Sushma, R. E. (2019). Aseptic packaging – a novel technology to the food industry. International Journal of Trend in Scientific Research and Development, 3, 307–310.

Smith, J. P., Zagory, D., & Ramaswami, H. S. (2004). Packaging of fruits and vegetables. In Processing fruits science and technology (2nd ed.). CRC Press, Boca Raton, FL.

Smolin, L., Grosvenor, M. B., & Gurfinkel, D. (2015). Nutrition: science and applications (2nd ed.). Wiley, Toronto, Canada.

Voicu, G., Constantin, G. A., Stefan, E. M., Tudor, P., Munteanu, M. G., & Żelaziński, T. (2019). Aspects regarding the aseptic packaging of food products. Acta Technica Corvininesis – Bulletin of Engineering, 12(1), 85–90.

Walsh, H., & Kerry, J. P. (2012). Packaging of ready-to-serve and retail-ready meat, poultry and seafood products. In Advances in meat, poultry and seafood packaging (pp. 406–436). Woodhead Publishing, Cambridge, UK.

Willhoft, E. M. A. (1993). Aseptic processing and packaging of particulate foods. In Aseptic processing and packaging of particulate foods (pp. 1–22). Springer, Boston, MA.

CHAPTER **17**

Foods with Edible Coatings
Science, Shelf Life, and Quality

Gurkirat Kaur, Swati Kapoor, and Neeraj Gandhi

CONTENTS

Email: Kaur Gurkirat, gurkirat@pau.edu

DOI: 10.1201/9781003091677-17

17.1 INTRODUCTION

The era of natural preservation of food and horticultural crops has taken up the pace in the present scenario. With respect to this, packaging of food products plays a pivotal role in extending the shelf life of produce. The major function of packaging is to provide protection to a food commodity from the time of harvest or production until consumption and maintain quality attributes throughout the distribution chain. The shortcomings associated with synthetic packaging material, particularly safety and non-biodegradable issues, have shifted the interest of researchers toward the application of more natural and edible grade films and coatings. However, the concept of edible coatings has long been acknowledged wherein fruits were preserved using wax or lipids for longer preservation and distant transportation. Edible coatings are considered a cost-effective alternate to low-temperature storage of food products.

In case of horticultural crops, as soon as the crop is harvested, it undergoes quality deterioration in terms of water loss, shriveling, wilting, spoilage, and decay, which negatively affects marketability and financial losses. Therefore, it becomes imperative to provide additional coatings to produce to minimize quality deterioration. It is pertinent to mention here that produce once spoiled or decayed once initiated could not be controlled further, irrespective of any amount or type of packaging or coatings applied. Hence, immediate application of post-harvest treatments in the form of coatings must be done to ensure good quality produce.

As per Pavlath and Orts (2009), edible films and coatings are defined as "any form of material used for enrobing (i.e., coating or wrapping) various food to extend shelf life of the product that may be eaten together with food with or without further removal is considered an edible film or coating." Edible films are basically thin and dried biopolymer layers applied separately on the food product, whereas coatings are applied directly to the food surface leading to film formation on the surface of the food (Krochta, 2002). Some of the quality characteristics of edible films and coatings are described in terms of mechanical strength, mass transfer, gas permeability, cost-effectiveness, and toxicity levels. Ultimately, the two most important marketable attributes of produce, i.e., color and texture, must be optimally maintained on application of eco-friendly coatings. None of the single coatings or film composition can extend the shelf life of variable horticulture produce or food commodity. Therefore, a vast amount of research is currently being undertaken to optimize each type of coating specific for the corresponding food commodity.

With respect to post-harvest treatments used to extend the shelf life of horticultural produce, low-temperature storage is mostly preferred. Each type of produce is stored at a different storage temperature as per their physiology and respiration rate. However, the cost associated with low-temperature storage is quite high. The concept of a cold chain is still in its infancy in developing countries due to lack of infrastructure, technology knowledge, and high cost. Similar issues are faced using modified atmosphere storage (MAS) and controlled atmosphere storage (CAS) techniques in food commodities. Therefore, the use of low-cost eco-friendly coatings and films to improve shelf life would give an edge to traders and producers in terms of financial benefits and reduction in post-harvest losses.

17.2 SCIENCE BEHIND THE MECHANISM OF COATINGS FOR ENHANCING SHELF LIFE OF FOODS

The foremost function of coatings is to provide a barrier between produce and environment to restrict quality loss of produce as depicted in Figure 17.1. There are a number of factors such as surface properties of fruits and vegetables, composition of coating material including emulsifiers, plasticizers, and antimicrobial compounds that determine the success of an edible coating on

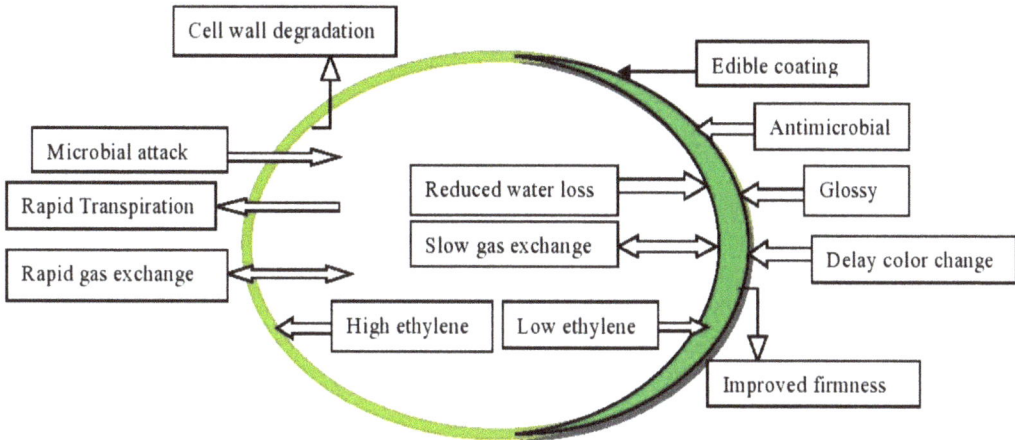

Figure 17.1 Effect of edible coatings on physiological properties of fruits and vegetables.

produce (Falguera et al., 2011). The coating layer must possess the following properties to ensure high-quality products:

- Provide protection against moisture evaporation/transpiration
- Slow down gaseous exchange/respiration in case of fresh fruits and vegetables
- Possess antimicrobial properties and prevent against microbial contamination
- Preserve volatile aromas and flavor components
- Carrier for potentially active compounds
- Enhance aesthetic appeal of produce for higher marketability
- Preserve color and textural properties of food products
- Should bear Generally Recognized as Safe (GRAS) status and does not react with the food commodity

17.2.1 Moisture/Transpiration

Fresh food produce is mostly composed of water, which largely determines its end life. Moisture loss or gain in food products is detrimental to its quality as the former is associated with weight loss and the latter is associated with microbial proliferation on produce. Thus, the impact of temperature and relative humidity on produce cannot be undermined and must be taken into account prior to application of coatings. Any type of barrier applied to food produce must be able to maintain optimum moisture levels in the produce and at the same time should possess antimicrobial properties to enhance the shelf life of the produce. Lipid-based coatings are hydrophobic in nature and are quite effective in reducing moisture loss.

17.2.2 Gaseous Exchange

Following harvest, most of fruits and vegetables continue to respire leading to faster oxygen depletion and hastening of the ripening process. However, when produce is coated with any barrier material the process of respiration slows down, lowering the oxygen depletion rate, and enhancing carbon dioxide accumulation within the produce as its difficult for the gases to escape the coated layer (Dhall, 2013). The restricted movement of gases due to coating leads to a shift from aerobic to partial anaerobic respiration, requiring less oxygen (McHugh and Senesi, 2000) for respiration. This phenomenon creates modified atmosphere inside the produce prolonging its shelf life;

however, care should be taken to not reduce oxygen concentration below 1–3% as it will lead to complete anaerobic respiration creating ethanol and acetaldehyde and affecting the sensory properties of the produce.

17.2.3 Ethylene Evolution Rate

Ethylene gas is a well-known compound that induces undesirable changes in the fresh produce after harvest as it hastens the ripening process leading to microbial growth, color loss, texture loss, etc. Application of edible coatings on fresh fruits and vegetables helps prevent oxygen entry inside the tissues, thereby interrupting ethylene production and reducing senescence.

17.2.4 Enzymatic

The inherent presence of defense enzymes in fruits and vegetables such as peroxidase, superoxide dismutase, and catalase help provide a natural defense against reactive oxygen species. These enzymes convert free radical oxygen to non-toxic forms such as water or oxygen (Gao et al., 2016). With respect to this, chitosan-based coatings on fruits and vegetables have shown synergistic effects to reduce free radicals, increase disease resistance, and induce defense laden enzymes in fresh produce (Badawy and Rabea, 2009; Xing et al., 2011).

17.2.5 Antimicrobial Activity

Most of the decay associated with fresh produce is produced by an attack by pathogenic fungi. Many attempts have been made to control fungal disease but more in the form of synthetic fungicides. As these synthetic fungicides do not qualify to be used in edible coatings, there are basically three categories that are considered, GRAS, essential oils, and antimicrobial antagonists, and they are used in edible coatings (Palou et al 2015). One of the potential base components used for developing antimicrobial films is chitosan. As per Andres et al. (2007) and Bravo-Anaya et al. (2016), the antimicrobial property associated with chitosan is due to interaction between negatively charged carboxylate functional groups (-COO-) on bacterial cell membranes and the positively charged amino group ($-NH_3$) of chitosan. This electrochemical interaction could possibly weaken and disrupt the cell membranes of microorganisms (Xing et al., 2016). Additionally, creation of abnormal osmotic pressure, higher lipid peroxidation, protein leakage, and amino protonation leading to cation formation are some of the molecular events caused by chitosan coating on fresh produce leading to microbial cell death (Lou et al., 2011; Wang et al., 2015). These antimicrobial properties could be further enhanced by adding active constituents such as essential oils, nanoparticles, etc., to chitosan base material used for coating on food products.

17.2.6 Texture

Firmness of produce is one of the quality indicators used to assess freshness of a product. Fresh fruits and vegetables on harvest start to ripen and loose there structural integrity due to reduction in adhesion between cell walls along with action of cell wall-degrading enzymes on pectin, the major cell wall component. On ripening, pectin becomes de-esterified leading to loss of cell integrity and reduced firmness of fresh produce. However, it has been well documented that use of calcium salts could help preserve cell wall structure as negatively charged de-esterified pectin can cross-link with positively charged calcium salts adding rigidity to the cell wall (Ortiz et al., 2010). Therefore, calcium pectate complex could help provide firmness to produce and could be incorporated into coatings for attaining the desired firmness.

17.3 TYPES OF EDIBLE FILMS/COATINGS

There are different coating techniques on the basis of type and quantity of the material that formulates edible films and coatings. Coating involves the application of a powder or a liquid over a film-forming base material. A simple coating process involves four general steps:

1. Coating material as a solution, suspension, or powder may be coated over the food surface following methods such as brushing, spraying, or casting.
2. Allow the base coating material to adhere properly with the surface of the food material.
3. The major film-forming step involves the coalescence of the coating material onto the surface of the food product.
4. The coating layer is stabilized over the food surface through the process of coacervation following the process of drying, cooling, or heating.

17.3.1 Commonly Used Coating Methods

Some of the commonly used coating methods include:

1. *Dipping:* The coating is done over the food surface by dipping the product in a coating solution considering the density, viscosity, surface tension, and withdrawal time of the product from the coating solution. Generally, the food is immersed within the coating solution for 5–30 seconds.
2. *Spraying:* This is considered to be a conventional technique involving application of edible coatings over the food surface. It is normally adopted if the film-forming solution is not thick.
3. *Brushing:* This is done to coat the fresh fruits and vegetables to improve their shelf life and is considered better than the dipping method, as the coating is more uniform over the food surface.
4. *Fluidized-bed coating:* This involves application of a very thin coating that has less density and smaller particle size over the dried materials. This process may be applied in either batch or continuous operations and causes adhesion of particles and drying of agglomerates.
5. *Solvent casting method:* This is one of the most common methods used to formulate edible films using hydrocolloids as a source material. Dispersions of the edible materials are formed using water and ethanol solutions in varying concentrations and are spread on the suitable substrate and finally drying the product.

17.3.2 Materials Utilized for the Formation of Edible Films and Coatings

Generally biopolymers with the potential to formulate edible films and coatings are categorized on the basis of type of forming material such as proteins, polysaccharides, lipids, and nanoemulsions incorporated into edible films; variable plasticizers to improve the film properties; and functional additives (Table 17.1). The films or coatings developed by using these variable source materials differ in mechanical, barrier, and tensile properties, which decide their applicability in different food products.

With protein-based coatings, proteins, lipids, and polysaccharides are mainly used for the formation of biopolymers. Although polysaccharides are easily available, cheap, non-toxic, and processed at higher temperatures, they are also a weak barrier to moisture due to their hydrophilic nature (Cazon et al., 2017). In contrast, lipids are hydrophobic and are commonly used to form food coatings that are comparatively inelastic (Hassan et al., 2018). Proteins possess better film-forming ability, abundant availability, and nutritional value because proteins are considered as a good source to develop biodegradable films and coatings (Kaewprachu et al., 2016). They convey better gas barrier and mechanical properties relative to the films formed by polysaccharides and lipids (Bourtoom, 2011).

Table 17.1   Base Materials Required for the Formulation of Edible Films and Coatings

Functional Compositions	Materials

Film-forming materials	**Protein:**
	  Animal proteins: Egg albumin casein, meat proteins, whey proteins
	  Plant proteins: Wheat proteins, soy proteins, corn zein

	Lipids: Waxes (beeswax, candelilla wax, jojoba beeswax, carnauba wax, etc.), non-hydrogenated vegetable oils. and fatty alcohols
	Polysaccharides: Chitin, starch, cellulose, agar, pectin, alginates, gums, etc.
Plasticizers	Water, glycerin, sorbitol, sucrose
Functional and other additives	Nanoemulsions, antioxidants, antimicrobials (essential oils), emulsifiers (tweens and lecithin)

Protein-based edible films use protein solutions containing solvent or a carrier, which is generally water, ethanol, or a mixture of ethanol and water. Exposure to conditions like heat, acids, and alkalis leads to protein denaturation, which is known to develop extended structures required for the formulation of edible films and coatings. Protein-based sources that are used to formulate the food coatings are depicted in Figure 17.2.

17.3.3 Types of Protein-Based Edible Films

17.3.3.1 Animal Proteins

17.3.3.1.1 Casein

Casein is known to be a considerable milk protein consisting of four subunits known as beta-casein, kappa-casein, alpha s1-casein, and alpha s2-casein, which constitute 36%, 13%, 38%, and 10% of the casein composition, respectively. Together these components form colloidal micelles within the milk, which are stabilized by the calcium-phosphate complex. Casein constitutes about 80% of the total milk protein, which has a tendency to precipitate on acidification of milk up to an isoelectric pH of casein, i.e., 4.6 (Audic et al., 2003). On acidification, the calcium-phosphate complex tends to solubilize, liberating casein molecules, which link together to form acid casein. This is insoluble in nature and the addition of an alkaline helps in converting the insoluble form of acid casein into the soluble and functional caseinates. The four protein subunits exhibit distinctive properties that

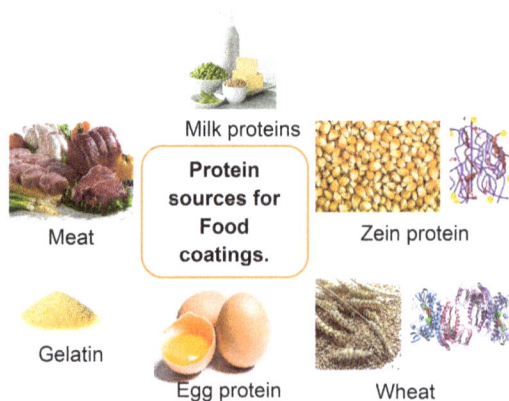

Figure 17.2   Materials used for protein-based (edible) films and coatings.

influence the ability of the casein in film forming. Casein has a tendency to form films easily from aqueous solutions without any processing because its random coil nature leads to strong interchain cohesion. It also possesses a lot of hydrophobic, intermolecular hydrogen and electrostatic bonds (Chambi and Grosso, 2006). Casein is easily available and has many properties. It is non-toxic, environmentally friendly, has good thermal stability, is capable of micelle formation, and has good water solubility and emulsification capacity, which makes it a good protein source for the preparation of edible biodegradable films (Belyamani et al., 2014). Most of the casein-formulated edible films have a serious limitation of poor mechanical and moisture barrier properties. Addition of several plasticizers cannot inculcate good mechanical properties and elasticity, although physical and chemical treatments can modify the polymer network and enhance the film's functionality (Bonnaillie et al., 2014). Casein cross-linking with polymers like gelatin and pectin using several enzymes like transglutaminase, glutaraldehyde, and other chemical agents may lead to improvements in film characteristics for their commercial applications (Shendurse et al., 2018).

17.3.3.1.2 Whey Protein

Whey is a milk protein and a by-product of cheese processing consisting of two fractions, whey protein concentrates (WPCs) (protein content 25–80%) and whey protein isolates (WPIs) (protein content ~90%). Alpha-lactalbumin, beta-lactoglobulin, bovine serum albumin, and immunoglobins are the four major proteins available in whey proteins (Ramos et al., 2012). Whey proteins have many specific and distinguishable features compared with other biopolymers such as electrostatic charge, amphiphilic nature, and conformational denaturation. Moreover, substantial self-association under a neutral environment does not take place, as in case of casein proteins. The film-forming properties mostly depend on various protein interactions that occur between different chains (Akcan et al., 2017). Whey proteins possess an advantage because they offer outstanding oxygen and oil barrier properties at low and intermediate relative humidity. Films or coatings developed from whey proteins are not good moisture barriers because whey protein itself is hydrophilic in nature (Hassan et al., 2018).

Temperature-induced denaturation of whey proteins in aqueous solution determines the formulation of edible film and coatings (Jauregi and Welderufael, 2010). Heating is done to modify the three-dimensional (3D) structure of whey proteins exposing the internal hydrophobic and sulfhydryl (S-H) groups. This ultimately promotes the oxidation of free sulfhydryl groups and interchange of disulfide (S-S) bonding on drying as the globular structure of β-lactoglobulin denatures at a temperature above 65°C, which leads to the formation of water-insoluble films or coatings. The films have a tendency to crack down on drying if they are prepared without following heat denaturation, whereas tensile strength and flexibility of the films and coatings derived from whey proteins can be improved by incorporating several plasticizers (e.g., glycerol, polyethylene glycol [PEG], and olive oil) in the denatured film solution (Shimada and Cheftel, 2002).

A study conducted by Wang et al. (2010) documented that a WPC-based packaging film may be developed. It was found that increasing the protein concentration enhanced the mechanical properties of the films. On the other hand, it also leads to the poor barrier properties of WPC films, which could be improved by heating the film at 80°C. So, films and coatings formulated with whey proteins are a feasible packaging alternate to be used as a biodegradable packaging film for preserving food.

17.3.3.1.3 Egg Albumen

Egg albumen is considered to be one of the significant sources of protein to be used for film or coating formation. There are five main protein fractions within the egg albumen: ovotransferrin, ovalbumin, ovomucin, ovomucoid, and lysozyme. Ovalbumin constitutes almost 50% of the total protein content of an egg and it consists of free sulfhydryl (S-H) groups, which are available for cross-linking (Mecitoglu et al., 2006). The other four protein fractions contain disulfide (S-S) bonds. Ovotransferrin is an iron-binding protein, whereas lysozyme is considered to be an important protein because of its antimicrobial activity against gram-negative bacteria. It is considered that the α-helix and β-pleated sheets (secondary structure) present in ovalbumin are random coiled polypeptides imparting good film-forming properties in egg white. Egg albumen proteins are easily denatured by heat and have the ability to form strong heat-set gels as they unfold revealing their internal sulfhydryl groups, which ultimately results in the formation of disulfide bonds enhancing their surface hydrophobicity (Dangaran et al., 2009). Denaturation of egg white is done to prepare the films and coatings made from egg albumen. It is done by regulating the pH around 10.5–11.5 and temperature at 40°C for 30 minutes. The temperature- and pH-induced denaturation leads to the unfolding of albumen proteins and exposing the internal sulfhydryl groups, which effects the formation of disulfide bond and increases the surface hydrophobicity.

17.3.3.1.4 Meat Proteins

There are generally three types of meat proteins: sarcoplasmic, stromal, and myofibrillar. Sarcoplasmic proteins consist of enzymes, myoglobulin, and cytoplasmic proteins; stromal proteins include collagen and elastin, and actin, myosin, and tropomyosin constitute the myofibrillar proteins. Of these, stromal and myofibrillar proteins are commonly used to develop the films and coatings. Collagen is extracted from the waste of the meat processing (skin, tendons, connective tissue, bones, and vascular system). It is fibrous in nature and is composed of three parallel alpha-chains that associate into a triple-stranded super helical structure (Haug et al., 2004). For the formation of gelatin, collagen is denatured by exposing it to mild pH conditions, and on cooling it again forms the triple helix structure and leads to the formation of gelatin (Badii and Howell, 2006). Gelatin is used as a thickener and texturizer and imparts better gelling properties. It exhibits good foaming properties, which makes it a good film-forming source material. Myosin and actin are a part of myofibrillar proteins and are soluble in salt solutions. Myofibrillar proteins are known to hold better applications in the food industry as they can be used in the formulation of food coatings.

17.3.3.2 Plant Proteins

17.3.3.2.1 Wheat Proteins

Gluten is the water-insoluble wheat protein consisting of globular proteins. It is extensively used for the preparation of edible films as it is easily and cheaply available and an abundant source (Kaushik et al., 2015). Gluten has the potential property of cohesiveness and elasticity, which gives integrity to the dough and allows the formation of film. Wheat gluten consists of two fractions, prolamin (gliadin) and glutelin (glutenin). Gliadin has solubility in 70% ethanol and gliadin is not soluble (Liu et al., 2015). Purity of wheat gluten significantly effects the appearance of film and its mechanical properties, which means gluten has a higher degree of purity resulting in stronger and transparent films. During the development of wheat gluten proteins, the native disulfide bonds break when the film-forming solutions are heated up and new disulfide bonds are formed when the films are dried. This phenomenon is considered to be very important for the formulation of films using wheat proteins along with the formation of hydrogen and hydrophobic bonds. The film flexibility can be enhanced using the appropriate plasticizer such as glycerin, wherein the low quantity of glycerol results in films with lower permeability against water vapor and oxygen but with improved tensile strength. Sorbitol can also be added as a plasticizer to enhance the film's flexibility, which has to be developed from wheat gluten, decreasing its film strength, elasticity, and barrier properties. The properties of films based on wheat proteins can be improved using glutaraldehyde, which acts as a cross-linking agent (Mohamed et al., 2019).

17.3.3.2.2 Soy Proteins

Soy proteins are extracted from soybeans, which are easily available in the market. Soy proteins have higher protein contents (38–44%) relative to cereal protein (8–15%) where soy flour (56% protein), soy concentrates (65% protein), and soy isolate (90% protein) are the major soya protein sources used for the formulation of food coatings (Cho and Rhee, 2004). Edible films developed from soy proteins possess excellent film-forming and functional properties. The higher molecular weights of soy proteins result in increased tensile strength and elongation properties. Soy proteins have polar and non-polar ends with strong intramolecular and intermolecular associations like dipole-dipole interactions, hydrogen bonding, and hydrophobic interactions. The side chains of soy protein molecules possess a higher charge and interactions between molecules resulting in the increase of film stiffness and tensile strength (Guerrero et al., 2011). The film or coatings can be developed from soy proteins following two steps: first, the protein solution is heated to break down the structure of proteins so that the disulfide bonds break, and secondly, sulfhydryl groups and hydrophobic bonds are exposed for the formation of a new disulfide, and hydrophobic and hydrogen bonds. On the other hand, films and coatings developed from soy proteins have weaker moisture barrier properties as soy protein is hydrophilic in nature. The addition of an appropriate amount of hydrophilic plasticizer, such as lipids, into the film-forming solution improves the barrier properties against water vapors. (Tian et al., 2018).Cross-linking of the protein chains through physical, chemical, and enzymatic treatments is another method to improve the film-forming characteristics of the soy protein-formulated food coatings. Films developed using soy proteins are proved to be much better than the films developed from lipids and polysaccharides in terms of their gas barrier properties.

17.3.3.2.3 Corn Zein

Zein accounts for 45–50% of corn proteins. They are easily soluble in alcohol but not in water because non-polar amino acids are present, which are responsible for imparting better water vapor barrier properties into the zein-based films (Soliman et al., 2009). Food coatings developed from zein are brittle in nature, which can be improved with the incorporation of polyols and fatty acids as plasticizers. Generally, films and coatings developed from zein possess moderate moisture and oxygen barrier properties along with better mechanical properties. Zein has variable properties compared with other biopolymer sources as it has a high ratio of non-polar amino acids and a low percentage of basic and acidic amino acids, which makes zein insoluble in water and barrier properties of such films (Shukla and Chheryan, 2001). Zein films are known to be glossy, tough, and grease proof with weak permeation to water vapor; hence, this makes them commercially available as a coating material for medical tablets and as a biodegradable packaging material.

- *Lipid-based coatings:* Hydrophobic substances used to prepare lipid-based films and coatings are categorized as waxes, non-hydrogenated vegetable oils, and fatty alcohols. Wax is a commonly used term for various non-polar substances and is preferred as a source material to formulate films and coatings as they can be kneaded at room temperature, and are brittle, crystalline, less viscous, and translucent to opaque in nature (Debeaufort and Voilley, 2009). Wax is an ester of a long chain aliphatic acid with an aliphatic alcoholic group. Their higher hydrophobicity is mainly due to their solubility in organic solvents as they have no polar components, hence, they cannot interact with water molecules exhibiting the hydrophobicity. Waxes may be natural or synthetic in origin; natural waxes include beeswax, candelilla wax, jojoba beeswax, carnauba wax, etc. (Aydin et al., 2017). It is an important consideration that the lipid-based edible waxes used as films or coatings must be compatible with the food with which they interact and should have a good sensory profile. Vegetable oils like palm oil, soybean oil, fish oil, and linseed oil; shortenings; and margarines are used to develop a resistant layer over the film or coating as they offer the benefit of refining the moisture

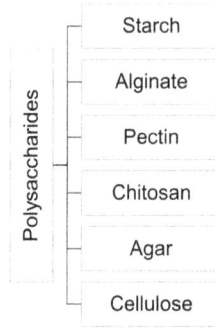

Figure 17.3 Materials used for polysaccharide-based (edible) films and coatings.

barrier properties of lipid-based edible films and coatings (Rhim and Shellhammer, 2005). Various emulsifiers and surfactants are incorporated into the coating medium to improve the surface adherence among the coating material and the surface of the food, which has to be coated. Lipid-based food coatings are commercially applied to coat various foods like bakery items (cakes, pizza, muffins, and biscuits) and minimally processed fruits and vegetables.

• *Polysaccharide-based coatings – types of polysaccharide-based edible films:* Most utilized sources of polysaccharide-based edible coatings are shown in Figure 17.3.

17.3.3.2.4 Chitin

Chitin is naturally present in abundance in cell walls of fungi and many other biological sources. Alkaline deacetylation of chitin results in the formation of chitosan, which is a copolymer of β-(1–4)-2-acetamido-D-glucose and β-(1–4)-2-amino-D-glucose units and possesses antimicrobial as well as film-forming properties (Nisha et al., 2016). Chitin is the most favorable biopolymer and serves as an alternate for synthetic food packages. The antimicrobial properties of chitosan-based films and coatings rely on their cationite properties, which exist in acidic media. Permeabilization and destruction of external membranes of bacterial cell walls occur at a pH value below pKa of chitosan protonated amino groups, which bind to the negatively charged carboxyl groups of lipopolysaccharides and peptidoglycans (de Azeredo et al., 2010). Chitosans are recognized as a potential source for food coatings as they are not toxic in nature, they have biocidal activity, and gas barrier properties. It also helps to limit senescence by checking the respiration rate and transpiration. Several characteristics of chitosan-based food coatings can be refined by adding glycerol or sorbitol as plasticizers. Various studies have shown the efficacy of chitosan-based films and coatings to enhance the shelf life of many perishable food commodities such as apples, bell peppers, pears, papaya, strawberries, cucumbers, plums, tomatoes, mangoes, meat, fish, etc. (Elsabee and Abdou, 2013).

17.3.3.2.5 Starch

Starch is one of the polysaccharide consisting of amylose and amylopectin, out of which amylose is known to have potential for film formation as it possess linear structure. The abundant bioavailability of starch, its thermoplasticity, and low cost makes it convenient to be a good source from which edible food coatings can be developed. Water is identified to reduce the tensile strength by making the intermolecular forces weaker. Plasticizers including several polyols like glycerol, sorbitol, xylitol, etc., are incorporated into the film-forming solution to restrict the brittleness of the developed films and coatings. Studies show that glycerol has proved to be a better plasticizer as it improves the physical and mechanical properties of the food coatings, whereas sorbitol is least effective due to the difference in their phase separation and crystallization properties.

Starch is not easily miscible in cold water; however, mixing it in hot water causes the crystalline structure to become distorted, resulting in partial solubility of the starch. Heating of starch in hot water in excess of 65–90°C results in the formation of hydrogen bonding and an irreversible gelatinization (Jiménez et al., 2012). Starch gelatinization is mandatory for film formation, which disrupts the amylopectin and liberates amylose as water enters inside the starch granules promoting the melting of starch contents. The amorphous structure of starch is mainly responsible for the poor mechanical properties and brittleness of the developed coatings, which can be improved by adding various plasticizers or by blending starch with another biopolymer or synthetic polymers.

17.3.3.2.6 Cellulose

Cellulose is a biodegradable biopolymer and is available in ample amounts in nature like in the cell walls of most plants and in some species of fungi, algae, and a few gram-negative bacteria. It is considered to be the most widely used source for edible film and coatings due to its renewability, low cost, biocompatibility, and chemical stability, and it is non-toxic, tasteless, odorless, and environmentally friendly (Leite et al., 2017). It involves units of D-glucose, which are linked with β-1,4 glycoside bonds. Methylcellulose (MC), hydroxypropyl methylcellulose (HPMC), and carboxy methylcellulose (CMC) are the common derivatives of cellulose mainly used for film and coating formulations (Tibolla et al., 2018). Cellophane is another cellulose derivative of that is developed from regenerated cellulose. It is known to be an interesting product as it is thin, and transparent, and has good oil, water, and gas barrier properties, making it suitable for food packaging applications mainly for fresh fruits and vegetables and bakery products. HPMC- and MC-based coatings are efficient in reducing the moisture loss and fat uptake in case of deep fried foods (Collazo-Bigliardi et al., 2018). Various antioxidant and antimicrobial substances are blended into the cellulose casting solutions to improve the functionality of developed films and coatings.

17.3.3.2.7 Agar

Many species of red algae (Rhodophyceae) are major sources from which to extract agar. Agar is derived from agarose, which is the gelling part of agar and is composed of repeating units of anhydro-α-L-galactose and β-D-galactose. It is a hydrocolloid involving a mixture of agarose and agaropectin and is able to form reversible gels by alternate cooling and heating cycles (Arham et al., 2016). Agar is applicable in most food packaging systems as a coating material because it has good gelling ability, good functional properties, and is stable at acidic pH conditions and elevated temperatures compared with other gelling systems. The films and coatings prepared from agar are known to be clear, strong, and flexible and hold better physical and mechanical strength, although they are not water soluble. Agar solution generally holds a tendency to gel at 90–103°C, hence, some methods such as solution or casting need a higher temperature for setting of agarose.

17.3.3.2.8 Pectin

Pectin constitutes one-third of the cell wall of certain fruit peels on a dry basis. It is an amorphous carbohydrate with high molecular weight and is mostly available in citrus peels and apples for extraction of pectin. It possess properties like thickening as well as emulsification to impose the ability of gel solidification (Valdés et al., 2015). Pectin polysaccharide is comprised of homogalacturonans, rhamnogalacturonan-I, and rhamnogalacturonan-II, and homogalacturonan is the main part of pectin. Carboxyl ends of units of galacturonic acid are esterified with methanol and the degree of esterification for pectin is classified as high methoxy pectin and low methoxy pectin

(De Cindio et al., 2016). Pectins have been proposed to be used as food coatings as they have good barrier properties against oxygen, oil, and aroma along with improved mechanical properties. Although they are hydrophobic in nature and not potent water barriers, several studies have shown that their application over fresh and minimally processed foods helped in prolonging their shelf life without adversely affecting their sensory attributes. Various plasticizers and emulsifiers are added to improve their mechanical properties and to increase their stability and better adhesion, respectively. Glycerol, acetylated monoglycerides, polyethylene, glycol, and sucrose are known plasticizers that cause significant improvement in film barrier properties and reduce their brittleness (Espitia et al., 2013).

17.3.3.2.9 Alginates

Brown seaweed is the major source for extraction of alginates and mainly the *Laminaria* sp. It is available in gel form in the form of calcium, sodium, and magnesium salts. An acidification is carried out to convert alginate into alginic acid following treatment with an alkali to obtain sodium alginate, which is water soluble (Cazon et al., 2017). The main important structural parameters of alginates include molecular weight, which is ultimately connected with the rheological behavior of alginate solutions. Edible films based on alginate are transparent and water soluble and the water solubility of these films can be enhanced by adding Ca^{2+} to cross-link the alginate, allowing the intermolecular interactions. Although alginates have very good film-forming properties, they still need to be plasticized with glycerol or sorbitol to reduce brittleness (Donati et al., 2005). Addition of glycerol and sorbitol has been reported to modify the water vapor and oxygen permeability of films developed from alginates. Several studies have reported that alginate food coatings improved the shelf life of sliced carrots for 5–7 days and nano-reinforcement agents like whisker cellulose helps in enhancing the mechanical strength, including its tensile strength (Li et al., 2017).

17.3.3.3 Nanoemulsion-Based Coatings

The development of foodborne pathogens on food surfaces (mostly fresh fruits or vegetables) is the leading reason for their short shelf life and several types of illnesses caused by consumption of such foods. Antimicrobial food packages have emerged as an alternative to reduce the issues of shelf life and health problems. There are many natural antimicrobials, antioxidants, and texture enhancers to improve the shelf life and overall quality of such perishable foods. The use of emulsion-based edible food coatings is the best known method to prolong the shelf stability of fresh foods as they help in associating the properties of lipophilic and hydrophilic functional compounds. Nanoemulsions are known to encapsulate various antioxidants, antimicrobials, nutraceuticals, and many other additives like color and flavor as nanocarriers (McClements and Rao, 2011; Salvia-Trujillo et al., 2017). They improve various interactions among the functional compounds that help improve their availability, metabolism, and absorption. Food-grade lipids such as corn oil, coconut oil, sunflower oil, and medium chain triglycerides serve as the source for lipophilic functional molecules to prepare the nanoemulsion-based edible coatings.

Plant-based natural essential oils such as lemongrass, oregano, clove, thyme, cinnamon, etc., possess strong antimicrobial properties against foodborne pathogens, although they cannot be used directly over the food surfaces as they are hydrophobic and volatile in nature. The development of nanoemulsions consisting of these essential oils can overcome these limitations and can be used in the food coatings to extend the shelf stability of fresh food commodities. Many essential oils are known to inhibit the activity and growth of foodborne bacteria such as *Listeria monocytogenes*, *Salmonella typhimurium*, *Escherichia coli* (O157:H7), and *Staphylococcus aureus*. Many studies have been conducted to evaluate the efficacy of nanoemulsion formulations as a coating material by dipping foods into them and concluded with positive results.

Nanoemulsions encapsulated with natural antioxidants (alpha-tocopherol and carotenoids) are freeze-dried and coated over the food surface as a nanometer-thin-sized layer to preserve them. This coating helps in retarding the gas and the fluid exchange among outer environments. The developed nanoedible coatings using nanoemulsions aid in enhancing the organoleptic and overall quality of the coated food products (Malnati et al., 2019).

Loss of texture and firmness during the storage period of fresh food commodities is another factor responsible for the decline in consumer acceptability. Food coatings based on nanoemulsions exhibit positive results as a texture modifier when used for fresh-cut fruits and vegetables. Xanthan gum, carnauba wax, pectin, and chitosan have been used as texture enhancers and found to be effective in maintaining the firmness and cell wall integrity of fresh fruits like apples, guava, and beans. Zambrano-Zaragoza et al. (2013) concluded in one of his studies that the shelf life of guava can be enhanced up to 30 days by glazing them with a blend of xanthan gum and carnauba wax-based nanoemulsion, retaining its texture when stored at 10°C.

17.4 APPLICATIONS OF EDIBLE FILM/COATINGS

Edible coatings have been considered as a traditional preservation technique for centuries, but this concept is new in the food processing industry and developed because of its potential to be utilized in the development of many future applications as per the need of ever-demanding customers. Many applications of films and coatings have been studied, and following is a summary of applications of edible coatings and films that have been successfully utilized (Figure 17.4).

17.4.1 Fruits and Vegetables

Fruits and vegetables contain 80–90% of moisture content and are highly perishable. Post-harvest losses are the major losses of fresh fruits and vegetables that occur between harvest and consumption.

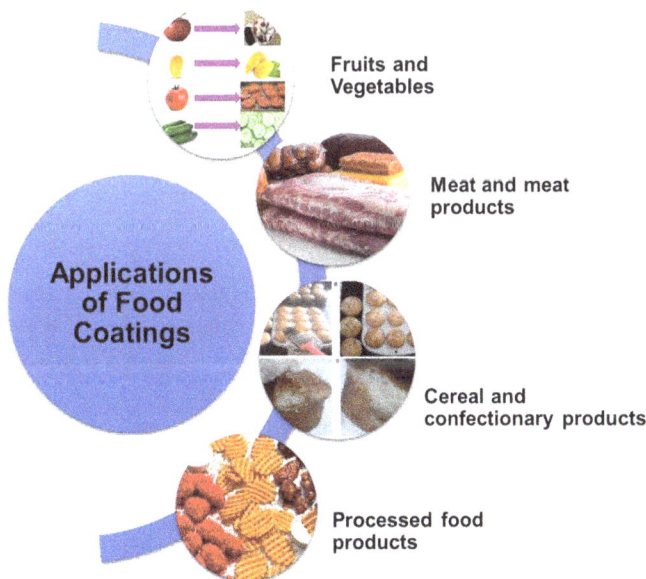

Figure 17.4 Areas of applications of food coatings.

The respiration rate increases on harvesting, which leads to metabolic loss, maturity, and senescence of crop. These changes depend on internal (cultivar, species, growth state) and external factors (Kluge et al., 2002). In fresh-cut fruits and vegetables, wounding of the fruit flesh can lead to disruption of texture, browning, and off-flavor production that ultimately affect the marketability and consumer preference. Therefore, CAS and MAS are presently used to control the post-harvest losses of fruit and vegetables. Edible coatings are a good alternative to these techniques, which replace natural waxy coatings of fruits and vegetables and provide a barrier to moisture, gas, and volatiles, thus decreasing respiration and senescence. This environmentally friendly coating effectively increases the shelf stability of fruits and vegetables while preserving color, texture, flavor, and moisture and improving mechanical handling properties of individual fruits/vegetables. Many functional ingredients such as antioxidants, nutrients, colors, flavors, anti-browning and anti-microbial compounds can be loaded within the edible coatings that can enhance product food quality and functionality (Ricardo et al., 2012). The edible coating material can be coated over the food surface by spraying, dipping, or glazing as a thin layer over the surface. A wide range of formulations have been developed that are used for fresh and minimally processed fruits and vegetables (Table 17.2). Carrageenan-coated cherries showed lower water loss (30%) compared with the control (Larotonda, 2007). Also, carrageenan-coated crops retained their textural integrity and glossy appearance during storage at refrigerated temperature (5°C).

Common coating-forming materials, i.e., lipids, polysaccharides, proteins, and resin can be used alone or in combination. The base material to be utilized for coating material is generally chosen on the basis of water solubility, hydrophobicity or hydrophilicity, mechanical properties, optical properties, barrier properties, cost, availability, ease in formation, and sensory properties. These coating characteristics depend on the type of material used as a structural matrix, the conditions under which coatings are prepared, and the type and concentration of additives used.(Dhall, 2013). The application and export of coated fruits and vegetables are governed by the legislation of each country.

17.4.2 Processed Foods

Edible coatings are applied for processed foods for the same reasons as for fruits and vegetables, i.e., to stabilize the food and thus extend the shelf life by reducing moisture loss, restricting oxygen entrance, sealing flavor volatiles, and retarding discoloration and microbial growth. Water activity and oxygen level are essential controlling factors in many processed foods, which directly affect texture and flavor. While packaging is a widely used practice to accomplish these objectives, edible coatings provide an alternative with several advantages. Plastic waste can be avoided with environmentally friendly edible coatings that continue to provide protection after opening the packaging. During frying, these coatings may reduce oil absorption by food or oil/shortening leakage from fried products. Edible coatings improve the texture and appearance of the product by reducing shrinkage during and after cooking. Further, edible coatings can act as carriers of bioactive components and preservatives, thus improving the functionality and shelf life of processed foods. Edible coatings are used in a variety of meat, poultry, seafood, nut, cereal-based, and confectionery products.

17.4.2.1 Meat, Poultry, and Seafood Products

Conventional methods were used in the last centuries to prevent shrinkage, microbial contamination, and surface discoloration of meat, which involves plastic wrapping and chemical applications. These methods are associated with several disadvantages, e.g., enhancement of cooling time and surface microbial counts due to plastic overwrapping. Quality of these products can be improved by using edible coatings as they retard loss of moisture, prevent fat oxidation, and reduce discoloration

Table 17.2   Application of Edible Coatings on Fresh Fruits and Vegetables

S. No.	Fruits and Vegetables	Edible Coating Used	References
1.	Apple (fresh cut)	Sodium alginate, gellan gum, and sunflower oil Whey protein concentrate, WPI, and whey protein peptide (WPP)	Rojas-Grau et al. (2008) Sonti (2003)
2.	Blueberry	Carboxy methylcellulose, sodium alginate, chitosan, monoglycerides and calcium caseinate	Duan et al. (2011)
3.	Tomato	*Aloe vera* gel	Athmaselvi et al. (2013)
4.	Grapes	*Aloe vera* gel	Chauhan et al. (2014a)
5.	Mango	Chitosan, *Aloe vera* Tapioca flour, sago flour, and oy protein Chitosan	Chauhan et al. (2014b) Salleh (2013) Abbasi et al. (2009)
6.	Apple	Neem oil, marigold flower extract, guar gum, and *Aloe vera* Soybean gum, paraffin wax, jojoba oil, and gum Arabic Whey protein concentrate, soy protein isolate, alginate, and carrageenan	Chauhan et al. (2014c) El-Anany et al. (2009) Ghavidel et al. (2013)
7.	Pineapple (fresh cut)	Pectin and alginate	Mantilla (2012)
8.	Strawberry	Sodium alginate and calcium alginate gel Gum Arabic and Arjan gum and psyllium mucilage Chitosan and tragacanth Linseed mucilage extract and chitosan Pectin, PVA, starch, soy protein, and gluten	Moadenia et al. (2010) Yarahmdi et al. (2014) Cobrera et al. (2010) Yossef (2014)
9.	Cantaloupe	*Aloe vera* gel Pectin and chitosan	Yulianingish et al. (2013) Garpas (2011)
10.	Banana	PVA, CMC, tannin	Magdy et al. (2014)
11.	Guava (fresh cut)	*Aloe vera* juice	Nasution et al. (2015)
12.	Pistachio	Chitosan and *Aloe vera* gel	Vanaei et al. (2014)
13.	Orange, tomato, mango, papaya, guava, and mushroom	Vegetable oil, Cellulose gum, and emulsifier	Nesperos-Carriedo et al. (1992)
14.	Potato	Chitosan, whey protein concentrate, and coconut oil	Saha et al. (2014)
15.	Papaya	Chitosan, *Aloe vera*, and papaya leaf extract	Marpudi et al. (2011)
16.	Plum	CMC and pectin	Panahirad et al. (2015)
17.	Sapota	*Aloe vera* juice	Padmja and Basco (2014)
18.	Pumpkin, carrot, radish, cantaloupe, and cucumber	Chitosan and calcium salt	Suwannark et al. (2015)
19.	Cucumber	Gum Arabic powder Beeswax and high-methoxy pectin Sorbitol and emulsifier Chitosan and *Aloe vera* Paraffin wax Sodium alginate, starch, glucose, sucrose, chitosan, sorbitol, and PVA	Junaimi et al. (2012) Moalemiyan and Ramaswamy (2012) Adetunji et al. (2014) Bahnasawy and Khater (2014) Zhang et al. (2004)
20.	Apple, potato, and cucumber	Potato starch, pea starch, and guar gum	Ghader et al. (2011)

Source: From Raghav et al. (2016).

of meat products. Edible coatings enhance the appearance of the product by sealing volatile flavors, eliminating dripping loss, acting as carriers of food additives/preservatives, and also reducing oil uptake during frying. Various types of edible coatings have been used including polysaccharides (carbohydrates and gums), proteins, lipids, or their combinations. The use of natural and safe preservatives, which have high antimicrobial, antioxidant, and functional properties, are mostly preferred for the preparation of coatings for meat, poultry, and fish products (Alsaggaf et al., 2017). Limited studies of edible coatings on red meats were reported, but their functionality to improve the quality of chicken and fish products has been well documented (Vital et al., 2016).

Polysaccharides including chitosan, alginates, carrageenan, agar, cellulose, starch, pectin, etc., are frequently used to develop edible coatings (Yousefi et al., 2018). These are usually colorless, non-greasy, and lower calories, which make them desirable coating material for meat, chicken, and fish products. They prevent dehydration, surface discoloration, and rancidity (Hassan et al., 2018). Chitosan has the largest application in bioactive edible coatings and films in fresh and processed meat, poultry, and seafood because of good antimicrobial and antioxidant activities (Cardoso et al., 2016). In case of steaks, a combination of chitosan-gelatin coatings reduced the purge losses and prevented oxidation by acting as an oxygen barrier when compared with uncoated steaks (Cardoso et al., 2016). Natural antimicrobial compounds (cinnamaldehyde and carvacrol) incorporated edible apple films that when applied to wrapped chicken breasts and ham surfaces exhibited better activity of carvacrol against microbes such as *Salmonella enteritidis* and *L. monocytogenes* compared with cinnamaldehyde (Ravishankar et al., 2009).

Protein coating materials used for meat, poultry, fish, and seafood products include various plant and animal protein sources like zein, whey proteins, collagen, gelatin, etc. Edible films formulated using whey proteins loaded with natural antioxidant extracts of laurel or sage significantly prohibited the oxidation of cooked meatballs when stored at temperature (−18°C) for a period of 2 months. Phenolic compounds present in laurel or sage extracts contributed to the oxygen barrier characteristics of whey protein-based edible films (Akcan et al., 2017). Refrigerated storage of raw beef meat surface coated with pectin-fish gelatin edible films (incorporating natural antioxidants present in olives) was studied and results indicated reduced lipid oxidation of products as measured by thiobarbituric acid reactive substances (TBARS) (Bermudez-Oria et al., 2019). Lipids are generally incorporated in the preparation of composite films, which are made from protein or polysaccharide materials to impart flexibility and impart moisture barrier properties because lipid-based films and coatings are devoid of structural integrity, hence, they are brittle in nature.

17.4.3 Nuts, Cereals, and Confectionery Products

Nuts are highly susceptible to oxidation and development of off-flavors due to their higher oil contents. Edible coatings may be used as an effective barrier that separates the nuts from atmospheric oxygen, thus preventing lipid oxidation. In addition coatings provide better adhesion of salt, spice mixtures, colors, flavors, and other additives. Studies showed that the emulsion-based coatings are more effective in preventing rancidity in pine nuts and walnuts (Mehyar et al., 2012). Similar results were also observed for pistachio kernels when coated with WPC–olive oil emulsion coatings (Javanmard, 2008). Nuts and almonds are coated with dextrins to preserve their texture and sensory attributes during storage (Murray et al., 1973).

Cereal-based bakery and extruded products are low moisture products and have crispy textures. Higher humidity causes an increase in moisture content during storage, which leads to loss of crispness and undesirable changes due to softness in texture. Crackers coated with emulsion coatings of corn starch, MC, and soybean oil stored at different relative humidity conditions (65, 75, and 85%) showed reduction in hydration kinetics at higher relative humidity (Bravin et al., 2006). Application of edible coatings based on chitosan or caseinates on partially dehydrated pineapple to be used in fruit cereal products was described by Talens et al. (2012). Emulsion coatings of sodium caseinate/beeswax/oleic acid, chitosan/oleic acid, and sodium caseinate/calcium caseinate/beeswax/oleic acid were analyzed. These layers protected pineapple fruits to avoid the cereals to reach the critical moisture content during the storage period. Results showed extension of shelf life of the pineapple-cereal-based product by using caseinate-based emulsion coatings. Edible coatings generally used for bakery products are generally chocolate or sugar based, and various preservatives can be added to extend the storage life and product safety.

Confectionery products are primarily coated to enhance the visual appearance and consumer appeal, e.g., shellac-based edible coatings provide a shiny coat on the surface of chocolate-paned products. Other functions of shellac coating include preventing cocoa butter from coming to the product's surface, which could stain the packaging, and acting as an oil barrier and thus preventing fat bloom of the product. Casein-based coatings comprised of vegetable oil and sodium caseinate were applied to coat chocolate products (Rice, 1994). Edible coatings of ready-to-eat food could enhance the safety, quality, and shelf life of products.

17.4.4 Flavor Encapsulation

Food coatings are successfully used for the encapsulation of flavors as these permit the production of dry and free-flowing flavor, act as a barrier for interaction with the food, prevent oxidative reactions, confinement during storage, and control release of flavors and thus enhance consumer satisfaction. The degree of encapsulation and release of flavor compounds mainly depends on the composition of the coating or film and the process used to form the coating around the flavoring. Various processes used for flavor encapsulation include the following:

- *Spray drying:* It is the most commonly used (90% usage) and cost-effective method. Spray drying involves the dispersion of the core particle into aqueous media and finally atomized in a during chamber (Santos et al., 2018).
- *Spray chilling:* This method involves the dispersion of core particle into the coating medium and the carrier particle is sprayed under a cold environment to solidify it around the core particle. This method is generally used to protect water-soluble cores; hence it is suitable for temperature-sensitive cores (Kiyomi Okuro et al., 2013).
- *Extrusion:* This is the preferred method to encapsulate the volatile flavors in glassy matrices. The dispersion is forced to pass through a die at high temperature and pressure into a bath for solidification of the particles (Risch, 1988).
- *Fluidized bed coating:* In this method, the particles remain suspended in air and the coating is applied over the surface, hence, it is used for applications in which very fine control of the core release properties is required (Dewettinck and Huyghebaert, 1999).
- *Inclusion complex formation:* This method is used for encapsulation of flavors and lipophilic nutrients (Martina et al., 2013).
- *Coacervation:* Biopolymers with opposite charges associate among themselves and lead to phase separation and formation of coacervates. They have the ability to entrap the core particles, and can be used in the encapsulation of flavors and many nutrients (Koupantsis et al., 2014).

The selection of wall material used for the encapsulation process is very important as it influences the encapsulation efficiency and stability of the capsule. The coating material should not react with other flavoring compounds and be able to hold the flavor within the capsule and protect flavor against adverse conditions during storage. Also, the material should be devoid of any unwanted and undesirable taste if it has to be used for any food application. Different coating materials used for encapsulation of functional food additives (Poornima and Sinthya, 2017) include the following:

- *Carbohydrates (starch, maltodextrins, chitosan, corn syrup solids, dextran, modified starch, cyclodextrins)*: This material is used for spray and freeze drying, extrusion, coacervation, and inclusion complexation.

- *Cellulose (CMC, MC, ethyl cellulose, cellulose acetate-phthalate, cellulose acetate butylate phthalate):* This is used for coacervation, spray-drying, and edible films.
- *Gums (gum acacia, agar, sodium alginate, carrageenan):* These are used or spray-drying and syringe method (gel beads).
- *Lipids (wax, paraffin, beeswax, diacylglycerols, oils, fats):* These are used for emulsion, liposomes, and film formation.
- *Proteins (gluten, casein, gelatin, albumin, peptides):* These are used for emulsion and spray-drying.

These materials are chosen on the basis of compatibility with the process, flavor used, cost, stability, functionality in food application, and legal and religious constraints. For example, for all-natural food flavoring, natural wall material such as gum acacia or pectin can be used. For water-soluble flavoring, a non-emulsifying wall material such as maltodextrin can be used. For water-insoluble flavoring, emulsion can be made using an emulsifying food polymer such as modified starch, gum acacia, or protein. Gelatin is not acceptable if the encapsulated flavoring is to be utilized in a kosher or halal product. Generally, water-soluble carbohydrates are the most commonly used wall materials in flavor encapsulation as each has a different functionality with respect to different encapsulation processes. Proteins (gelatin, fish, and dairy proteins) are minimally used in the coacervation process.

The emulsifying properties of coatings depend on the type of flavoring encapsulated, encapsulation process, and final food application. Water-soluble flavoring, generally a mixture of alcohol or propylene glycol (at around 0.1% usage levels), does not require emulsions. An emulsifying matrix is required in the system in case insoluble flavors are used for encapsulation. Flavor losses can be minimized using emulsification during the encapsulation process (spray drying and extrusion process). Various studies recorded that good quality emulsion substantially improved the retention of water-insoluble flavorings during the encapsulation process (Reineccius, 1998; Soottitantawat et al., 2004). Also, it is not always necessary that small average particle sizes for more soluble flavors improve the retention. Dry products such as savory mixes, bakery products, and confectionery goods do not require the emulsification properties of edible films or coatings but require dry flavorings in substantial portions, for example, in dry beverage mixes. Encapsulated flavor based on maltodextrins and corn syrup solids cannot be used for dry beverage flavor application due to their high viscosity and lack of emulsion stability, which leads to the requirement of a secondary emulsifier. Modified starch or gum acacia are excellent emulsifiers that may be used in combination with maltodextrins or corn syrup solid flavor carrier. These secondary emulsifiers are not commonly used with spray-dried flavors; however, they provide stability to extruded flavorings. Use of modified starches in flavor encapsulation provides a stable flavor emulsion.

For the retention of flavor compounds during drying process, the ability of an edible coating to hold flavor compounds is important. More volatile components get lost during the encapsulation process resulting in dry flavor. Loss of volatiles such as in the spray dryer during exit through the drier or into the environment is also another area of concern. These volatiles must be removed from the environment using a costly scrubbing process, which leads to high processing cost and low product quality in case of poor retention encapsulating materials. As modified food starches act as excellent emulsifiers and yield good emulsions, their emulsion quality significantly affects the flavor retention during drying (Baranauskiene et al., 2007). Higher levels of infeed solids of modified starches and gum acacia contributed the best flavor retention property (Reineccius et al., 2003). Simple sugars and their alcohols, maltodextrins, and corn syrup solids have poor emulsification properties at high levels of infeed solids, which leads to poor flavor retention during drying. During drying the solids content of the infeed slurry is the major factor for flavor retention, which is a totally different ranking of carriers and retention can be obtained if the experimental design uses the carriers at constant viscosity. A retention of water-soluble components-volatile model flavor system was evaluated by Dronen (2004) at constant viscosity basis during spray drying. Carrageenans, in the form of beads, gels, and films were found to be more efficient in retention of polar aroma compounds when compared with the usual lipid-based matrices. Carrageenan films are capable of reducing the volatility of encapsulated flavors (Chakraborty, 2017) and preserving the volatile flavors during film-process formation, and gradually releasing these flavors as required with time (Marcuzzo et al., 2010).

Similar to flavor retention, flavor release from the encapsulated material is also very important. Depending on the processing conditions of encapsulation and final food product application, controlled and slow or delayed release of flavor can be required, e.g., delayed release in case of thermal processing prevents less flavor loss due to heat processing and rapid release in case of dry beverage mix during reconstitution. Extrusion, coacervation, and inclusion complexation processes directly provide controlled release properties to flavorings. As spray-dried particles are water soluble, application of secondary coatings (e.g., with fat, oils, and shellac) may be required to impart controlled flavor release properties. Because application of secondary coatings is costly and problematic, use of appropriate encapsulation methods is suggested to accomplish controlled release of flavoring constituents.

Oxidation is the most common problem encountered during storage of flavors. Encapsulated coatings/films provide varying protection to flavorings from oxygen, e.g., the higher the dextrose equivalent (DE) in maltodextrin coating, the better the protection against oxidation (Arvisenet et al., 2002). Higher DE materials are more hygroscopic and lead to poor drying and poor flavor retention. Therefore, optimum emulsifying materials can be blended with higher DE maltodextrins, corn syrup solids, or simple sugars to impart required emulsifying properties and oxidative stability at lower product cost. Because modified starches provide excellent flavor retention and emulsion stability to encapsulated flavorings but are not effective against oxidation, inexpensive starch hydrolysates may be added for desired quality attributes (Reineccius et al., 2003).

17.4.5 Delivery of Flavors and Active Ingredients

Food coatings can be used to incorporate flavors and active ingredients to the matrix such as antioxidants, antimicrobial agents, anti-browning components, colors, and bioactive compounds or nutraceuticals to enhance the functionality. Active compounds can be incorporated directly into the polymer matrix or can be encapsulated to better protect their activity and properties. Pullulan-based Japanese films are commercially available in a variety of colors, spices, and seasonings (Guilbert and Gontard, 2005), and can be used to enhance sensory properties of food in addition to their excellent oxygen barrier properties. Plasticizers are generally added for preparation of food coatings as the structure of coatings (especially based on polysaccharide- and protein-based coatings) is often brittle and stiff because of extensive interactions between polymer components (Krochta, 2002). Glycerol, sucrose, acetylated monoglyceride, and PEG are commonly used plasticizers that decrease glass transition temperature of the polymers and increase flexibility of the coating (Guilbert and Gontard, 1995). Plasticizers attract water molecules because of their hygroscopic nature, therefore providing flexibility. Emulsifiers are surface active agents that are used to reduce the surface tension between two different phases (water-oil or water-air interfaces). Because of their amphiphilic nature, their addition increases the hydrophilicity and coatability of the food surface, thus improving the oxygen barrier properties of coatings.

Several studies have reported the incorporation of nutraceuticals such as minerals, vitamins, and fatty acids into edible food coating formulations to enhance the nutritional value of food. The concentration of nutrient must be carefully added because it affects the basic functionality of the coating such as barrier and mechanical properties. Park and Zhao (2004) studied the water vapor barrier property of the chitosan-based film, which improved with an increased concentration of mineral (5–20% w/v zinc lactate) or vitamin E in the film matrix. Tensile strength of coatings/film was also reported to be affected by the incorporation of high concentrations of calcium or vitamin E, irrespective of the change in other mechanical properties, such as film elongation, puncture strength, and puncture deformation.

Recent studies of incorporation of probiotics into edible good coatings/films and coatings show a new strategy toward biopreservation and healthier foods. In most of countries, species of *Lactobacillus* and *Bifidobacterium* are most commonly used as probiotics, whereas *E. coli*, some *Bacillus* species, and the yeast *Saccharomyces boulardii* are also used successfully (Pavli et al., 2018).

Various antioxidant and antimicrobial substances such as ascorbic acid, potassium sorbate, benzoic acid, sodium benzoate, lysozyme, calcium ascorbate, and calcium lactate, and herbal preservatives such as essential oils from plants are utilized in food coatings to increase the shelf life of food products. Essential oils and their components extracted from plants effectively act as natural preservatives as antimicrobial substances and are therefore studied widely. An additional benefit of using essential oils in food coatings is their compatibility with food sensory properties. Ayala-Zavala et al. (2012) developed pectin and cinnamon leaf oil (36.1 g/L)-based emulsified films. The application of these antioxidant and antibacterial coatings to fresh-cut peaches resulted in decreased bacterial growth against *S. aureus*, *L. monocytogenes*, and *E. coli* (O157:H7) and increased antioxidant level and sensory acceptability. Low water solubility of essential oils limits their application in food products, therefore, nanoemulsions are made to improve the water dispersion and prevent degradation of oils. Essential oils with polysaccharides are used to prepare nanosized emulsions with required functional properties. A comparison storage study of nanoemulsion coating with conventional emulsion coating containing lemongrass essential oil showed a higher rate of inactivation of *E. coli* (O157:H7) in the case of nanoemulsion-based coatings (Salvia-Trujillo et al., 2015). Also, these nanoemulsion-based coatings with lemongrass essential oil completely inhibited the natural microflora of fresh-cut Fuji apples when used at concentrations of 0.5 and 1.0% within 2 weeks. Antioxidants are added generally to emulsified food coatings to prevent the food from rancidity, discoloration, and degradation. Ascorbic acid is a natural antioxidant that is used extensively to prevent enzymatic browning or discoloration of fruits and vegetables. Studies showed retardation of browning of pear wedges when ascorbic acid and potassium sorbate incorporating MC and stearic acid-based emulsified food coatings were applied to fruits (Olivas et al., 2003). Similarly in the case of apricots and green peppers, food coating based on ascorbic acid, MC, PEG, and stearic acid decreased weight and vitamin C losses of crops (Ayranci and Tunc, 2004). Therefore, food coatings can be utilized for innovative purposes rather than their current uses by not only improving functional properties but also by carrying active substances that can extend the storage life of food by preventing surface microbial growth. Studies on hydrocolloid films are widely reported, but studies on application of active emulsified coatings/films and specifically their industrial implementation is still limited.

17.5 FUNCTIONAL PROPERTIES

So far numerous edible coating formulations have been investigated by many researchers as per produce requirements. It has been observed that the success of any coating solely relies on its barrier properties as it is able to create a micro-modified atmosphere around the produce that helps to retard food deterioration. Effectiveness of any coating material also depends on product wettability that determines its adherence over the produce surface. Some of the significant parameters to be considered during evaluation of films or coatings on shelf life extension of produce has been shown in Figure 17.5. To achieve optimal performance edible film/coating, composites are to be considered where each and every component contributes to enhanced performance of the coating.

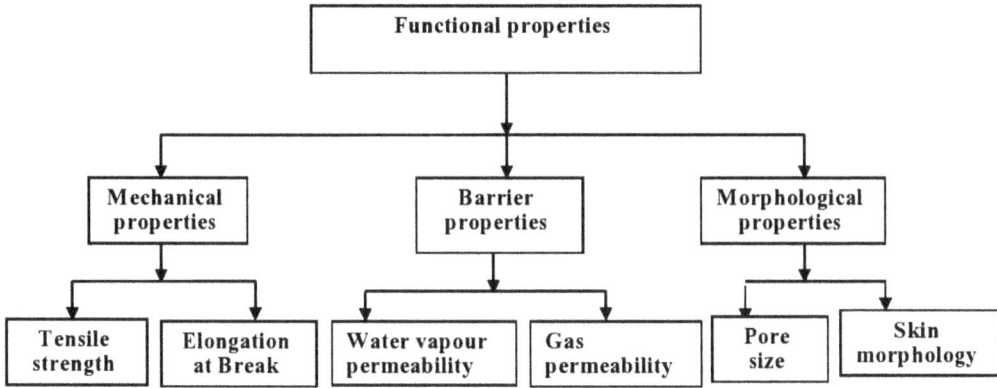

Figure 17.5 Functional properties of edible films and coatings.

17.5.1 Morphological Properties

17.5.1.1 Pore Size

Porosity is one of the essential features that determine permeability properties of films. The proportion of pores blocked by coating solution on the produce surface influence the gaseous exchange pattern and water vapor permeability (WVP) of the produce, therefore, individual coating applications vary with respect to different fruits as reported by Bai et al. (2003). The authors found that shellac-based coating is excellent for a Red Delicious apple but not suitable for Braeburn and Granny Smith apples as it causes anaerobiosis. Also, Banks et al. (1993) noted that inherent differences in skin's resistance to diffusion and respiration leads to variable responses of individual produce to a coating treatment as differing proportions of pores are blocked by the coating solution. Therefore, the authors suggested that surface coatings of fruits should be added to fresh produce to be used for processing rather than fresh produce used for direct marketing because of anaerobic conditions.

17.5.1.2 Surface Morphology

To ensure coating coverage over the product surface, a spreading coefficient is usually considered. This is a function of work of adhesion and cohesion (Sapper and Chiralt, 2018). The spreading coefficient equation (Equation 17.1), as described by Rulon and Robert (1993), is reported either in negative or zero. The work of adhesion (W_a), as shown in Equation (17.2), assists in liquid extension over a product surface, whereas work of cohesion (W_c), as shown in Equation (17.3), facilitates contraction; a balance between the two determines the wetting behavior of the coating solution. As per the authors, a value of W_s nearing zero will result in better coating of a surface.

$$W_s = W_a - W_c \tag{17.1}$$

$$W_a = \gamma LV + \gamma SV - \gamma SL \tag{17.2}$$

$$W_c = 2\gamma LV \tag{17.3}$$

where γLV, γSV, and γSL are designated as liquid-vapor, solid-vapor, and solid-liquid interfacial tension, respectively. These are calculated from the surface tension of the coating solution and contact angle of liquid drop on the solid surface. Additives such as calcium salts are known to increase cohesion in cell walls of fresh produce.

17.5.2 Mechanical Properties

17.5.2.1 Tensile Strength

Edible films are usually characterized by their tensile strength and elongation properties, which are important to bear physical impact caused by impact, vibrations, and pressure. The film/coating must be mechanically durable to prevent it from fracture and breakage. Physical techniques such as irradiation, ultrasound, heating and pressure applications, chemical techniques, crosslinking, acid-alkali reactions, and enzymatic techniques are used to improve resistance of films (Conca, 2002) As per Letendre et al. (2002), protein-polysaccharide interactions generate a 3D network with improved mechanical properties on autoclaving. This behavior could be possibly explained as high temperature generates less ordered structure making more functional groups available for cross-linking (Thakur et al., 1997). Other physical treatments such as homogenization, which is used to reduce particle size, is positively associated with producing stronger whey protein-based films due to increased interfacial area (Perez-Gago and Krochta, 2001). Additives such as citric acid as a cross-linking agent forms cross-links with the hydroxyl group of polysaccharides and improves their compatibility to improve tensile strength (Da et al., 2011; Olivato et al., 2012). To increase film thickness, glycerol is usually used as glycerol molecules fill up the void in the polymer matrix and interact with the polymer, thus increasing thickness (Sudaryati et al., 2010).

17.5.2.2 Elongation at Break

Elongation at break is defined as the ability of films to extend before breaking and indirectly relates to film plasticity and integrity when applied to food products (Galus and Lenart, 2013). Some of the factors such as base material type, concentration, and solvent used in the film formation affect elongation and tensile strength of the film (Herliany et al., 2013). Glycerol is known to increase elongation at break and flexibility by reducing intermolecular interactions between polymer chains. As a result, mobility between molecular chains increases leading to increase in elongation (Kokoszka et al., 2010; Katili et al., 2013).

17.5.3 Barrier Properties

17.5.3.1 Water Vapor Permeability

Among mass transfer properties, WVP of designed edible films or coatings plays a huge role in reducing water loss and maintaining firmness of the produce. Use of hydrophilic components in edible film formulation does not provide an adequate water vapor barrier due to moisture sensitivity, whereas the addition of lipids (non-polar) in the formulation enhances the WVP of the films. However, the type of lipid used also affects the final water vapor transmission (WVT) properties such as solid lipids (palm oil, beeswax, stearic acid) result in lower WVP compared with liquid lipids (oleic acid) (Wu et al., 2002). In case of protein-based films, pH value plays an important role in determining the WVP of films and it has been observed that at neutral pH the best water barrier properties are achieved for edible films (McHugh, 2000). Additionally, use of cross-linking agents like citric acid could help improve water barrier properties (Ma et al., 2018). Theoretically, resistance of water vapor transmission through films is governed by Fick's first law of diffusion and Henry's law of solubility as per the following mathematical equation (ASTM, 1990):

$$WVP = \frac{j_{AZ}\Delta z}{\rho_{A1} - \rho_{A2}}$$
(17.4)

Δz = Film thickness
j_{AZ} = Water vapor transmission rate
$\rho_{A1} - \rho_{A2}$ = Water vapor partial pressure gradient

17.5.3.2 Gas Barrier Property

Among gas barrier properties, oxygen and carbon dioxide are the two major gases that influence the respiration of fresh fruits and vegetables and lipid peroxidation of fat-rich foods. To measure gas permeability properties, techniques such as isostatic concentration, manometric, volumetric, gas chromatographic, bioluminescence, and infrared detection have been used (Chen, 1995). Plasticizers are known to reduce intermolecular interactions between polymer chains, thereby increasing mobility of film and facilitating diffusion of gas molecules from the polymer network (Sabbah et al., 2017). It has been seen that edible films prepared from proteins usually have lower oxygen permeability compared with polyethylene and polyvinyl chloride but close to polyester film and lower than cellulose films as well (Park and Chinnan, 1995). As per the authors, the CO_2/O_2 ratio plays a significant role in designing edible film or coatings for fresh horticultural produce, and it has been documented that higher ratio will allow les accumulation of carbon dioxide gas. Therefore, gas barrier properties must be characterized for commodity-specific food products and designed in a way to maximize the shelf life of the respective product.

17.6 FUTURE PROSPECTS AND CONCLUSION

Edible coatings used on food products has various implications such as to maintain sensory properties, avoid allergic reactions, be cost-effective, be easy to use, be food grade and have positive interactions with the produce on which applied. To meet these objectives, elaborate scientific studies are required as each type of additive used to formulate edible coatings/films can have a dramatic effect on the final quality of the produce. Also, lack of knowledge about an application method of edible films/coatings also poses a hurdle to achieve the desired effect in the final produce. Today, use of nanoparticles as an active delivery system for antimicrobial, anti-browning, or antioxidant additives is widely practiced and much of the research is focused on developing nanoscale additives with greater activity. Electrospray application of edible coatings on food products is also garnering attention due to uniform and thin layers formed around the produce. However, the research literature pertaining to these studies is so wide and has not been structured in a way to provide concise knowledge about their use and application. Moreover, laboratory-scale experiments are proven successful but there are limitations in the use of these technologies at commercial scale. Therefore, taking into account factors such as food safety legislation, edible film/coating formulation, mode of application, and target food product, literature should be published that can be followed globally rather than individually. Lastly, efforts must be made to commercialize edible film/coatings in developing countries and create awareness among the masses regarding the same.

REFERENCES

Abbasi, N.A., Iqbal, Z., Maqbool, M., & Hafiz, I.A. (2009). Postharvest quality of mango (*M. indica*) fruits as affected by chitosan coating. *Pakistan Journal of Botany*, *41*(1), 343–357.

Adetunji, C.O., Fadiji, A.E., & Aboyeji, O.O. (2014). Effect of chitosan coating combined *Aloe vera* gel on cucumber postharvest quality during ambient storage. *Journal of Emerging Trends in Engineering and Applied Science*, *5*(6), 391–397.

Akcan, T., Estevez, M., & Serdaroglu, M. (2017). Antioxidant protection of cooked meatballs during frozen storage by whey protein edible films with phytochemicals from *Laurus nobilis L.* and *Salvia officinalis*. *LWT – Food Science and Technology*, *77*, 323–331.

Alsaggaf, M.S., Moussa, S.H., & Tayel, A.A. (2017). Application of fungal chitosan incorporated with pomegranate peel extract as edible coating for microbiological, chemical and sensorial quality enhancement of Nile tilapia fillets. *International Journal of Biological Macromolecules*, *99*, 499–505.

Andres, Y., Giraud, L., Gerente, C., & Le Cloirec, P. (2007). Antibacterial effects of chitosan powder: mechanisms of action. *Environmental Technology*, *28*, 1357–1363.

Arham, R., Mulyati, M.T., Metusalach, M., & Salengke, S. (2016). Physical and mechanical properties of agar based edible film with glycerol plasticizer. *International Food Research Journal*, *23*, 1669–1675.

Arvisenet, G., Voilley, A., & Cayot, N. (2002). Retention of aroma compounds in starch matrices: competitions between aroma compounds toward amylose and amylopectin. *Journal of Agricultural and Food Chemistry*, *50*, 7345–7349.

ASTM (American Society for the Testing of Materials). (1990). *Standard test methods for water vapour transmission of materials*. Philadelphia, PA: ASTM.

Athmaselvi K.A., Sumitha, P., & Reathy, B. (2013). Development of *Aloe vera* based edible coating for tomato. *International Agrophysics*, *27*, 369–375.

Audic, J.L., Chaufer, B., & Daufin, G. (2003). Non-food applications of milk components and dairy co-products: a review. *Lair*, *83*, 417–438.

Ayala-Zavala, J.F., Silva-Espinoza, B.A., Cruz-Valenzuela, M.R., Leyva, J.M., Ortega-Ramírez, L.A., Carrazco-Lugo, D.K., Pérez-Carlón, J.J., Melgarejo-Flores, B.G., González-Aguilar, G.A., & Miranda, M.R.A. (2012). Pectin-cinnamon leaf oil coatings add antioxidant and antibacterial properties to fresh-cut peach. *Flavour and Fragrance Journal*, *28*, 39–45.

Aydin, F., Kahve, H. I., & Ardic, M. (2017). Lipid based edible films. *Journal of Scientific and Engineering Research*, *4*, 86–92.

Ayranci, E., & Tunc, S. (2004). The effect of edible coatings on water and vitamin C loss of apricots (*Armeniaca vulgaris* Lam.) and green peppers (*Capsicum annuum* L.). *Food Chemistry*, *87*, 339–342.

Badawy, M.E.I., & Rabea, E.I. (2009). Potential of the biopolymer chitosan with different molecular weights to control postharvest gray mold of tomato fruit. *Postharvest Biology and Technology*, *51*, 110–117.

Badii, F., & Howell, N.K. (2006). Fish gelatin: structure, gelling properties and interaction with egg albumen proteins. *Food Hydrocolloids*, *20*, 630–640.

Bahnasawy, A.H., & Khater, El-S.G. (2014). Effect of wax coating on the quality of Cucumber fruit during storage. *Journal of Food Processing & Technology*, *5*–6.

Bai, J., Hagenmaier, R.D., & Baldwin, E.A. (2003). Coating selection for "Delicious" and other apples. *Postharvest Biology and Technology*, *28*, 381–390.

Banks N.H., Dadzie, B.K., & Cleland, D.J. (1993). Reducing gas exchange of fruits with surface coatings. *Postharvest Biology and Technology*, *3*, 269–284.

Baranauskiene, R., Bylaite, E., Zukauskaite, J., Venskutonis, R.P. (2007). Flavor retention of peppermint (*Mentha piperita* L.) essential oil spray-dried in modified starches during encapsulation and storage. *Journal of Agricultural and Food Chemistry*, *55*, 3027–3036.

Belyamani, I., Prochazka, F., & Assezat, G. (2014). Production and characterization of sodium caseinate edible films made by blown-film extrusion. *Journal of Food Engineering*, *121*, 39–47.

Bermudez-Oria, A., Rodríguez-Gutierrez, G., Rubio-Senent, F., Fernández-Prior, Á., Fernández-Bolaños, J. (2019). Effect of edible pectin-fish gelatin films containing the olive antioxidants hydroxytyrosol and 3, 4-dihydroxyphenylglycol on beef meat during refrigerated storage. *Meat Science*, *148*, 213–218.

Bonnaillie, M.L., Zhang, H., Akkurt, S., Yam, L.K., & Tomasula, M.P. (2014). Casein films: the effects of formulation, environmental conditions and the addition of citric pectin on the structure and mechanical properties. *Polymers*, *6*, 2018–2036.

Bourtoom, T. (2011). Edible protein films: properties enhancement. *International Food Research Journal*, *16*, 1–9.

Bravin, B., Peressini, D., & Sensidoni, A. (2006). Development and application of polysaccharide-lipid edible coating to extend shelf-life of dry bakery products. *Journal of Food Engineering*, *76*(3), 280–290.

Bravo-Anaya, L.M., Soltero, J.F.A., & Rinaudo, M. (2016). DNA/chitosan electrostatic complex. *International Journal of Biological Macromolecules*, *88*, 345–353.

Cardoso, G.P., Dutra, M.P., Fontes, P.R., Ramos, A.L.S., Gomide, L.A.M., & Ramos, E.M. (2016). Selection of a chitosan gelatin-based edible coating for color preservation of beef in retail display. *Meat Science*, *114*, 85–94.

Cazon, P., Velazquez, G., Ramírez, J.A., & Vazquez, M. (2017). Polysaccharide-based films and coatings for food packaging: a review. *Food Hydrocolloids*, *68*, 136–148.

Chakraborty, S. (2017). Carrageenan for encapsulation and immobilization of flavor, fragrance, probiotics, and enzymes: a review. *Journal of Carbohydrate Chemistry*, *36*, 1–19.

Chambi, H., & Grosso, C. (2006). Edible films produced with gelatin and casein cross-linked with transglutaminase. *Food Research International*, *39*, 458–466.

Chauhan, S., Gupta, K.C., & Agrawal, M. (2014a). Development of *A. vera* gel to control postharvest decay and longer shelf life of grapes. *International Journal of Current Microbiology and Applied Sciences*, *3*(3), 632–642.

Chauhan, S., Gupta, K.C., & Agrawal M. (2014b). Efficacy of chitosan and calcium chloride on postharvest storage period of mango with the application of hurdle tech. *International Journal of Current Microbiology and Applied Sciences*, *3*(5), 731–740.

Chauhan, S., Gupta, K.C., & Agrawal, M. (2014c). Efficacy of natural extract on the storage quality of apple. *International Journal of Current Microbiology and Applied Sciences*, *3*(3), 706–711.

Chen, H. (1995). Functional properties and applications of edible films made of milk proteins. *Journal of Dairy Science*, *78*, 2563–2583.

Cho, S. Y., & Rhee, C. (2004). Mechanical properties and water vapor permeability of edible films made from fractionated soy proteins with ultrafiltration. *Lebensmittel-Wissenschaft und-Technologie*, *37*, 833–839.

Cobrera, L.E.P., Narvez, G.C.D., Coronel, A.T., & Martinez, C.G. (2010). Effects of edible chitosan-linseed mucilage coating on quality and shelf life of fresh cut strawberry. National Autonomous University of Mexico, Mexico City, Mexico. https://www.semanticscholar.org/paper/Effects-of-edible-chitosan-linseed-mucilage-coating-Cabrera-N%C3%A1rvaez/6764f81e158195adbd193d1044fc417b557d9b58

Collazo-Bigliardi, S., Ortega-Toro, R., & Boix, A.C. (2018). Isolation and characterisation of microcrystalline cellulose and cellulose nanocrystals from coffee husk and comparative study with rice husk. *Carbohydrate Polymers*, *191*, 205–215.

Conca, K.R. (2002). Protein-based films and coating for military packaging applications. In: Gennadios, A, editors. *Protein-based films and coatings*. Boca Raton, FL: CRC Press, pp. 551–577.

Da, R.A., Zambon, M.D., Aas, C., & Ajf, C. (2011). Thermoplastic starch modified during melt processing with organic acids: the effect of molar mass on thermal and mechanical properties. *Industrial Crops and Products*, *33*, 152–157.

Dangaran, K , Tomasula, P. M., & Qi, P. (2009). Structure and function of protein-based edible films and coatings. In: *Edible films and coatings for food applications*. New York: Springer, pp. 25–56.

De Azeredo, H.M.C., de Britto, D., & Assis, O.B.G. (2010). Chitosan edible films and coatings – A review. In: MacKay, RG, Tait, JM, editors. *Handbook of chitosan research and applications*. Hauppauge, NY: Nova Publishers, pp. 179–194.

De Cindio, B., Gabriele, D., & Lupi, F. R. (2016). Pectin: properties determination and uses. In: *Encyclopedia of food and health*. Oxford, UK: Academic Press, pp. 294–300.

Debeaufort, F., & Voilley, A. (2009). Lipid-based edible films and coatings. In: *Edible films and coatings for food applications*. New York: Springer, pp. 135–168.

Dewettinck, K., & Huyghebaert, A. (1999). Fluidized bed coating in food technology. *Trends in Food Science & Technology*, *10*(4–5), 163–168.

Dhall, R.K. (2013). Advances in edible coatings for fresh fruits and vegetables: a review. *Critical Reviews in Food Science and Nutrition*, *53*, 435–450.

Donati, I., Holtan, S., Mørch, Y.A., Borgogna, M., Dentini, M., Skjåk-Bræk. (2005). New hypothesis on the role of alternating sequences in calcium-alginate gels. *Biomacromolecules*, *6*, 1031–1040.

Dronen, D. (2004). Characterization of volatile loss from dry food polymer materials. PhD dissertation, University of Minnesota, Minneapolis, MN.

Duan J., Wu R., Strick B.C., & Zhao Y. (2011). Effect of edible coating on the quality of fresh blueberry under commercial storage condition. *Postharvest Biology and Technology*, *59*, 71–79.

El-Anany, A.M., Hassan, G.F.A., & Ali, M.R. (2009). Effect of edible coating on the shelf life and quality of Anana Apple (*Malus domestica Borkh*) during cold storage. *Journal of Food Technology*, *7*(1), 5–11.

Elsabee, M. Z., & Abdou, E. S. (2013). Chitosan based edible films and coatings: a review. *Materials Science and Engineering*, *33*, 1819–184.

Espitia, P.J.P., Du, W.X., Avena-Bustillos, R.J., Soares, N.F.F., & McHugh, T.H. (2013). Edible films from pectin: physical-mechanical and antimicrobial properties – A review. *Food Hydrocolloids*, *35*, 1–10.

Falguera, V., Quintero, J.P., Jiménez, A., Muñoz, J.A., & Ibarz, A. (2011). Edible films and coatings: structures, active functions and trends in their use. *Trends in Food Science and Technology*, *22*, 292–303.

Galus, S., & Lenart, A. (2013). Development and characterization of composite edible films based on sodium alginate and pectin. *Journal of Food Engineering*, *115*, 459–465.

Gao, H., Zhang, Z.K., Chai, H.K., Cheng, N., Yang, Y., Wang, D.N., Yang, T., & Cao, W. (2016). Melatonin treatment delays postharvest senescence and regulates reactive oxygen species metabolism in peach fruit. *Postharvest Biology and Technology*, *118*, 103–110.

Garpas, M.E.M. (2011). Development for fresh-cut cantaloupe, Master's thesis, Texas A& M University, College Station, Texas.

Ghader, F.M., Hamzah, M.A.Q., Hifzi, A.A.B., & Barey, G.S. (2011). Antifungal effectiveness of potassium sorbate incorporation edible coating against spoilage moulds of apple, cucumber and tomato during refrigerated storage. *Journal of Food Science.* doi:10.1111/1750-3841-2011.02059x.

Ghavidel, R.A., Davoodi, M.G., Adibasl, A.F., Tanoori, T., & Sheykholeslami, Z. (2013). Effect of selected edible coating to extend shelf life of fresh cut apples, *International Journal of Agriculture and Crop Science*, *616*, 1171–1178.

Guerrero, P., Stefani, P.M., Ruseckaite, R.A., de la Caba, K. (2011). Functional properties of films based on soy protein isolate and gelatin processed by compression molding. *Journal of Food Engineering*, *105*, 65–72.

Guilbert, S., & Gontard, N. (1995). Edible and biodegradable food packaging. In: Ackermann P, Jägerstad M, Ohlsson T, editors. *Foods and packaging materials-chemical interactions*. Cambridge, UK: The Royal Society of Chemistry, pp. 159–168.

Guilbert, S., & Gontard, N. (2005). Agro-polymers for edible and biodegradable films: review of agricultural polymeric materials, physical and mechanical characteristics. In: Han JH, editor. *Innovation in food packaging*. San Diego, CA: Elsevier, pp. 263–276.

Hassan, B., Ali, S., Chatha, S., Hussain, A.I., Zia, K.M., & Akhtar, N. (2018). Recent advances on polysaccharides, lipids and protein based edible films and coatings: a review. *International Journal of Biological Macromolecules*, *109*, 1095–1107.

Haug, I.J., Draget, K. I., & Smidsrod, O. (2004). Physical and rheological properties of fish gelatin compared to mammalian gelatin. *Food Hydrocolloids*, *18*, 203–213.

Herliany, N.E., Santoso, J., & Salamah, E. (2013). Characteristics of biofilm-based on carrageenan. *Jurnal Akuatika*, *4*(1), 10–20.

Jauregi, P., & Welderufael, F.T. (2010). Added-value protein products from whey: extraction, fractionation, separation, purification. *Nutrafoods*, *9*, 13–23.

Javanmard, M. (2008). Shelf life of whey protein-coated pistachio kernels (*Pistacia Vera L.*). *Journal of Food Process Engineering*, *31*, 247–259.

Jiménez, A., Fabra, M.J., Talens, P., & Chiralt, A. (2012). Edible and biodegradable starch films: a review. *Food and Bioprocess Technology*, *5*, 2058–2076.

Juhaimi, F., Ghafoor, K., & Babika, E.E. (2012). Effect of gum Arabic edible coating on weight loss, firmness and sensory characteristics of cucumber (*Cucumis sativus* L.) fruit during storage. *Pakistan Journal of Botany*, *44*(4), 1439–1444.

Kaewprachu, P., Osako, K., Benjakul, S., Tongdeesoontorn, W., & Rawdkuen, S. (2016). Biodegradable protein-based films and their properties: a comparative study. *Packaging Technology and Science*, *29*, 77–90.

Katili, S., Harsunu, B.T., & Irawan, S. (2013). Effect of plasticizer concentration of glycerol and chitosan compositions in the solvent on the physical properties of chitosan edible film. *Journal Teknologi*, *6*, 29–38.

Kaushik, R., Kumar, N., Sihag, M.K., & Ray, A. (2015). Isolation, characterization of wheat gluten and its regeneration properties. *Journal of Food Science and Technology*, *52*, 5930–5937.

Kiyomi Okuro, P., de Matos Junior, F.E., & Favaro-Trindade, C.S. (2013). Technological challenges for spray chilling encapsulation of functional food ingredients, spray chilling encapsulation. *Food Technology and Biotechnology*, *51*(2), 171–182.

Kluge, R.A., Nachtigal, J.C., Fachinello, J.C., & Bilhalva, A.B. (2002). *Fisiologia e manejop´os-colheita de frutas de clima temperado.* Sao Paulo, Brazil: Livraria e Editora Rural, Campinas, pp. 214.

Kokoszka, S., Debeaufort, F., Hambleton, A., Lenart, A., & Voilley, A. (2010). Protein and glycerol contents affect physico-chemical properties of soy protein isolate-based edible films. *Innovative Food Science and Emerging Technologies*, *11*, 503–510.

Koupantsis, T., Pavlidou, E., & Paraskevopoulou, A. (2014). Flavour encapsulation in milk proteins – CMC coacervate-type complexes. *Food Hydrocolloids*, *37*, 134–142.

Krochta, J.M. (2002). Proteins as raw materials for films and coatings: definitions, current status, and opportunities. In: Gennadios A, editor. *Protein-based films and coatings.* Boca Raton, FL: CRC Press, pp. 1–41.

Larotonda, F.D.S. (2007). *Biodegradable films and coatings obtained from carrageenan from Mastocarpus stellatus and starch from Quercus suber.* PhD dissertation, University of Porto, Porto, Portugal.

Leite, A.L.M.P., Zanon, C.D., & Menegalli, F.C. (2017). Isolation and characterization of cellulose nanofibers from cassava root bagasse and peelings. *Carbohydrate Polymer*, *157*, 962–970.

Letendre, M., D'Aprano, G., Lacroix, M., Salmieri, S., & St-Gelais, D. (2002). Physicochemical properties and bacterial resistance of biodegradable milk protein films containing agar and pectin. *Journal of Agricultural and Food Chemistry*, *50*, 6017–6022.

Li, L., Sun, J., Gao, H., Shen, Y., Li, C., Yi, P., He, X., Ling, D., Sheng, J., Li, J., Liu, G., Zheng, F., Xin, M., Li, Z., & Tang, Y. (2017). Effects of polysaccharide-based edible coatings on quality and antioxidant enzyme system of strawberry during cold storage. *International Journal of Polymer Science*, 2017, 9746174.

Liu, D., Nikoo, M., Boran, G., Zhou, P., & Joe, M. (2015). Regenstein, collagen and gelatin. *Annual Review on Food Science and Technology*, *6*, 527–557.

Lou, M.M., Zhu, B., Muhammad, I., Li, B., Xie, G.L., Wang, Y.L., Li, H.Y., & Sun, G.C. (2011). Antibacterial activity and mechanism of action of chitosan solutions against apricot fruit rot pathogen *Burkholderia seminalis. Carbohydrate Research*, *346*, 1294–1301.

Ma, W., Rokayya, S., Xu, L., Sui, X., Jiang, L., & Li, Y. (2018). Physical-chemical properties of edible film made from soybean residue and citric acid. *Journal of Chemistry*, *2018*, 1–8.

Magdy, M.H.M., Al-Shamrani, Khalid M., & Al-Arifi, A.S. (2014). Edible coating of shelf life extension of fresh banana fruit on gamma irradiated plasticized poly (VA)/CMC/tannin/composite. *Materials Sciences and Applications*, *5*, 395–415.

Malnati, R.M.E.J., Du-Pont, M.X.A., & Morales, D.A.O. (2019). Method for producing a nanoemulsion with encapsulated natural antioxidants for preserving fresh and minimally processed foods, and the nanoemulsion thus produced. Patent. WIPO WO2019039947A1.

Mantilla, N.V. (2012). Development of an alginate based on antimicrobial edible coating to extend the shelf life of fresh cut pineapple. Master's thesis, Texas A&M University, College Station, TX.

Marcuzzo, E., Sensidoni, A , Debeaufort, F., & Voilley, A. (2010). Encapsulation of aroma compounds in biopolymeric emulsion based edible films to control flavour release. *Carbohydrate Polymers*, *80*(3), 984–988.

Marpudi, S.L., Abirami, L.S.S., Pushkala, R., & Srividya, N. (2011). Enhancement of storage life and quality of Papaya fruits using *Aloe vera* based microbial coating. *Indian Journal of Biotechnology*, *1*, 83–89.

Martina, K., Binello, A., Lawson, D., Jicsinszky, L., & Cravotto, G. (2013). Recent applications of cyclodextrins as food additives and in food processing. *Current Nutrition & Food Science*, *9*, 167–179.

McClements, D.J., & Rao, J. (2011). Food-grade nanoemulsions: formulation, fabrication, properties, performance, biological fate, and potential toxicity. *Critical Reviews in Food Science and Nutrition*, *51*, 285–330.

McHugh, T.H. (2000). Protein-lipid interactions in edible films and coatings. *Nahrung*, *44*, 148–151.

McHugh, T. H., & Senesi, E. (2000). Apple wraps: a novel method to improve the quality and extend the shelf life of fresh-cut apples. *Journal of Food Science*, *65*, 480–485.

Mecitoglu, C., Yemenicioglu, A., Arslanoglu, A., Elmaci, Z.S., Korel, F., & Cetin, A.E. (2006). Incorporation of partially purified hen egg white lysozyme onto zein films for antimicrobial food packaging. *Food Research International*, *39*, 12–21.

Mehyar, G.F., Al-Ismail, K., Han, J.H., & Chee, G.W. (2012). Characterization of edible coatings consisting of pea starch, whey protein isolate, and carnauba wax and their effects on oil rancidity and sensory properties of walnuts and pine. *Journal of Food Science*, *77*(2), E52–E59.

Moadenia, N., Ehsani, M.R., Emamdejomeh, Z., Asadi, M.M., Misani, M., & Mazahari, A.F. (2010). A note on the effect of calcium alginate coating on quality of refrigerated strawberry. *Irish Journal of Agricultural and Food Research*, *49*, 165–170.

Moalemiyan, M., & Ramaswamy, H.S. (2012). Quality retention of shelf life of extension in Mediterranean cucumber coated with a pectin based film. *Journal of Food Research*, *1*(3), 159–168.

Mohamed, S.A.A., El-Sakhawy, M., Nashy, S.H.A., & Othman, A.M. (2019). Novel natural composite films as packaging materials with enhanced properties. *International Journal of Biological Macromolecules*, *136*, 774–784.

Murray, D.G., Luft, L.R., & Low D.E. (1973). Corn starch hydrolysates. *Food Technology*, *27*, 32–33, 36, 38, 40.

Nasution, Z., Wei, J.N., & Hamzah, Y. (2015). Characteristics of fresh cut Guava coated with *Aloe vera* gel as affected by different additives. *Kasetsart Journal (Natural Science)*, *49*, 111–121.

Nesperos-Carriedo, M.O., Baldwin, E.A., & Shaw, P.E. (1992). Development of an edible coating for extending postharvest life of selected fruits and vegetables. *Proceedings of Florida State Horticultural Society*, *104*, 122–125.

Nisha, V., Monisha, C., Ragunathan, R., & Johney, J. (2016). Use of chitosan as edible coating on fruits and in micro biological activity - An ecofriendly approach. *International Journal of Pharmaceutical Science Invention*, *5*, 7–14.

Olivas, G.I., Rodrigues, J.J., & Barbosa-Cánovas, G.V. (2003). Edible coatings composed on methylcellulose, stearic acid, and additives to preserve quality of pear wedges. *Journal of Food Processing and Preservation*, *27*, 359–366.

Olivato, J.B., Grossmann, M.V., Bilck, A.P., & Yamashita, F. (2012). Effect of organic acids as additives on the performance of thermoplastic starch/polyester blown films. *Carbohydrate Polymers*, *90*, 159–164.

Ortiz, A., Graell, J., & Lara, I. (2010). Cell wall-modifying enzymes and firmness loss in ripening 'Golden Reinders' apples: a comparison between calcium dips and ULO storage. *Food Chemistry*, *128*, 1072–1079.

Padmja, N., & Basco, J.D. (2014). Preservation of Sapota (*M. zapota*) by edible *Aloe vera* gel coating to maintain its quality. *Food Science*, *3*(3), 177–179.

Palou, L., Valencia-Chamorro, S., & Pérez-Gago, M. (2015). Antifungal edible coatings for fresh citrus fruit: a review. *Coatings*, *5*, 962–986.

Panahirad, S., Nasser, M., Hassani, R.N., Ghanbarzadeh, B., & Nahandi, F.Z. (2015). Plum shelf life enhancement by edible coating based on pectin and carboxymethyl cellulose. *Journal of Biodiversity and Environmental Sciences*, *7*(1), 423–430.

Park, H.J., & Chinnan, M.S. (1995). Gas and water vapor barrier properties of edible films from protein and cellulosic materials. *Journal of Food Engineering*, *25*, 497–507.

Park, S., & Zhao, Y. (2004). Incorporation of a high concentration of mineral or vitamin into chitosan-based films. *Journal of Agricultural and Food Chemistry*, *52*, 1933–1939.

Pavlath, A.E., & Orts, W.J. (2009). Edible films and coatings: Why, What, and How? In: Embuscado, ME, Huber KC, editors. *Edible films and coatings for food applications*. New York: Springer Science+Business Media, pp. 1–23.

Pavli, F., Tassou, C., Nychas, G.J.E., & Chorianopoulos, N. (2018). Probiotic incorporation in edible films and coatings: bioactive solution for functional foods. *International Journal of Molecular Sciences*, *19*, 150.

Perez-Gago, M., & Krochta, J. (2001). Lipid particle size effect on water vapor permeability and mechanical properties of whey protein/beeswax emulsion films. *Journal of Agricultural and Food Chemistry*, *49*, 996–1002.

Poornima, K., & Sinthya, R. (2017). Application of various encapsulation techniques in food industries. *International Journal of Latest Engineering Research and Applications*, *2*(10), 37–41.

Raghav, P.K, Agarwal, N., & Saini, M. (2016). Edible coating of fruits and vegetables: a review. *International Journal of Scientific Research and Modern Education*, *1*(1), 2455–5630.

Ramos, O.L., Fernandes, J.C., Silva, S,I., Pintado, M.E., & Malcata, F.X. (2012). Edible films and coatings from whey proteins: a review on formulation, and on mechanical and bioactive properties. *Critical Reviews in Food Science and Nutrition*, *52*, 533–552.

Ravishankar, S., Zhu, L., Olsen, C.W., McHugh, T.H., & Friedman, M. (2009). Edible apple film wraps containing plant antimicrobials inactivate foodborne pathogens on meat and poultry products. *Journal of Food Science*, *74*(8), M440–M445.

Reineccius, G.A. (1998). Kinetics of flavor formation during Maillard browning. In: Teranishi R, Wick EL, Hornstein L, editors. *Flavor chemistry: thirty years of progress*. New York: Springer, pp. 345–352.

Reineccius, G.A., Liardon, R., & Luo, Z. (2003). The retention of aroma compounds in spray dried matrices during encapsulation and storage. In: LeQuere JL, Etievant PX, editors. *Flavour research at the dawn of the twenty first century*. Cachan, France: Lavoisier, pp. 3–9.

Rhim, J.W., & Shellhammer, T.H. (2005). Lipid-based edible films and coatings. In: *Innovations in food packaging*. Academic Press, pp. 362–383.

Ricardo, D., Andrade, O.S., & Osorio, F.A. (2012). Atomizing spray systems for application of edible coatings. *Comprehensive Reviews in Food Science and Food Safety*, *113*, 323–337.

Rice J. (1994). What's new in edible films? *Food Processing*, *55*(7), 61–62.

Risch, S.J. (1988). Encapsulation of flavors by extrusion. In: *Flavor encapsulation. ACS Symposium Series*. New York and Columbus, OH, ACS Publications, Vol. 370, pp. 103–109.

Rojas-Grau, M.A., Tapia, M.S., & Belloso, O.M. (2008). Effect of polysaccharide based edible coating on shelf life of fresh cut Fuji apple. *LWT – Food Science and Technology*, *41*(1), 139–147.

Rulon, J., & Robert, H. (1993). Wetting of low-energy surfaces. In: Berg JC, editor. *Wettability*. New York: Marcel Deckker, pp. 4–73. ISBN 0824790464.

Sabbah, M., Di Pierro, P., Giosafatto, C.V.L., Esposito, M., Mariniello, L., Regalado-Gonzales, C., & Porta, R. (2017). Plasticizing effects of polyamines in protein-based films. *International Journal of Molecular Science*, *18*, 1026.

Saha, A., Gupta, R.K., & Tyagi, Y.K. (2014). Effect of edible coating of the shelf life and quality of Potato (*S. tuberosam*) tubers during storage. *Journal of Chemical and Pharmaceutical Research*, *6*(12), 802–809.

Salleh, N.S.M. (2013). Development of starch and Soy Protein edible coating and its effect on the postharvest life of mango (*Mangiferaindica*). Universiti Teknologi MARA, Selangor, Malaysia, Doctoral dissertation.

Salvia-Trujillo, L., Rojas-Graü, M.A., Soliva-Fortuny, R., & Martín-Belloso, O. (2015). Use of antimicrobial nanoemulsions as edible coatings: impact on safety and quality attributes of fresh-cut Fuji apples. *Postharvest Biology and Technology*, *105*, 8–16.

Salvia-Trujillo, L., Soliva-Fortuny, R., Rojas-Graü, M.A., McClements, D.J., & Martín-Belloso, O. (2017). Edible nanoemulsions as carriers of active ingredients: a review. *Annual Reviews in Food Science and Technology*, *8*, 439–466.

Santos, D., Colette Maurício, A., Sencadas, V., Santos, J.D., Fernandes, M.H., & Gomes, P.S. (2018). Spray drying: an overview. In: *Biomaterials – physics and chemistry*, IntechOpen. http://dx.doi.org/10.5772/intechopen.72247

Sapper, M., & Chiralt, A. (2018). Starch-based coatings for preservation of fruits and vegetables. *Coatings*, *8*, 152.

Shendurse, A.M., Gopikrishna, G., Patel, A.C., & Pandya, A.J. (2018). Milk protein based edible films and coatings–preparation, properties and food applications. *Journal of Nutrition and Health Food Engineering*, *8*, 219–226.

Shimada, K., & Cheftel, J.C. (2002). Sulfhydryl group/disulfide bond interchange reactions during heat-induced gelation of whey protein isolate. *Journal of Agriculture and Food Chemistry*, *37*, 161–168.

Shukla, R., & Chheryan, M. (2001). Zein: the industrial protein from corn. *Industrial Crops and Products*, *13*, 171–192.

Soliman, E., Eldin, M.M., & Furuta, M. (2009). Biodegradable zein-based films: influence of gamma-irradiation on structural and functional properties. *Journal of Agricultural and Food Chemistry*, *57*, 2529–2535.

Sonti, S. (2003). Consumer perception and application of edible coating on fresh cut fruits and vegetables. M.Sc. thesis., Osmania University College of Technology, Telangana, India.

Soottitantawat, A., Yoshii, H., Furuta, T., Ohgawara, M., Forssell, P., Partanen, R., Poutanen, K., & Linko, P. (2004). Effect of water activity on the release characteristics and oxidative stability of d-limonene encapsulated by spray drying. *Journal of Agricultural and Food Chemistry*, *52*, 1269–1276.

Sudaryati, H.P., Mulyani, S.T., & Hansyah, E.R. (2010). Physical and mechanical properties of edible film from porang (*Amorphopallus oncophyllus*) flour and carboxymethyl-cellulose. *Jurnal Teknologi Pertanian, 11*, 196–201.

Suwannark, J., Phanumong, P., & Rattanapanone, N. (2015). Combined effect of calcium salt treatment and chitosan coating on quality and shelf life of caused fruits and vegetables. *CMU Journal of Natural Sciences, 14*(3), 269–284.

Talens, P., Pérez-Masía, R., Fabra, M.J., Vargas, M., & Chiralt, A. (2012). Application of edible coatings to partially dehydrated pineapple for use in fruit-cereal products. *Journal of Food Engineering, 112*(1–2), 86–96.

Thakur, B.R., Singh, R.K., & Handa, A.K. (1997). Chemistry and uses of pectin - a review. *Critical Reviews in Food Science & Nutrition, 37*, 47–73.

Tian, H., Guo, G., Fu, X., Yao, Y., Yuan, L., & Xiang, A. (2018). Fabrication, properties and applications of soy-protein-based materials: a review. *International Journal of Biological Macromolecules, 120*, 475–490.

Tibolla, H., Pelissari, F.M., Martins, J.T., Vicente, A.A., & Menegalli, F.C. (2018). Cellulose nanofibers produced from banana peel by chemical and mechanical treatments: characterization and cytotoxicity assessment. *Food Hydrocolloids, 75*, 192–201.

Valdés, A., Burgos, N., Jiménez, A. & Garrigós, M.C. (2015). Natural pectin polysaccharides as edible coatings. *Coatings, 5*, 865–886.

Vanaei, M., Sedaghat, N., Abbaspour, H., Kaviani, M., & Azarbad, H.R. (2014). Novel edible coating based on *Aloe vera* gel to maintain pistachio quality. *International Journal of Scientific Engineering and Technology, 3*(8), 1016–1019.

Vital, A.C.P., Guerrero, A., Monteschio, J.D.O., Valero, M.V., Carvalho, C.B., De Abreu Filho, B.A., Madrona, G.S., & Do Prado, I.N. (2016). Effect of edible and active coating (with rosemary and oregano essential oils) on beef characteristics and consumer acceptability, *PLoS One, 11*(8), e0160535.

Wang, H. J., Sun, C., & Huang, L.Q. (2010). Preparation and properties of whey protein packaging film. In: Proceedings of the 17th IAPRI world conference on packaging, 259–264.

Wang, Q., Zuo J.H., Wang, Q., Na, Y., & Gao, L.P. (2015). Inhibitory effect of chitosan on growth of the fungal phytopathogen, *Sclerotinia sclerotiorum*, and sclerotinia rot of carrot. *Journal of Integrative Agriculture, 14*, 691–697.

Wu, Y., Weller, C.L., Hamouz, F., Cuppett, S.L., & Schnepf, M. (2002). Development and applications of multicomponent edible coatings and films: a review. *Advances in Food and Nutrition Research, 44*, 347–394.

Xing, Y., Li, X., Xu, Q., Yun, J., Lu, Y., & Tang, Y. (2011). Effects of chitosan coating enriched with cinnamon oil on qualitative properties of sweet pepper (*Capsicum annuum* L.). *Food Chemistry, 124*, 1443–1450.

Xing, Y., Xu, Q., Li, X., Chen, C., Ma, L., Li, S., Che, Z., & Lin, H. (2016). Chitosan-based coating with antimicrobial agents: preparation, property, mechanism, and application effectiveness on fruits and vegetables. *International Journal of Polymer Science, 2016*, 4851730.

Yarahmdi, M., Azizi, M., Morid, B., & Kalatejari, S. (2014). Postharvest application of gum and mucilage as edible coating on postharvest life and quality of strawberry fruit. *International Journal of Advanced Biological and Biomedical Research, 2*(4), 1279–1286.

Yossef, M.A. (2014). Composition of different edible coating material for improvement of quality and shelf life of perishable fruits. *Middle-East Journal of Applied Science, 4*(2), 416–424.

Yousefi, M., Azizi, M., Mohammadifar, M.A., & Ehsani, A. (2018) Antimicrobial coatings and films on meats: a perspective on the application of antimicrobial edible films or coatings on meats from the past to future. *Bali Medical Journal, 7*, 87.

Yulianingish, R., Maharani, D.M., Hawa, L.C., & Sholikhan, L. (2013). Physical quality observation of edible coating made from *Aloe vera* on cantaloupe minimally processed. *Pakistan Journal of Nutrition, 12*(9), 800–805.

Zambrano-Zaragoza, M.L., Mercado-Silva, E., Ramirez-Zamorano, P., Cornejo-Villegas, M.A., Gutierrez-Cortez, E., & QuintanarGuerrero, D. (2013). Use of solid lipid nanoparticles (SLNs) in edible coatings to increase guava (*Psidium guajava* L.) shelf-life. *Food Research International, 51*, 946–953.

Zhang, M., Xiao, G., Luo, G., Peng, J., & Salokhe, V.M. (2004). Effect of coating treatments on the extension of the shelf life of minimally processed cucumber. *International Agrophysics, 18*, 87–102.

CHAPTER 18

Instrumental Methods for Food Safety and Shelf Life Assessment

Haamiyah Sidiq, Aiman Zehra, Sabreena Yousuf, and Sajad Mohd Wani

CONTENTS

Email: Sajad Mohd Wani, wanisajad82@gmail.com

DOI: 10.1201/9781003091677-18

18.1 INTRODUCTION

Instrumental methods are carried out with instruments that are designed to have automated sample introduction and data processing. The most suitable instrumental technique depends on the physicochemical characteristics of the analytes, the detection limit, and the resolution required. Due to the complex nature of the food matrix, it often becomes difficult to accurately analyze one component in the presence of others using the classical method of analysis. Interferences are encountered during the measurement of minor components in the presence of the components present in bulk quantities. This may lead to inaccurate and unreliable results and sometimes erroneous and false results because of the lack of specificity and sensitivity of the classical methods. Requirements for objectivity and efficiency in quality control lead to the necessity of the determination of quality parameters by methods that are fast, low-cost, and sufficiently precise (Nachev et al., 2019). This chapter is dedicated to providing information on various instrumental methods that have become a backbone for food safety and shelf life assessment.

18.2 HIGH-PERFORMANCE LIQUID CHROMATOGRAPHY

18.2.1 Principle

High-performance liquid chromatography (HPLC) is a technique used to separate the components of the chemical mixture and is modified column chromatography. The components are first dissolved in liquid solvents and then forced to flow through a chromatographic column under high pressure. The mixture is resolved into its components in the column. The amount of resolution depends on the extent of interactions between solute components and the stationary phase (Figure 18.1) (Kupiec, 2004).

18.2.2 Applications of HPLC

There is a wide range of applications related to HPLC in the field of food. The major topic of concern mainly for humans is our health as we consume processed and more versatile food than ever. To ensure the safety of food material, various analytical methods have been used.

The qualitative and quantitative analysis is done by HPLC. It is one of the most appropriate fast, precise, accurate, and highly sensitive techniques, and it can be used even for compounds present in very low quantities and within complex matrixes.

The HPLC technique is usually preferred for analysis of aromatic hydrocarbons, polymers, and drugs and has been used to detect substances that are low volatile and have poor thermal stability(Li et al., 2010). HPLC has been used in the determination of polycyclic aromatic hydrocarbons in fried and roasted meat products. It has been used to determine the aromatic amines that are highly toxic

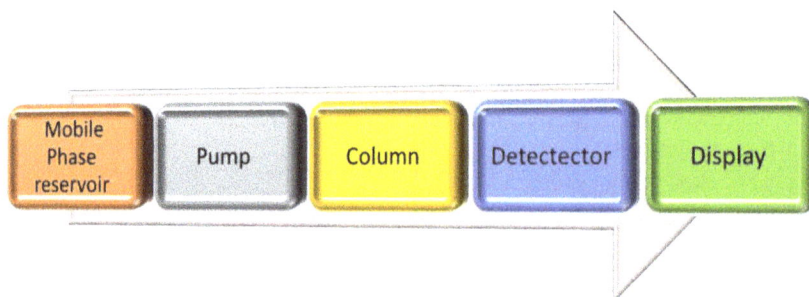

Figure 18.1 Flow diagram of high-performance liquid chromatography.

and carcinogenic. These compounds are mostly present in artificial pigments. The determination time of analysis in HPLC is only 5 minutes compared with thin-layer chromatography (TLC) analysis, which takes 50 minutes to complete (Gebauer et al., 2017). Application of HPLC is seen in the determination of pesticide residues (Cai et al., 2017). Meng et al. (2016) reported vitamin B_1 and B_2 can be effectively determined by HPLC. Synthetic dyes in foods and beverages have been determined by HPLC (Prado and Godoy, 2002), and is used to determine the major mycotoxin (patulin) in apple and apple juice concentrate, which is the major cause of apple rot (Wu et al., 2008). HPLC is a rapid and simple technique used to analyze vitamin K in food, tissues, and blood (Jakob and Elmadfa, 2000). Vitamins C, B, and E are determined in a wide range of food products using HPLC (Fontannaz et al., 2006; Sund et al., 2007; Sasaki et al., 2020). HPLC is also used to separate phenolic compounds in food (Bonoli et al., 2003; Sies, 2010; Krystyna and Sentkowska, 2015).

18.3 GAS CHROMATOGRAPHY (GC)

18.3.1 Principle

When the mobile phase comes in contact with an adsorbent (stationary phase), certain amounts of it get adsorbed on the solid support according to the laws of Freundlich or Langmuir. If the gas comes in contact with a liquid (stationary phase), some amount of it gets dissolved in the liquid according to Henry's law of partition. Different species of molecules (which can be volatilized without decomposition below 300°C and are stable with respect to isomerization) spend different periods in the stationary phase and the mobile phase and this affects the time it takes the gas to pass through the column. Various compounds emerge from the column at different intervals of time to form a chromatogram. This is the basis of separation in gas chromatography (GC) (Chatwal and Arnand, 2016).

The capabilities of integrated GC-mass spectrometry (GCMS) are almost unique in meeting the requirements for analytical methods, which are not only highly sensitive but also specific and reliable in providing information on specific compounds as a function of their concentrations. In GCMS, GC separates the components of a mixture and MS provides the identification on the structural components. Sample volatility is the basic requirement of GC; therefore, MS ionization techniques that require gas-phase analytes are ideally suited to GC/MS.

18.3.2 Applications of Gas Chromatography

Some of the applications of GC are fatty acid profiling; separation and identification of proteins, carbohydrates, preservatives, flavor, colorants, vitamins, and pesticide residues in food samples; analysis of fatty acids produced by bacteria that enables the identification of the bacteria; analysis of dairy products for aldehydes and ketones, which cause rancidity; analysis of butter for butterfat content, colors, and flavors; the GC profiling of the essential volatile oils; and detection of steroid administered to animals (Chatwal and Arnand, 2016).

Comparison of volatile trapping techniques for the comprehensive analysis of food flavoring can be carried out by GCMS (Diez-Simon et al., 2020). Active flavor compounds of rambutan (*Nephelium opossum* L) seed fat is also determined by solid-phase microextraction (SPME) GCMS (Khairy et al., 2017). Moreover, the flavoring compounds, which might be used as adulterants (imitation of commercially valuable fragrant varieties) in rice, can be detected by using headspace SPME coupled with the GCMS selected ion monitoring technique (Peng et al., 2020). Volatile non-intentionally added substances coming from a starch-based biopolymer (which is intended for use as a food contact material) can also be evaluated by GCMS (Osorio et al., 2019). Quantification of targeted pesticide residues and additional substances of concern in fresh-food commodities is done using GC-MS (Vargas-Pérez et al., 2020). Bisphenols (which are endocrine disruptors and can lead to serious health issues) in canned meat have also been evaluated by this technique (Cunha et al.,

2020). A novel approach for rapid identification of *Staphylococcus aureus*, *Vibrio parahaemolyticus*, and *Shigella sonnei* in foods by SPME coupled with GCMS has been used (Wang et al., 2018).

18.4 LIQUID CHROMATOGRAPHY-MASS SPECTROMETRY

Liquid chromatography-mass spectrometry (LC-MS) is a rapid and cost-effective technique used for quantitative measurements of organic molecules (Taylor, 2005).

18.4.1 Applications of LC-MS

LC-MS has been used to determine pesticide residues, such as N-methyl carbamate from spinach, tomatoes, potatoes, apples, cucumbers, and mandarins. There have been 160 pesticides (different classes) determined from tomato, pear, and orange by using LC-MS. Phenoxy acid residues found in food have been determined in rice. LC-MS has been used to determine carbosulfan and its metabolites in oranges, rice, potatoes, and strawberries. Nitenpyram, isocarbophos, and isofenphos-methyl have been determined in peppers. Target pesticides and non-target metabolites are estimated by LC-MS in oranges, lemons, grapes, and olive oil. LC-MS has been used to determine Thiacloprid Thiamethoxam 11 organophosphorus in honey. Veterinary drugs such as avermectins, benzimidazoles, -agonists, -lactams, corticoids, macrolides, nitroimidazoles, quinolones, sulfonamides, and tetracyclines have been determined in raw milk by LC-MS (Malik et al., 2010).

In foods, a large number of toxic substances such as antibiotics, pesticides, and toxins of various origins are present in very small quantities, which can be very toxic to human health. LC-MS can detect these sensitive compounds in low concentrations. LC-MS has been used for the determination of mycotoxin in food supplements (Mavungu et al., 2009). Spanjer et al. (2008) used the LC-MS/MS multimethod for mycotoxin determination after a single extraction, with validation data, for peanuts, pistachios, wheat, maize, cornflakes, raisins, and figs. Gentili (2007) has used LC-MS methods for analyzing anti-inflammatory drugs in animal-food products. Tsutsumiuchi (2003) used LC/MS/MS for the determination of acrylamide in processed foods.

18.5 PASTING PROPERTIES

Viscosity is measured by the process of heating and cooling at a known period (Gamel et al., 2012). A gel is formed due to the swelling of starch granules, as a result of the stirring and heating of water and starch/flour slurry.

The Rapid Visco Analyzer (RVA) test is performed in five stages:

1. Addition of water to starch/flour
2. Heating
3. Holding at a maximum temperature
4. Cooling
5. Final holding stage

RVA starts with an initial temperature set at 50°C, with a holding time of 1 minute at 50°C and then heating again for about 3 minutes 42 seconds at 95°C; holding at 95°C for 2 minute 30 seconds and again cooling it to 50°C over 3 minutes 48 seconds; and, finally, holding at 50°C for 2 minutes.

18.5.1 Application of the RVA in Cereal Analysis

The most important and useful property of starch is its pasting property, which is significantly influenced by many factors such as nature and source of starch additives, isolation methods, particle size, waxy or non-waxy contribution, and its interaction with starches and processing techniques. The most versatile natural food component is starch, which is composed of two glucosidic macro-molecules with entirely different properties and structures. Starch molecules undergo gelatinization during which they absorb water and lose their crystallinity, resulting in swelling up to several times their original size. The pasting property of flours during cooking and cooling undergo many pro-cesses including swelling, deformation, fragmentation, and solubilization (Zhu, 2015).

Currently, the RVA (originally developed by CSIRO, Australia, to measure sprout damage in wheat) and the visco-amylograph (VAG) are the most used equipment for characterizing starch paste viscosity by measuring the gelatinization properties of flour and native or modified starch. Ahmed and Al-Attar (2015) and Ahmed et al. (2016) studied the pasting properties of various starches (e.g., rice, chestnuts) using a Brabender Micro-Visco-AmyloGraph.

The pasting properties of starch and protein are better understood during the processing of whole grain products. Slavin et al. (2000) studied the pasting properties of starch and protein of barley, pearl millet, rye, and sorghum whole grain meals, using an RVA. Bason et al. (1993) and Batey et al. (1997) detected the damage of barley during its sprouting. The recent application includes the investigation of the quality parameters of starch pasting properties of noodle texture in wheat (He et al., 2004, 2006). The quality of barley malt was estimated by Zhou and Mendham (2005) and Izydorczyk (2008).

18.6 FOURIER TRANSFORM INFRARED SPECTROSCOPY

Infrared (IR) spectroscopy is one of the most powerful analytical techniques that offers the pos-sibility of chemical identification. This technique, when coupled with intensity measurements, may be used for quantitative analysis. The IR spectrum of a chemical substance is a fingerprint for its identification. In Fourier transform infrared spectroscopy (FTIR), the sample is not subjected to thermal effects, whereas in dispersive IR spectroscopy, the sample is subjected to the thermal effect from the focused beam.

18.6.1 Applications of FTIR

FTIR spectroscopy has become a leading technique for the analysis of sugars in food samples. Quantification of sugars in mango juice as a function of ripening is done by FTIR spectroscopy (Kova and Wilson, 2001; Duarte et al., 2002). FTIR spectroscopy can also be applied to rapidly pre-dict the quality of fish fillets (Hernández et al., 2014). Coupled with an attenuated total reflectance (ATR) device, the time-consuming procedure for sample preparation is reduced and a small quan-tity of solid, liquid, or concentrated analyte solutions is required without using additional reagents or preparative processes. This technique is rapid, non-destructive, and suitable to estimate the con-stituents, composition, or pureness of samples. FTIR-ATR is an important tool to analyze potential adulterated samples of food products (Miaw et al., 2018). This technique is also used to detect adulteration in different varieties of honey (Gallardo-Velazquez et al., 2009).

18.7 DIFFERENTIAL SCANNING CALORIMETRY

18.7.1 Principle

Calorimetry is a universal method of investigating processes that are connected with the genera-tion or consumption of heat (Höhne et al., 2003). In differential scanning calorimetry (DSC), the thermal properties of a sample are recorded as a function of temperature during a predetermined temperature scan (Garden et al., 2018).

18.7.2 Applications of DSC

DSC coupled with machine learning has been used to detect adulteration of raw bovine milk. It has also been used to find the impact of wheat bran on wheat starch gelatinization (De Bondt et al., 2020). This technique has been effectively used to study and analyze the thermal properties of polymer materials. Melting temperatures, polymorphic transformations, and decompositions can be determined using this technique (Drzeżdżon et al., 2018).

18.8 POLYMER CHAIN REACTION

18.8.1 Principle

Polymer chain reaction (PCR) is a three-step cyclic in vitro procedure based on the ability of the DNA polymerase to copy a strand of DNA (Uyttendaele et al., 2014). When two primers bind to complementary strands of target DNA, the sequence in between is amplified exponentially with each cycle, making the technique a very sensitive tool. The presence of even one copy of the original template within the reaction mixture can be detected within a couple of hours, as about a billion copies are created. The results of PCR are traditionally (in conventional PCR) detected by agarose gel electrophoresis and stained with a nonspecific (i.e., sequence-independent) DNA-intercalating dye such as ethidium bromide.

18.8.2 Applications

Rapid, quantitative, and precise detection by real-time PCR (RTi-PCR) methods is crucial for the prevention and risk assessment of food poisoning (Akyol, 2018). Several simplex and multiplex PCR and RTi-PCR protocols have been developed and used for the identification of desired pathogens. Reverse transcriptase-PCR-denaturing gradient gel electrophoresis (DGGE) has been used as a rapid method for routine determination of *Vibrio* spp. in foods (Chahorm and Prakitchaiwattana, 2017). The detection and characterization of *Listeria monocytogenes* and *Listeria ivanovii* in foods and environmental sources is also done using PCR (Chen et al., 2017). For the visual detection of *L. monocytogenes* in food, DNAzyme-based PCR signal cascade amplification has been developed (Liu et al., 2018). Duplex PCR-enzyme-linked immunosorbent assay (ELISA) has been developed for simultaneous detection of *Salmonella* spp. and *Escherichia coli* O157: H7 in food (Hu et al., 2018). *Salmonella typhi* bacteria in a contaminated egg can also be detected using this technique (Nurjayadia et al., 2019). The multifaceted yeast *Kazachstania servazzii* in food can be quickly detected using this assay (Spanoghe et al., 2016). A fast RTi-PCR assay based on the TaqMan probe has been developed for the identification of edible rice grasshopper (*Oxya chinensis*) in processed food products (Kim et al., 2018). Porcine DNA in processed food samples has been rapidly detected using a streamlined DNA extraction method combined with the SYBR Green RTi-PCR assay (Lee-Tan et al., 2019).In addition, bovine and fish DNA in gelatin admixture, food products, and dietary supplements are also determined (Sultana et al., 2020). Furthermore, transgenic rice line TT51-1 in processed foods can be detected using conventional PCR, RTi-PCR, and droplet digital PCR (Wang et al., 2019).

18.9 COLOR

18.9.1 Principle

Color arises from the presence of light in greater intensities at some wavelengths than at others (Pomeranz and Meloan, 2004). In practice, it is limited to the part of the electromagnetic spectrum that is visible to the human eye from 380 to 770 nm. Color is the stimulus that results from the detection of light after it has interacted with an object. Three factors are involved: a light source, an object, and a receiver-detector.

18.9.2 Applications

A colorimeter equipped with illuminant D65 and based on the CIELAB color space has been used to monitor the color changes that take place due to dehydration of salmon, beef, and apples (Franco et al., 2019). Moreover, the color change kinetics of shrimp due to drying temperature and drying medium velocity has also been investigated (Hosseinpour et al., 2013). RGB (red, green and blue) color imaging has been used for the grading of dates (Manickavasagan et al., 2014). The potential of RGB digital imaging and hyperspectral imaging (900–1700 nm) is also evaluated for discriminating maturity level in apples under different storage conditions along with the shelf life (Garrido-Novell et al., 2012). Furthermore, the color stability of anthocyanins extracted from fermented purple sweet potato culture shows that the purple sweet potato anthocyanins are more stable under acid conditions (pH 2.0–4.0) than sub-acid conditions (pH 5.0–6.0) as per UV–Vis absorption spectra and CIELAB color coordinates (Fan et al., 2008).

18.10 TEXTURE ANALYSIS

Food texture measuring devices have five essential elements: driving mechanism, probe element, force, sensing element, and read-out system. The mechanical textural characteristics can be divided into primary parameters of hardness, cohesiveness, viscosity, elasticity and adhesiveness, and into derived parameters of brittleness, chewiness and gumminess.

The load-deformation relationship gives a measure of hardness. The rate at which a material disintegrates under mechanical action is a measure of cohesiveness and it is generally tested in terms of secondary parameters. The ease with which the material yields under an increasing compression load is a measure of brittleness, crunchiness, and crumbliness. The energy required to masticate a food quantifies tenderness, chewiness, and toughness (Pomeranz and Meloan, 2004). A characteristic of semisolid food is gumminess, which has a low degree of hardness and a high degree of cohesiveness. The rate at which a food returns to its original condition after the removal of a deforming force is an index of a food's elasticity. The work required to overcome the attractive forces between the surface of food and the surface of other materials with which the food comes in contact is a measure of adhesiveness.

18.10.1 Applications

A texture analyzer has been used to relate the quality characteristics of aged eggs and fresh eggs to vitelline membrane strength (Kirunda and McKee, 2000). In addition, a texture analyzer has been used to trace rigor mortis and subsequent resolution of chicken breast muscle (Li et al., 2010). Texture profile analysis has also been used for quince sponge cake (Salehi and Kashaninejad, 2018); shelled sunflower seed caramel snack (Gupta et al., 2007); and peach, nectarine, and date flesh (Rahman and Al-Farsi, 2005). Mechanical and acoustic evaluation of potato chip crispness can also be performed using this technique (Taniwaki and Kohyama, 2012). Tensile tests have been used to monitor bread aging (Abu-Ghoush et al., 2007) and to detect the effect of preservatives on the shelf life and quality of flatbread. Firmness testing of fruits is used to describe the mechanical properties of the fruit tissue and provide information on the storability and resistance to injury of the product during handling. Instrumental measurements of texture are preferred rather than sensory evaluation, because instruments may reduce variation among measurements due to human factors and are, in general, more precise (Concha-Meyer et al., 2015).

18.11 MICROBIOLOGICAL TECHNIQUES

1. *Pour plate technique:* This technique is used to count the number of colony-forming bacteria present in a liquid specimen. The principle of this technique is to focus on a specific number of bacteria. It is usually used to isolate the bacteria from different colonies and to demonstrate its characteristics such as color, texture, size, etc. In this method, a fixed amount of inoculum (generally 1 mL) from a broth/sample is placed in the center of a sterile Petri dish using a sterile pipette. Molten cooled agar (approximately 15 mL) is then poured into the Petri dish containing the inoculum and mixed well. After the solidification of the agar, the plate is inverted and incubated at 37°C for 24–48 hours in a BOD incubator. Microorganisms will grow both on the surface and within the medium.
2. *Spread plate technique:* This technique is used for enumeration, enrichment, screening, and selection of microorganisms. The culture is spread uniformly on the surface of the agar plate, which results in the formation of isolated colonies that are distributed evenly across the agar surface.
3. *Streaking technique:* This method is used to obtain a pure culture from the mixed culture. Quadrant streaking is done in the Petri plate in such a way that all four corners are used for isolating a single bacterial colony. The streak plate method, being a rapid qualitative isolation method, is a dilution technique that involves spreading a loopful of culture over the surface of an agar plate.

18.11.1 Applications

Microbiological tests in food products are used to identify and ensure safety and to check whether the food has been contaminated by the presence of spoilage and pathogenic organisms. Apart from this, microbiological tests are used to detect the presence of yeast and molds. The major objective of the microbiological risk assessment in food is the definition of the likelihood of a foodborne pathogen to provoke illness in humans following the consumption of a given food product. A 2015 study reported that there have been 600 million cases of foodborne diseases and 420,000 deaths recorded and estimated each year worldwide. The source of this is food contamination by pathogenic and spoilage causing bacteria (Zacharski et al., 2018); 32 diseases are caused by foodborne hazards in which 11 diarrheal disease agents are included. *E. coli, Salmonella* spp., *Campylobacter* spp., and some *Bacillus* strains are pathogenic bacteria that represent the greatest issue for fresh products. On the other hand, the major concern for processed and long-term storage foods are foodborne pathogens such as *L. monocytogenes* and some fungal strains (Giacometti et al., 2013).

18.12 ENZYME-LINKED IMMUNOSORBENT ASSAY

ELISA is a technique designed to detect and quantify soluble substances such as peptides, proteins, antibodies, and hormones. In ELISA, the antigen is immobilized on a solid surface and then complexed with an antibody that is linked to a reporter enzyme. Detection is accomplished by measuring the activity of the reporter enzyme via incubation with the appropriate substrate.

18.12.1 Applications

ELISA is based on antibody-antigen interaction wherein the detection may take from 2 to 3 hours. A wide range of applications for ELISA has been found in the food industry including to detect food allergens in peanuts, milk, walnuts, almonds, and eggs. It has also been used to detect celiac disease (Dossi et al., 2007).

ELISA is mainly used to detect pesticides, heavy metals, and pollutants in foods. It has also been used to determine nutrient content in a food sample. ELISA can detect the concentration of contaminants in the range of parts per million (ppm) (Berg and Otley, 2002).

18.13 CONCLUSION

Various instrumental methods are used for food safety and shelf life assessment. In the broadest sense, an instrument for analysis does not generate quantitative data; instead, it simply converts information to a form that is more readily observable. Thus, the instrument can be visualized as a communication device that performs the function of generating a signal, transduction, amplification, and presentation of the signal. Instrumental methods have become a backbone in food analysis; therefore, analysts must have an understanding of the principles, applications, and limitations of these techniques.

REFERENCES

Abu-Ghoush, M., Herald, T. J., Dowell, F., Xie, F., Aramouni, F. M., & Madl, R. (2007). Effect of preservatives addition on the shelf-life extensions and quality of flat bread as determined by near-infrared spectroscopy and texture analysis. *International Journal of Food Science and Technology*, 43(2):357–364.

Ahmed, J., & Al-Attar, H. (2015). Effect of drying method on rheological, thermal, and structural properties of chestnut flour doughs. *Food Hydrocolloids*, 51: 76–87.

Ahmed, J., Al-Attar, H., & Arfat, Y.A. (2016). Effect of particle size on compositional, functional, pasting and rheological properties of commercial water chestnut flour. *Food Hydrocolloids*, 52:888–895.

Akyol, I. (2018). Development and application of RTi-PCR method for common food pathogen presence and quantity in beef, sheep and chicken meat. *Meat Science*, 137:9–15. doi:10.1016/j.meatsci.2017.11.001

Bason, M.L., Ronalds, J.A., Wrigley, C.W., & Hubbard, L.J. (1993).Testing for sprout damage in malting barley using the Rapid Visco-Analyzer. *Cereal Chemistry*, 70(3):269–272.

Batey, I.L., Curtin, B.M., & Moore, S.A. (1997). Optimization of Rapid-Visco Analyser test conditions for predicting Asian noodle quality. *Cereal Chemistry*, 74(4):497–501.

Berg, D., & Otley, C. C. (2002). Skin cancer in organ transplant recipients: epidemiology, pathogenesis, and management. *Journal of the American Academy of Dermatology*, 47(1):1–20.

Bonoli, M., Pelillo, M., Toschi, T.G., & Lercker, G. (2003). Analysis of green tea catechins: a comparative study between HPLC and HPCE. *Food Chemistry*, 81(4):631–638.

Cai, W., Guan, Y., & Zhou, Y. (2017). Detection and characterization of the metabolites of rutaecarpine in rats based on ultra-high-performance liquid chromatography with linear iontrap-Orbitrap mass spectrometer. *Pharmaceutical Biology*, 55:294–298.

Chahorm, K., & Prakitchaiwattana, C. (2017). Application of reverse transcriptase-PCR-DGGE as a rapid method for routine determination of *Vibrio* spp. in foods. *International Journal of Food Microbiology*, 264:46–52. doi:10.1016/j.ijfoodmicro.2017.10.014

Chatwal, G.R., & Anand, S.K. (2016). *Instrumental methods of chemical analysis* (5th ed.). Mumbai: Himalaya Publishing House.

Chen, J., Healey, S., Regan, P., Laksanalamai, P., & Hu, Z. (2017). PCR-based methodologies for detection and characterization of *Listeria monocytogenes* and *Listeria ivanovii* in foods and environmental sources. *Food Science and Human Wellness*, 6(2):39–59. doi:10.1016/j.fshw.2017.03.001

Concha-Meyer, A., Eifert, J. D., Williams, R. C., Marcy, J. E., & Welbaum, G. E. (2015). Shelf life determination of fresh blueberries (*Vaccinium corymbosum*) stored under controlled atmosphere and ozone. *International Journal of Food Science*, 2015:164143.

Cunha, S.C., Inácio, T., Almada, M., Ferreira, R., & Fernandes, J.O. (2020). Gas chromatography-mass spectrometry analysis of nine bisphenols in canned meat products and human risk estimation. *Food Research International*, 135:109293. doi:10.1016/j.foodres.2020.109293

De Bondt, Y., Liberloo, I., Roye, C., Goos, P., & Courtin, C.M. (2020). The impact of wheat (*Triticum aestivum* L.) bran on wheat starch gelatinization: A differential scanning calorimetry study. *Carbohydrate Polymers*, 241: 116262. doi:10.1016/j.carbpol.2020.116262

Diez-Simon, C., Ammerlaan, B., van den Berg, M., Van Duynhoven, J., Jacobs, D., Mumm, R., & Hall, R.D. (2020). Comparison of volatile trapping techniques for the comprehensive analysis of food flavourings by gas chromatography-mass spectrometry. *Journal of Chromatography A*, 1624:461191. doi:10.1016/j.chroma.2020.461191

Dossi, N., Piccin, E., Bontempelli, G., Carrilho, E., & Wang. J. (2007). Rapid analysis of azo-dyes in food by microchip electrophoresis with electrochemical detection. *Electrophoresis*, 28(22):4240–4246.

Drzeżdżon, J., Jacewicz, D., Sielicka, A., & Chmurzyński, L. (2018). Characterization of polymers based on differential scanning calorimetry based techniques. *Trends in Analytical Chemistry*, 110:51–56. doi. org/10.1016/j.trac.2018.10.037

Duarte, I.O., Barros, A.I., Delgadillo, I., & Almeid, A. (2002). Application of FTIR spectroscopy for the quantification of sugars in mango juice as a function of ripening. *Journal of Agricultural and Food Chemistry*, 50(11):3104–3111.

Fan, G., Han, Y., Gu, Z., & Gu, F. (2008). Composition and colour stability of anthocyanins extracted from fermented purple sweet potato culture. *LWT*, 41(8):1412–1416. doi:10.1016/j.lwt.2007.09.003

Fontannaz, P., Kilinc, T., & Heudi, O. (2006). HPLC-UV determination of total vitamin C in a wide range of fortified food products. *Food Chemistry*, 94(4):626–631.

Franco, S., Jaques, A., Pinto, M., Fardella, M., Valencia, P., Núñez, H., Ramírez, C., Simpson, R. (2019). Dehydration of salmon (Atlantic salmon), beef, and apple (Granny Smith) using Refractance window™: Effect on diffusion behaviour, texture, and color changes. *Innovative Food Science and Emerging Technologies*, 52:8–16. doi:10.1016/j.ifset.2018.12.001

Gallardo-Velazquez, T., Osorio-Revilla, G., Zuniga-de Loa, M., & Rivera-Espinoza, Y. (2009). Application of FTIR-HATR spectroscopy and multivariate analysis to the quantification of adulterants in Mexican honeys. *Food Research International*, 42(3):313–318.

Gamel, T.H., Abdel-Aal, E.S.M., Wood, P.J., Ames, N.P., & Tosh, S.M. (2012). Application of the Rapid Visco Analyzer (RVA) as an effective rheological tool for measurement of ß-glucan viscosity. *Cereal Chemistry*, 89(1):52–58.

Garden, J.L., Moiroux, G., Dignac, C., & Chaussy, J. (2018). Derivative scanning calorimetry: A new highly sensitive method in differential calorimetry. *Thermochimica Acta*, 670:202–210. doi:10.1016/j. tca.2018.10.020

Garrido-Novell, C., Pérez-Marin, D., Amigo, J.M., Fernandez-Novales, J., Guerrero, J.E., & Garrido-Varo, A. (2012). Grading and colour evolution of apples using RGB and hyperspectral imaging vision cameras. *Journal of Food Engineering*, 113(2):281–288. doi:10.1016/j.jfoodeng.2012.05.038

Gebauer, S., Friebe, S., & Scherer, G. (2017). High performance liquid chromatography on calixarene-bonded silica gels. III. Separations of cis/trans isomers of proline-containing peptides. *Journal of Chromatographic Science*, 36:388–394.

Gentili, A. (2007). LC-MS methods for analyzing anti-inflammatory drugs in animal-food products. *Trends in Analytical Chemistry*, 26(6):595–608.

Giacometti, J., Tomljanovic, A.B., & Josic, D. (2013). Application of proteomics and metabolomics for investigation of food toxins. *Food Research International*, 54(1):1042–1051. https://doi.org/10.1016/j. foodres.2012.10.019.

Gupta, R.K., Sharma, A., & Sharma, R. (2007). Instrumental texture profile analysis (TPA) of shelled sunflower seed caramel snack using response surface methodology. *Food Science and Technology International*, 13(7):455–460. doi: 10.1177/1082013207088369

He, Z., Xu, Z., Xia, L., Xia, X., Yan, J., Zhang, Y., & Chen, X. (2006). Genetic variation for waxy proteins and starch properties in Chinese winter wheats. *Cereal Research Communications*, 34(2):1145–1151.

He, Z.H., Yang, J., Zhang, Y., Quail, K.J., & Pena, R.J. (2004). Pan bread and dry white Chinese noodle quality in Chinese winter wheats. *Euphytica*, 139:257–267.

Hernández, M., Gallardo-Velazquez, T., Osorio-Revilla, G., Almaraz-Abarca, N., & Castaneda-Perez, E. (2014). Application of MIR-FTIR spectroscopy and chemometrics to the rapid prediction of fish fillet quality. *CyTA – Journal of Food*, 12(4):369–377.

Höhne, G.W.H., Hemminger, W.F., & Flammersheim, H.J. (2003). *Differential scanning calorimetry* (2nd ed.). Berlin: Springer-Verlag.

Hosseinpour, S., Rafiee, S., Mohtasebi, S.S., & Aghbashlo, M. (2013). Application of computer vision technique for online monitoring of shrimp color changes during drying. *Journal of Food Engineering*, 115(1):99–114. doi:10.1016/j.jfoodeng.2012.10.003

Hu, J., Huang, R., Wang, Y., Wei, X., Wang, Z., Geng, Y., Jing, J., Gao, H., Sun, X., Dong, C., & Jiang, C. (2018). Development of duplex PCR-ELISA for simultaneous detection of *Salmonella* spp. and *Escherichia coli* O157: H7 in food. *Journal of Microbiological Methods*, 154:127–133. doi:10.1016/j.mimet.2018.10.017

Izydorczyk, M. (2008). *Using RVA to measure pre-germination in barley and predict germination energy after storage.* Winnipeg, Canada: Canadian Grain Commission.

Jakob, E., & Elmadfa, I. (2000). Rapid and simple HPLC analysis of vitamin K in food, tissues and blood. *Food Chemistry*, 68(2):219–222.

Khairy, H.L., Saadoon, A.F., Zzaman, W., Yang, T.A., & Easa, A.M. (2017). Identification of flavor compounds in rambutan seed fat and its mixture with cocoa butter determined by SPME-GCMS. *Journal of King Saud University Science*, 30(3):316–323. doi:10.1016/j.jksus.2017.03.001

Kim, S.Y., Kim, M.J., Jung, S.K., & Kim, H.Y. (2018). Development of a fast real-time PCR assay based on TaqMan probe for identification of edible rice grasshopper (*Oxya chinensis*) in processed food products. *Food Research International*, 116:441–446. doi:10.1016/j.foodres.2018.08.059

Kirunda, D.F.K., & McKee, S.R. (2000). Relating quality characteristics of aged eggs and fresh eggs to vitelline membrane strength as determined by a texture analyzer. *Poultry Science*, 79(8):1189–1193.

Kova, K., & Wilson. (2001). Developments in mid-infrared FT-IR spectroscopy of selected carbohydrates. *Carbohydrate Polymers*, 44(4):291–303.

Krystyna, P., & Sentkowska, A. (2015). Recent developments in the HPLC separation of phenolic food compounds. *Critical Reviews in Analytical Chemistry*, 45:41–51.

Kupiec, T. (2004). Quality-control analytical methods: High-performance liquid chromatography. *International Journal of Pharmaceutical Compounding*, 8(3):223–227.

Lee-Tan, L., Ahmed, S.A., Kit-Ng, S., Citartan, M., Raabe, C.A., Rozhdestvensky, T.S., & Tang, T.H. (2019). Rapid detection of porcine DNA in processed food samples using a streamlined DNA extraction method combined with the SYBR Green real-time PCR assay, *Food Chemistry*, 309:125653. doi:10.1016/j.foodchem.2019.125654

Li, C., Shi, P., Xu, C., Xu, X., & Zhou, G. (2010). Tracing processes of rigor mortis and subsequent resolution of chicken breast muscle using a texture analyzer. *Journal of Food Engineering*, 100(3):388–391. doi:10.1016/j.jfoodeng.2009.10.040

Liu, Z., Yao, C., Yang, C., Wang, Y., Wan, S., & Huang, J. (2018). Development of DNAzyme-based PCR signal cascade amplification for visual detection of *Listeria monocytogenes* in food. *Analytical Biochemistry*, 553:7–11. doi: 10.1016/j.ab.2018.05.015

Malik, K.A., Blasco, C., & Picó, Y. (2010). Liquid chromatography–mass spectrometry in food safety. *Journal of Chromatography*, 1217:4018–4040.

Manickavasagan, A., Al-Mezeini, N.K., & Al-Shekaili, H.N. (2014). RGB color imaging technique for grading of dates. *Scientia Horticulturae*, 175:87–94. doi:10.1016/j.scienta.2014.06.003

Mavungu, J., Monbaliu, S., Scippo, L.S., Maghuin-Rogister, G. Schneider, Y.J., Larondell, Y., Callebaut, A., Robbens, J., Peteghem, C.V., & de Saeger, S. (2009). LC-MS/MS multi-analyte method for mycotoxin determination in food supplements. *Food Additives and Contaminants*, 26(6):885–895.

Meng, X.F., Geng, N.W., & Yang, K. (2016). Molecularly imprinted polymer-high performance liquid chromatography for the determination of tetracycline drugs in animal derived foods. *Food Control* 69: 171–176.

Miaw, C.S.W., Assis, C., Silva, A.R.C.S., Cunha, M.L., Sena, M.M., & de Souza, S.V.C. (2018). Determination of main fruits in adulterated nectars by ATR-FTIR spectroscopy combined with multivariate calibration and variable selection methods. *Food Chemistry*, 254:272–280

Nachev, V.G., Titova, T, P., & Kosturkov, R.D. (2019). Instrumental data fusion for food analysis application. *IFAC PapersOnLine*, 52(25):58–63.

Nurjayadia, M., Pertiwi, Y.P., Islami, N., Azizah, N., Efrianti, U.R., Saamia, V., Wiranatha, I.M., Nastassya, L., & El-Enshasyee, H.A. (2019). Detection of the *Salmonella typhi* bacteria in contaminated egg using realtime PCR to develop rapid detection of food poisoning bacteria. *Biocatalysis and Agricultural Biotechnology*, 20:101214. doi:10.1016/j.bcab.2019.101214

Osorio, I., Dreolin, N., Aznar, M., Nerin, C., & Hancock, P. (2019). Determination of volatile non intentionally added substances coming from a starch-based biopolymer intended for food contact by different gas chromatography-mass spectrometry approaches. *Journal of Chromatography A*, 1599:15–222. doi:10.1016/j.chroma.2019.04.007

Peng, J., Yang, Y., Zhou, Y., Hocart, C.H., Zhao, H., Hu, Y., & Zhang, F. (2020). Headspace solid-phase microextraction coupled to gas chromatography mass spectrometry with selected ion monitoring for the determination of four food flavoring compounds and its application in identifying artificially scented rice. *Food Chemistry*, 313:126136. doi:10.1016/j.foodchem.2019.126136

Pomeranz, Y., & Meloan C.E. (2004). *Food analysis theory and practice* (3rd ed.). New Delhi: CBS Publishers.

Prado, M.A., & Godoy, H.T. (2002). Validation of the methodology to determine synthetic dyes in foods and beverages by HPLC. *Journal of Liquid Chromatography & Related Technologies*, 25(16): 2455–2472.

Rahman, M.S., & Al-Farsi, S.A. (2005). Instrumental texture profile analysis (TPA) of date flesh as a function of moisture content. *Journal of Food Engineering*, 66:505–511. doi:10.1016/j.jfoodeng.2004.04.022

Salehi, F., & Kashaninejad, M. (2018). Texture profile analysis and stress relaxation characteristics of quince sponge cake. *Journal of Food Measurement and Characterization*, 12(2):1203–1210. doi:10.1007/s11694-018-9734-3

Sasaki, K., Hatate, H., & Tanaka, R. (2020). Determination of 13 vitamin B and the related compounds using HPLC with UV detection and application to food supplements. *Chromatographia*, 83:839–851.

Sies, H. (2010). Polyphenols and health: Update and perspectives. *Archives of Biochemistry and Biophysics*, 501:6–9.

Slavin, J.L., Jacobs, D., & Marquardt, L. (2000). Grain processing and nutrition. *Critical Reviews in Food Science and Nutrition*, 4:309–326.

Spanjer, M.C., Rensen, P.M., & Scholten, J.M. (2008). LC–MS/MS multi-method for mycotoxins after single extraction, with validation data for peanut, pistachio, wheat, maize, cornflakes, raisins and figs. *Food Additives and Contaminants*, 25(4):472–489.

Spanoghe, M., Jara, M.G., Rivière, J., Lanterbecq, D., Gadenne, M., & Marique, T. (2016). Development and application of a quantitative real-time PCR assay for rapid detection of the multifaceted yeast *Kazachstania servazzii* in food. *Food Microbiology*, 62:133–140. doi:10.1016/j.fm.2016.10.015

Sultana, S., Hossain, M.A.M., Azlan, A., Johan, M.R., Chowdhury, Z.Z., & Ali, M. E. (2020). TaqMan probe based multiplex quantitative PCR assay for determination of bovine, porcine and fish DNA in gelatin admixture, food products and dietary supplements. *Food Chemistry*, 325:126756. doi:10.1016/j.foodchem.2020.126756

Sund, I., Murkovic, M., Bandoniene.D., & Brigitte, M. (2007). Vitamin E content of foods: Comparison of results obtained from food composition tables and HPLC analysis. *Clinical Nutrition*, 26:145–153.

Taniwaki, M., & Kohyama, K. (2012). Mechanical and acoustic evaluation of potato chip crispness using a versatile texture analyzer. *Journal of Food Engineering*, 112:268–273. doi:10.1016/j.jfoodeng.2012.05.015

Taylor, P.J. (2005). Matrix effects: The Achilles heel of quantitative high-performance liquid chromatography-electrospray-tandem mass spectrometry. *Clinical Biochemistry*, 38:328–334.

Tsutsumiuchi, K., Hibino, M., Kambe, M., Oishi, K., Okada, M., Miwa, J., & Taniguchi, H. (2003). Application of Ion-trap LC/MS/MS for determination of acrylamide in processed foods. *Journal of the Food Hygienic Society of Japan*, 45(2):95–99.

Uyttendaele, M., Rajkovic, A., Ceuppens, S., Bart, L., Coillie, E.V., Herman, L., Jasson, V., & Imberechts, H. (2014). PCR applications in food microbiology. *Encyclopedia of Food Microbiology*, 1033–1041. doi:10.1016/B978-0-12-384730-0.00246-9

Vargas-Pérez, M., Domínguez, I., González, F., & Frenich, A.G. (2020). Application of full scan gas chromatography high resolution mass spectrometry data to quantify targeted-pesticide residues and to screen for additional substances of concern in fresh-food commodities. *Journal of Chromatography A*, 1622:461118. doi:10.1016/j.chroma.2020.461118

Wang, X., Tang, T., Miao, Q., Xie, S., Chen, X., Tang, J., Peng, C., Xu, X., Wei, W., You, Z., & Xu, J. (2019). Detection of transgenic rice line TT51-1 in processed foods using conventional PCR, real-time PCR, and droplet digital PCR. *Food Control*, 98:380–388. doi:10.1016/j.foodcont.2018.11.032

Wang, Y., Liu, S., Pu, Q., Li, Y., Wang, X., Jiang, Y., Yang, D., Yang, Y., Yang, J., & Sun, C. (2018). Rapid identification of *Staphylococcus aureus*, *Vibrio parahaemolyticus* and *Shigella sonnei* in foods by solid phase microextraction coupled with gas chromatography–mass spectrometry. *Food Chemistry*, 262: 7–13. doi:10.1016/j.foodchem.2018.04.088

Wu, R.-N., Dang, Y.-L., Niu, L., & Hu, H. (2008). Application of matrix solid-phase dispersion–HPLC method to determine patulin in apple and apple juice concentrate. *Journal of Food Composition and Analysis*, 21(7):582–586.

Zacharski, K.A., Southern, M., Ryan, A., & Adley, C.C. (2018). Evaluation of an environmental monitoring program for the microbial safety of air and surfaces in a dairy plant environment. *Journal of Food Protection*, 81(7):1108–1116. https://doi.org/10.4315/0362-028X.JFP-17-464

Zhou, M.X., & Mendham, N.J. (2005). Predicting barley malt extract with a Rapid Viscoanalyser. *Journal of Cereal Science*, 41:31–36.

Zhu, F. (2015). Isolation, composition, structure, properties, modifications, and uses of yam starch. *Comprehensive Reviews in Food Science and Food Safety*, 14: 357–386.

CHAPTER 19

Hygienic Design and Cleaning

Nadira Anjum, Insha Zahoor, Mohd Aaqib Sheikh, Arshied Manzoor,
Meenakshi Trilokia, and Anju Bhat

CONTENTS

19.1 INTRODUCTION

Foodborne diseases affect a large number of people throughout the world. A food processing plant is abundant in nutrients required for the growth of microorganisms. Moreover, the microbes multiply when given favorable conditions like temperature, pH, moisture, etc. Thus, the food contact surfaces of equipment used for processing in food industries are prone to the risk of contamination (Somers and Wong, 2004). Sanitary design can be defined as the use of clearly defined methods and specifications for the design, fabrication, and installation of facilities and equipment, which

Email: Nadira Anjum, nadiraanjum1736@gmail.com

DOI: 10.1201/9781003091677-19

when integrated, allow prevention of contamination as well as help in timely and effective cleaning of the entire manufacturing asset. Hygienic design is considered a prerequisite in the process of manufacturing food products to ensure quality and safety. Hygienic considerations are important in all phases of processing like the selection of proper equipment, plant design, equipment specifications, and material handling and training of staff at all stages of processing (Kuo et al., 2001). Proper hygienic design results in the prevention of contamination of food products from various physical, biological, and microbial agents to maintain consumer health. Faults in the implementation of hygienic design of both equipment and the plant may lead to incidents of food poisoning and spoilage (Faille et al., 2002). For hygienic design, product contact surface and non-contact surfaces where the product is splashed should be taken into consideration. Several food safety programs like Hazard Analysis and Critical Control Point (HACCP) are used to maintain hygienic conditions in a food business.

On the other hand, cleaning can be defined as the complete removal of all types of soil using a proper cleaning agent under recommended conditions. This is considered an essential measure to prevent contamination. Cleaning and sanitation procedures are carried out not only for the removal of soil but for the removal of microorganisms that may otherwise lead to spoilage, disease, and poor hygienic conditions. Cleaning and sanitation operations are necessary in the food industry as the microbial count present in any food product at the time of consumption should be below the acceptable limit. The amount of microorganisms present in a product also depends on the conditions in storage and distribution. Moreover, the period between production and consumption of a product will also influence the microbial load of any product. The type of microbe depends on the type of food production as well as the composition of the particular product (Verrips et al., 1980). The surface to be cleaned should be smooth and free from cracks, crevices, and sharp corners. Any equipment that is difficult to be cleaned requires aggressive cleansing agents, longer cleaning time, and it is recommended to clean that equipment frequently. There is no reason to build lines of processing that do not meet hygienic guidelines (Holah and Thorpe, 1990). Cleanability also depends on the depth of defects present on the surface. In the food, beverage, and pharmaceutical industries, highly exacting hygienic demands are imposed on the surfaces that come into close contact with the product to be manufactured. These hygienic demands are described in the form of standards that need to be followed by food businesses to maintain hygienic conditions on the premises as well as in the equipment used for manufacturing products.

The purpose of hygienic design is to

- Provide contamination-free product contact surfaces in the food processing line that are easily cleanable to remove any adherent soil,
- Minimize the occurrence of dead areas in the processing line where chemical as well as microbial contamination may take place,
- Provide maximum protection to the raw materials as well as to the finished product, and
- Give access to the procedures of cleaning, maintenance, training, and inspection.

19.2 LEGISLATION, STANDARDS, AND GUIDELINES FOR HYGIENIC EQUIPMENT DESIGN

To ensure hygienic fabrication, design, and construction of any food processing equipment, different regulatory bodies carry out equipment inspections on a routine basis. Various standards organizations have created different standards for hygienic equipment design, and they are different for each type of industry. Today, a modern food business operator must meet the standards from national as well as local regulations and by retail customers (Schmidt and Erickson, 2017). Following the standards and regulations given by different regulatory bodies, food business operators can minimize the risks to food safety and quality by physical, chemical, and biological hazards. Apart from the standards and regulations for hygienic equipment, there are some regulations for cleaning equipment. An efficient cleaning process needs to be properly supported by the management, as it first defines a particular schedule and then communicates the prepared schedule to the staff dealing with the process. Legislations, standards, and guidelines for hygienic equipment design and cleaning are outlined in the following sections.

19.2.1 Legislation

Legislation includes Annex I of the Machine Directive 2006/42/EC and 98/37/EC and Annex V of Council Directive 93/43/EEC on the Hygiene of Foodstuffs. This legislation demands that the equipment, as well as the packaging material, should be constructed and used in such a way that minimizes the overall risk of contamination. The materials used for construction should enable proper cleaning and disinfestation procedures. EU Regulation (EC) No. 852/2004 on the Hygiene of Foodstuffs deals with cleaning and hygiene of premises of food businesses.

19.2.2 Standards and Guidelines

The American Meat Institute (AMI), United States Department of Agriculture (USDA), and 3-A are responsible for hygienic design in the United States; whereas in Europe the European Hygienic Engineering & Design Group (EHEDG) is considered as the most important organization in the area of hygienic design.

19.3 MATERIALS USED FOR CONSTRUCTION

The most important step in the hygienic design of any equipment is the selection of appropriate construction materials. The food contact surfaces should be inert and non-toxic under the conditions of use. There should not be any chance of chemical reaction of the material with the sanitizing and cleansing agents. Apart from this, the materials used should be non-porous and smooth to prevent contamination as there is the possibility of the presence of microbes and insect eggs on uneven surfaces like cracks and crevices. The original finish of the material of construction should remain maintained throughout the processing and the materials should be resistant to deformation and pitting or corrosion (Milledge and Jowitt, 1980). The material should facilitate the complete removal of residues of food during the process of cleaning to prevent microbial growth. The coating used for food contact surfaces should be resistant to all the conditions of processing. Several materials are used in the construction of any food equipment, and all of these materials have different properties (Shapton and Shapton, 1998). Before selecting any material, the food business operators should be well aware of the recent developments in materials and should take advice from the material suppliers whenever required. The materials that are not recommended for construction and low-grade materials should be avoided when constructing any food equipment (Moerman and Partington, 2014). The availability of construction material should be considered while designing any equipment. The most commonly used materials for food equipment are given in the following sections.

19.3.1 Metals and Alloys

Because carbon steel is extremely sensitive to corrosion, particularly to salt and chlorine-containing solutions, is not recommended for use in food contact surfaces. The most used materials for food equipment in the category of metals and alloys are as follows.

19.3.1.1 Stainless Steel

Stainless steel is mostly used for the fabrication of food equipment. The use of stainless steel in the construction of equipment is considered safe as it is inert and resistant to corrosion under all conditions to which it is subjected. Because of the longer durability of stainless steel, it is widely used in food processing equipment. Various types of stainless steel used in designing equipment differ in their properties depending on their composition. Mostly properties are related to the percentage of chromium and nickel. Chromium is considered responsible for higher corrosion resistance, whereas nickel strengthens the alloy (Hall, 2000). The commonly used American Iron and Steel Institute (AISI) grades of stainless steel in food processing equipment are 304, 304L, 316, 316L, and 321.

On continued use, the properties of stainless steel can change as the layer of chromium oxide is affected by some cleaners that are abrasive and incompatible with the material of construction. To avoid this, the process of passivation is highly recommended. The metals mostly used for this purpose are gold, titanium, and platinum. Titanium has high corrosion resistance, especially in an acidic and salty medium, which makes its use possible for citrus products. Platinum, like titanium, has excellent corrosion resistance but the cost of both these metals is so high that it often limits their use. Gold is also recommended for use in food contact surfaces as it prevents abrasion and is compatible with materials like glass. Molybdenum prevents corrosion but should be incorporated in cost-effective levels (EHEDG, 2007).

19.3.2 Plastic Materials

A wide range of plastic materials is used for the fabrication of food contact surfaces. The most widely used materials are high- and low-density polyethene, rigid forms of polyvinyl chloride, glass-reinforced polyester, and acrylonitrile butadiene styrene. One of the most important advantages of plastics over metals is their low cost. Plastics are resistant to abrasion and are suitable for use over a wide range of temperatures. They can be cleaned easily so accumulation of dirt is minimized. The most important application of plastic materials is in the manufacture of pipelines that transport materials.

19.3.3 Elastomers

Rubbers or elastomers used for food contact surfaces should be resistant to elongation and shrinkage when subjected to temperature changes. Examples of widely used rubbery materials are fluoroelastomers, neoprene, natural rubber, nitrile or nitrile/butyl rubber silicone, and ethylene propylene diene monomer rubber. Chloroprene is applied over a wide range of temperatures. Before selecting any elastomer for construction of any part of processing equipment, a detailed study about its resistance to various chemicals and to certain cleaning and sanitizing agents should be undertaken (Plett and Graßhoff, 2006).

19.3.4 Glass

Glass has the potential to break and chip, hence, the application of glass as a construction material for food contact surfaces is limited. It is used in vessels for the light and sight opening. A special form of glass known as Pyrex has been successfully used for food contact surfaces.

Apart from the materials listed earlier, other materials like ceramics, paper, and wood are used to a lesser extent in the construction of food processing equipment. Papers are mostly recommended for single-use, especially in pipeline gaskets that can be disassembled regularly, whereas ceramics are recommended for use in membranes meant for filtration. The use of wood is restricted to hardwood cutting boards and tight grain butcher blocks (John et al., 2017).

19.4 HYGIENIC DESIGN OF FOOD PROCESSING EQUIPMENT

As discussed previously, microbes grow and multiply in favorable conditions like certain temperatures, moisture, time, and nutrients. For the hygienic design of both equipment and plant, consideration of several factors is important: compatibility of the equipment with the product, environment surrounding the equipment, maintenance of equipment, and the fluids used to clean the equipment. A hygienic design of any equipment determines the efficiency of any cleaning method used. Equipment with poor hygienic design is difficult to clean, and the complete removal of microorganisms from it is impossible as there is always a chance of multiplication of the microbes surviving the cleaning process (Lelieveld, 1994). Before starting the process of production again, the microbes present in the processing line must be inactivated. The presence of dead areas in the processing line must be avoided because microbes reside in these areas and have time to multiply. Cracks and crevices on the food contact surfaces are also to be avoided to prevent the multiplication of microbes in the processing line (Matuszek, 2012). Moreover, after using the equipment, it should be free from moisture as the presence of moisture supports the growth of microbes. Food business operators must undertake a detailed study about whether the equipment is cleanable with a variety of cleaning agents like steam, hot water, or chemicals and whether the equipment in the processing line facilitates hygienic operation. The hygienic design of equipment prevents food contact surfaces from corrosion that may result from chemical cleaning and sanitizing agents. Improperly installed equipment in food businesses results in difficulty in cleaning and poor sanitation practices. Poor hygienic design of equipment has resulted in a decrease in the production of particular food products, product recalls, and even closure of food businesses (Lelieveld et al., 2003). Educating and training people and staff about the safe and hygienic operation of equipment is the most important factor and needs to be given as much attention as possible by food business operators. After a thorough inspection and testing of the equipment and inspection of the equipment construction material, EHEDG certifies the hygienic design. According to EHEDG guidelines (Document 8, EHEDG, 2004), the requirements for hygienic design of equipment are listed as follows:

- *Surface finish:* To facilitate a hygienic and easier cleaning operation the food contact surfaces should be smooth. Rough food contact surfaces must be avoided, as these surfaces must undergo wear and tear during the processing operation and rough surfaces make the process of cleaning more difficult (Riedewald, 2006).
- *Joints:* Whether permanent (welded), dismountable, or flanged, the type of joint should prevent the possibility of contamination. The surface of joints should be smooth and free from crevices as uneven surfaces may harbor microbes. Also, the joints should prevent ingress of microbes through leaks.
- *Fasteners:* According to Statutory Instrument No. 3073 (Machinery Regulations 1992), in designing any equipment the use of fasteners like rivets, nuts and screws, etc., must be avoided unless it is necessary. In any complex food processing equipment, rotating parts when fixed by fasteners can retain a part of the product in the slots. This is difficult to remove and leads to a potential health hazard by making the conditions favorable for the growth of harmful microorganisms.
- *Drainage:* The accumulation of cleaning and rinsing fluids in the equipment and in the pipelines meant for drainage leads to the growth of microorganisms, leading to contamination of the food product. To avoid this, self-draining equipment and pipelines are recommended.
- *Internal angles and corners:* Machinery regulations suggest that the radius of internal angles and corners should facilitate an easier, efficient, and thorough cleaning operation. Because the cleaning of internal angles is difficult, technically giving a proper value of radius is not possible, so it is recommended to have a radius as large as possible.
- *Dead spaces:* During the process of designing and installation of any equipment, care should be taken to prevent the formation of dead spaces. Dead spaces are those areas in the processing line in which product or any cleaning fluid could accumulate, leading to the growth of microbes and contamination of the product.

- *Bearings and shaft seals:* We know that bearings require lubricants for proper functioning; thus, it is recommended that bearings should be lubricated by the product itself. If the bearings are not lubricated by the product, then lubricants of an edible nature should be used. Also, the seals used should be easily cleanable.
- *Controls*: Because controls are frequently touched by the staff, they should be designed to avoid the chances of contamination. Also, the controls should be easy to clean.
- *Instrumentation:* The materials used for the manufacturing of instruments should be appropriate and desirable. If any instrument contains fluid, like some pressure gauges, then the fluid should be approved for use on the food contact surface. Installation of the instruments should also be done hygienically. For installation of any equipment in the food processing line, care should be taken in maintaining a proper distance between the equipment and between the equipment and walls.
- *Doors, covers, and panels:* These should be designed to restrict the entry of contaminants and prevent the accumulation of dust, dirt, and soil. Whenever required, doors, covers, and panels should be removed to facilitate proper cleaning.

19.5 PRINCIPLES OF HYGIENIC DESIGN

Principles of sanitary equipment design are developed to meet the expectations of food business operators. These principles are designed in proper consultation with manufacturers, the government, and other organizations involved in the process of certification. While following the principles of hygienic design, the possibility of product recall is prevented. These principles help the equipment manufacturers and food business operators, where the equipment is to be installed and used, to work together and find the problems of common concern (EHEDG, 2004). These principles of hygienic design help food business operators think about the aspects of safety and quality before installing any equipment. Apart from this, the principles of sanitary equipment design help to identify and prevent all types of potential risks (Matuszek, 2011b). Hygienic design of equipment initially involves the higher cost of purchasing new equipment, but it is beneficial in the long run and prevents the reputation of the food business. It also results in a reduction in the amount of water and chemicals used and prevents dirt, debris, and microbial accumulation (Cossen et al., 2005). One of the most important reasons for developing principles of hygienic design was the complexity of the available standards. Also, it was noticed that the design for the same kind of equipment meant for a common purpose in the processing line differed from processor to processor. These 10 principles of hygienic design were created by the equipment design task force of the AMI in 2001:

1. Cleanable to a microbiological level
2. Made of compatible materials
3. Accessible for inspection, maintenance, cleaning, and sanitation
4. No product or liquid collection
5. Hollow areas should be hermetically sealed
6. No niches
7. Sanitary operational performance
8. Hygienic design of maintenance enclosures
9. Hygienic compatibility with other plant systems
10. Validated cleaning and sanitizing protocols

19.6 CLEANING

Cleaning primarily refers to a series of combined operations aimed at eliminating adherent impurities (dust, food residues, dirt, grease, germs, organic material, microorganisms, and debris) from the product contact surfaces or surface areas of equipment through mechanical means like sweeping (dry cleaning) or using detergent, surfactants, enzymatic cleaners, and water (wet cleaning) to prevent the possible cross-contamination of hazards. Soils, dirt, food waste, and other debris can be a potential source of hazards and can attract pests that can contaminate the production environment. Maintaining a predefined level of hygiene on product contact surfaces or clean work environment is critical in any food service premise to prevent malfunctions, and reduce the microbial load or contamination that would alter the safety, identity, quality, and purity of the product (Otto et al., 2011). Effective regular cleaning to remove pollutants from food premises is an important operation in food processing because of the significant contribution to product hygiene and food safety. Before starting a cleaning operation in the food industry, it is necessary to have a clear understanding of the type of cleaning agent used, the type of soil to be removed, and the method of cleaning used. It has been studied that the amount of soil adherent to the equipment is not uniformly distributed (Costa et al., 2013). Soil adhered to the food processing equipment can be of organic and inorganic. There are three basic elements that together ensure cleaning within the food premise is of a high standard: chemicals, equipment, and techniques. Inappropriate cleaning methods may increase contamination around the plant rather than decreasing it from the surface. To ensure cleaning is conducted correctly, a defined and systematic approach is required to mitigate the spread of diseases (Stanfield, 2003). Frequent systematic cleaning minimizes the risk of food contamination and is key to controlling the spread of infection. The process of cleaning is considered an integral part of the overall processing operation. Today care is taken to prevent excessive usage of water during cleaning. The water used for cleaning equipment, raw materials, and premises should be recycled (Watson, 1908). The friction of cleaning minimizes most contamination and exposes any remaining germs to the effects of a sanitizer or disinfectant later. A well-planned cleaning program should include

- Cleaning method to be used
- Equipment and areas to be cleaned
- Specific standards
- Frequency of cleaning
- Chemicals or processes to be used
- Time and temperature specifications
- Staff responsible for each task

We know that soil is considered an undesirable material present on food contact surfaces. The main source of soil is the food being processed, but the water used for cleaning and processing can also lead to the formation of soil due to the minerals present in it. There are different types of soil that vary in composition, and various factors affect the formation of soil on food contact surfaces. Before choosing any detergent for the removal of soil in a cleaning operation, it is recommended understanding the nature of the soil to be removed. The selection of an inappropriate detergent can have a harmful effect and, instead of removing the soil, it may precipitate the soil (Costa et al., 2013). Based on solubility, soil can be classified into the following types:

- Acid-soluble soil
- Alkali-soluble soil
- Water-soluble soil
- Acid-, alkali-, and water-soluble soil

Protein-based soils are usually removed with an alkalizing agent. Fat is present in the form of an emulsion and can be removed with the help of hot water. Carbohydrates and simple sugars also are easily soluble in hot water. Complex carbohydrates like starches can be removed with the help of mild detergents. Compared with other types of soil, minerals deposits are difficult to remove, hence, acidic detergents are used to efficiently clean them. Cleaning is imperative to adhere to legal and fundamental requirements of global food standards and ensure food hazards are under control (Tidswell, 2005). Cleaning is a prerequisite to ensure equipment remains hygienic and operating under optimal conditions to produce high-quality and safe products, in addition to minimizing the risk of contamination. The basic eight elements that must be considered while developing a cleaning program are described by the acronym TACT WINS.

T = Time
- Contact time is defined by how long the cleaner is in contact with the surface being cleaned (food residues, dirt, grease, and the food contacting surface).

A = Action
- What type of cleaning method (clean in place [CIP] or cleaning out of place [COP] or manual or high-pressure or power washing or gelling, or foaming or agitation, circular or ultrasonic) will be used and what type of physical force will be exerted onto the surface.

C = Cleaner/Concentration
- What type (alkaline based or acid based or foaming or non-foaming) and amount of cleaner (concentration of detergent)will be most successful?
- The optimum concentration of cleaning solution is a crucial aspect regarding the effectiveness of cleaning.

T = Temperature
- Choosing the right amount of energy as in heat in the cleaning solution is key to dissolving the soil molecule on the surface.

W = Water
- Water comprises 90–95% of the cleaning solution, and impurities in the water can affect the effectiveness of the cleaning solution.
- Understanding the level of impurities in water used to prepare a cleaning solution is crucial in the selection of the right cleaner.

I = Individual
- Effective cleaning depends on the cleaning crew performing the clean-up operation.
- All workers performing the cleanup operation should follow the standard operating procedures (SOPs) to ensure maximum cleaning effectiveness.
- Staff should be provided with the tools needed to complete the task with the least amount of effort and time.
- The cleaning crew must be aware of personal safety (safety first).

N = Nature
- Properly identifying the composition of the soil that needs to be removed allows the right chemical to be used.

S = Surface
- The proper cleaning agents must be selected so that they do not harm the equipment surfaces being cleaned.
- The effectiveness of the cleaning method is also dependent on the design and integrity of the equipment.
- The ideal standard material should be smooth, non-porous, abrasion-resistant, and inert.

19.7 PROCEDURE OF CLEANING

Cleaning is a necessary step for food manufacturing sectors because one cannot sanitize a dirty surface, so cleaning is of paramount importance in the food industry. Cleaning methods should be assessed for adequacy through valuation and inspection protocols. The correct sequence of a general cleaning procedure for surfaces in a food plant (Lelieveld et al., 2005) includes the following:

1. *Gross clean:* This involves the removal of all ingredients (food pieces, packaging materials, organic materials, etc.) from the production area or equipment surfaces before gross cleaning.

2. *Pre-rinse:* This step is used to remove deposits (food residues, dirt, and debris present) that cannot be easily removed by picking, sweeping, or other manual forms of gross cleaning. This may be accomplished by flushing the equipment surface with cold or warm water under moderate pressure. A considerable amount of residue is removed in the pre-rinsing step. It helps to reduce the amount of detergent used.

3. *Detergent application:* This lowers the surface tension of water so that soil or remaining food residues, grease, dirt, and other food deposits that remain on food contact surfaces may be dislodged or loosened and to suspend them for subsequent flushing away. Chemically active compounds (caustic soda, caustic potash, and surfactants) tweak the nature of soil (food residues) to make it more soluble and easier to efface and carry away surface contamination. Detergents used on the site/equipment to dislodge food residues, dirt, grease, germs, organic material, microorganisms, and debris must be appropriate for the task and undergo a thorough risk assessment before purchase and application (Watkinson, 2008). Cleaning compound requirements vary according to the surface to be cleaned, and the factors that can affect their effectiveness are dilution rate, concentration, contact time, temperature, and mechanical action. A wide range of detergents is used in the cleaning of food contact surfaces like inorganic alkalis and inorganic and organic alkalis. Sodium hydroxide, sodium bicarbonate, and sodium metasilicate, are the most used alkalis. Inorganic acids include nitric acid, sulfuric acid, and hydrochloric acid, whereas organic acids include citric acid, tartaric acid, and gluconic acid (Holah, 2003).

4. *Post rinsing:* Post rinsing is done to remove all traces of the cleaning solution and remaining food deposits with clean potable water and to minimize the amount of splash or aerosol formed that may recontaminate surfaces. This step can be accomplished by applying hot or cold water. Sometimes an acidic or alkaline solution may also be used for post rinsing (Matuszek, 2011a).

5. *Disinfection:* This reduces or destroys the number of infectious agents and other kinds of microorganisms (including bacteria, viruses, fungi, and parasites) to an acceptable level by direct exposure to chemicals or physical agents. Disinfectant refers to a chemical substance that destroys infectious agents (not necessarily the bacterial spores) and is intended to be used in disinfection. Halogens, peroxides, and surfactants are the most used disinfectants. Halogens used include chlorine dioxide, hypochlorous acid, and iodine. Commonly used peroxides are peracetic acid and hydrogen peroxide. Cationic and amphoteric surfactants are also used for cleaning food contact surfaces (Wallace, 1997).

6. *Terminal rinsing:* This is the thorough cleaning of product contact surfaces so that the disinfected area is not recontaminated. Terminal rinsing is usually done with cold water but hot water at a temperature less than 100°C can also be used. The main purpose of final rinsing is to completely remove the disinfectant as well as food residues from food contact surfaces.

19.8 METHODS OF CLEANING

A proper and well-defined cleaning program and its application should be the first concern of a food business operator for safety maintenance and the production of quality food products. Cleaning should be done according to global food standards. The selection of a proper cleaning method is an important step in the cleaning program. Food business operators should maintain a record of all the activities related to cleaning (Chiarello-Ebner, 2006). Based on parts disassembled at the time of cleaning, the cleaning methods are usually classified into three types: mechanical cleaning, COP, and CIP. All three types are discussed in the following sections.

19.8.1 Mechanical Cleaning

Mechanical cleaning requires a trained and well-qualified staff to disassemble the equipment completely for proper cleaning. Labor and time requirements are high, which increases cost. Because the parts are reassembled after cleaning, there are more chances of contamination. Equipment parts are prone to damage during the process of assembling and disassembling.

19.8.2 Cleaning out of Place

The COP method cleans only some parts of the equipment. The removed parts are then cleaned either manually or in specially designed pressure tanks. In pressure vessels, the parts are cleaned by a combination of turbulence and pressure from the cleaning liquid. Usually, hot solutions are used to clean the equipment parts. The equipment parts are disassembled as much as needed to expose the adherent dirt. It involves the use of high-pressure nozzles for the application of detergents, hot water, and sanitizers.

19.8.3 Clean in Place

The CIP method is suitable for food processing systems that are used to process liquid foods. Processing lines for liquid foods are characterized by less accumulation of soil that can be easily removed. This type of cleaning does not require disassembling the parts of the processing line. It is most commonly practiced in dairy industries and industries for soft drinks and mineral water. A system with closed parts like valves, exchangers, pumps, vessels, etc., which cannot be disassembled, is usually cleaned by this method. Because the parts are not disassembled at the time of cleaning, this method is called clean in place. Detergents, water, and sanitizing solutions are circulated through the system in a predetermined manner (Jennings et al., 1957).

19.9 CONCLUSION

Hygienic design and cleaning are the primary concerns of the proper functionality of food processing equipment areas. It is the responsibility of food business operators to meet the demands of people and to ensure the quality and safety of food. They must collaborate with the manufacturers of equipment and engineers to implement a food safety management system at an early stage to prevent contamination and ensure quality and safety. Hygienic design and cleaning help prevent product recalls from the market, maintain the reputation of the food business operator, and help the food business owners survive the competition in the market. Any food business operator must undertake a study regarding the materials of construction, methods of cleaning, the cleaning agents used, and the standards and principles to be followed. Finally, it can be concluded that there is no reason to develop food processing lines that do not facilitate hygienic design and cleaning.

REFERENCES

Chiarello-Ebner, K. (2006). Pursuing efficiency: new developments in cleaning technology. *Pharmaceutical Technology*, *30*(3), 52–64.
Cossen, H., Kastelien, J., & van der Kamp, J.W. (2005). New era for hygienic food manufacturing. *New Food*, *1*, 65–69.
Costa, C.A., Luciano, M.A., & Pasa, A.M. (2013). Guiding criteria for hygienic design of food industry equipment. *Journal of Food Process Engineering*, *36*(6), 753–762.
EHEDG. (2004). *Document 8. Hygienic Design Criteria* (2nd ed.). European Hygienic Engineering and Design Group, Frankfurt Germany.
EHEDG. (2007). Materials of construction for equipment in contact with food. *Trends in Food Science and Technology*, *18*, S40–S50.
Faille, C., Jullien, C., Fontaine, F., Bellon-Fontaine, M. N., Slomianny, C., & Bénézech, T. (2002). Adhesion of *Bacillus* spores and *Escherichia coli* cells to inert surfaces: role of surface hydrophobicity. *Canadian Journal of Microbiology*, *48*, 728–738.
Hall, K. (2000). Hygienic surfaces. *Food Processing*, 19–20.

Holah, J.T. (2003). Cleaning and disinfection. In: Lelieved, H.L.M., Mostert, M.A., Holah, J., White, B. (Eds.), *Hygiene in Food Processing*. Cambridge, UK, Woodhead Publishing.

Holah, J.T., & Thorpe, R.H. (1990). Cleanability in relation to bacterial retention on unused and abraded domestic sink materials. *Journal of Applied Microbiology, 69*, 599–606.

Jennings, W.G., McKillop, A.A., & Luick, J.R. (1957). Circulation cleaning. *Journal of Dairy Science, 40*, 1471.

John, B.K., Martin, K., & Andrew, A. (2017). Materials selection and fabrication practices for food processing equipment manufacturers in Uganda. *International Journal of Scientific and Technology Research, 6*(8), 338–346.

Kuo, T., Huang, S.H., & Zhang, H. (2001). Design for manufacture and design for "X": concepts, applications, and perspectives. *Computers and Industrial Engineering, 41*, 241–260.

Lelieveld, H.L.M. (1994). HACCP and hygienic design. *Food Control, 5*, 140–144.

Lelieveld, H.L.M., Mostert, M.A., & Curiel, G.J. (2003). Hygienic equipment design. In: Lelieveld, H.L.M., et al. (Eds.), *Hygiene in Food Processing*. Cambridge, Woodhead Publishing.

Lelieveld, H.L.M., Mostert, M.A., & Holah, J.T. (2005). *Handbook of Hygiene Control in the Food Industry*. Cambridge, UK, Woodhead Publishing.

Machinery Regulations. (1992). *The Supply of Machinery (Safety) Regulations 1992, Statutory Instrument No. 3073*, London, HMSO.

Matuszek, T. (2011a). Cleaning effectiveness and environmental criteria. *Journal of Hygienic Engineering and Design, 1*, 44–46.

Matuszek, T. (2011b). Food production quality and risk assessment on machinery design. *Journal of Hygienic Engineering and Design, 1*, 66–71.

Matuszek, T. (2012). Food equipment hygienic design philosophy. In: *EHEDG Yearbook 2011/2012* (pp. 62–65). Frankfurt, Germany, EHEDG.

Milledge, J.J, and Jowitt, R. (1980). The cleanability of stainless steel used as a food contact surface. *Proceedings of the Institute of Food Science and Technology, 13*, 57–62.

Moerman, F., & Partington, E. (2014). Materials of construction for food processing equipment and services: requirements, strengths and weaknesses. *Journal of Hygienic Engineering and Design, 6*, 1–37.

Otto, C., Zahn, S., Rost, F., Zahn, P., Jaros, D., & Rohm, H. (2011). Physical methods for cleaning and disinfection of surfaces. *Food Engineering Reviews, 3*, 171–188.

Plett, E.A., & Graßhoff, A. (2006). Cleaning and sanitation. In: Heldman, D.R., Lund, D.B. (Eds.), *Handbook of Food Engineering* (pp. 929–975). Boca Raton, FL, CRC Press/Taylor & Francis.

Riedewald, F. (2006). Bacterial adhesion to surfaces: the influence of surface roughness. *PDA Journal of Pharmaceutical Science and Technology, 60*(3), 164–171.

Schmidt, R.H., & Erickson, D.J. (2017). Sanitary design and construction of food equipment. FSHN0409, 1–7.

Shapton, D.A., & Shapton, N.F. (1998). *Principles and Practices for the Safe Processing of Foods* (p. 472). Cambridge, UK, Woodhead Publishing.

Somers, E.B., & Wong, A.C.L. (2004). Efficacy of two cleaning and sanitizing combinations on *Listeria monocytogenes* biofilms formed at low temperature on a variety of materials in the presence of ready-to-eat meat residue. *Journal of Food Protection, 67*(10), 2218–2229.

Stanfield, P. (2003). Retail foods sanitation: prerequisites to HACCP. In: Hui, Y.H., et al. (Eds.), *Food Plant Sanitation*, 563. New York, Marcel Dekker, Inc.

Tidswell, E.C. (2005). Bacterial adhesion: considerations within a risk-based approach to cleaning validation. *PDA Journal of Pharmaceutical Science and Technology, 59*,10–32.

Verrips, C.T., Smid, D., & Kerkhol, A. (1980). The intrinsic microbial stability of water-in-oil emulsions II. Experimental. *European Journal of Applied Microbiology and Biotechnology, 10*, 73.

Wallace, L.A. (1997). Human exposure and body burden for chloroform and other trihalomethanes. *Critical Reviews in Environment Science and Technology, 27*(2), 113–194.

Watkinson, W.J. (2008). Chemistry of detergents and disinfectants. In: Tamime, A.E. (Ed.), *Cleaning-In-Place: Dairy, Food and Beverage Operations* (pp. 56–80). Oxford, UK, Blackwell.

Watson, H.E. (1908). A note of the variation of the rate of disinfections with the change of the concentration of disinfectant, *Journal of Hygiene, 8*, 53.

CHAPTER 20

Food Quality and Safety Regulation Issues

Shemilah Fayaz, Farah Naqash, Spurti Morab, and Basharat Nabi Dar

CONTENTS

20.1 INTRODUCTION

Globalization has brought the whole world together like never before and made nations interdependent for mutual growth and development. Food has had a major impact in shaping human civilizations across history. The importance of food as a subset of international trade and commerce has gained special attention due to its importance in our daily lives. Foodborne disease outbreaks in the past resulted in laws and regulations against adulteration, spoilage, etc., which were mostly punitive. The liberalization of world trade in recent decades culminated in local, regional, and international regulations and standards that focus on food safety and quality aspects and harmonization of global food trade with the prevention of outbreaks and illnesses as the major goal. The shift from a punitive approach to foodborne illnesses and outbreaks to a preventive approach forms the crux of the evolution of the regulatory framework in the last century. This chapter covers a broader understanding, background, and framework of food safety regulations and issues thereof.

Email: Shemilah Fayaz, sheebushah8@gmail.com

DOI: 10.1201/9781003091677-20

The food safety framework includes laws, regulations, standards, and regulatory authorities that may be local, regional or global, private or governmental, enforcement agencies, etc. The early food laws and regulations were a direct outcome of public outcry due to malpractices resulting in foodborne illnesses and outbreaks. For example, an early instance can be traced to the 13th century when the adulteration in bread resulted in the enactment of the Assize of Bread & Ale Law in England (Davis, 2004). Similarly, the rampant food adulteration in the United States and Canada resulted in the Massachusetts Act (also called the U.S. Food Safety Act) in 1785 (Janssen, 1975). With the advancements in science and technology, liberalization of global trade, changes in food habits, the rising awareness among masses, and a better understanding of foods and food safety hazards, food laws and regulations have evolved with time to the point that the U.S. Food and Drug Administration (USFDA) initiated the "New Era of Smarter Food Safety Blueprint" in 2020, leveraging technology for food safety.

20.2 FOOD SAFETY VERSUS FOOD QUALITY

It is necessary to understand the difference in food safety versus food quality as it affects public policy and practical outcomes. Food safety deems a food fit or unfit for human consumption. It deals with the parameters that ensure whether food has the potential to cause harm to its consumer or not. There is an increasing trend of setting a basic benchmark of food safety standards including Good Manufacturing Practices (GMP), Good Agricultural Practices (GAP), Hazard Analysis and Critical Control Point (HACCP), traceability, etc., by various countries for imported food and even foods produced with the country. On the other hand, food quality has much broader and relative connotations and may involve attributes like color, taste, labeling, size, ingredients, packaging, etc. Although food safety ensures that food is fit for consumption, food quality determines its acceptability and value in the market.

20.3 UNDERSTANDING FOOD SAFETY HAZARDS

As per the Food Safety Inspection System (FSIS), a food safety hazard is "any biological, chemical or, a physical property that may cause a food to be unsafe for human consumption". Food hazards may be, intentionally or unintentionally, introduced to foods anywhere from farm to fork. https://www.fsis.usda.gov/

Biological hazards: They include bacteria, viruses, parasites, prions, etc., and/or their products. Foodborne illnesses can occur due to ingestion of these hazards, which may result in infection or intoxication. They occur due to bacteria like *Bacillus cereus*, *Clostridium botulinum*, *Clostridium perfringens*, *Escherichia coli*, *Salmonella*, *Vibrio cholera*, etc.; viruses like hepatitis A virus, norovirus, rotavirus, etc.; parasites like *Taenia* spp., *Toxoplasma gondii*, *Giardia* spp., *Entamoeba histolytica*, *Cryptosporidium parvum*, etc.; and prions, which are believed to cause bovine spongiform encephalopathy (BSE) or mad cow disease.

Chemical hazards: There are a plethora of chemical hazards that can enter food through the environment (mycotoxins, marine toxins, arsenic, cadmium), human interventions (pesticide and veterinary drug use and abuse), manufacturing processes (acrylamide, ethyl carbamate, furan, etc.) and addition of food additives (colors, enzymes, emulsifiers, etc.).

Physical hazards: Any extraneous material foreign to a food other than biological and chemical hazards is called a physical hazard. They are usually non-toxic but indicate mishandling and unhygienic conditions during the various levels like processing, storage, marketing, etc., of the products. They can pose serious issues like wounds, lacerations and perforations due to their shape, size, sharpness, etc., if left unchecked. Some examples of physical hazards are metal and wood pieces, nails, insects, hair, plastic, etc.

Along with the previously mentioned hazards, certain foods or their ingredients may serve as allergens to some people causing allergies and in some cases anaphylactic shock. People across the world have been reported to be allergic to certain foods like peanuts, eggs, milk and milk products, sesame seeds, soy, etc. Research is ongoing to understand the possible food safety ramifications of genetically modified or novel foods.

20.4 WORLDWIDE FOOD SAFETY ISSUES

The human population is grappling with threats and worrisome scenarios brought to the fore by issues about food safety. According to the World Health Organization (WHO), unsafe food is the prime cause of diseases as mild as diarrhea to as deadly as cancer. Contaminated food is known to be the cause of illness of about 600 million people in the world, leading to about 420,000 deaths every year, estimating the loss of about 33 million healthy life years (disability adjusted loss years[DALYs]). https://www.who.int/. Unsafe food also negatively affects the global economy in terms of medical expenses arising out of unsafe food consumption in countries belonging to low- and middle-income groups. Unsafe food and the associated diseases thus hamper the development of all socioeconomic circles by putting burdens on healthcare, national economies, trade, and tourism. Given the fact that food chains cross diverse national borders, the scares do not remain confined to a particular geographical location, even if they are born in a specific geography. This could be exemplified by the fact that in 2017/2018 contaminated meat from South Africa was exported to 15 countries resulting in 216 deaths and 1600 outbreaks of listeriosis.

The primary concern associated with unsafe food has been seen in the face of widespread foodborne illnesses, resulting from biological agents like bacteria, parasites, viruses, or chemical entities making their way to the human body through food or water. Bacteria such as *Salmonella* and *E. coli* are known to cause diseases such as fever, vomiting, diarrhea, and abdominal pain. *Listeria monocytogenes*, found in ready-to-eat foods and unpasteurized dairy products, are implicated in miscarriages in pregnant women and even the death of newborn babies. *V. cholerae* also causes abdominal pain, vomiting, severe dehydration, and even death, when ingested through contaminated food or water. Similar kinds of diseases are attributed to viruses and parasites as well. Chemicals or toxins occurring naturally in food products such as mycotoxins, aflatoxin, and cyanogenic glycosides can potentially damage the immune system and lead to the development of cancer. Heavy metal contamination also is responsible for causing neurological damage and often kidney failure. The average per case economic cost of *E. coli* O157:H7 is estimated to be $9606 (2010) and that of *Salmonella* is $4312, indicating the economic burden associated with countering illnesses caused by them (Scharff, 2012). A report from World Bank in 2018 reported that the loss in productivity observed with the outbreaks of foodborne illnesses in low- and middle-income countries amounted to about $95.2 billion per year, along with the annual cost of treatment estimated at $15 billion. Almost all food products are associated with the risk of producing foodborne illnesses and subsequent food scares. For instance, milk and its products, despite being nutritious and wholesome, are associated with grave issues of safety and quality. Milk harbors microbes that could be pathogenic as well as saprophytic and bacteria as well as viruses, including but not limited to *E. coli*, *L. monocytogenes*, *Salmonella* spp., *Staphylococcus aureus*, *Clostridium* spp., *Cryptosporidium parvum*, and so on. These microorganisms are transferred to milk either from the cattle or by contamination due to handling and water supply (Fox and Cogan, 2004). Staphylococci growing in milk secrete a heat-resistant toxin causing food poisoning. Likewise, other pathogenic microbes are known to cause food poisoning, gastroenteritis, and allied infections such as typhoid fever, diphtheria, tuberculosis, etc. (Singhal et al., 2020).The European Union Rapid Alert System for Food and Feed (EURASFF), which is an online portal providing free access to safety information about food and feed in EU member states, lists the hazards and safety issues associated with milk and milk products. From a period of January 1, 2015, to December 31, 2019, the database provided an account of a wide range of biological hazards found in milk and milk products such as *Pseudomonas*, *Brucella*, *B. cereus*, *Cronobacter*, *Salmonella*, *E. coli*, and their associated outbreaks (Montgomery et al., 2020). Other than biological contaminants, physical and chemical contaminants as well as inadequate control measures also pose a significant health hazard. Other hazards associated with milk products include foreign bodies, additives, mycotoxins, allergens, labeling issues, and defective packaging. Other than the contaminants, milk fraud issues such as the melamine scandal and adulteration are also widely reported (Handford et al., 2016).

Meat and its products are preferred in the diet due to their nutritional adequacy and desirable properties. However, these products are rendered unacceptable due to spoilage caused by improper handling and unhygienic processing (Das et al., 2019). Consumption of such meat is a cause of meat-borne illnesses attributed to the pathogens carried (Biswas et al., 2008). Contaminants enter the meat processing cycle at the pre-harvest, harvest, and post-harvest stages. Various physical hazards rendering meat unsafe include needles, glass, wires, insects, pests, packaging materials, etc. Chemical hazards include pesticides, fertilizers, drugs, oils, lubricants, and environmental contaminants such as mercury, lead, arsenic, aflatoxins, and intentionally added chemicals beyond permissible limits. The category of biological hazards include bacteria, viruses, parasites, and prions. Various spoilage and pathogenic organisms, which are a matter of potential concern, are listed in Table 20.1. Due to these microorganisms, various meat products have been recalled across different global markets. Shang and Tonsor (2017) provided a wide account of the effect of meat product recalls on consumer meat demand in the United States. The recalls identified by *E. coli* and non-*E. coli* recalls have been shown to impact the overall meat consumption and mar the interests of the consumers, causing negative consequences economically.

Table 20.1 Spoilage and Pathogenic Bacteria Associated with Meat and Meat Products

	Bacteria	Types of Meat	References
Spoilage microorganisms	*Pseudomonas*, *Acinetobacter*, *Alcaligenes*, and *Moraxella* spp.	Unpreserved meat products stored at chilled temperatures (4°C ± 1°C)	Brown (2000) Doulgeraki and Nychas (2013)
	Enterobacter spp.	Refrigerated meat product	Dave and Ghaly (2011)
	Enterobacter and *Pseudomonas* spp.	Modified atmosphere packed meat (especially pork)	Dave and Ghaly (2011)
	The lactic acid bacteria, *Enterococci*, *Micrococci*, and yeasts	Raw, salted, and cured products such as corned beef, uncooked hams, and bacon	Brown (2000); Dave and Ghaly (2011)
Pathogenic microorganisms	*Campylobacter jejuni* and *Campylobacter coli*	Raw poultry products Beef, pork, and poultry products	Mataragas et al. (2008); Sofos (2014)
	C. botulinum *Escherichia* spp.	Undercooked poultry-meat products, non-intact meat products such as beef	Akhtar et al. (2009); Sofos (2014)
	L. monocytogenes		Sofos and Geornaras (2010)
	Salmonella spp.	Ready-to-eat pork or poultry-meat products, or reheated meat products, partially cooked pork products, and poultry foods	Sofos (2014) Mataragas et al. (2008); Sofos (2014)
	Y. enterocolitica	Processed pork or poultry-meat products (ready-to-eat or to be reheated) and partially cooked pork products	Sofos and Geornaras (2010); Sofos (2014) Mataragas et al. (2008); Sofos (2014)
		Cooked, reheated pork products or improper cooking products, contaminated water	

Source: Adapted from Das et al. (2019).

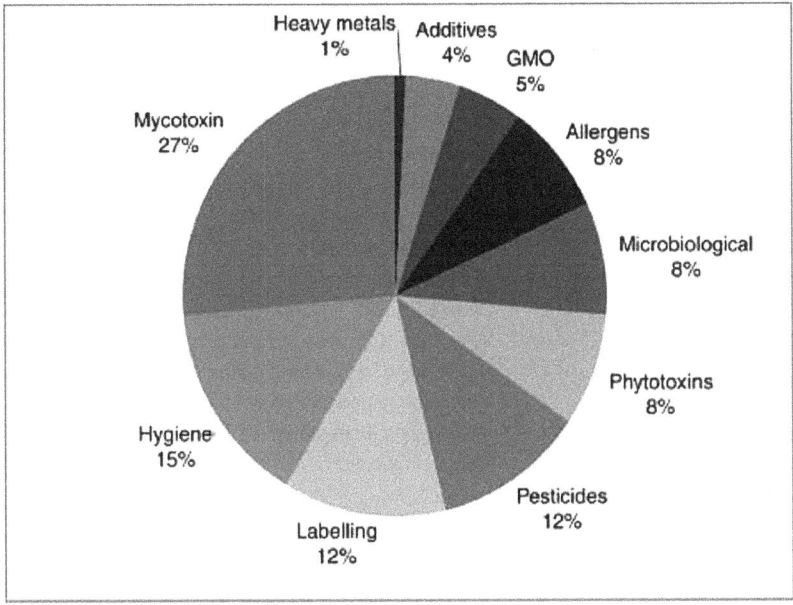

Figure 20.1 Analysis of notifications (n = 122) concerning cereals and bakery products communicated through the European Rapid Alert System for Food and Feed during the calendar year 2015. (Adapted from Alldrick, 2017, with permission from Elsevier.)

Grain and cereal-based foods are consumed across the world by a large expanse of the population as part of the staple diet. However, as with any foodstuff, cereal-based products also have chemical, physical and microbiological hazards associated with them that form issues of potential concern. Safety concerns observed in cereal-based products include those that are inherently produced such as phytotoxins, heavy metals such as arsenic, mycotoxins due to post-harvest and agronomic practices, microbiological contaminants, pesticides, and process-induced toxicants. Criminal acts involving the fraudulent addition of adulterants such as melamine and methamidophos are also included in risk factors associated with cereal products. Figure 20.1 provides an overview of the safety issues associated with cereal products as notifications made to EURASFF (Alldrick, 2017).

Seafood is a favorite food relished by a huge chunk of the population world over, including products obtained from wild capture and aquaculture. A wide range of food safety hazards is related to seafood including ciguatoxin, scombroid poisoning, microbial contamination, biotoxins from algae, and environmental pollutants. Enormous outbreaks recorded in the literature have been linked to the consumption of seafood. From 2005 to 2015, about 308 cases of illnesses and 45 hospitalizations linked to seafood consumption were reported in Australia (New South Wales). A breakdown of the data further revealed 65.11% of outbreaks were due to finfish consumption, 23.35% due to shellfish, and 11.63% due to crustaceans. About 25.58% of the total outbreaks were a result of scombroid poisoning. These outbreaks highlight the failures in ensuring food safety as a result of the complex food chain and factors such as poor sanitation and improper temperature (Hussain et al., 2017).

Fruits and vegetables are consumed principally for their nutritional and bioactive reserve, which promotes health and also helps to combat various health issues. Contamination of fresh produce pre-or post-harvest causes fruits and vegetables to produce to harbor spoilage and pathogenic microorganisms causing foodborne illnesses such as have been widely reported in the literature. In the recent past the number of disease outbreaks attributed to ready-to-eat fruits and vegetables has increased and are mainly caused by *E. coli* and *L. monocytogenes* (Paramithiotis et al., 2017).

Tables 20.2 and 20.3 highlight various food recalls and public health warnings issued by FSIS, the U.S. Department of Agriculture (USDA) and the USFDA. The recalls are often done based

Table 20.2   Recalls and Alerts by FSIS

Product Recall	Reason of Recall	Date
Blueberry yogurt	Potential for mold	Jan 11, 2021
Ice cream products	Contaminated with extraneous material	Jan 10, 2021
Chopped salad kit	Undeclared eggs	Jan 01, 2021
Ready-to eat Polish sausage	Misbranding, undeclared allergen	Jan 02, 2021
Baked chicken meal	Foreign matter contamination	Dec 19, 2020
French silk pie	Undeclared pecans	Dec 28, 2020
Frozen meat and poultry	Without inspection	Dec 19, 2020
Ready-to-eat chicken breasts	Undercooked	Nov 26, 2020
Frozen snack	Misbranding, undeclared allergen	Nov 11, 2020
Ready-to-eat chicken salad	Misbranding, undeclared antigen	Sep 16, 2020
Smoked sausage products	Misbranding, undeclared antigen	Aug 22, 2020
Canned soup	Misbranding, undeclared antigen	Aug 18, 2020
Italian meatball, beef ravioli, pepperoni pizza	Without inspection	Jul 24, 2020
Samosas containing chicken	Without inspection	Jun 17, 2020
Refrigerated beef products	*E. coli* O157:H7 contamination	Jun 13, 2020
Frozen not-ready-to-eat chicken and turkey bowl products	Foreign matter contamination	May 22, 2020
Ready-to-eat meat stick	Misbranding, undeclared allergen	Feb 08, 2020
Stuffed pizza sandwich	Misbranding, undeclared allergen	Jan 29, 2020
Salad products	*E. coli* O157:H7 contamination	Nov 21, 2019
Meat and poultry products	Insanitary conditions	Oct 18, 2019
Ready-to-eat poultry products	*Listeria* contamination	Sep 28, 2019

Source: Retrieved from FSIS, USDA usda.gov, and USFDA https://www.fda.gov/.

Table 20.3   Public Health Alerts Issued by FSIS

Public Health Alert	Reason	Date
Frozen, fully cooked, not shelf-stable chicken	Misbranding, undeclared antigen	Dec 31, 2020
Raw, fresh goat meat	Without inspection	Dec 8, 2020
Chicken and beef samosa containing pastry ingredients	Misbranding	Nov 21, 2020
Chicken and pork tamales containing diced tomatoes (recalled)	Misbranding	Nov 15, 2020
Spaghetti and meatballs	Foreign matter contamination	Nov 4, 2020
Ready-to-eat sausage	*Listeria* contamination	Aug 7, 2020
Ready-to-eat meat and poultry products containing onions (recalled)	*Salmonella* contamination	Aug 5, 2020
Raw beef ravioli	*E. coli* O157:H7 contamination	May 12, 2020
Romaine lettuce products	*E. coli* O157:H7 contamination	Nov 22, 2019
Raw pork products	Misbranding, undeclared antigen	Nov 2, 2019
Beef products	*E. coli* O157:H7 contamination	Oct 16, 2019
Raw chicken products	*Salmonella* contamination	Aug 24, 2018
Products containing whey powder (recalled)	*Salmonella* contamination	Jul 20, 2018
Chicken nuggets meal products	*Salmonella* contamination	Apr 20, 2017

Source: Retrieved from FSIS and USDA.

on the severity of the associated hazard and are classified into three categories: (1) class I recall, wherein the use of product consumption would cause serious health consequences including death; (2) class II recall, wherein a remote probability of adverse health effects exist; and (3) class III recall, wherein the product consumption will not bring adverse effects. Where a recall cannot be issued if the source of the outbreak has not been identified or improper handling gives rise to illness, FSIS may issue public health alerts for safe food handling.

Given the severity of these risks, food has to be presented to the consumer table so there is no shadow of a doubt about its safety and wholesomeness. This is where the need for regulation of food production, processing, handling, storage, and distribution comes into play. Reiterating the need for ensuring food safety, the International Conference on Food Safety held in Addis Ababa in February 2019 and International Forum on Food Safety and Trade in Geneva held in 2019 enshrined food safety to be a key parameter for attaining sustainable global development. Rigorous implementation and not mere framing is, however, imperative and a prerequisite for meeting such sustainable development.

20.4.1 Why Are Food Safety Regulations Necessary?

1. To ensure food safety and quality
2. To regulate the production, processing, packaging, distribution, marketing, and trade of food products
3. To set standards and legislation for foods from farm to fork based on established scientific knowledge
4. To regulate the import and export of foods
5. To prevent foodborne illnesses and outbreaks
6. To control malicious or unintended food hazards
7. To create awareness among consumers
8. To harmonize the quality of foods as per global standards for ease of trade
9. To settle disputes involving foods at the local, regional, and global level
10. To enforce foolproof product recall in case of an outbreak or eventuality
11. To set standard operating procedures (SOPs), and guidelines for sanitation and sanitation operating procedures(SSOPs), etc., for operations in the food industry
12. To create a framework for researching, understanding, and adapting to dynamics of foods, food habits, technology, and trade
13. To build an environment of trust among customers regarding the foods available in the market
14. To bridge the gaps between governments, businesses, and consumers

20.5 HISTORY OF FOOD REGULATIONS

Food has had a huge influence on the existence and evolution of human civilization. The earliest advances in food safety could have been the knowledge of identifying a poisonous plant and a nutritious one. There are inferences of regulations on the manufacture of beer in the Code of Hammurabi that date back to the Babylonian era. Chanakya, the Indian philosopher and economist, mentioned "food adulteration" in his book *Arathshastra* around 375 BC. However, food safety regulations began to gain traction after food started being prepared on a commercial scale. In Table 20.4 are some notable developments in the evolution of food safety regulations across history and the world.

20.6 FOOD STANDARDS

The International Organization for Standardization (ISO) defines a standard as "a document that provides requirements, specifications, guidelines, or characteristics that can be used consistently to ensure that materials, products, processes, and services are fit for their purpose." The standards

Table 20.4   Major Milestones Regarding Food Safety Regulations across the World

S. No.	Year	Milestone	Organization	Country
1	1202	Assize of Bread and Ale Law was enacted	King of England (King John)	UK
2	1785	To curb the rampant food adulteration, The Massachusetts Act (also called the U.S. Food safety Act) was enacted	U.S. Congress	U.S.
3	1862	Charles M. Wetherill, a chemist, was appointed by President Abraham Lincoln to head the Department of Agriculture. He founded the Bureau of Chemistry, which culminated in the U.S. Food and Drug Administration (USFDA)	USDA, USFDA	U.S.
4	1872	British Food & Drug Act dealt with gross adulterations of cereals and pulses, spices, milk, etc. The law had provisions of seizure of foods and punishments in the courts	British government	India (The country was under British rule until 1947)
	1875	Inland Revenue Act is considered the first food law after its confederation. It prohibited adulteration and covered foods, drinks, drugs, and even alcohol	Canada government	Canada
4	1897	Tea Importation Act was enacted in the United States, which made the inspection of all the imported tea at ports mandatory	U.S. Congress	U.S.
5	1906	President Theodore Roosevelt signs the Food, Drug, and Cosmetics Act, which banned the trade of adulterated foods, drinks, and drugs between states, passed by U.S. Congress	U.S. Congress	U.S.
6	1913	The Gould Amendment made it mandatory to mention content of the food package "plainly and conspicuously" in the United States	U.S. Congress	U.S.
7	1938	The Federal Food, Drug and Cosmetic (FDC) Act, mandating safe tolerance limits for unavoidable poisonous substances, authorizing inspections of factories, etc., was passed by Congress	U.S. Congress	U.S.
8	1939	The first Food Standards were issued for canned tomatoes, puree, etc., in the United States	U.S. Congress	U.S.
	1954	Several laws were implemented to ensure food safety in India like the Prevention of Food Adulteration Act (1954), Essential Commodity Act (1955), The Fruit Products Order (1955), The Meat Food Products Order (1973), etc.	Indian government	India
9	1961	Codex Alimentarius or the Food Code, established in November 1961 by the Food and Agriculture Organization of the United Nations (FAO), was joined by the World Health Organization (WHO) in June 1962 and held its first session in Rome In October 1963. The texts of Codex Alimentarius are developed and maintained by the Codex Alimentarius Commission (CAC). The main goal of the commission is to protect the health of consumers and ensure fair practices in the international food trade. As of 2020, there are 189 members of the CAC	CAC/FAO/WHO	Italy
10	1962	The Hazard Analysis Critical Control Point (HACCP) concept was developed by NASA and the Pillsbury Company jointly to provide safe food to astronauts. In the following decades, HACCP gained popularity in the food regulatory mechanism. It was made mandatory for seafoods in 1997 and, subsequently, extended to meat and poultry manufacturers. The year 2002 saw HACCP made mandatory in the juice industry	NASA	U.S.

(Continued)

Table 20.4 Major Milestones Regarding Food Safety Regulations across the World (*Continued*)

S. No.	Year	Milestone	Organization	Country
	1972	The Foodstuffs, Cosmetics, and Disinfectants Act, 1972 is a major legislation in South Africa. From forbidding sale of foods harmful to consumers to protecting them from misbranding and misinformation regarding foods, it covers various aspects of food safety and quality	Directorate: Food Control (Department of Health)	South Africa
11	1983	WHO-Europe recommended the implementation of HACCP in the region	WHO-Europe	Europe
12	1991	The year 1991 saw the first joint organization, Food Standards Australia New Zealand (FSANZ), between two countries regarding food systems	FSANZ	Australia/New Zealand
13	1997	In this year, the Canadian Food Inspection Agency Act established the Canadian Food Inspection Agency(CFIA). Its legal guidelines had existed in the confederation of Canada for more than 137 years	CFIA	Canada
14	2002	European Food Safety Authority (EFSA) came into in 2002 by the General Food Law	ESFA	Europe
	2003	Saudi Arabia consolidated all the agencies ensuring food and drug safety in the country into a Saudi Food & Drug Authority	Saudi Food & Drug Authority	Saudi Arabia
	2003	The outbreak of bovine spongiform encephalopathy triggered the enactment of "The Food Safety Basic Law" in 2003 in Japan. Later, it led to the establishment of the Food Safety Commission, which deals with risk assessment based on established scientific practices	Food Safety Commission (Cabinet Office)	Japan
15	2006	In a major development, the government of India in 2006 established the Food Safety and Standards Authority of India (FSSAI) as a regulatory authority to accomplish the functions assigned to it in the Food Safety and Standards Act passed by the Indian parliament. The objectives of FSSAI are not only confined to act as a regulatory or law implementing agency but also to promote the production and supply of safe and healthy food to Indian population	FSSAI	India
16	2011	The FDA Food Safety Modernization Act (FSMA) was enacted in the United States	FDA	U.S.
17	2020	FDA Commissioner S. Hahn launched the New Era of Smarter Food Safety Blueprint, which is a way to leverage modern technology for food safety	FDA	U.S.

are usually set by international agencies like Codex Alimentarius and are adopted by various other regulatory authorities around the globe, but countries may have their own standards based on local requirements and sensibilities. However, for international trade, they must be set as per the established scientific norms.

Standards serve the following important functions:

1. Act as a guide for manufacturers to produce products as per the required benchmark of safety and quality
2. Ensure uniformity and prevent undue wastage of resources in the products and services
3. Serve as references in dispute resolution arising in trade
4. Provide a competitive advantage and help build trust among consumers

Standards may be "Mandatory/Compulsory," which means the minimum safety and quality attributes a product must have to gain access to the markets, or "Voluntary," which means they are used by producers/manufacturers over and above the mandatory standards to have a competitive advantage.

As discussed, the last century saw a remarkable development in food safety regulations throughout the world. Although efforts were being made for hassle-free international trade and cooperation by establishing organizations like the World Trade Organization (WTO),ISO, Codex Alimentarius Commission (CAC), Global Food Safety Initiative (GFSI), etc., many national and regional regulatory authorities also came into being like the USFDA, European Food Safety Authority, Food Safety and Standards Authority of India, etc. These bodies are in constant pursuit of safeguarding the health, economic, and environmental well-being of the planet.

The following section is intended to inform the reader regarding a number of major regulatory authorities and organizations, their history, jurisdictions, and modus operandi.

20.6.1 Codex Alimentarius

Codex Alimentarius or the "Food Code" is an amalgamation of internationally accepted standards, guidelines, codes of practice, etc., regarding foods, whether raw, semi-processed, or processed, to ensure food safety, quality, and harmonization of the international food trade. The tagline on the Food and Agriculture Organization (FAO) website reads "Codex Alimentarius is about safe, good food for everyone-everywhere."

History: Several developments in the past century led to the evolution of Codex Alimentarius into what it is now. The development of global standards for milk and milk products by the International Dairy Federation is believed to be the precursor of the Codex Alimentarius. The formation of FAO in 1945, ISO in 1947, WHO in 1948, and the European Codex Alimentarius in 1954 led to the establishment of the CAC in 1961 at the 11th FAO conference. The joint FAO/WHO Food Standards conference urged the CAC to create Codex Alimentarius, thus establishing it in 1963.

The Codex texts are categorized into three themes:

1. *Standards:* These are about product characteristics. They are either general like the maximum residual limits for pesticides or veterinary drugs or specific to a food product (i.e., there are separate detailed standards for quick frozen broccoli and quick frozen cauliflower).
2. *Guidelines:* They determine the course of action for various food-related activities, such as guidelines on "nutrition labeling" or "general principles for the addition of essential nutrients to foods."
3. *Codes of practice:* These are established practices that help streamline the various steps involved in the food industry to ensure food safety and quality.

These standards, guidelines. and codes are modified by the respective committees from time to time. There are 189 members in the Codex as of 2020.

The Codex has played a remarkable role in establishing trust between sellers and buyers or exporters and importers involved in food trade internationally. FAO and WHO jointly ensure that the standards, guidelines, and procedures underlined in the Codex Alimentarius are based on established science. Due to its authenticity, food laws and legislations of several countries are based on Codex Alimentarius. With billions of tonnes of food exchanged globally, the Codex Alimentarius has been used as a reference in dispute resolution arising in the trade. The provisions in it range from what is in the food (ingredients, residues, food additives, contaminants, etc.), how it is tested (methods of analysis and sampling), and how it is packaged (labeling, presentation, etc.) to how it is distributed (import and export inspection and certification) http://www.fao.org/fao-who-codexalimentarius/en/.

20.6.2 The International Organization for Standardization (ISO)

ISO is an independent, non-governmental and international organization that develops and publishes standards for various kinds of products covering diverse fields that touch our lives like manufacturing, agriculture, foods, chemicals, services, pharma, healthcare, technology, etc. To date, ISO has developed over 23,592 standards, which are included in their catalogue and it has a membership of 165 national standard organizations.

History: It was in 1946 that 65 delegates from 25 countries discussed the future of International Standardization to ensure safety, reliability, and quality of products for consumers in London, which culminated in the formation of ISO in 1947.

The organization pools expertise from various member organizations to create science-based standards that are relevant to the changing global trends. It ensures a proper benchmark in reliability, quality, and safety.

Along with developing and publishing standards, the organization is involved in conducting research, capacity building of its members, and raising public awareness regarding economic, societal, and environmental aspects of standardization.

ISO has been instrumental in

1. Creating a conducive environment for businesses to flourish by helping decrease the costs and improving access to new markets.
2. Enabling better national and international regulations by providing an authentic basis that is supported by teams of experts in the respective fields.
3. Bridging the gap between the expectations of consumers and the specifications of products and services by evolving mechanisms that make customer satisfaction an indispensable variable in the development of standards.

ISO's conformity assessment, which comprises certification, testing, and inspection, is a tool to ascertain whether a business/company is conforming to the requirements and specifications underlined in the standard for the production of the product. If a product produced by a company is ISO certified, it translates into confidence and trust for the consumer and authenticity and a competitive advantage for the company. ISO's Committee on Conformity Assessment (CASCO) deals with such issues.

ISO Standards: ISO has separate standards for various products and services that are published in the ISO catalog. Some of the standards are copyrighted and about 300 are freely available. For example, ISO 9000 (Quality Management System) is a set of standards that deal with statutory and regulatory requirements related to products and services. ISO 22000 (Food Safety Management System) standards deal specifically with food safety. It can be applied independently or integrated with other management systems. https://www.iso.org/standards.html

20.6.3 World Trade Organization (WTO)

The WTO is the largest economic organization that regulates trade and commerce between participating nations. It provides a platform for the trading nations to negotiate and reach agreements and a legal and institutional framework to implement and monitor those agreements to ensure a hassle-free international trade. WTO has played a major role in harmonizing global trade since the inception of its predecessor organization, the General Agreements on Tariffs and Trades (GATT), in 1948.

History: WTO was created on January 1, 1995 replacing GATT. The difference between both organization is that while GATT mainly dealt with the international trade of goods, the sphere of influence of WTO was widened to encompass the trade of goods, services, and intellectual property (IPR).

WTO is engaged in a plethora of activities ranging from negotiating and removing obstacles involved in international trade; administering and monitoring the application of WTO agreed rules for trade in goods, services, and IPR; reviewing trade policies; settling disputes between trading countries; capacity building of government officials of developing countries; conducting economic research; raising general awareness among masses, etc. https://www.wto.org/

The establishment of WTO also resulted in the following binding agreements that had a far-reaching impact on the international trade of agricultural products and food:

1. The Agreement on the application of Sanitary and Phytosanitary (SPS) measures
2. The Agreement on Technical Barriers to Trade (TBT)

20.6.3.1 Sanitary and Phytosanitary (SPS) Measures

SPS measures are basic rules that envisage how food safety and animal and plant health are ensured while regulating international trade. The agreement aims to ascertain that the food being traded is "safe" as per the established requirements while ensuring that unrealistic and unnecessary safety requirements are not used to protect local producers. The measures provide the participating nations the discretion to use their standards if they are based on scientific proof.

20.6.3.2 Technical Barriers to Trade (TBT)

TBT measures ensure that technical regulations, standards, and conformity assessment procedures do not create unnecessary hurdles in the trade while safeguarding the interests of participating nations regarding the safety and protection of the environment. The Technical Barriers to Trade Committee is a forum for members to deliberate on the issues concerning regulations and their implementation. They cover all kinds of products including agricultural products and foods excluding food standards and issues covered under SPS.

20.6.4 Global Food Safety Initiative (GFSI)

GFSI is a business-driven private initiative of an international business network, the Consumer Goods Forum (CGF), which was created in 2000 to overcome the trust deficit between consumers and the food industry. The organization works in the following four major domains:

1. *Harmonizing of food trade:* The organization works on the principle of "once certified, recognized everywhere" to avoid unnecessary duplications in audits by businesses. The Benchmarking Requirements are a built-in consensus with experts and members while consulting internationally accepted standards like Codex Alimentarius and ISO.
2. *Capacity building for businesses:* GFSI has devised a set of guidance tools referred to as the Global Markets Programme. It is a step-by-step program meant for capacity building at various levels of business operation.
3. *Public-private collaborations*: The organization has been working on bringing various stakeholders like government regulators and private businesses on a common platform to deliberate on the current challenges, opportunities, and prospects in the food industry. They host "Government to Business" (G2B) meetings, which are attended by intergovernmental organizations, national governments, national regulators, and various companies.
4. *Knowledge sharing:* GFSI has been successful in bringing multiple stakeholders on one platform to share knowledge and expertise in issues related to food safety. https://mygfsi.com/

20.6.5 U.S. Food and Drug Administration (USFDA)

The USFDA is a federal agency under the Department of Health and Human Services in the United States that regulates food products, drugs, medical devices, radiation-emitting devices, vaccines, veterinary and animal products, tobacco products, cosmetics, etc. The legal framework of the FDA comes from the Food, Drug, and Cosmetic Act (FD&C) and some other federal laws enacted by the

U.S. Congress. The FDA has the authority to regulate quality and claims regarding substances sold as foods or food supplements

History: The Pure Food & Drugs Act of 1906 paved the way for the establishment of a federal agency that would take care of consumer interests and prevent malpractice in the U.S. markets of the time. Harvey Washington Viley, the chief Chemist of the Bureau of Chemistry (Department of Agriculture) is said to have vociferously advocated for the federal regulation.

The USFDA regulates about 78% of the food supply in the country leaving meat, poultry, and some egg products, which are regulated by the Food Safety and Inspection Service of the USDA. Its regulations influence about 35,000 farms, 300,000 restaurant chains, and its products are handled at around 270,000 registered facilities around the world. It ensures that the information regarding the products it approves is available in the public domain for everyone to benefit from. In case of a food product being mislabeled or posing a health threat or causing an outbreak, food producers are required to recall their food products. In case of an outbreak, the FDA's Coordinated Outbreak Response and Evaluation (CORE)network handles the activities involved in the mitigation of the issue. Also, the FDA, in collaboration with the Centers for Disease Control and Prevention (CDC), provides information on safety during public health emergencies like natural disasters. https://www.fda.gov/

20.6.6 European Food Safety Authority

To ensure access to safe and wholesome food by the population of Europe, in the backdrop of many food safety issues reported in the 1990s, the European Commission rolled out general principles and rules regarding food and feed, commonly known as the General Food Law Regulation. It underlined the principles, procedures, and requirements at all levels of food and feed production, processing, transport, marketing, and sale. It was through this regulation that the European Food Safety Authority was conceived as an independent agency meant for scientific advice and support. The European Food Safety Authority provides expert advice based on established scientific practices, which in turn are provided by the "Scientific Panel and Scientific Committee" as and when required by the European Commission, European Parliament, and its members in matters concerning food safety. https://www.efsa.europa.eu/en/about/howwework

20.7 CONCLUSION

With the changing world dynamics of economy, polity, society, and technology, food safety concerns will evolve with time. There is a need for a much-improved collaboration among nations to address the concerns and build a stronger and foolproof regulatory framework rooted in science so that a consumer in one corner of the world feels safe and comfortable in taking the food grown, processed, packaged, and transported across borders from another corner. This can only be ensured when farmers, businesses, regulatory authorities, governments, intergovernmental, and independent agencies work in unison toward the common goal of ensuring food safety and quality from farm to fork.

REFERENCES

Akhtar, S., Paredes-Sabja, D., Torres, J.A., Sarker, M.R. (2009). Strategy to inactivate *Clostridium perfringens* spores in meat products. Food Microbiology 26 (3), 272–277.

Alldrick, A.J. (2017). Food Safety Aspects of Grain and Cereal Product Quality. In: Wrigley, C., Batey, I., Miskelly, D. (Eds.), Cereal Grains: Assessing and Managing Quality, London, Elsevier Academic Press, pp. 393–424.

Brown, M.H. (2000). Processed meat products. In: Lund, B., Baird-Parker, A.C., Gould, G.W. (Eds.), The Microbiological Safety and Quality of Food. Gaithersburg, Aspen Publishers.

Biswas, A., Kondaiah, N., Bheilegaonkar, K., Anjaneyulu, A., Mendiratta, S., Jana, C., et al. (2008). Microbial profiles of frozen trimmings and silver sides prepared at Indian buffalo meat packing plants. Meat Science, 80(2), 418–422.

Dave, D., Ghaly, A.E. (2011). Meat spoilage mechanisms and preservation techniques: a critical review. American Journal of Agricultural and Biological Sciences, 6(4), 486–510.

Das, A.K., Nanda, P.K., Das, A., Biswas, S. (2019). Hazards and Safety Issues of Meat and Meat Products. In: Singh, R.L., Mondal, S. (Eds.), Food Safety and Human Health, London, Elsevier Academic Press, pp. 145–168.

Davis, J. (2004). Baking for the common good: a reassessment of the assize of bread in Medieval England 1. The Economic History Review, 57(3), 465–502

Doulgeraki, A.I., Nychas, G.J.E. (2013). Monitoring the succession of the biota grown on a selective medium for pseudomonads during storage of minced beef with molecular-based methods. Food Microbiology 34 (1), 62–69.

Fox, P.F., Cogan, T.M. (2004). Factors That Affect the Quality of Cheese. In: Fox, P.F., McSweeney, P., Cogan, T.M., Guinee, T. (Eds.), Cheese: Chemistry, Physics and Microbiology (3rd ed.), Vol. 1: General Aspects. London, Elsevier Academic Press, pp. 583–608.

Handford, C.E., Campbell, K., Elliott, C.T. (2016). Impacts of milk fraud on food safety and nutrition with special emphasis on developing countries. Comprehensive Reviews in Food Science and Food Safety, 15(1), 130–142. https://doi.org/10.1111/1541-4337.12181

Hussain, M.A., Saputra, T., Szabo, E.A., Nelan, B. (2017). An overview of seafood supply, food safety and regulation in New South Wales, Australia. Foods, 6, 52. https://doi.org/10.3390/foods6070052

Janssen, W.F. (1975).America's first food and drug laws. Food Drug Cosmetic Law Journal, 30, 665–672.

Mataragas, M., Skandamis, P., Drosinos, E. (2008). Risk profiles of pork and poultry meat and risk ratings of various pathogen/product combinations. International Journal of Food Microbiology, 126, 1–12.

Montgomery, H., Haughey, S.A., Elliott, C.T. (2020). Recent food safety and fraud issues within the dairy supply chain (2015–2019). Global Food Security, 26, 100447.

Paramithiotis, S., Drosinos, E.H., Skandamis, P.N. (2017). Food recalls and warnings due to the presence of foodborne pathogens –a focus on fresh fruits, vegetables, dairy and eggs. Current Opinion in Food Science, 18, 71–75.

Scharff, R. L. (2012). Economic burden from health losses due to foodborne illness in the United States. Journal of Food Protection, 75(1), 123–131.

Shang, X., Tonsor, G.T. (2017). Food safety recall effects across meat products and regions. Food Policy, 69, 145–153.

Singhal, P., Kaushik, G., Hussain, C.M., Chel, A. (2020). Food Safety Issues Associated With Milk: A Review. In: Grumezescu, A., Holban, A.M. (Eds.), Safety Issues in Beverage Production (1st ed.), Vol. 18: The Science of Beverages. London, Elsevier Academic Press, pp. 399–427.

Sofos, J.N. (2014). Meat and Meat Products. Food Safety Management. Elsevier, pp. 119–162.

Sofos, J.N., Geornaras, I. (2010). Overview of current meat hygiene and safety risks and summary of recent studies on biofilms, and control of *Escherichia coli* O157:H7 in non-intact, and *Listeria monocytogenes* in ready-to-eat, meat products. Meat Science, 86 (1), 2–14.

Web Addresses Consulted

Codex Alimentarius, FAO/WHO Food Standards, http://www.fao.org/fao-who-codexalimentarius/en/

European Food Safety Authority, https://www.efsa.europa.eu/en/about/howwework

Global Food Safety Initiative (GFSI), The Coalition of Action on Food Safety, https://mygfsi.com/

International Organization for Standardization, Standards, https://www.iso.org/standards.html

U.S. Department of Agriculture Food Safety and Inspection Service, https://www.fsis.usda.gov/sites/default/files/media_file/2021-03/Guidebook-for-the-Preparation-of-HACCP-Plans.pdf

U.S. Department of Agriculture Food Safety and Inspection Service, Recalls & Public Health Alerts, https://www.fsis.usda.gov/currentrecalls

U.S. Food and Drug Administration, Fact Sheet: FDA at a glance, https://www.fda.gov/about-fda/fda-basics/fact-sheet-fda-glance

U.S. Food and Drug Administration, FDA History, https://www.fda.gov/about-fda/history-fdas-fight-consumer-protection-and-public-health

U.S. Food and Drug Administration, Recalls, Market Withdrawals, & Safety Alerts, https://www.https://www.fda.gov/safety/recalls-market-withdrawals-safety-alerts

U.S. Food and Drug Administration, What Does FDA Regulate, https://www.fda.gov/about-fda/fda-basics/what-does-fda-regulate

United States Department of Agriculture Food Safety and Inspection Service, Public Health Alerts (usda.gov)

United States Department of Agriculture Food Safety and Inspection Service, Current Recalls and Alerts (usda.gov)

World Health Organization, Food safety (who.int)

World Health Organization, Food Safety, https://www.who.int/news-room/fact-sheets/detail/food-safety

World Trade Organization, Overview, https://www.wto.org/english/thewto_e/whatis_e/wto_dg_stat_e.htm

World Trade Organization, Understanding the WTO Agreement on Sanitary and Phytosanitary Measures, https://www.wto.org/english/tratop_e/sps_e/spsund_e.htm

Index

Pages in italics and bold refer to figures and tables, respectively.

For Product Safety Concerns and Information please contact our EU
representative GPSR@taylorandfrancis.com
Taylor & Francis Verlag GmbH, Kaufingerstraße 24, 80331 München, Germany

www.ingramcontent.com/pod-product-compliance
Lightning Source LLC
Chambersburg PA
CBHW082104220326
41598CB00066BA/5212

9780367550370